Achieving sustainable cultivation of wheat

Volume 2: Cultivation techniques

It is widely recognised that agriculture is a significant contributor to global warming and climate change. Agriculture needs to reduce its environmental impact and adapt to current climate change whilst still feeding a growing population, i.e. become more 'climate-smart'. Burleigh Dodds Science Publishing is playing its part in achieving this by bringing together key research on making the production of the world's most important crops and livestock products more sustainable.

Based on extensive research, our publications specifically target the challenge of climate-smart agriculture. In this way we are using 'smart publishing' to help achieve climate-smart agriculture.

Burleigh Dodds Science Publishing is an independent and innovative publisher delivering high quality customer-focused agricultural science content in both print and online formats for the academic and research communities. Our aim is to build a foundation of knowledge on which researchers can build to meet the challenge of climate-smart agriculture.

For more information about Burleigh Dodds Science Publishing simply call us on +44 (0) 1223 839365, email info@bdspublishing.com or alternatively please visit our website at www.bdspublishing.com.

Related titles:

Achieving sustainable cultivation of wheat Volume 1: Breeding, quality traits, pests and diseases
Print (ISBN 978-1-78676-016-6); Online (ISBN 978-1-78676-018-0, 978-1-78676-019-7)

Achieving sustainable cultivation of maize Volume 1: From improved varieties to local applications
Print (ISBN 978-1-78676-008-1); Online (ISBN 978-1-78676-010-4, 978-1-78676-011-1)

Achieving sustainable cultivation of maize Volume 2: Cultivation techniques, pest and disease control
Print (ISBN 978-1-78676-012-8); Online (ISBN 978-1-78676-014-2, 978-1-78676-015-9)

Achieving sustainable cultivation of rice Volume 1: Breeding for higher yield and quality
Print (ISBN 978-1-78676-024-1); Online (ISBN 978-1-78676-026-5, 978-1-78676-027-2)

Achieving sustainable cultivation of rice Volume 2: Cultivation, pest and disease management
Print (ISBN 978-1-78676-028-9); Online (ISBN 978-1-78676-030-2, 978-1-78676-031-9)

Achieving sustainable cultivation of sorghum Volume 1: Genetics, breeding and production techniques
Print (ISBN 978-1-78676-120-0); Online (ISBN 978-1-78676-122-4, 978-1-78676-123-1)

Achieving sustainable cultivation of sorghum Volume 2: Sorghum utilisation around the world
Print (ISBN 978-1-78676-124-8); Online (ISBN 978-1-78676-126-2, 978-1-78676-127-9)

Chapters are available individually from our online bookshop: https://shop.bdspublishing.com

climate•SMART•publishing
IN AGRICULTURAL SCIENCE

BURLEIGH DODDS SERIES IN AGRICULTURAL SCIENCE

NUMBER 06

Achieving sustainable cultivation of wheat

Volume 2: Cultivation techniques

Edited by Professor Peter Langridge, University of Adelaide, Australia

burleigh dodds
SCIENCE PUBLISHING

Published by Burleigh Dodds Science Publishing Limited
82 High Street, Sawston, Cambridge CB22 3HJ, UK
www.bdspublishing.com

Burleigh Dodds Science Publishing, 1518 Walnut Street, Suite 900, Philadelphia, PA 19102-3406, USA

First published 2017 by Burleigh Dodds Science Publishing Limited

Library of Congress Control Number: 2017932202

British Library Cataloguing in Publication Data
A catalogue record for this book is available from the British Library

ISBN 978-1-78676-020-3 (print)
ISBN 978-1-78676-022-7 (online)
ISBN 978-1-78676-023-4 (online)
ISSN 2059-6936 (print)
ISSN 2059-6944 (online)

Typeset by Deanta Global Publishing Services, Chennai, India
Printed by Lightning Source

Contents

Series list ix

Introduction xiii

Key priorities in wheat research: the Wheat Initiative's Strategic Research Agenda xix

Part 1 Wheat cultivation techniques

1 Variety selection in wheat cultivation 3
Arun Kumar Joshi, International Maize and Wheat Improvement Center (CIMMYT); Vinod Kumar Mishra, Banaras Hindu University, India; and Simanchal Sahu, Orissa University of Agriculture and Technology, India
- 1 Introduction 3
- 2 Wheat variety selection methods: natural and traditional selection 4
- 3 Wheat variety selection methods: modern molecular breeding 8
- 4 Variety selection by plant breeders 12
- 5 Variety selection by farmers 17
- 6 Conclusion 18
- 7 Where to look for further information 18
- 8 References 18

2 Establishment and root development in wheat crops 25
Peter J. Gregory and Christina K. Clarke, University of Reading, UK
- 1 Introduction 25
- 2 Root development and growth during establishment 26
- 3 Crop establishment practices 28
- 4 Case study: the effects of embryo size and seed mass on early vigour 33
- 5 Research contributions to improved wheat establishment and production 35
- 6 Future trends in research 36
- 7 Where to look for further information 38
- 8 Acknowledgements 38
- 9 References 38

3 Conservation tillage for sustainable wheat intensification: the example of South Asia 41
Vijesh Krishna, Georg-August University of Göttingen, Germany; Alwin Keil, International Maize and Wheat Improvement Center (CIMMYT), India; Sreejith Aravindakshan, Wageningen University, The Netherlands; and Mukesh Meena, Indian Institute of Soil and Water Conservation, India
- 1 Introduction 41
- 2 Factors affecting the diffusion of CT wheat in South Asia 43
- 3 Recent evidence of the agronomic and economic impacts of CT wheat in South Asia 45

4 Constraints to the diffusion of CT practices in wheat in South Asia 49
5 Recent developments in CT wheat 53
6 Concluding remarks 55
7 Where to look for further information 56
8 Acknowledgements 56
9 References 56

4 Improving water management in winter wheat 63
Q. Xue, J. Rudd, J. Bell, T. Marek and S. Liu, Texas A&M AgriLife Research and Extension Center at Amarillo, USA
1 Introduction 63
2 Winter wheat yield 68
3 Yield determination under water-limited conditions 69
4 The role of measuring evapotranspiration (ET) 70
5 Water-use efficiency 70
6 Wheat yield, evapotranspiration (ET) and water-use efficiency (WUE) relationships 71
7 Case studies 73
8 Future trends and conclusion 80
9 Where to look for further information 80
10 References 81

5 Post-harvest wheat losses in Africa: an Ethiopian case study 85
Tadesse Dessalegn, Tesfaye Solomon, Tesfaye Gebre Kristos, Abiy Solomon, Shure Seboka and Yazie Chane, Ethiopian Institute of Agricultural Research, Ethiopia; Bhadriraju Subramanyam and Kamala A. Roberts, Kansas State University, USA; and Fetien Abay and Rizana Mahroof, South Carolina State University, USA
1 Introduction 85
2 Methods 88
3 Demographics 88
4 Crop production information 90
5 Post-harvest issues 91
6 Preventing post-harvest losses 99
7 Information flow and training requirements 99
8 Gender and food security issues 100
9 Conclusion 101
10 Acknowledgements 101
11 Where to look for further information 102
12 References 102

Part 2 Wheat crop management

6 Wheat crop modelling to improve yields 107
J. R. Guarin and S. Asseng, University of Florida, USA
1 Introduction 107
2 Extrapolation from crop models in time and space 111

3 Experiments for testing crop models 114
4 Adapting crop models for climate change impact 117
5 Natural resource management 121
6 Managing seasonal variability 124
7 Decision support for farmers 126
8 Future trends 127
9 Conclusion 129
10 Where to look for further information 130
11 Acknowledgements 130
12 References 130

7 Integrated crop management of wheat 143
Brian L. Beres, Reem Aboukhaddour and Haley Catton, Agriculture and Agri-Food Canada, Lethbridge Research and Development Centre, Canada
1 Introduction 143
2 Impact of agronomists on agriculture 145
3 Case study: ICM for wheat stem sawfly (*Cephus cinctus* Norton) 153
4 Future trends and conclusion 159
5 Where to look for further information 160
6 References 160

8 Organic production of wheat and spelt 167
T. F. Döring, Humboldt-Universität zu Berlin, Germany
1 Introduction 167
2 Wheat in organic rotations 169
3 Organic wheat in species mixtures 179
4 Diversification of organic wheat at the genetic level 182
5 Organic wheat and farm biodiversity 187
6 Future trends and conclusion 189
7 Where to look for further information 189
8 Acknowledgements 190
9 References 190

9 Durum wheat: production, challenges and opportunities 199
J. M. Clarke, K. Nilsen, D. Khitiri, X. Lin and C. J. Pozniak, University of Saskatchewan, Canada; and K. Ammar, International Maize and Wheat Improvement Center (CIMMYT), Mexico
1 Introduction 199
2 Durum wheat yield and agronomic traits 200
3 Abiotic stresses on durum wheat 202
4 Major diseases of durum wheat 205
5 Insect pests of durum wheat 210
6 End-use quality of durum wheat 212
7 Durum wheat breeding technologies 215
8 Future trends and conclusion 218
9 Where to look for further information 220
10 References 220

Part 3 Improving wheat cultivation in the developing world

10 Supporting smallholders in improving wheat cultivation 237
Tinashe Chiurugwi, Simon Kerr, Ian Midgley and Lesley Boyd, National Institute of Agricultural Botany (NIAB), UK; Johnson Kamwaga, Food Crops Research Centre – Njoro, Kenya; Peter Njau, Highlands Agri-consult Services Ltd, Kenya; Terry van Gevelt, University of Cambridge, UK; and Claudia Canales and Max Marcheselli, the Malaysian Commonwealth Studies Centre (MCSC) and the Cambridge Malaysian Education and Development Trust (CMEDT), UK
1 Introduction: smallholder wheat cultivation 237
2 Limitations to smallholder wheat cultivation 242
3 How research addresses limitations to smallholder wheat cultivation 246
4 Case study: Agri-transfer 250
5 Future trends 253
6 Where to look for further information 254
7 References 255

11 Improving wheat cultivation in Asia 261
Rajiv Kumar Sharma, Global Wheat Improvement Program – CIMMYT, India
1 Introduction 261
2 Improving and exploiting genetic diversity 262
3 Improving breeding techniques 264
4 Dealing with biotic stresses 266
5 Dealing with abiotic stresses 267
6 Improving aspects of quality 269
7 Improving cultivation practices 270
8 Case study: improving cultivation in India 271
9 Summary 274
10 Where to look for further information 274
11 References 275

12 Improving wheat production in the Central and West Asia and North Africa (CWANA) region 287
W. Tadesse, A. Amri, M. Sanchez-Garcia, M. El-Bouhssini, M. Karrou, S. Patil, F. Bassi and M. Baum, International Center for Agricultural Research in the Dry Areas, Morocco; and T. Oweis, International Center for Agricultural Research in the Dry Areas, Jordan
1 Introduction 287
2 Major wheat production environments 290
3 Challenges to wheat production 291
4 Improved technologies to enhance wheat production 292
5 Intensification of wheat technologies 301
6 Future trends 303
7 Where to look for further information 305
8 References 305

Index 309

Series list

Title	Series number
Achieving sustainable cultivation of maize - Vol 1 From improved varieties to local applications Edited by: Dr Dave Watson, CGIAR Maize Research Program Manager, CIMMYT, Mexico	001
Achieving sustainable cultivation of maize - Vol 2 Cultivation techniques, pest and disease control Edited by: Dr Dave Watson, CGIAR Maize Research Program Manager, CIMMYT, Mexico	002
Achieving sustainable cultivation of rice - Vol 1 Breeding for higher yield and quality Edited by: Prof. Takuji Sasaki, Tokyo University of Agriculture, Japan	003
Achieving sustainable cultivation of rice - Vol 2 Cultivation, pest and disease management Edited by: Prof. Takuji Sasaki, Tokyo University of Agriculture, Japan	004
Achieving sustainable cultivation of wheat - Vol 1 Breeding, quality traits, pests and diseases Edited by: Prof. Peter Langridge, University of Adelaide, Australia	005
Achieving sustainable cultivation of wheat - Vol 2 Cultivation techniques Edited by: Prof. Peter Langridge, University of Adelaide, Australia	006
Achieving sustainable cultivation of tomatoes Edited by: Dr Autar Mattoo, USDA-ARS, USA & Prof. Avtar Handa, Purdue University, USA	007
Achieving sustainable production of milk - Vol 1 Milk composition, genetics and breeding Edited by: Dr Nico van Belzen, International Dairy Federation (IDF), Belgium	008
Achieving sustainable production of milk - Vol 2 Safety, quality and sustainability Edited by: Dr Nico van Belzen, International Dairy Federation (IDF), Belgium	009
Achieving sustainable production of milk - Vol 3 Dairy herd management and welfare Edited by: Prof. John Webster, University of Bristol, UK	010
Ensuring safety and quality in the production of beef - Vol 1 Safety Edited by: Prof. Gary Acuff, Texas A&M University, USA & Prof.James Dickson, Iowa State University, USA	011
Ensuring safety and quality in the production of beef - Vol 2 Quality Edited by: Prof. Michael Dikeman, Kansas State University, USA	012
Achieving sustainable production of poultry meat - Vol 1 Safety, quality and sustainability Edited by: Prof. Steven C. Ricke, University of Arkansas, USA	013
Achieving sustainable production of poultry meat - Vol 2 Breeding and nutrition Edited by: Prof. Todd Applegate, University of Georgia, USA	014
Achieving sustainable production of poultry meat - Vol 3 Health and welfare Edited by: Prof. Todd Applegate, University of Georgia, USA	015

Achieving sustainable production of eggs - Vol 1 016
Safety and quality
Edited by: Prof. Julie Roberts, University of New England, Australia

Achieving sustainable production of eggs - Vol 2 017
Animal welfare and sustainability
Edited by: Prof. Julie Roberts, University of New England, Australia

Achieving sustainable cultivation of apples 018
Edited by: Dr Kate Evans, Washington State University, USA

Integrated disease management of wheat and barley 019
Edited by: Prof. Richard Oliver, Curtin University, Australia

Achieving sustainable cultivation of cassava - Vol 1 020
Cultivation techniques
Edited by: Dr Clair Hershey, formerly International Center for Tropical Agriculture (CIAT), Colombia

Achieving sustainable cultivation of cassava - Vol 2 021
Genetics, breeding, pests and diseases
Edited by: Dr Clair Hershey, formerly International Center for Tropical Agriculture (CIAT), Colombia

Achieving sustainable production of sheep 022
Edited by: Prof. Johan Greyling, University of the Free State, South Africa

Achieving sustainable production of pig meat - Vol 1 023
Safety, quality and sustainability
Edited by: Prof. Alan Mathew, Purdue University, USA

Achieving sustainable production of pig meat - Vol 2 024
Animal breeding and nutrition
Edited by: Prof. Julian Wiseman, University of Nottingham, UK

Achieving sustainable production of pig meat - Vol 3 025
Animal health and welfare
Edited by: Prof. Julian Wiseman, University of Nottingham, UK

Achieving sustainable cultivation of potatoes - Vol 1 026
Breeding, nutritional and sensory quality
Edited by: Prof. Gefu Wang-Pruski, Dalhousie University, Canada

Achieving sustainable cultivation of oil palm - Vol 1 027
Introduction, breeding and cultivation techniques
Edited by: Prof. Alain Rival, Center for International Cooperation in Agricultural Research for Development (CIRAD), France

Achieving sustainable cultivation of oil palm - Vol 2 028
Diseases, pests, quality and sustainability
Edited by: Prof. Alain Rival, Center for International Cooperation in Agricultural Research for Development (CIRAD), France

Achieving sustainable cultivation of soybeans - Vol 1 029
Breeding and cultivation techniques
Edited by: Prof. Henry Nguyen, University of Missouri, USA

Achieving sustainable cultivation of soybeans - Vol 2 030
Diseases, pests, food and non-food uses
Edited by: Prof. Henry Nguyen, University of Missouri, USA

Achieving sustainable cultivation of sorghum - Vol 1 031
Genetics, breeding and production techniques
Edited by: Prof. Bill Rooney, Texas A&M University, USA

Achieving sustainable cultivation of sorghum - Vol 2 032
Sorghum utilisation around the world
Edited by: Prof. Bill Rooney, Texas A&M University, USA

Achieving sustainable cultivation of potatoes - Vol 2 033
Production and storage, crop protection and sustainability
Edited by: Dr Stuart Wale, Potato Dynamics Ltd, UK

Achieving sustainable cultivation of mangoes 034
Edited by: Professor Víctor Galán Saúco, Instituto Canario de Investigaciones Agrarias (ICIA), Spain & Dr Ping Lu, Charles Darwin University, Australia

Achieving sustainable cultivation of grain legumes - Vol 1 035
Advances in breeding and cultivation techniques
Edited by: Dr Shoba Sivasankar et al., CGIAR Research Program on Grain Legumes, ICRISAT, India

Achieving sustainable cultivation of grain legumes - Vol 2 036
Improving cultivation of particular grain legumes
Edited by: Dr Shoba Sivasankar et al., CGIAR Research Program on Grain Legumes, ICRISAT, India

Achieving sustainable cultivation of sugarcane - Vol 1 037
Cultivation techniques, quality and sustainability
Edited by: Prof. Philippe Rott, University of Florida, USA

Achieving sustainable cultivation of sugarcane - Vol 2 038
Breeding, pests and diseases
Edited by: Prof. Philippe Rott, University of Florida, USA

Achieving sustainable cultivation of coffee 039
Breeding and quality traits
Edited by: Dr Philippe Lashermes, Institut de Recherche pour le Développement (IRD), France

Achieving sustainable cultivation of bananas - Vol 1 040
Cultivation techniques
Edited by: Prof. Gert Kema, Wageningen University, The Netherlands & Prof. André Drenth, University of Queensland, Australia

Global Tea Science 041
Current status and future needs
Edited by: Dr V. S. Sharma, Formerly UPASI Tea Research Institute, India & Dr M. T. Kumudini Gunasekare, Coordinating Secretariat for Science Technology and Innovation (COSTI), Sri Lanka

Integrated weed management 042
Edited by: Emeritus Prof. Rob Zimdahl, Colorado State University, USA

Achieving sustainable cultivation of cocoa - Vol 1 043
Genetics, breeding, cultivation and quality
Edited by: Prof. Pathmanathan Umaharan, Cocoa Research Centre – The University of the West Indies, Trinidad and Tobago

Achieving sustainable cultivation of cocoa - Vol 2 044
Diseases, pests and sustainability
Edited by: Prof. Pathmanathan Umaharan, Cocoa Research Centre – The University of the West Indies, Trinidad and Tobago

Water management for sustainable agriculture 045
Edited by: Prof. Theib Oweis, Formerly ICARDA, Lebanon

Improving organic animal farming 046
Edited by: Dr Mette Vaarst, Aarhus University, Denmark & Dr Stephen Roderick, Duchy College, Cornwall, UK

Improving organic crop cultivation 047
Edited by: Prof. Ulrich Köpke, University of Bonn, Germany

Managing soil health for sustainable agriculture - Vol 1 048
Fundamentals
Edited by: Dr Don Reicosky, USDA-ARS, USA

Managing soil health for sustainable agriculture - Vol 2 049
Monitoring and management
Edited by: Dr Don Reicosky, USDA-ARS, USA

Rice insect pests and their management 050
E. A. Heinrichs, Francis E. Nwilene, Michael J. Stout, Buyung A. R. Hadi & Thais Freitas

Improving grassland and pasture management in temperate agriculture 051
Edited by: Prof. Athole Marshall & Dr Rosemary Collins, University of Aberystwyth, UK

Precision agriculture for sustainability 052
Edited by: Dr John Stafford, Silsoe Solutions, UK

Introduction

The chapters in this volume, *Achieving sustainable cultivation of wheat Volume 2: Cultivation techniques*, pick up on the core themes identified in the Wheat Initiative's Strategic Research Agenda:

- Core theme 1: increase wheat field potential
- Core theme 2: protect yield potential
- Core theme 3: protect the environment and increase the sustainability of wheat production systems
- Core theme 4: ensure the supply of high quality, safe wheat

They cover key issues such as improvements in individual cultivation techniques to make wheat production more resource efficient and sustainable, as well as the best ways of combining such techniques in systems such as integrated or organic crop management. The book also reviews research on the key challenges in adopting new varieties, methods and technologies in practice in the developing world.

Part 1 Wheat cultivation techniques

Chapter 1 provides an overview of key issue in wheat variety selection. It looks first at relevant developments in breeding, including pedigree, modified bulk, selected bulk, non-selected bulk and back cross selection methods, before discussing more recent developments such as marker-assisted selection, genomic selection and next generation phenotyping. It then focuses on key criteria for breeders in selecting new varieties, from targeting a particular set of environmental conditions or mix of crop management methods to addressing specific biotic or abiotic stresses, or focusing on quality requirements for end users. The chapter compares these with farmer requirements which may differ from those of breeders. As an example, hybrid varieties may only be suitable for farmers who have access to funds or credit to purchase hybrid seed every year and are able to practice high input agriculture to obtain a sufficient return. Small-scale subsistence farmers often rely on seeds saved by farmers the previous years and thus prefer pure line varieties, which can be grown for a number of years from the same seed source. The chapter shows how breeder and farmer objectives can be better aligned through techniques such as participatory varietal selection (PVS).

Chapter 2 reviews recent research on wheat crop establishment, the period encompassing plant development and growth prior to the elongation of the main stem. As the chapter points out, good establishment lays the foundations for wheat yield. The chapter describes current research on how the seedling root system establishes and then outlines how key agronomic practices (time and depth of sowing, seeding rate and soil moisture conditions) affect the initial growth of the plant. The chapter includes a case study from Australia on how embryo size affects seed mass which affects early growth, and concludes with some recent ideas as to how root systems might be optimised in different soils to increase production.

Chapter 3 explores the use of conservation agriculture techniques for sustainable wheat intensification. Conservation agriculture emphasises the importance of soil health through minimal soil disturbance, residue retention and permanent soil cover, combined with crop rotations to enhance soil moisture and nutrition. A key element is conservation tillage (CT) which encompass zero-tillage, reduced-tillage, and ridge-tillage practices in wheat production. CT has been seen as an important way of addressing stagnant wheat yields in regions such as the Indo-Gangetic Plains of South Asia. The chapter reviews factors affecting the diffusion of CT in South Asia and the latest evidence of its agronomic and economic impact. It also assesses constraints to the diffusion of conservation tillage practices in wheat, taking India as a case study. The chapter shows the benefits of CT in reducing energy, fertilizer and especially irrigation input costs whilst maintaining yields, but shows that its diffusion has been variable, particularly amongst smallholders. The chapter identifies problems such as insufficient support networks for CT wheat technology, variation of yield impacts across some farming systems and locations, and the fact that tractor and CT drill ownership is unaffordable for many poorer smallholders. To improve adoption, it argues for developments such as customised hiring services to reduce transaction costs for small farms, farmer group- or cooperative-based investments in CT machinery, and the development and deployment of more location-specific and scale-appropriate machinery.

Chapter 3 emphasised the role of conservation agriculture in increasing soil water retention. Chapter 4 looks in more detail at improving water management in wheat cultivation, focussing on winter wheat. Globally, winter wheat is mainly grown in Eurasia, China, Iran, and the USA. Drought stress can significantly reduce winter wheat yields, even in high precipitation environments. Since water is the most important factor affecting crop production, development of crop management practices to conserve and optimize water use and improve crop water use efficiency (WUE) becomes essential, particularly under changing climate conditions. Chapter 4 reviews progress in winter wheat water management and WUE, drawing on long-term field experiments in the USA's southern Great Plains (which have a long history of winter wheat research). The chapter discusses yield determination under water-limited conditions, as well as the relationships between yield, evapotranspiration and WUE. It also summarises best management practices based on case studies. These practices range from soil and water conservation techniques (such as no-till, residue mulching and rotation associated with conservation agriculture), irrigation scheduling and deficit irrigation, to genetic improvement of drought tolerance.

According to estimates provided by the African Postharvest Losses Information System (APHLIS), physical grain losses range from 10-20%. It has been estimated that overall post-harvest grain losses for sub-Saharan Africa could be as high as US$4 billion/year, around 15% of total production. This compares to the findings reported for post-harvest losses of wheat in Ethiopia discussed in Chapter 5. These suggest high average losses of around 17%, with losses ranging from 14 to as high as 23%. Chapter 5 explores the range of issues underlying post-harvest losses through a case study highlighting the problem in Ethiopia. It emphasises the importance of investigating the experience and views of farmers themselves. Farmers indicated post-harvest losses to be greatest at harvest and during storage, primarily due to insects, molds, and vertebrate pests, especially rodents. Rain during harvest also greatly increased post-harvest losses.

To resolve these problems, farmers identified the need for more information about weather conditions (particularly during the harvest period), better moisture measurement, better storage techniques (including using pesticides during storage), and better market information. Storage methods can be improved by using hermetic storage or polypropylene bags (such as Purdue Improved Cowpea Storage (PICS) bags), metal or plastic drums, or metal silos. It is also important that storage methods are cost effective and accessible to remote or poor households. Finally, providing continued training on post-harvest loss mitigation methods will help greatly reduce post-harvest losses and increase food security of Ethiopian wheat farmers.

Part 2 Wheat crop management

Crop simulation models are robust tools for scientists, farmers, and policymakers that can be used to develop practices and strategies to increase agricultural productivity and sustainability on a local, regional, or global scale under conditions and treatments difficult or impossible to test using solely field experimentation. Crop models are increasingly being used to explore solutions to issues such as global food security, climate change impact assessment and adaptation, nitrogen fertilization, irrigation, and seasonal variability management.

Chapter 6 examines the experimental datasets, dynamic modeling processes, validation and fundamental applications of current wheat crop models. It reviews research on ways of using crop models, for example in responding to the challenge of climate change by modelling the effects of changing temperatures and rainfall patterns. As an example, it refers to studies that show where changing weather patterns could make new regions suitable for wheat cropping. Models can also highlight ways of balancing yield improvement with more efficient and sustainable use of inputs whilst conserving natural resources. Crop modeling also enables exploration of the benefits from seasonal forecasts to manage seasonal variability to improve efficiency and profitability. One way to manage the increased risks of less predictable weather is through the use of in-season decision support systems that incorporate crop models, such as Yield Prophet, Crop Prophet or AgroClimate. Yield Prophet, for example, is a web interface and mobile application built around the APSIM crop modeling system that produces a comprehensive crop report based on simulated growing conditions using farm location, cultivar details, soil data, hundred-year weather data, and seasonal climate forecasts. This crop report gives the latest seasonal weather forecasts and estimated impacts on crop yields, including updates on growth stage, yield potential, available soil water and nitrogen, water and nitrogen stress indices, and risks of exposure to heat or frost stress at critical crop developmental stages. This allows farmers to make more informed decisions about the timing, placing and amount of irrigation to apply, for example.

As Chapter 7 explains, integrated crop management (ICM) is the practice of using multiple agronomic and pest control methods to maximize yield and net returns, minimize inputs while maintaining economic and environmental sustainability. ICM is a holistic framework concerned with optimally and sustainably managing abiotic and biotic production components to allow actual crop yield to approach the genetic yield potential. This is accomplished by maximizing potential yield through improved crop breeding, mitigating production losses by optimizing nutrients, moisture and agronomic practices, and using integrated pest management to reduce losses from weeds, diseases and insects. This chapter describes the role of agronomists in developing ICM strategies. It also shows how ICM principles can be applied. As an example, adoption of no-till and preservation of crop residues can result in colder and often wetter soils at planting, which can also delay seedling emergence. Integration of crop protection inputs such as dual fungicide and insecticide seed treatments can help offset weak systems vulnerable to poor stand establishment and low yield performance. The chapter provides a detailed case study of the use of ICM to control the wheat pest wheat stem sawfly (WSS), combining pre-crop strategies to reduce the local number of adult female WSS available to lay eggs; planting and in-crop strategies to reduce the number of eggs and to reduce egg or larval survival prior to the cutting stage; and harvest management strategies to reduce or prevent damage from cutting and to enhance biocontrol of WSS.

Wheat plays a major role in organic agriculture around the globe. Chapter 8 reviews the current state of organic cultivation of wheat and its close relative, spelt, with a focus on crop diversification, through rotational design, use of species mixtures containing wheat, and diversification at the genetic level, for example through variety mixtures. The chapter shows that there are constraints associated with diversification in organic wheat cultivation systems. Diversifying rotations may dilute wheat production within the rotation; intercrops containing wheat are often characterised by unpredictability and technical difficulties; and diversification of wheat through evolving crop populations may be hampered by seed-borne diseases. The chapter concludes that optimization of organic wheat cultivation systems across the world through diversification depends on a continuous and reliable, systematic and comprehensive collection of agronomic data under conditions of increased crop diversity.

Durum wheat is principally used for the manufacture of pasta, couscous and, to a lesser extent, in the production of bulgur and bread. Chapter 9 explores the challenges and opportunities of durum wheat production in the twenty-first century. After discussing research on durum wheat yields, it covers the major abiotic stresses facing durum wheat cultivation, as well as the principal diseases and insect pests. It describes approaches used to mitigate production constraints and the progress realized through these approaches in breeding in recent years. The chapter also examines end use quality of durum wheat and future trends in the form of emerging breeding technologies.

Part 3 Improving wheat cultivation in the developing world

Chapters 10 to 12 review ways of improving cultivation in the developing, including case studies from Africa, Central and South Asia. It has been estimated that just over half of global wheat production comes from about 150 million hectares cultivated in 75 low and middle income countries, mainly in South, South East and West Asia, North Africa, Middle East, sub-Saharan Africa and Latin America. Wheat production in these regions is mainly carried out by smallholder farmers. Chapter 10 summarises some of the main problems that constrain smallholder wheat cultivation and discuss ways in which these can be addressed. It reviews the key research on the range of constraints facing smallholders, from biophysical constraints to socioeconomic limitations. There is a particular focus on women given their central role in smallholder farming. It reviews how each of these constraints can be overcome. To illustrate what this means in practice, it includes a case study from Kenya. As the chapter points out, new, improved and disease-resistant wheat varieties developed by the Kenya Agricultural & Livestock Research Organisation (KALRO) do not always find their way into the fields of smallholder farmers. Part of the problem is: a lack of awareness and understanding by smallholder farmers on how to grow a good wheat crop, poor access to affordable, genuine, certified seed, and inadequate dissemination platforms to educate and engage smallholder farmers. It describes the two-year Agri-Transfer technology transfer project. This has used measures such as farmer-managed trials and demonstrations, a wheat production handbook and farmer-managed certified seed production schemes to improve adoption rates.

Wheat is the principal staple food in most countries of the Central and West Asia and North Africa (CWANA) region, accounting for 45% of the region's per capita calorie intake with an average wheat consumption of about 200 kg/capita/year. Wheat production in the region has increased in recent decades but most of the countries in the region are still unable to meet their national demand. After reviewing the major wheat production environments in the region, Chapter 11 reviews the main challenges of wheat production, the available wheat production technologies and the ways forward to improve wheat production while conserving the natural resource base in the region. Areas for improvement discussed include the use of: modern varieties (MV), sustainable nutrient and water management, crop rotation systems and other conservation agriculture techniques as well as integrated pest management.

The chapter then shows how these techniques were effectively deployed in sustainable intensification programmes. Different approaches were followed in each country to disseminate improved wheat packages (including improved varieties, fertilizers, irrigation frequency and raised-bed management). Yield results of participating farmers in all countries showed that increases under all production systems (i.e. irrigated, supplementary irrigated or rain-fed) were achievable through the use of improved technologies as compared to the use of farmers' own practices. An average increase of 27% over a five year period was achieved in the fields of participating farmers across all countries involved.

It has been estimated that wheat production in South Asia needs to grow at the rate of 2–2.5% annually until the middle of 21st century to feed its population. As Chapter 12 points out, China exemplifies both the successes and the continuing challenges of improving wheat yields. Chinese wheat production experienced rapid growth after the beginning of hybridization. However, more intensive cultivation has led recently to serious problems such as decline in soil fertility, environmental pollution and concerns about remaining water resources. These problems have resulted in significant yield variations and slowing yield improvement. The chapter reviews a number of key ways of addressing these challenges and improving cultivation in Asia. These include: enhancing and exploiting genetic diversity, improving breeding techniques, methods for dealing with biotic and abiotic stresses, improving aspects of grain quality and developing more efficient and sustainable cultivation techniques such as the use of conservation agriculture (a theme also highlighted in Chapter 3). The chapter includes a case study showing how improvements have been implemented in India.

Summary

The chapters in Volume 2 summarise a wealth of research on ways of improving wheat cultivation. They discuss a range of individual techniques from variety selection, tillage and planting through to water management and reducing post-harvest losses. They also show the broader systems which make it possible to select, combine and apply individual techniques, from crop modelling to conservation agriculture (CA). They also show how a holistic approach, as advocated in integrated crop management, can be used, for example, to offset problems in applying CA techniques in certain conditions. Finally, the chapters highlight how important it is to understand the often complex web of constraints which prevent smallholders from adopting new varieties, methods and technologies and deploying them successfully. They highlight the key importance of engaging with farmers and involving them from the start, for example in participatory breeding programmes, if initiatives to support them are to be appropriate, and innovations designed to help them are to be understood, accepted and adopted.

Key priorities in wheat research: the Wheat Initiative's Strategic Research Agenda

Introduction

Wheat is the most widely grown crop in the world and provides 20% of daily protein and food calories. It is the second most important food crop in the developing world after rice. The strong rise in wheat yields in the post-war period has been associated with genetic improvements in yield potential, resistance to diseases and pests, adaptation to abiotic stresses, and advances in agronomic practises from the green revolution in the 1960s to the early 1990s. However, the rate of increase in wheat productivity has slowed, with large variations of yield between countries. Genetic yield increases of up to 1% per year have been achieved historically, but global 'on-farm' wheat yields are beginning to plateau in many wheat producing areas.

As a result, in recent years wheat production levels have not satisfied demand, triggering price instability, hunger riots, and government instability. With a predicted world population of over 9 billion by 2050 (UN Population Division), the demand for wheat will increase by 60% compared with 2010. To meet this demand, global annual yield increases must rise from the current level of 1% per year (2001–2010) to 1.6% per year (2011–2050).

All wheat producing countries share an urgent need to increase the rate of genetic gains for yield, tolerance to abiotic stresses, resistance to pathogens and pests, as well as to improve input use efficiency for sustainable wheat production, while ensuring high quality, safe and nutritious food products. To take full advantage of wheat genetic potential, improved agronomic practises and the development of innovative cropping systems are paramount. These challenges can be met through an unprecedented effort from the wheat research community to increase and coordinate funding at the international level.

The Wheat Initiative and its Strategic Research Agenda

To meet the expected 60% increase in demand for wheat by 2050 in a sustainable and resilient way, more research is needed through significant investment and collaboration between public and private partners. The Wheat Initiative (www.wheatinitiative.org) was established in 2011 under the aegis of the G20 Ministries of Agriculture to address this challenge by coordinating international research efforts on wheat and by providing opportunities for an increased and more efficient utilisation of resources through the alignment of national and regional activities and pooling of resources.

In order to increase wheat production by at least 60% by 2050, several targets were identified in the Wheat Initiative Vision Document of 2013:

- Increasing the yield potential of wheat cultivars;
- Closing the yield gap on under-performing land and increasing the sustainability of cropping systems;
- Monitoring wheat diseases and developing wheat cultivars with durable resistance;
- Increasing resource use efficiency and tolerance to abiotic stress;

- Improving the nutritional and processing quality and safety of wheat cultivars
- Tailoring wheat cultivars and types to diverse agro-ecosystems and production systems;
- Supporting implementation of modern breeding methods by all breeders;
- Ensuring access to shared platforms and standards;
- Benefiting from a Wheat Information System to provide easy access to data and information.

The Wheat Initiative Strategic Research Agenda (SRA) was developed in 2015 with the aim of achieving these targets in 2030 and beyond through partnerships between private and public organisations. The SRA has been designed to set out clear policy-relevant research priorities for the short (1–5 years), medium (5–10 years) and long term (>10 years), and to list the strategic actions needed to address these priorities.

The SRA is organised into four thematic core themes and two cross-cutting themes:

1 Increase wheat yield potential,
2 Protect yield potential,
3 Protect the environment and increase the sustainability of wheat production systems,
4 Ensure the supply of high quality, safe wheat.

The two cross-cutting themes are:

5 Enabling technologies and shared resources,
6 Knowledge exchange and education

The following sections explore the four main themes of the SRA in more detail, looking at aspirations, challenges, aims and research needs. The text is taken from the main SRA document (http://www.wheatinitiative.org/news-media/wheat-initiative/wheat-initiative-strategic-research-agenda). It has been reproduced with the permission of the Wheat Initiative.

Core theme 1: Increase wheat yield potential

Aspiration: Wheat cultivars with yield potential enhanced by up to 50% in 2040

Yield increases of up to 1% per year have historically been achieved in some regions of the world, but not all. This global increase was obtained by a combination of increased genetic yield potentials and improved crop management practices. Yield potential is defined as the yield of a cultivar when grown in environments to which it is adapted, with non-limiting nutrients and water and with pests, diseases, weeds, lodging and other stresses effectively controlled. Wheat yield potential differs according to the environment in which it is grown. The challenge will be to further increase the yield potential in wheat cultivars adapted to diverse environments and cropping systems by improving the efficiency of light interception and conversion into biomass by photosynthesis and of partitioning into grain.

The current ceiling of genetic yield potential is likely to be the result of constraints in many processes in the scheme of carbon capture, biomass accumulation and redistribution into grain. The biology of yield is very complex and therefore needs to be tackled by a diversity of approaches including:

- Increasing our understanding of the genetic and physiological basis of yield potential;
- Modelling yield potential in different regional areas/conditions and defining adapted ideotypes;
- Improving light interception in the canopy by modifying plant and ear architecture;
- Enhancing solar energy conversion efficiency in wheat by optimisation of photosynthesis;
- Optimising plant phenology, development and partitioning to maximise grain yield;
- Transferring discoveries and technology breakthroughs obtained in other species into wheat.

Breeders can deliver new cultivars with higher yield potential if provided with the germplasm, resources and the tools to do it, as shown by the success that has been achieved to date by private and public breeders in the developed and developing world. However, the current rate of gain is insufficient and significant investment to develop more robust programmes and large international and public-private partnerships are needed. Fundamental and applied research will contribute to enhancing genetic gain for yield potential and is required to translate it to final products (cultivars) and information on how to grow them. These requirements can be translated into four key research activity areas that cover the topics identified above:

- Increasing the knowledge base for improving yield potential, including the genetic control of yield components, physiological and biochemical processes that can increase biomass and harvest index, the optimisation of phenology and plant architecture, the identification or generation of variation in the efficiency of photo synthesis in wheat germplasm;
- Utilising game-changing breeding methods and the latest technologies to improve the effectiveness of breeding for yield potential, such as the use of hybrids, application of 'omics', bioinformatics, and genomics selection;
- Developing new genetic resources for breeders with tools such as interspecific crosses, targeted mutation and genetic engineering;
- Exploiting research results from other genera to improve yield in wheat.

Core theme 2: protect yield potential

Core theme 2 consists of two sub-topics:

- Subtopic 2.1: controlling wheat diseases and pests
- Subtopic 2.2: improving tolerance of wheat to abiotic stress

These are discussed below.

Subtopic 2.1: controlling wheat diseases and pests

Aspiration: Wheat cultivars with durable resistance to most major pests and diseases.

Up to 20% of the global wheat production is estimated to be lost each year due to diseases and pests. This amounts to 140 million tonnes (435 billion dollars at $250/ton) of estimated wheat crop loss in 2012 (FAOSTAT 2014). Globally, current wheat yields would not be achievable without the application of chemical protectants, in particular in developed countries. However, concerns about health and environment call for reduction of chemicals when possible, and for their replacement by cultivars resistant to diseases and pests, which are the prerequisite for environmentally and consumer friendly wheat production. Many farmers in developing countries depend on resistant cultivars. However, most resistance genes, as well as chemical strategies, can be overcome by rapidly evolving pathogens. The occurrence of the new stem rust race Ug99, virulent on 90% of all cultivars grown world-wide when it was detected, yellow rust races adapted to higher temperatures than previous races, potential emerging disease like wheat blast or the rapidly increasing importance of *Fusarium* species causing head scab but also crown and root rot call for more investments in developing wheat cultivars with durable resistance and/or in managing available resistances.

Major wheat pathogens include rusts (yellow, brown and stem rust), powdery mildew, *Fusarium* species producing mycotoxins in the grain leading to its unsuitability for human or animal consumption, and necrotrophic pathogens developing upon dead tissues such as some *Septoria* and *Pyrenophora* species.

While sustained increases in wheat yields are needed to meet future demands, the remarkable ability of pests and pathogens to mutate, produce novel toxins, or otherwise adapt to overcome resistance makes breeding for durable resistance a considerable challenge. Many diseases and pests impact wide international wheat growing zones while others cause local or regional epidemics.

Deployment of host resistance to diseases and pests is the most effective and environmentally sound way of plant protection, as has been demonstrated recently by the Borlaug Global Rust Initiative (BGRI) in the case of resistance against the stem rust race Ug99 which is threatening wheat production in Africa and Asia. Indeed, global coordination of efforts are needed to develop management and gene stewardship deployment strategies with the ultimate aim of achieving broad spectrum and durable disease resistance in wheat crops. This will involve:

- Continued identification of novel sources of disease and pest resistance including the newest technologies of phenotyping to broaden the genetic base of resistance against primary wheat pathogens and pests;
- Development of diagnostic genetic markers to assist breeding programmes in the development of disease resistant cultivars;
- Deep knowledge on wheat-pathogen interactions, in relation with the environment and management systems, and use of this knowledge for the development of new resistance strategies;
- Monitoring of pathogen populations for their ability to overcome resistance genes;

- Understanding the impact of climate change on pathogens in order to assess potential new epidemics;
- Deployment of cis/transgenes to augment strategies towards durable disease resistance;
- Coordinated efforts in pre-breeding and breeding programmes to optimise strategies for the efficient use of resistance genes in wheat and to ensure their sustainable deployment by developing resistance management systems.

To improve resistance of wheat to pests and diseases, an efficient collaboration between phytopathologists, geneticists, molecular biologists and breeders is needed to cover the whole chain from the identification of new races of pathogens, the development of new resistant cultivars to integrated pest management.

There are two key factors for improving resistance of wheat to diseases and pests: pathogen surveillance and the identification of new sources of resistances in the wheat gene pool. Completed in parallel, it will be possible to develop novel wheat management regimes that resist the most economically important wheat pathogens and pests.

Identified resistance sources will be analysed in terms of the genetics of the respective resistances followed by the development of molecular markers and the development of efficient (pre-breeding) strategies for marker-based introgression ideally as multiple gene combinations into adapted cultivars.

Genomic tools currently available in wheat and developments in the near future will pave the way for resistance breeding at the allele level. To achieve this, special emphasis has to be given to the isolation of resistance genes from wheat and wild relatives on a large scale, especially for those proven to confer durable resistance.

Knowledge of the sequence of resistance genes will facilitate the detection of the natural allelic diversity and the estimation of the effectiveness of different alleles and provide an opportunity for the creation of new, functional alleles by site directed mutagenesis. With the development of genome editing techniques, it will also be possible to perform targeted editing of genes conferring susceptibility with the goal of generating novel resistant phenotypes. Furthermore, isolated genes will facilitate a better understanding of genetic networks leading to resistance in order to improve resistance and to create durable resistances.

Besides research on the host's natural resistance, a deeper understanding of the biology of wheat pathogens and their evolution is needed with respect to wheat-pathogen interactions, effector diversity and the epidemiology of diseases. This also holds true for knowledge concerning the effect of the wheat microbiome on resistance.

Subtopic 2.2: improving tolerance of wheat to abiotic stress

Aspiration: Wheat cultivars better adapted to diverse environmental conditions in terms of yield and yield stability under stress.

Environment stresses, particularly heat and drought but also salinity, soil acidity, frost and many other factors severely limit wheat production globally. Wheat yield models indicate that a 1°C temperature increase has the potential to restrict wheat yields by 10% in some parts of the world and that the wheat producers in South Asia and North Africa will be hit hardest by climate change. Wheat yield in 2050 could decline by 27% compared with 2000 in some regions under the predicted climate change scenarios.

In most production environments, multiple abiotic stresses challenge wheat plants simultaneously. While tolerance to some stresses is under simple genetic control, others are regulated by multiple interacting genetic mechanisms. Managing tolerance to abiotic stresses will require:

- Characterisation of the environmental and abiotic stresses that limit wheat production in different regions of the world, including assessing the impact of climate change, and development of conceptual models of adaptation to different mega-environments;
- Development of techniques for assessing stress adaptive characteristics of germplasm under field conditions as well as under controlled conditions where appropriate;
- Identification of sources of tolerance to individual stresses and combinations of stress from a broad base of genetic resources;
- Genetic, physiological and biochemical analysis of stress response pathways and processes in wheat;
- Refinement of genetic and physiological models for enhanced stress tolerance;
- Building resilience into cultivars so that they can cope with variable environmental conditions;
- Isolation of genes and alleles underlying stress tolerance loci.

Research needs to address the complexity of stresses faced by wheat producers across highly diverse production environments. Higher plants have evolved multiple, interconnected strategies that enable them to survive unpredictable environmental fluctuations. However, these strategies are not always well developed in the cultivars grown by wheat producers and most of the strategies used in model non-crop species are focused on plant survival at the expense of yield.

The Intergovernmental Panel on Climate Change has predicted that rising temperatures, drought, floods, desertification and weather extremes will severely affect agriculture. For wheat, the genetic control of traits determining yield in water limited and low yielding environments are generally expected to be of low heritability, polygenic and many of the key loci will show epistatic rather than additive effects. Breeding and genetic techniques need to detect and select for these types of loci while also accounting for factors such as maturity, height, resistance or tolerance to soil borne diseases. Tolerance to related stresses such as boron, acidity, salinity and nutrient deficiencies must also be taken into account.

While breeders have been successful in improving performance through direct selection for lines under a range of environmental condition, the opportunity now exists to integrate other approaches. Areas where greater research activity will offer significant advances include:

- Climate modelling and monitoring, including the characterisation of the stresses and probability of stress in different wheat production environments;
- Statistical tools for the analysis of field and controlled environment trials;
- Comprehensive models of physiological and biochemical responses to stresses and stress combinations;

- Field phenotyping methods and well-characterised field sites that have been validated in breeding and gene discovery work;
- A wide range of genomics technologies;
- Studies of the impact of abiotic stress on grain quality;
- Information from other plant systems both crop and non-crop;
- Impact of cropping systems on managing stressful environments.

Core theme 3: protect the environment and increase the sustainability of wheat production systems

Core theme 3 consists of two subtopics:

- Subtopic 3.1: nutrient use efficiency
- Subtopic 3.2: agronomy and crop management

These are discussed below.

Subtopic 3.1: nutrient use efficiency

Aspiration: Increase the efficiency of nutrient use so that the applied fertiliser is used by the crop, for example increase N use of over 60% of the applied amount.

Adequate nutrition, and particularly nitrogen availability, is essential for optimal yield and quality of wheat crops. However, given economic and environmental costs, efficient use is essential. More than 25 Mt fertilisers are used annually on wheat with a global nitrogen (N) use efficiency as low as 30% and high discrepancies between regions. It is, for instance, estimated that N use efficiency is twice as high in Western Europe compared with India and China, and 50% more efficient than in the USA. Micronutrients, particularly iron (Fe) and zinc (Zn), are important for both plant health and the nutritional value of the grain for humans. Rising CO_2 levels is expected to decrease levels of these nutrients in grain and higher yields are frequently associated with decreases in nutrient density.

There is a major opportunity for improvement in both the genetic and the agronomic components of wheat nutrient use efficiency, including:

- Enhancing use efficiency for all the macro- and micro-nutrients important for optimal crop production. Special focus should be given to nutrients that are either costly to produce and supplied in excess of plant requirements in some agro-ecosystems (e.g. N), derived from finite rock mineral resources (e.g. phosphorus), have negative environmental impacts if inappropriately used (e.g. N), and are essential elements of human and animal diets (e.g. Zn, Fe);
- Combining cultivars adapted to optimal agricultural practices (Subtopic 3.2) and optimising nutrient uptake by the roots, nutrient utilisation to produce biomass and nutrient translocation to the grain to ensure quality;

- Integration of genetics, ecophysiology, rhizosphere microbiology and symbiotic interactions to identify traits and chromosomal regions relevant for nutrient use efficiency improvement;
- Taking into account global change impacts on nutrient use for production and quality.

Research is needed to maximise capture, partitioning and remobilisation of nutrients in the canopy to the grain through genetic improvement. Research will be targeted primarily to:

- Improve the capacity to phenotype nutrient use efficiency in a standardised way in terms of definitions and practical aspects such as controlling the environment to create defined nutrient availability and measurement technologies to evaluate the responses of the plant/crop at different physiological levels;
- Develop pre-breeding programmes to facilitate the use of genetic resources (landraces, synthetics, wild relatives) for traits that will enhance nutrient use efficiency. This may include further research on increasing leaf/canopy photosynthesis per unit of N using the existing genetic variability;
- Favour beneficial interactions with soil microorganisms and enhance the capability to capture N and other nutrients (root architecture, mycorrhizal/diazotroph associations, chemical and biological nitrification inhibitors, root exudation of organic acids);
- Identify loci and alleles involved in enhanced nutrient use efficiency. This approach may be combined with the development of crop simulation modelling, sensitivity analysis for nutrient use efficiency traits as well as the identification of chromosomal regions associated to model parameters.

Subtopic 3.2: agronomy and crop management

Aspiration: Develop highly stable wheat production systems that integrate novel genetic traits, deliver attainable yield goals and quality targets, enhance environmental sustainability, and mitigate production threats.

The yield gap, as defined by the difference between the yield potential and the average farmers' yield in a given environment and time, varies enormously between regions of the world with different climate and cropping systems.

Crop management plays an important part in the expression of the genetic potential and in the sustainability of crop production systems:

- Roughly one-half of yield potential is realised through agronomic factors; thus, attainable wheat yield and improved system stability can only be achieved or possibly exceeded when genetics are coupled with innovative integrated crop management strategies, including those based on principles of conservation agriculture;
- If wheat used more of the annual rainfall (not just that from the growing season) grain yield could be increased. Enhanced water efficiency combined with emerging water use technologies and precision tools provide opportunities to reduce the gap;

- In addition to water, nitrogen is the major limiting factor in wheat. Management strategies, including placement and foliar feeding should aim to enhance use efficiency and/or limit losses to groundwater or atmosphere;
- Agronomics must include components to minimise losses from diseases, weeds, or insect pests.
- The best combinations of agronomic practices and genotypes will therefore improve the soil characteristics (pH, organic matter, water infiltration rate, water holding capacity), irrigation efficiency, drought tolerance, and reduce the chemical load brought in by fertilisers, pesticides and herbicides.

Bringing together experts from a broad range of disciplines will contribute to the refinement of wheat production systems to meet the global challenges facing wheat growers and end-users today and well into the future through:

- Site-specific technologies tailored to the needs of different production systems. An understanding of the trade-offs as agronomics are manipulated is needed so that the proposition of solutions is meaningful;
- Integration of new cultivars adapted to innovative and diversified management systems;
- Advancements in farming equipment (e.g. seeding equipment) and precision agriculture tools as key components of agronomic solutions and decision support systems.
- These knowledge-based tools and next generation precision agriculture approaches will be key to enabling farmers to reach optimal wheat yield and quality with minimum resources irrespective of agro-ecological production areas.

Research is needed to develop highly productive and stable cropping systems with enhanced environmental sustainability. This will include:

- Development of holistic and integrated agronomic strategies to mitigate the negative impacts of climate change and to enhance environmental sustainability;
- Optimal crop residue management, no-till and cover crops;
- Development of site-specific resource optimisation, integrated nutrient management and wheat rotational strategies for the long-term sustainability and stability of wheat production systems;
- Integration of novel genetic traits and precision agriculture tools into innovative cropping systems;
- Intensified manipulation of agronomic factors to fully exploit wheat yield potential and technological quality to meet end-use requirements;
- Development of decision support systems to mitigate insect, weed and disease related threats to wheat production;
- Agronomics to improve both crop water use and efficiencies in rain-fed and irrigated moisture regimes;
- A knowledge transfer strategy to ensure uptake of new knowledge and innovations at the farm gate and to update scientists on changing field realities.

Core theme 4: ensuring the supply of high quality, safe wheat

Aspiration: Production of nutritious and healthy wheat grains with adequate end-use quality to meet the demand from subsistence users to industry and produce the great variety of food products made from wheat.

Wheat produces 20% of our daily protein and food calories. Wheat grain proteins are an important determinant of the quality of end-products such as bread and pasta but may also cause allergenic reactions. Grain carbohydrates affect food processing conditions, as well as the texture, shelf life, and digestibility of end products. Non-starch carbohydrates are of interest in reducing obesity and the incidence of cardiovascular diseases, type 2 diabetes and some cancers. Wheat is also an important source of dietary fibre, minerals and vitamins (notably B vitamins) in the human diet and also contains a wide range of bioactive phytochemicals, many of which have been proposed to have health benefits. Understanding variation in the grain content and composition is necessary to enhance their beneficial health effects and to reduce negative effects. In addition, grain processing plays an important role in determining the availability, functionality and physiological effects of the grain components.

Improvement of quality and safety necessitates the integration of knowledge from multiple disciplines like cereal chemistry, microbiology, human nutrition, toxicology, agronomy and genetics. This complex task will involve studies considering:

- Gluten and starch composition and grain hardness;
- Human nutrition and biofortification;
- Food safety;
- Food technology and chemistry;
- Local breeding and environmental stability, combined with the extremely diverse end-uses of bread and durum wheat;
- Developing genomic resources for breeding for end-use quality;
- Tuning protein composition to maintain end-use properties while reducing the protein content, resulting in improved nitrogen use efficiency;
- Specific traits for specific end-uses, including engagement with processors and nutritionists.

Future research on wheat quality and safety should aim at:

- Standardising methods to examine gluten proteins and improve understanding of the role of gluten proteins on different dough processing conditions and end-product properties, including unifying the nomenclature of gluten alleles;
- Germplasm screening for the identification of sources of variation for various quality component traits;
- A deep understanding of the inheritance of grain bioactive compounds, and the genetics of the bioavailability of micronutrients and high dietary fibre to improve the nutritional value of wheat and increase healthy cereal-based food consumption;

- A deep understanding of the nature and content of proteins of wheat showing allergen and toxic reactions and developing low-allergen wheat suitable for patients suffering in wheat related food disorders;
- Understanding the effects of food manufacturing processes on the digestibility of wheat proteins, bio-availability of nutrients, and the interaction with gut micro-organisms;
- Fine tuning gluten, starch properties and grain hardness according to specific (and diverse) end-uses by understanding Genotype x Environment x Management interaction;
- Reducing mycotoxins (see core theme 2, subtopic 2.1) and toxic minerals in wheat and wheat products;
- Development of low cost biomarkers for the above determinants of wheat quality and safety.

Cross-cutting theme 5: enabling technologies and shared resources

Cross-cutting theme 5 consists of three subtopics:

- Subtopic 5.1: enabling technologies and methods
- Subtopic 5.2: shared platforms and technologies
- Subtopic 5.3: genetic resources

These are discussed below.

Subtopic 5.1: enabling technologies and methods

Aspiration: Wheat to become a preferred research target for scientists as a result of freely available technology, germplasm and resource base.

Wheat has a large and complex genome, is grown in diverse production environments, and has broad germplasm resources. The very best tool-kit of crosscutting technologies and methodologies, together with increased characterisation, use, and sharing of available genetic diversity of wheat, is needed to support the challenges faced by wheat researchers and breeders.

There have been rapid advances on techniques supporting biological research, many coming out of medical research or studies in model crops. Supporting the rapid transfer of these technologies to wheat and developing new methods and resources based on wheat represents a key aim of the Wheat Initiative. The main aims of this core theme are:

- Development of cross-cutting enabling technologies in support of core themes 1–4 to address shared platforms and methodologies for breeding, genotyping and phenotyping;

- Completion of a high quality reference sequence and full annotation of the wheat genome to serve as a resource for marker development and gene discovery in wheat;
- Completion of a 'Wheat Information System' (WheatIS) to provide the international wheat community a single, comprehensive access point for wheat data and informatics tools;
- Development, characterisation and deployment of genetic resources, and technologies to enhance the rate of genetic gain in wheat breeding programs.

Research needs include a reference sequence of the wheat genome which will build a foundation to enable gene isolation, functional analyses, new allele discovery for pre-breeding, polymorphism discovery for marker-assisted selection, and an increased understanding of the impact of epigenetic modifications and transposable elements on gene regulation. Generation of a reference sequence anchored to genetic and phenotypic maps is the focus of the International Wheat Genome Sequencing Consortium (IWGSC). IWGSC is coordinating research to complete the bread wheat genome sequence by focussing on:

- Completion of genetically anchored physical maps for all 21 wheat chromosomes;
- Sequencing of the minimum tiling-path of BAC clones derived from the physical maps;
- Assembly and annotation of pseudo-molecules for each of the 21 wheat chromosomes;
- Integration of the sequence data into the WheatIS.

Parallel research activities to integrate whole genome shotgun sequencing and improve the definition of the wheat gene space as well as support for the annotation of the reference genome sequence are also a priority. Sequencing of the durum wheat genome, using the hexaploid wheat genome reference sequence as a foundation, is a high priority for the durum wheat genetics and breeding community. In addition to the genome sequence, large RNAseq data sets are being produced for wheat as well as proteomic and metabolomics data. These will be related to the genome sequence and phenotypic traits, and integrated into the WheatIS. Additional genomic resources, necessary to answer post-genomic challenges such as diversity, population genetics, regulation and epigenetics, also need to be integrated into the WheatIS.

In addition, research into novel methods, tools and technologies will also facilitate the development of widely-adapted elite genotypes and accelerate selection cycles to reduce time from discovery to the target market. Predictive science has the potential to improve the efficiency and sophistication of breeding by phenotyping only those genotypes with the greatest potential for cultivar release. Research to develop new algorithms and predictive models for genome wide-association mapping studies and implementation of genomic selection in applied breeding programmes is needed.

Recombination underpins the generation of new genetic combinations through crossing and provides the raw material for selective breeding. Recombination is mostly limited to distal regions of wheat chromosomes, restricting new combinations of alleles to these regions in breeding populations. Modulation of recombination along the chromosomes should allow new genotypes to emerge as well as facilitate the introgression of useful genes from wheat related species.

Deployment of hybrid wheat will likely contribute to wheat productivity and yield stability in the future. Research focusing on defining heterotic pools to maximise hybrid vigour and yield, and to develop effective pollination control systems for cost-effective hybrid seed production systems are needed.

Genetic engineering through mutation, genome editing and transformation is a critical tool for the identification and characterisation of wheat genes and has the potential to generate new genetic variation that will be valuable in breeding new cultivars. International coordination of well-characterised TILLING (Targeted Induced Local Lesions in Genomes) populations, and research programs to allow full implementation of gene and genome editing technologies is needed.

Although the techniques for producing and evaluating transgenic wheat lines are now well established, the technology is closely regulated and the regulations vary across jurisdictions. In many regions, field evaluation of transgenic wheat is not permitted or is complex and expensive. Currently, no transgenic wheat is under commercial production. Other technologies, such as genome editing, are now being applied to wheat but it is still unclear how this technology will be viewed by different regulatory agencies. The Wheat Initiative supports the use of genetic engineering to advance wheat research and looks forward to its responsible application to wheat improvement. Given the importance of genetic engineering to wheat research and the variation in regulatory requirements, the Wheat Initiative encourages partnerships and collaborations that assist in the development and evaluation of genetically engineered wheat breeding lines and genotypes, while respecting jurisdictional requirements and sensitivities.

Finally, as the global climate is changing, the interactive effects of plant traits, environment and management must be understood under region-specific climate change scenarios. The ability of process-based crop models to simulate the interactive effects of plant traits, environment and management makes such models attractive tools for plant breeding. However, research is needed to improve model responses to disease and pests, climate and soil factors, including high temperature, elevated atmospheric CO_2 and variable water and nutrient supply as well as their interaction with specific crop traits, and to develop a framework to handle crop model uncertainties and enable model applications to:

- Characterise and define target environments including climate change 'hotspots';
- Test the potential impact of specific morphophysiological traits on yield, quality and resource use for key environments and climate change hotspots;
- Explore possible ecophysiological mechanisms affecting traits;
- Assist in understanding interactions of traits, environment and management;
- Improve the characterisation of traits in genetic studies and breeding nurseries.

In recent years, several international projects have led to an explosion of phenotypic data and detailed characterisation of wheat molecular variation, differential transcriptome profiles, metabolite profiling and proteomic characterisation. Research focusing on a systems biology approach is a high priority, where integrating multiple -omics platforms and sophisticated bioinformatics systems can be used to unravel the underlying mechanisms of plant growth and development, adaptation to harsh environments, and resistance to biotic stresses. This is a growing area in model plant research and it is important that it is also captured for addressing the needs of wheat improvement. Incorporation of major adaptation gene information into wheat models is a first step in this process.

Subtopic 5.2: shared platforms and technologies

Integrated research approaches require combining data and information from all areas of wheat research. We need to develop an integrated information system that provides access to information that can then be applied to wheat research, practical breeding, crop modelling and wheat management strategies. Currently, wheat data are independently dispersed among several databases around the world which limits their full potential for exploitation by the wheat research and breeding community.

The development of a single-access web-based WheatIS (www.wheatis.org) linking existing and new databases will allow sharing and integration of data resources as well as provide a common platform housing available bioinformatics tools. To achieve the full potential of the Wheat Information System, efforts are needed to:

- Define standards, protocols and processes for sharing and integrating wheat data in keeping with other international initiatives;
- Develop a web platform allowing the exchange of standardised data files and associated meta data between different repositories;
- Develop a distributed index search engine to allow dynamic search of different databases through a single entry point;
- Develop a robust and fast distributed infrastructure;
- Provide user-friendly bioinformatics tools to use the data efficiently;
- Establish links with the wheat users' community through outreach, training, and dissemination activities.

Novel high throughput phenotyping platforms for the field and controlled environment are also allowing precise measurement of wheat traits, either for phenotypic selection or for genetic studies designed to dissect complex traits. To reduce the gap between genomics and phenotyping for wheat improvement, it is necessary to characterise existing infrastructure and to strengthen international initiatives aimed at developing new technologies for wheat phenotyping, including protocols for simulating target environments and for trait phenotyping and/or identification. Priorities include:

- Improving the precision of current phenotyping techniques;
- Establishing common protocols and reference germplasm to facilitate data comparison and exchange;
- Developing low cost, high throughput tools for field-level screening and for specific traits;
- Assessing new technologies to study currently inaccessible characteristics, e.g. non-invasive detection of root architecture and function.

High-throughput phenotyping facilities and the use of modern sensor technologies is expected to result in the generation of large volumes of data. Therefore, a concerted effort is needed to develop efficient tools for the analysis, storage and publication of high-resolution phenotyping data. There is also a critical need for research to evaluate and improve the prediction of trait effects measured in controlled conditions relative to performance in the field. Defining global sets of reference cultivars and defining standards for phenotypic and environmental data collection are a priority to allow for comparison and statistical integration of phenotypic datasets between locations and years.

Finally, high throughput genotyping methods are providing an unprecedented supply of molecular markers for anchoring genomic sequences that will allow the genetic dissection of complex wheat traits, for use in wheat breeding programs. The number of Single Nucleotide Polymorphism (SNP) markers in wheat is no longer a limitation for application into breeding and several SNP detection platforms based on different technologies are already available to the community. However, such technologies are not always accessible to breeders with limited resources and it is important that training and access to resources are supported within the wheat community. Further research into common high throughput Genotyping by Sequencing (GbS) and other genotyping/sequencing strategies, as well as association of genetic variation with plant phenotypes is needed to facilitate wheat genetic improvement.

Subtopic 5.3: genetic resources

Utilisation of genetic resources is the basis of wheat breeding and is therefore fundamental to sustaining global wheat production. However, part of the genetic diversity present in landraces of wheat and in ancestral species was not captured or was lost during wheat domestication and subsequent selective breeding. Introgression of novel functional variation found in wheat landraces, old cultivars and related species into adapted backgrounds will be critical to enhancing yield potential of wheat and to provide new sources of resistance and tolerance to biotic and abiotic stresses. Diversity plays also a central role in a hybrid wheat approach. However, the use of genetic resources in breeding is currently limited, largely because there has not been a global assessment and characterisation of the hundred thousands of accessions conserved in genebanks, leading to scarce genotypic and phenotypic information. Additionally, gene transfer from related species has been hampered by linkage to deleterious traits and by biological aspects such as differences in ploidy level or genomes involved. In addition, information is dispersed among genebanks, which makes systematic access from the wheat research community difficult.

Ex-situ collections of wheat and related species held in genebanks need to be managed securely in a cost-effective manner and accessed easily and ideally freely by a broad range of users. While great advances have been made in recent years, further support is needed to help genebank managers link and operate individually held collections within a global system and to support the wheat global ex-situ conservation and utilisation strategy through efficient coordination, capacity building, integrated information systems, and by rectifying key gaps in existing collections including genomic and phenotypic data. Specifically, efforts to update and implement the global wheat conservation strategy should be supported with an emphasis on the use of wheat and related species germplasm in breeding.

Current work on the development of a single access web-based system enabling cross-searching of multiple gene banks must be supported to facilitate sharing of genetic material. This will be further enhanced by the inclusion of bioinformatics tools linking them with systematic genomic and phenotypic data (for example through the WheatIS).

To facilitate utilisation of genetic resources, new methods for their effective incorporation and uptake into breeding programs are required. This involves developing reliable phenotypic and genotypic data on accessions in gene bank collections as well as pre-breeding efforts. The wheat genetic resources community needs to operate as a global partnership to ensure the cohesion, security, functionality and use of germplasm collections following the recommendations of the International Treaty on Plant Genetic Resources.

Crosscutting theme 6: knowledge exchange and education

Aspirations:

- *A well-connected public-private community of wheat researchers sharing knowledge and information throughout the world, self-perpetuating and expanding by means of training, education, networking and mentorship*
- *A broad understanding of wheat, wheat science and of the role of research and technology in addressing the challenges of facing global wheat production*

Exchanging knowledge between wheat researchers globally and building capacity The need to exchange knowledge to strengthen wheat research globally and to enhance public-private partnership appeared in all the areas covered by the previous scientific core themes of the SRA. Education in wheat research was pointed out as a necessity by scientists and policy makers. For example, industry has difficulties recruiting wheat breeders because there are not enough being trained. In addition, capacity building in agriculture in developing countries was identified as a means to address more efficiently the needs faced in specific regions of the world. Finally, a major limitation to the use of modern technology is variable community support and understanding of the potential of these technologies to solve food insecurity issues.

Mentoring and networking are critical components for success in all disciplines of science, and even more important in smaller scientific communities, such as wheat improvement, where resources are limited but the need for impact is great. Mentoring is considered so important for the development of scientists that the US National Science Foundation now requires its grantees to provide information on how they will mentor postdoctoral fellows.

The working world of scientists is changing. For the same reasons that the Wheat Initiative must work to strengthen public-private partnerships in research and product development, we also must acknowledge and incorporate all of the roles that researchers must play to be successful in their careers – such as manager, mentor, administrator, and scientist – which also in turn contribute to the successful delivery of Wheat Initiative strategic objectives.

Education, training, networking and mentorship are especially important for engaging and retaining early career researchers, especially from the developing world, as well as female scientists. The Wheat Initiative's strategic goals will only be achieved if a vibrant community of scientists is fostered and supported well into the future. They must be supported to establish and maintain quality standards for research, data collection, and knowledge sharing, but also to develop a collective voice to be able to advocate for needed funds, policy support, or public attention. Such a community will likely further contribute to the Wheat Initiative's strategic goals by engaging with the public on occasion for the purposes of dialogue and learning in areas of interest and importance to them.

Addressing the challenges of communication, training and education will be the responsibility of people involved or connected to wheat research and industry. The Wheat Initiative will aim to support these activities, help develop generic tools and programmes and facilitate the development of training and communication partnerships. This will be achieved through:

- Developing networking activities to build communities around common research objectives;
- Encouraging and facilitating mobility of researchers globally and at all levels to foster transfer of knowledge;
- Organising seminars and conferences to foster creative thinking and cross-disciplinary exchanges;
- Organising training workshops around specific areas of research and technology for the scientific community, wheat growers and end-users;
- Fostering education and training of early career scientists in wheat science, in developed and developing countries;
- Advertising training opportunities and fostering social networks of scientists, both virtually and face-to-face;
- Supporting scientists' efforts to engage with public audiences and the media to contribute to dialogue and learning.

Summary

The Wheat Initiative SRA emphasises the importance of research in areas such as:

- Better understanding of plant physiology, genetics and genetic diversity
- The combination of this understanding with advances in breeding techniques to produce higher-yielding, more resilient cultivars
- The development of cultivars with greater nutrient use efficiency, resistance to pests and diseases as well as resilience to abiotic stresses such as drought, together with enhanced nutritional quality
- The development of more efficient, sustainable and 'climate smart' methods of cultivation to make the most of new cultivars

Above all, the SRA emphasises the critical role of collaboration and partnership, of shared resources, knowledge exchange and education in deploying research effectively.

The two volumes of *Achieving sustainable cultivation of wheat* both share these aspirations and seek to play their part in helping to achieve these goals. Volume 1 addresses key issues set out in Core Themes 1, 2 and 4, whilst Volume 2 explores aspects of cultivation in Core Theme 3. The bringing together of an international team of expert contributors to review and summarise the wealth of research in wheat science exemplify the goals set in Cross-cutting Themes 5 and 6. It is hoped these volumes will provide a helpful foundation on which future research can build.

Part 1

Wheat cultivation techniques

Chapter 1

Variety selection in wheat cultivation

Arun Kumar Joshi, International Maize and Wheat Improvement Center (CIMMYT); Vinod Kumar Mishra, Banaras Hindu University, India; and Simanchal Sahu, Orissa University of Agriculture and Technology, India

1 Introduction
2 Wheat variety selection methods: natural and traditional selection
3 Wheat variety selection methods: modern molecular breeding
4 Variety selection by plant breeders
5 Variety selection by farmers
6 Conclusion
7 Where to look for further information
8 References

1 Introduction

The development of new varieties of crop plants largely occurs through variation and selection. Variation, defined as the emergence of differences between individuals in a population, serves as the raw material for varietal development or plant breeding. Using important traits as the criteria, superior plants suited to particular environments, management methods or markets can then be selected. For example, if a variable population is exposed to selection in the presence of a commonly grown variety (known as a check variety), some genotypes may perform better than the check. A breeder or farmer may then be able to select a genotype with greater potential than the commonly grown variety. The superiority of an identified line is normally validated through a series of trials, which often take the form of a multi-year multi-location evaluation. Once the superiority of a new variety has been established, it is registered and released for cultivation by a competent authority. In many countries, a committee is responsible for identifying new crop varieties developed by the public sector. In India, this is the task of the Varietal Identification Committee, which usually takes decisions during an annual national workshop. Following this, the government again verifies the superiority through another committee and then issues an official notification for the release of variety for cultivation by farmers. There are typically also standard identification procedures for varieties developed by the private sector, but these differ between countries. Seeds are then produced following a defined procedure, so as to maintain the superior variety's physical and genetic purity and to meet minimum seed standards. They are then disseminated among farmers and other stakeholders.

http://dx.doi.org/10.19103/AS.2016.0004.11

2 Wheat variety selection methods: natural and traditional selection

Varieties are now selected for cultivation according to a well-organized procedure; however, even before humans started experimenting with improving native plant populations, natural selection played its role in the evolution of different crop species. During prehistoric times, selection occurred in nature and man chose desirable crops or plants. Artificial selection became a fundamental crop improvement process when crops were first domesticated for agriculture. Humans then started modifying within individual crop populations, guided by their own needs. The selection procedures adopted during prehistory were not based on sound scientific knowledge, but gradually the scientific principles of plant science were discovered. A turning point was the discovery of the laws of inheritance by Gregor Mendel in 1866. Currently, leading-edge techniques such as molecular tools have entered the selection arena to further enhance our ability to choose the desired genotype.

2.1 Natural selection

Natural selection is inherent in a genetically heterogeneous population at any place or time, and is the only known process that can produce adaptive evolutionary changes (Silvertown and Doust, 1993). It manifests itself through differences in genotypic fitness, which is a complex parameter encompassing adaptation, survival ability and viability. *On the Origin of Species* by Charles Darwin (1859) is a classic work on the role of natural selection in the evolution of crop plants. Darwin argued that though there were morphological species types, these arose by natural selection among the variable members of the previous species. Francis Galton was the first to apply statistical methods to the study of differences in human populations and believed that nature, and not nurture, determined hereditary (Gillham, 2001).

Before domestication, crop species were subject only to natural selection. The basis for natural selection was adaptation to the prevailing environment. The plant types more adapted to their environment produced more progeny than others, hence were selected by nature. However, the role of natural selection has been greater for some plant characteristics than others. Photoperiod adaptations, for example, are considered the most important plant characteristic because day neutral crops (such as wheat and rice) are more widely adapted than day sensitive crops. Other important characteristics that have influenced natural selection are tolerance to biotic and abiotic stresses, dormancy, quality and seed shattering. Natural selection generally favours heterogeneity as it has more buffering capacity. Present-day agriculture, however, has moved towards growing homogeneous populations of crops for morphological features, quality and also for disease resistance. In remote areas of many parts of the world, such as the hills of Bhutan and Nepal, landraces have emerged, which are more genetically diverse than other varieties, probably due to the larger role of natural selection in subsistence farming, compared to more industrialized agriculture.

2.2 Selection by humans

The selection of varieties by humans, termed selective breeding, is the result of a conscious decision by a farmer or plant breeder to keep the progeny of a particular parent in preference to others in order to retain or reject a particular plant characteristic. For example, in wheat, selection may be for large heads and seeds, attractive grain colour, synchronous flowering, early maturity, pleasing taste or superior chapati or bread quality. In fact, selection by humans has brought about changes in almost all of the characteristics of wheat. Arnold (1985) has suggested that all crops were developed by humans through conscious selection, whereas Jenkins (1966) had argued the opposite. Different phases in the human improvement of crops by variety selection can be distinguished: the era of domestication, the era of pre-Mendelian selection, the post-Mendelian era and the modern era of crop breeding. The rate of change has accelerated recently due to rapid advances in knowledge in agricultural science.

Role of domestication

Domestication, the process of bringing wild species under human management, was the first step in crop improvement by humans, which led to the beginnings of agriculture and the development of cultivated plants. The domestication of crop plants began around 10 000 years ago (Murray, 1970; Renfrew, 1973). Over years of cultivation, man has selected, knowingly or unknowingly for the characteristics that make plant populations more suited to his needs. Hence, there are now large differences between cultivated plants and wild populations surviving in nature. For wheat, domestication is believed to have occurred about 10 000 years ago in Asia, in the Fertile Crescent of the Middle East. The staple crops in this region used to be einkorn (*Triticum monococcum* L.) and emmer (*T. turgidum* sp. Dicoccum L.) wheat, both non-brittle types. Natural crossing among different species followed by allopolyploidization and mutations in genes governing threshability and other traits, however, led to the formation of today's economically important bread wheat *Triticum aestivum* (Faris, 2014).

Variety selection in the pre-Mendelian era

Choosing the right kind of wheat planting material has long been a significant agricultural practice. There are references, for example, to the selection during Roman times of the biggest ears and the largest kernels (Jenkins, 1966). Prior to 1800, very little plant research could be considered plant breeding; however, some of the botanical studies of that time laid the foundations for modern plant breeding. The Vilmorin Company, established in France in 1727, contributed extensively to the development of plant breeding knowledge and the production of improved varieties. Then, in 1819, Patrick Shirreff began his experiments into wheat and oat hybridization and selection with the purpose of obtaining superior cultivars (Stoskopf et al., 1993), and successfully developed a new oat variety in 1824 and a new wheat variety in 1832. He was among the first cereal breeders to develop the principle of selecting pure lines. His work was based on selecting individual plants from heterogeneous populations, in which pure line selection had never previously been practised (Stoskopf et al., 1993). Red May, dating from 1830, was perhaps the first wheat variety produced in the United States by selection. Then in 1837, Henry Zemmerman was one of the earliest individuals in North America to select a pure line wheat cultivar (Ball, 1930). In the same year, Le Couteur, a farmer from the Isle of Jersey, selected wheat based on form difference and subjected the isolates to progeny tests (catalogue.nla.gov.au/Record/813341). Red Fife wheat, a selection made in 1842, set a new performance standard for yield and quality. The pedigree method of selection was developed in 1890 and was in use in Great Britain, Sweden and other European countries before the turn of the century. It was developed when it was realized that genetically superior plants are lost when selected plants are cultivated in bulk. The rediscovery of Mendel's laws of inheritance (dating from 1866) in 1900 by Correns (Germany), De Vries (Holland) and Tschermak (Austria) led people to understand the laws of heredity and especially how hybridization and mutation could create totally new variations for developing and selecting a variety.

Variety selection in the post-Mendelian era

Natural populations of wheat are heterogeneous mixtures of several homozygous genotypes, thus several pure lines can be selected from such populations. In 1910, W. H. Nilsson was the first to realize that a pure line can be produced from a population through descent from the self-fertilization of an individual homozygous plant (Stoskopf et al., 1993). Earlier, in 1903 Johannsen has proposed the concept of pure line based on his studies with beans (*Phaseolus vulgaris*). He explained the genetic basis of pure lines, which was subsequently popularized by Vilmorin in France and Nilsson Ehle in Sweden. In pure line selection, a large number of plants are selected from a wheat population (naturally available or created through hybridization or any other means) and are harvested individually. Their individual progenies are evaluated, and the best progeny is released as a pure line variety. In this method, selection is based on progeny testing. It is used for improvement of local varieties and also for handling segregating populations derived from an organized crossing programme.

Mass selection, considered as the oldest method of breeding, has played a significant role in variety selection. This is a simple approach in which a large number of plants of similar phenotype are selected and seeds bulked to derive a new variety. Selection is based on the appearance of visible superiority of the plants and the selected plants are not subjected to progeny testing. This method had been used in wheat for improving land races or purification of a mixed variety (Walker, 1969). There are studies indicating its use to improve yield and tillering in wheat (Derera and Bhatt, 1972; Redden and Jensen, 1974).

Combination breeding involves making crosses to transfer one or more characteristics, usually governed by few genes, from one variety into another. In making a cross, one can develop a hybrid (F_1) or a pure line variety. To develop pure line varieties, the F_1 is selfed and selection is made in subsequent segregating generations. Three important selection procedures for selection in the segregating generation are pedigree, bulk and backcross selection. Pedigree selection, which predated the rediscovery of Mendel's laws, involves the maintenance of detailed pedigree records. In other words records of progenies are kept, so that each progeny can be traced back to the F_2 plant from which it originated (Allard, 1960). When applying the method to wheat, the population size in segregating generations should be large because genes for traits like yield are located in almost all chromosomes (Shebeski, 1967). It is often used to correct weaknesses in an established variety through combination breeding and is also useful in the selection of new superior recombinant types (transgressive breeding) (Jenson, 1988). However, this method suffers from the disadvantage of the ineffectiveness of single plant selection in an early generation. It is also expensive, limits the amount of material carried (Harrington, 1937) and involves detailed and precise record keeping.

Various modifications have been proposed to increase the efficiency of the pedigree method. Several experts began to think of handling segregating generations as bulks and then handling a larger number of crosses. Nilsson-Ehle (1908) at Swalof was the first to use the bulk method for selection in segregating populations to screen winter wheat progeny for high yield and winter hardiness (Newman, 1912; Allard, 1960). Harlan et al. (1940) proposed the bulk method to handle selection in a large number of crosses. In this method, which has the considerable merit of simplicity, the F_2 and subsequent generations are harvested in mass or as bulks, while the semi-natural selection imposed by environment changes gene/allele frequencies (Walker, 1969). At the end of the bulking period, individual plants are selected and evaluated in the same manner as in the pedigree method. Large periods of bulk handling are required for character fixation because homozygous types are fixed by the F_6–F_7 generations (Jain, 1961; Allard and Jain, 1962). The bulk method is advantageous for maintaining substantial variability over an extended period during which selection of desirable types can be made (Suneson and Wiebe, 1962). Though natural selection in bulk breeding may be useful in selecting subtle differences in different traits (Suneson and Stevens, 1953), in many cases it might eliminate desirable plants such as dwarf genotypes in crosses involving tall and dwarf (Sharma, 1989; Jennings and Aquino, 1968). Therefore, artificial selection may be used as a support at suitable stages (Jain, 1961).

Several modifications of the bulk method have been suggested, among which the mass pedigree method (Harrington, 1937) and single seed descent (SSD) (Johnson and Bernard, 1962; Grafius, 1965; Brim, 1966) are important. The mass pedigree method is based on bulking the populations when the environment is unfavourable for selection and doing single plant selection when environment becomes favourable. SSD, which involves the quick advancement of generation by taking one seed from each of the 200–250 F_2 plants and carrying out selection when lines become homozygous, was first suggested by Goulden (1939, 1941). SSD is quite convenient, time saving and advantageous from the point of maintaining considerable variability (Brim, 1966; Empig and Fehr, 1971; Qualset, 1975).

Singh et al. (1998) investigated the crossing and selection schemes (pedigree, modified bulk, selected bulk and non-selected bulk) in wheat at CIMMYT, Mexico, and established that there was little difference among schemes in terms of grain yield and other traits. The selection of parents was the most important feature for yield potential and other agronomic traits irrespective of cross type and selection scheme (Singh et al., 1998). According to these authors, owing to genetic gains and cost efficiency, selected bulks is the most attractive selection method (Singh et al., 1998). In CIMMYT, Mexico, pedigree selection was primarily used from 1944 to 1985. From the mid-1980s to the mid-1990s the main selection method was modified pedigree/bulk and this was replaced in the late 1990s by the selected bulk method. Due to genetic gains and cost efficiency (Singh et al., 1998), selected bulk is regarded as the most attractive selection scheme for CIMMYT's wheat breeding.

The backcross method, proposed by Harlan and Pope (1922) and studied in detail by Briggs (1930, 1935, 1938, 1941, 1958), was designed to transfer a desirable allele in place of an undesirable one through repeated backcrossing with the recipient (recurrent) parent. The donor parent is used only once in the hybridization. This method requires a good recurrent parent, a donor parent and proper expression of the character (Briggs and Allard, 1953), and is more practicable in the case of simply inherited traits. For polygenic traits, each backcross progeny needs to be selfed for 2–3 generations with simultaneous selection. Selection followed by single backcrossing has been found very useful in obtaining higher yielding genotypes at CIMMYT and is commonly used in their global wheat programme (Singh et al., 2008).

3 Wheat variety selection methods: modern molecular breeding

3.1 Marker-assisted selection

From the 1980s onwards, there has been considerable research into various molecular-based selection tools. Marker-assisted selection (MAS), in particular, is having a growing impact on wheat improvement programmes (Gupta et al., 2008, 2014; Tyagi et al., 2014). While MAS for selection on major genes proved straightforward, it has been more challenging to apply it for selection on quantitative traits. It has been estimated that by the end of first decade of this century, at least 60 genes/quantitative trait loci (QTL) had been handled using MAS around the world, mainly in developed countries (Gupta et al., 2010). Among the developing countries, there has been remarkably successful implementation of MAS in India to improve several cultivars, proving the potential of the tool (Gupta et al., 2011). Backcrossing strategies are particularly suited to a MAS approach, with markers providing the means for either foreground selection, or a combination of foreground and background selection. The latter allows an accelerated recovery of the recipient parent's genome, which is the purpose of backcrossing. Some of the molecular markers developed for various traits are listed in Table 1.

Table 1 Some successful examples of marker-assisted selection and gene pyramiding in wheat

No.	Target trait(s)	Target loci	Marker type	Effect of selection	Reference
1	Powdery mildew resistance	3 genes combinations (*Pm1+Pm3a, Pm1+Pm10, Pm3a+Pm10*)	RFLP	Powdery mildew resistance in introgression lines	Liu et al. (2000)
2	HMW-glutenins	Major genes (*Glu-A1* and *Glu-D1*)	AS-PCR	Improvement in glutenin quality	de Bustos et al. (2001)
3	Scab resistance	One QTL	SSR	Scab resistance in $F_{2:3}$	Zhou et al. (2003)
4	Leaf rust resistance	2 genes (*Lr19, Lr24*)	STS	Successful pyramiding in F_3 lines	Singh et al. (2004)
5	FHB resistance, orange blossom wheat midge resistance and leaf rust resistance	6 QTL for *FHB, Sm1* and *Lr21*	SSR	Successful introduction of FHB, *Sm*1 and *Lr*21 resistance genes	Somers et al. (2005)
6	Increased GPC	*Gpc-B1* gene	SSR	Improved GPC in BC_2F_4 plants	Davies et al. (2006)
7	Pre-harvest sprouting (PHS)	2 QTL	SSR	Increased grain dormancy in white-grained wheat	Kottearachchi et al. (2006)

Table 1 (*Continued*)

No.	Target trait(s)	Target loci	Marker type	Effect of selection	Reference
8	Fusarium head blight	3 QTL	SSR	Maximum grain fill phenotypic selection following the marker-based selection	Miedaner et al. (2006)
9	Powdery mildew resistance	3 QTL	SSR	Effective selection for powdery mildew resistance in both greenhouse and field experiments	Tucker et al. (2006)
10	Cereal cyst nematode resistance	2 genes (*CreX*, *CreY*)	SCAR	Higher resistance in pyramided line	Barloy et al. (2007)
11	Dough properties, durable rust resistance and height	*Rht-B1*, *Rht-D1*, *Rht8*, *Lr24/Sr24*, *Lr34/Yr18*, *Glu-A3*, *Glu-B3*	SSR	Increased genetic improvement for specific target genes, particularly at the early stage of a breeding programme	Kuchel et al. (2007)
12	Leaf rust resistance	4 genes (*Lr1*, *Lr9*, *Lr24*, *Lr47*)	STS, SCAR, CAPS	Effective selection for resistance gene	Nocente et al. (2007)
13	FHB resistance and DON content	3 QTL	SSR	Increased gain for major QTL only	Wilde et al. (2007)
14	Spot blotch resistance	4QTLs	SSR	Enhanced spot blotch resistance	Kumar et al. (2009)
15	Spot blotch resistance	4QTLs	SSR	Enhanced spot blotch resistance	Kumar et al. (2010)
16	Leaf rust resistance	*Lr24* + *Lr28*	SCAR	Better yield potential than recipient parent	Chhuneja et al. (2011)
17	Adult plant powdery mildew resistance	*QPm.caas-1A* + *QPm.caas-4DL* + *QPm.caas-2BS* + *QPm.caas-2BL* + *QPm.caas-2DL*	–	Expressed better APR to powdery mildew than the more resistant parent	Bai et al. (2012)

(*Continued*)

Table 1 (*Continued*)

No.	Target trait(s)	Target loci	Marker type	Effect of selection	Reference
18	Glutenin	*Glu-A1x, Glu-B1x* and *Glu-D1x*	SSR	Improved baking quality	Izadi-Darbandi and Yazdi-Samadi (2012)
19	Terminal heat tolerance	2QTLs	SSR	Enhanced heat tolerance	Paliwal et al. (2012)
20	Heat tolerance	14 QTL	SSRs	Validation of QTL	Sadat et al. (2013)
21	Spot blotch resistance	Co-location with leaf rust resistance gene	SSR	Spot blotch resistant genes was co-located with *Lr34* and *Lr46*	Lillemo et al. (2013)
22	Terminal heat tolerance	7 QTLs	SSR	Enhanced heat tolerance	Tiwari et al. (2013)
23	Zn and Fe content in grain	10 QTLs	SNP and DArT	Increased Zn and Fe content in grain	Srinivasa et al. (2014)
24	Grain protein content	*Gpc-B1* gene	SSR	Enhanced grain protein content in a well-adopted variety HUW234	Vishwakarama et al. (2014)
25	Grain protein content	*Gpc-B1*gene	SSR	Enhanced >3% grain protein content as compared to recipient variety	Mishra et al. (2015)
26	Stem rust resistant genes	*Sr25, SrWeb, Sr50*	SSR	All these genes pyramided in HUW234, a mega variety of NEPZ, India	Yadav et al. (2015)
27	Spot blotch resistance	2 QTLs	SSR	Pyramided 2 QTLs for spot blotch resistance in to well-adopted variety	Vasistha et al. (2015)

3.2 Genomic selection

Genomic selection is an emerging technique of selection (Meuwissen et al., 2001), in which joint merit of a large number of mapped markers across the genome is used (de Koning and McIntyre, 2012). The major difference between conventional breeding and genomic selection approaches is that in the former, selection is based on the phenotypic performance, whereas in the latter, genetic makeup of plants/lines is used as the deciding factor. The successful utilization of this approach became possible due to major advances in the development of statistical tools such as BayesB and the GBLUP (Gianola et al., 2009; Gianola, 2013; de los Campos et al., 2013; Daetwyler et al., 2013). Burgueño et al. (2012) first used GBLUP for assessing G × E under genomic selection, while Heslot et al. (2014) used crop-modelling data for same assessment. Jarquin et al. (2014) proposed a random effect GBLUP model where random variance–covariance structures of markers and environmental covariables were used to determine the effects of markers and environmental covariates.

To perform genomic selection, a population (referred to as the training population), is genotyped and phenotyped, and used to produce a statistical model. This model is then used on another population (called the breeding population) which is genotyped, but not phenotyped. The performances for various traits of the breeding population are predicted using allelic identity with loci that were found to be associated with the phenotype in the training population (Würschum et al., 2013). Selections of new breeding parents are made based on the genomic estimated breeding value (GEBV) which is derived from the combination of useful loci that occur in the genome of each individual in the breeding population. Compared to conventional breeding approach, genomic selection is much faster as it is not necessary to wait for late filial generations (F_6 or even beyond) to phenotype quantitative traits such as grain yield. In GS approach, GEBV of a third set of individuals, called validation population (VP), is also determined whose correlation with phenotypic evaluation is used to estimate the 'accuracy' of the genomic selection model. The recent experiences on wheat breeding suggest that genomic selection is a promising approach to improve complex traits like grain yield (Massman et al., 2013).

Major advances bring 'genomic selection' closer to its inclusion in the standard variety selection toolbox; however, a higher level of success has been achieved in animal breeding than in plant breeding (de Koning and McIntyre, 2012; Jonas and de Koning, 2013). Dairy cattle improvement per generation by genomic selection is double that of those achieved by traditional breeding (de Koning, 2016), but results for plant breeding have not been as impressive. Nevertheless, GBS is under development for wheat as a powerful genotyping approach capitalizing on 'next-generation' DNA sequencing technology advancements (Poland et al., 2012a). It has been successfully applied to CIMMYT breeding germplasm to develop genomic selection (GS) prediction models (Poland et al., 2012a). In initial studies, the cross-validation prediction accuracy with genotype by sequencing (GBS) was greater than that achieved with an established marker platform (DArT) and sufficiently high that GS could be applied in breeding programmes (Poland et al., 2012b).

3.3 Next-generation phenotyping

Selection based on physical appearance of the plant or characteristics of its traits (phenotype) used to be considered an easy method, and has been utilized extensively for the development of varieties. It is interesting to note that with increasing use of genomic tools for high-resolution linkage mapping, the necessity of precision phenotyping is in no way less than the conventional breeding (Cobb et al., 2013). The experiences drawn over past few years suggest that designing effective next-generation phenotyping is not simple and requires coordination among different subjects such as biologists, computer scientists, statisticians and engineers (Houle et al., 2010). In fact precise phenotyping is being seen as a major limitation in adopting genomic selection approach. It is believed that in the next two decades, phenotyping will depend more and more on increasing automation and throughput for improved precision, fast data recording and genetic analysis (Weber and Broman, 2001; Spalding and Miller, 2013). Advances in phenotyping technology can provide us with tools to accurately measure plants' characteristics on a much larger scale and thus will narrow the divisions between genomics, plant function and agricultural traits (Breccia and Nestares, 2014).

A new area which combines precision with automation in the recording of phenotypic observations is known as 'phenomics'. Phenomics entails phenotyping at multiple levels of organization (ranging from cellular components to whole plants and canopy) and encompasses structural, physiological and performance-related traits (Dhondt et al., 2013). High-throughput techniques for whole plant-level as well as at organ-level phenotyping are being developed. Data collection is now very fast, and the precision of the data has been enhanced manifold using advanced imaging technologies, including visible imaging (machine vision), imaging spectroscopy (multispectral and hyperspectral remote sensing), thermal infrared imaging, fluorescence imaging, 3D imaging and tomographic imaging (MRT, PET and CT). A useful list of available image-analysing tools can be found at www.plant-image-analysis.org (Lobet, 2013).

There are various examples of successful automated systems in controlled conditions for measuring aboveground traits (Pereyra-Irujo et al., 2012; Tisne et al., 2013), root architecture (Famoso et al., 2010; Clark et al., 2011) and both shoot and root growth (Ruts et al., 2013; Camargo et al., 2014; Yang et al., 2014). Field-phenotyping systems have also been developed using proximal sensing devices (Busemeyer et al., 2013). Infrared imagery, stereo image analysis, acoustic-based distance sensing, non-contact measurement of chlorophyll fluorescence, laser distance sensing and near-infrared spectroscopy are potential tools to obtain phenomic data in field conditions (White et al., 2012). There are imaging techniques that are used to characterize the plant temperature responses to the water status and transpiration rate and detect difference in stomatal conductance (Chen et al., 2014b). Large-scale platforms for tissue or cell-level phenotyping are still undeveloped, however. Advances in the robotic sampling of plants grown in field or greenhouse platforms combined with automatic analysis and proper conservation of the sample for further analysis will facilitate biochemical and histological characterization in large-scale phenotyping.

4 Variety selection by plant breeders

There is no one ideal wheat variety for all situations: different varieties are required depending on the selection objectives. Breeders and farmers may have the same end goals in mind; however, farming conditions vary across countries and regions and due to differences in management practices and level of industrialization. Breeders therefore have to develop varieties for a range of conditions, while farmers only require varieties suitable for their own environment or market.

4.1 Environment and location

The growing environment or location and size of the breeding operation influence the traits a breeder desires for a new or improved variety. Wheat is grown from the equator to 67° N in Scandinavia and 45° S in Argentina, Chile and New Zealand (Trethowan et al., 2005), so many different varieties are required due to the large-scale variations in conditions over this huge area. The chosen varieties must be able to cope with the prevailing biotic and abiotic constraints of the intended region, have the resilience to tolerate the relevant abiotic stresses and harness the positive potential of the given environment. In addition, different parts of the world have different cropping patterns and different end users have different food habits. For example, some parts of world have fallow wheat cropping systems, some rice–wheat, and others potato–wheat. While *chapati* is popular in South Asia, noodles in China and South East Asia, people from Europe prefer bread. No one wants red wheat in South Asia while it is very popular in Europe and America.

Based on the diversity of soil, temperature, rainfall and other environmental factors, CIMMYT wheat breeders categorized the global wheat growing areas into 12 mega-environments (MEs; Rajaram et al., 1994). A ME was defined as a broad, not necessarily contiguous area, occurring in more than one country and frequently transcontinental, defined by similar biotic and abiotic stresses, cropping system requirements, consumer preferences and, for convenience, by volume of production (Rajaram et al., 1994). MEs were defined for all three kinds of cultivated wheats – spring, winter and facultative. The first six MEs (ME1–ME6) were defined for spring wheat, next three (ME7–ME9) for facultative and the remaining three (ME10–ME12) for winter wheats. The location of MEs are given in Fig. 1 while the characteristic features in Table 2.

Selection of variety for a given ME takes care of major stresses, although there may be variation in secondary stresses. Some of the varieties are referred to as mega varieties since they have covered a large area after their release. The Green Revolution variety Sonalika was the first such mega variety, covering more than 14 million hectares in South Asia. Subsequently, several mega varieties were released, the most recent being PBW343 and HD 2967 in India; the first one covered around 7–8 million ha during early part of this century while the latter was grown in around 7 million hectares in the just concluded 2016–17 crop cycle.

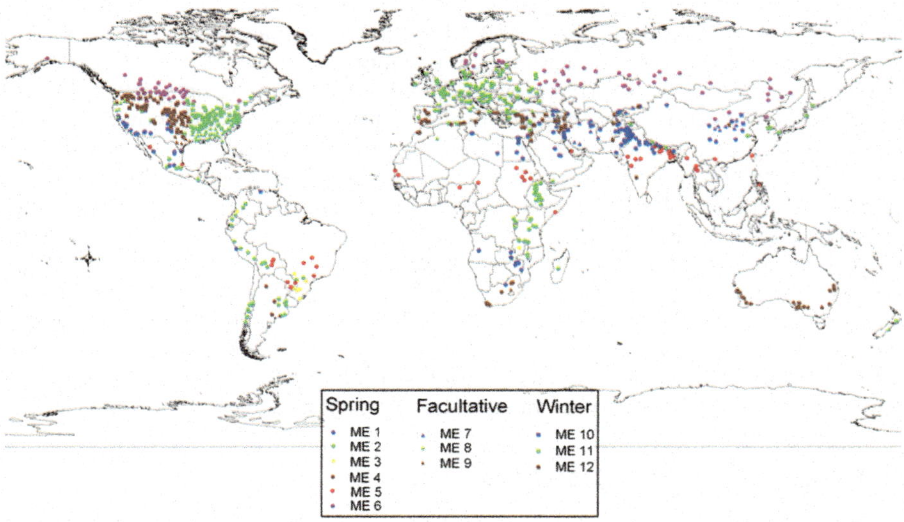

Figure 1 Global distribution of mega-environments of wheat (Source: http://wheatatlas.org/megaenvironments).

Table 2 Summary of characteristic features of 12 mega-environments of wheat (Source: http://wheatatlas.org/megaenvironments)

	Wheat types and temperature range					
	Spring wheat				Facultative wheat	Winter wheat
Rainfall/Irrigation	16°C >Tmin* > 11°C	16°C >Tmin* ≥ 3°C	11°C >Tmin* ≥ 3°C	−13°C >Tmin*	3°C >Tmin* ≥ −2°C	−2°C >Tmin* ≥ −13°C
Irrigated[1]					7	10
Low/irrigated[1]			1			
Low		4C	4A, 4B		9	12
Moderate/low				6B		
High/irrigated[1]	5				8	11
High		2A, 2B, 3		6A		

*Average minimum for 3 consecutive coolest months); [1] >5% of 5 arc min grid cell equipped for irrigation; ME2A Tmin judged in wettest quarter (3 consecutive wettest months) with = 250 mm precipitation, elevation = 1400 m; ME2B coolest quarter precipitation = 150 mm, elevation <1400 m; ME3 identical to ME2 (merged A&B) except soil pH < 5.2; ME4 wettest quarter precipitation = 100 mm, <400 mm; ME6 only at high latitudes (>45° N or S), Tmin = 9°C for warmest quarter.

Most wheat breeders in general work for a limited geographical region, but in some countries like India and China, the target region may be fairly big, such as 11 m ha in the North Western Plains Zone of India encompassing the states of Punjab, Haryana and western Uttar Pradesh. Likewise, wheat breeders at CIMMYT breed wheat varieties for the entire of the developing world. Wheat is grown on about 110 million hectares in more than 70 developing countries, so CIMMYT breeders must have a thorough knowledge of soil type and environmental factors such as temperature, rainfall, diseases and pests. This helps in understanding and deploying different traits for different parts of the world so that wheat production is not negatively affected by growing conditions in the various environments. Such a breeding programme requires an intelligent plan for selection of parents to make appropriate crosses by focusing on key characteristics for each ME. For example, farmers in high temperature and high humidity regions will need tolerance to terminal heat stress and spot blotch disease, while a low temperature region will demand resistance against yellow rust and powdery mildew.

Once the actual breeding process has come to an end, experimental wheats destined for a certain ME are evaluated repeatedly under those conditions. Superior lines are selected for additional improvement only if they can withstand the particular stresses predominating in that environment. MEs therefore are useful in deciding research and breeding priorities and selection of varieties before they are released for cultivation.

4.2 Crop management practices

Management practices for wheat cultivation differ from country to country and often from region to region within a country. Strategies employed in the hills are different from those used on the plains and are also altered by the available infrastructure. The current world is witnessing a change in cultural practices, for instance, reduced tillage is becoming a popular trend in many Latin American and Asian countries. One such region is the rice–wheat cropping system areas in South Asia (Hobbs, 2001; Joshi et al., 2007a). Wheat management conditions can be broadly divided into the following categories: i) rainfed; ii) irrigated-high fertility or low fertility; iii) conventional management, for example, surface seeding; iv) modern management for example, machine planting; and v) reduced tillage or conservation agriculture.

The rainfed areas will require varieties that carry drought tolerance; however, rainfed areas in plains, such as those in South Asia and Africa, also suffer from higher temperatures, so varieties with combined tolerance to drought and heat are required. In rainfed hilly and temperate regions, wheat will need cold tolerance and the duration of crop cycle is much longer. In such regions, generally winter wheat or facultative wheats are grown in contrast to warmer locations where the wheat grown is spring type. In milder locations, such as in Bhutan or the higher hills of Nepal, facultative wheat varieties are grown. Winter wheat requires vernalization and cannot be grown in warmer areas. The rainfed wheat varieties are generally taller, as a water-stressed environment demands a robust wheat with more biomass to survive under stress and produce grain. Such varieties do have more vigorous and deep root system. Absence of lodging, which is so frequent in irrigated wheat regions, also allow taller varieties to be selected and grown in rainfed areas.

Wheat varieties selected for irrigated management system are different from those selected for rainfed regions. Owing to the lodging-prone conditions, these varieties are dwarf or semi-dwarf. Irrigated management can be high fertility or low fertility depending on the ability of farmers to apply fertilizers. In high-fertility irrigated regions, the varieties selected are generally shorter than those in low-fertility regions, since they are more responsive to fertilizer application and therefore, have much higher yield potential and may carry much higher risk of lodging. However, in low-fertility conditions where farmers apply lower doses of fertilizer, the varieties are generally of the semi-dwarf type with the ability to perform well under conditions of low nutrient availability.

Conservation tillage requires cultivars suited to a particular resource conservation practice (Joshi et al., 2007b). Therefore, suitable germplasm adapted to conservation tillage practices is necessary for the farmers where this technique is gaining popularity. For surface seeding and zero-till planting, the cultivar should possess faster root development to quickly establish the crop (Trethowan and Reynolds, 2005). Ecologies or locations having more weed problems and where tillage is not used to control weeds will also need such cultivars. In the Indo-Gangetic plains of India, zero till has been found highly beneficial in reducing the population of *Phalaris minor* (Joshi et al., 2007b).

Selection of varieties may also depend on the level of mechanization (Joshi et al., 2007b). One reason could be that the sowing depth of seeds may vary under conventional and reduced tillage. In surface seeding, seeds are dispersed on the surface, while in machine planting the depth will be greater. Hence, cultivars displaying better germination and growth under shallow or surface seeding are more desirable for surface-seeded conditions (Joshi et al., 2007b). The variability available in synthetic wheat lines has been exploited to develop and select such varieties for farmers (Mujeeb-Kazi et al., 1996).

4.3 Biotic and abiotic stresses

Two of the most important wheat stresses are heat and drought and they are expected to increase in coming years due to global warming and increasing water scarcity. Terminal heat stress is defined by Fischer and Byerlee (1991) as occurring when mean daily temperatures are greater than 17.5°C in the coolest month. Stress-adaptive traits are of high importance for heat-stressed locations. Suitable varieties for such regions require traits such as an extensive root system (under irrigated non-saline environments), medium tall to tall plant stature, optimum tillering capacity and low canopy temperature, which indicate the potential for uptake of nutrients and water (Blum, 1996; Reynolds et al., 2001; Joshi et al., 2007b). In addition to terminal heat, early heat tolerance is also becoming important. For instance, in many regions, such as the rice–wheat cropping system of eastern Indo-Gangetic plains of South East Asia, short-maturing rice varieties (such as hybrids) and resource conservation technologies are enabling farmers to sow wheat earlier than normal when temperature remains somewhat higher than the optimum. In these varieties, mild vernalization might be useful as it may help to maintain cultivars in the vegetative growth phase when temperatures are warmer. In the 1990s and early years of this century, the CIMMYT-derived cultivar PBW 343 was adopted in around 8 m ha in India due to its adoptability in early as well as normal sowing conditions.

Crops are often grown in environments where water is a limiting factor. Hence, taking advantage of available moisture and reducing evaporation from soil are important factors. Any increase in early seedling vigour should reduce evaporative losses from the soil surface (Richards, 1992). If crop duration is short, greater vigour is likely to increase final biomass and yield and it may be an effective way to reduce weed growth in most environments. Among traits that contribute to increased seedling vigour, coleoptile length is the most important (Fick and Qualset, 1976; Whan, 1976). Short coleoptiles result in poor emergence, which leads to poor crop establishment. This is true particularly when seeds of semi-dwarf wheat cultivars and lines are sown deeply to seek moisture or are sown into stubble (Richards et al., 2001). In dry environments, farmers sowing into a receding moisture profile following rainfall often sow at greater depth (8–12 cm) to ensure seed contact with available soil moisture (Paulsen, 1987). Better emergence is achieved by sowing wheat with long coleoptiles. Allan (1980) suggested that the accumulation of modifier genes that favour emergence could be important in breeding and selection for better emerging semi-dwarf wheat cultivars. There are reports suggesting the possibility of selection within families of *Rht1* and *Rht2* dwarfing genes for plants with potentially longer coleoptiles having the ability to emerge from deep sowing; likewise, non-*Rht1* or non-*Rht2* hexaploid bread wheat (using Seri 82 and Culiacan 89) can have better emergence characteristics under deep sowing in rainfed conditions (Trethowan et al., 2001).

In addition to heat and drought, many areas suffer from stresses such as high pH. In high pH soils, seed contact with an alkali crust is harmful for seed germination under conventional as well as reduced tillage (Joshi et al., 2007b). Therefore, selecting cultivars tolerant to saline soils is important. Already there are a number of varieties recommended for saline-alkali soils in different countries.

4.4 Demands of consumers or industry

For a long time, most of the important wheat-based foods were specific to their own regions, for instance, Arabic flat breads were prevalent in the Middle East; oriental noodles in China, Japan and Korea; and chapati in South Asia. However, a number of reasons including migration, industrialization and increased communication have permitted the exchange of cultural and dietary habits across regions and countries (Montague, 1998). At present, the world's burgeoning urban populations are looking for healthier, more nutritious foods which are convenient to prepare, and have increased income to spend on food (particularly in Asia, Latin America and Africa). There is therefore increased demand for wheat-based convenience foods that are ready-to-eat. Newly marketed wheat-based foods, such as noodles and flat breads in Europe, the Americas and Australia, leavened breads and wheat-based fast foods in whole of Asia are easily accepted by the urban populations. As an example, the 'flour tortilla', a wheat flat bread from Mexico, is considered one of the fastest growing products of the baking industry in the United States (Kohn, 2000).

There are large differences in grain composition among cultivars of the same wheat species, and both quantitative and qualitative compositional differences can make a cultivar suitable or unsuitable for a given baking process or food type. In view of the global changes described above, farmers need to grow varieties which match the quality parameters required by the market and industry. Hence, wheat breeders are increasingly selecting for the specific quality traits that satisfy market and processing demands for those producing food for elite urban populations. The demand for wheat in rural communities is often different, though, so farmers in those locations will select varieties more suited to their markets.

5 Variety selection by farmers

5.1 Participatory varietal selection

For a long time, variety selection has been carried out at research stations. The conditions on farms, however, may be entirely different from those at the research stations, for a variety of reasons. Approaches in which farmers participate in the selection process, such as participatory varietal selection (PVS), have therefore been found to be useful (Witcombe et al., 1996, 2001). The PVS approach generally involves mother-baby trials. The mother trial consists of a set of lines (6–10 or maybe more) grown in standard plots in the farmers' fields alongside one or two of the best local checks. The baby trial is a field trial of one or two of the best lines selected from the mother trial by the farmers. The plot used for the baby trial is larger (1/4–1 acre in size); its aim is to compare the selected line with the local checks at a much larger scale. In the PVS method, a group of farmers are shown the crops grown in the mother trial and the general features of the lines (including traits that cannot be seen by human eye such as nutrient characteristics in the grain) are explained to them. The farmers are then asked to rate the varieties according to their preference. This generally takes place before the crops reach physiological maturity; however, if the farmers are asked to express a preference for grain traits, this takes place at maturity. The varieties the farmers prefer are then planted in baby trials in subsequent years. There may be a large number of baby trials in a village or in a cluster of villages.

Through the PVS approach, farmers get ample opportunity to judge varietal options together with their associated crop production technologies. The seed sector is strengthened by the creation of a network of seed villages and farmers also benefit in terms of marketing due to the creation of links with other sectors. PVS trials are encouraged in countries like India (in eastern Gangetic plains), Nepal, Bangladesh, Bhutan and also in Africa, as they allow farmers to use their own judgement to assess and select superior lines (Pandit et al., 2011), which ensures that trade is based on distinct crop traits (Joshi et al., 2005). The PVS approach also fast tracks the dissemination of varieties, since lines enter seed production on day 1 of their introduction to farmers' fields. Ug99 resistant lines were released and disseminated in South Asia and Africa through participatory approaches (Joshi et al., 2011). Although several cultivars were made available to farmers, only a few were adopted, indicating that farmers have their own methods of selecting varieties based on a package of traits which improve profitability (Singh and Huerta-Espino, 2004; Pandit et al., 2011).

5.2 Seed cost

Variety selection may also depend upon the cost of the seed. This can be illustrated by examining the case of hybrid versus pure line wheat varieties, where seed of the former is more costly than that of the latter. The economic acceptance of hybrid wheat is complex and rests on a number of factors including the level of heterosis, yield stability, cost of seed, expected return on investment and the technology of hybrid seed production (Cisar and Cooper, 2002). Hybrid varieties are selected by only those farmers whose businesses are more advanced, or who have access to funds or credit to purchase hybrid seed every year and are able to practice high-input agriculture. Small-scale subsistence farmers generally stick to seeds saved by farmers the previous years and thus prefer pure line varieties, which can be grown for a number of years from the same seed source. Hybrid seed would be more attractive to farmers if it were more affordable, or if it increased yields (and therefore profits) significantly under existing crop management regimes.

6 Conclusion

Variety selection in wheat cultivation depends upon a number of factors, mainly biotic and abiotic stress conditions prevailing at the location where crop is to be planted. Other factors determining variety selection are the type of variety to be developed (hybrid of pure line), market requirement, management practices, cropping system and the socio-economics of the farming community. This century is witnessing a shift from traditional breeding to modern molecular breeding with increased emphasis on marker-assisted selection, genomic selection and next-generation phenotyping and genotyping. It is being believed that variety selection will soon be depending more and more on molecular approaches but precision phenotyping will continue to play an important role. However, in order to be commercially successful, a varietal selection process must lead to varieties that are acceptable to farmers and the consumers.

7 Where to look for further information

For further information, the authors suggest to read the references cited in this chapter. Other chapters of the book could be equally useful. Further sources could be the websites of CIMMYT (www.cimmut.org), ICARDA (www.icarda.org), FAO (FAO.org) and wheat atlas (http://wheatatlas.org). Some information can also be found at Wikipedia (https://en.wikipedia.org).

8 References

Allan, R. E. 1980. Influence of semi dwarfism and genetic background on stand establishment of wheat. *Crop Sci.*, 20: 634–8.

Allard, R. W. 1960. *Principal of Plant Breeding*. John Wiley and Sons, Inc.: New York.
Allard, R. W. and Jain, S. K. 1962. Population studies in predominantly self-pollinating species II. Analysis of quantitative genetic changes in a bulk population of barley. *Evolution*, 16: 90–101.
Arnold, M. H. 1985. Sugar beet. *Outlook Agr.*, 14: 109–14.
Bai, B., He, Z. H., Asad, M. A., Lan, C. X., Zhang, Y., Xia, X. C., Yan, J., Chen, X. M. and Wang, C. S. 2012. Pyramiding adult-plant powdery mildew resistance QTLs in bread wheat. *Crop Past. Sci.*, 63(7): 606–11.
Ball, C. R. 1930. The history of American wheat improvement. *Agr. Hist.*, 4(2): 48–71.
Barloy, D., Lemoine, J., Abelard, P., Tanguy, A. M., Rivoal, R. and Jahier, J. 2007. Marker-assisted pyramiding of two cereal cyst nematode resistance genes from *Aegilops variabilis* in wheat. *Mol. Breed.*, 20: 31–40.
Blum, A. 1996. Crop responses to drought and the interpretation of adaptation. *Plant Growth Reg.*, 20(2): 135–48.
Breccia, G. and Nestares, G. 2014. Next-generation phenotyping in plants: Old problems, new promises. *BAG*, 25: 5–8.
Briggs, F. N. 1930. Breeding wheat resistant to bunt by the backcrossing method. *J. Am. Soc. Agron.*, 22: 239–44.
Briggs, F. N. 1935. The back cross method in plant breeding. *J. Am. Soc. Agron.*, 27: 971–3.
Briggs, F. N. 1938. The use of back cross in crop improvement. *Am. Nat.*, 72: 285–92.
Briggs, F. N. 1941. The use of back cross in plant breeding. Proceedings of the Seventh International Genetics Congress, Edinburgh and London, Oliver & Boyd, pp. 81–2.
Briggs, F. N. 1958. Back crossing: Its development and present application. In Proceedings of the First International Wheat Genetics Symposium. University of Manitoba: Winnipeg, pp. 8–11.
Briggs, F. N. and Allard, R. W. 1953. The current status of back cross method of plant breeding. *Agron. J.*, 45: 131–8.
Brim, C. A. 1966. A modified pedigree method of selection in soybeans. *Crop Sci.*, 6: 220.
Burgueño, J., de los Campos, G., Weigel, K. and Crossa, J. (2012). Genomic prediction of breeding values when modeling genotype × environment interaction using pedigree and dense molecular markers. *Crop Sci.*, 52, 707–19.
Busemeyer, L., Ruckelshausen, A., Möller, K., Melchinger, A. E., Alheit, K. V., Maurer, H. P., Hahn, V., Weissmann, E. A., Reif, J. C. and Würschum, T. 2013. Precision phenotyping of biomass accumulation in triticale reveals temporal genetic patterns of regulation. *Sci. Rep.*, 3: 2442.
Camargo, A., Papadopoulou, D., Spyropoulou, Z., Vlachonasios, K., Doonan, J. H. and Gay, A. P. 2014. Objective definition of rosette shape variation using a combined computer vision and data mining approach. *PLoS One*, 9: e96889. doi: 10.1371/journal.pone.0096889.
Chen, D., Neumann, K., Friedel, S., Kilian, B., Chen, M., Altmann, T. and et al. 2014b. Dissecting the phenotypic components of crop plant growth and drought responses based on high-throughput image analysis. *Plant Cell*, 26, 4636–55. doi: 10.1105/tpc.114.129601.
Chhuneja, P., Vikal, Y., Kaur, S., Singh, R., Juneja, S., Bains, N. S., Berry, O., Sharma, A., Gupta, S. K., Charpe, A., Prabhu, K. V. and Dhaliwal, H. S. 2011. Marker-assisted pyramiding of leaf rust resistance genes *Lr24* and *Lr28* in wheat (*Triticum aestivum*). *Indian J. Agri. Sci.*, 81(3): 214–18.
Cisar, G. and Cooper, D. B. 2002. Hybrid wheat. In B. C. Curtis, S. Rajaram, H. Gomez Macpherson (Eds), *Bread Wheat: Improvement and Production*. Food and Agriculture Organization: Rome, Italy, pp, 157–74.
Clark, R. T., MacCurdy, R. B., Jung, J. K., Shaff, J. E., McCouch, S. R., Aneshansley, D. J. and Kochian, L. V. 2011 Three-dimensional root phenotyping with a novel imaging and software platform. *Plant Physiol.*, 156: 455–65.
Cobb, J. N., De Clerck, G., Greenberg, A., Clark, R. and McCouch, S. 2013. Next-generation phenotyping: Requirements and strategies for enhancing our understanding of genotype–phenotype relationships and its relevance to crop improvement. *Theor. Appl. Genet.*, 126: 867–87.
Daetwyler, H. D., Calus, M. P.L., Pong-Wong, R., de los Campos, G. and Hickey, J. M. 2013. Genomic prediction in animals and plants: Simulation of data, validation, reporting, and benchmarking. *Genetics*, 193: 347–65.

Davies, J., Berzonsky, W. A. and Leach, G. D. 2006. A comparison of marker-assisted and phenotypic selection for high grain protein content in spring wheat. *Euphytica*, 152: 117–34.

de Bustos, A., Rubio, P., Soler, C., Garcia, P. and Jouve, N. 2001. Marker assisted selection to improve HMW-glutenins in wheat. *Euphytica*, 119: 69–73.

de Koning, D. J. 2016. Meuwissen et al. on Genomic Selection. *Genetics*, 203(1): 5–7.

de Koning, D. J. and McIntyre, L. 2012. Setting the standard: A special focus on genomic selection in Genetics and G3. *Genetics*, 190: 1151–2.

de Los Campos, G., Hickey, J. M., Pong-Wong, R., Daetwyler, H. D. and Calus, M. P. 2013. Whole-genome regression and prediction methods applied to plant and animal breeding. *Genetics*, 193(2): 327–45.

Derera, N. F. and Bhatt, G. M. 1972. Effectiveness of mechanical mass selection in wheat (*Triticumaestivum* L.). *Aust. J. Agr. Res.*, 23: 761–8.

Dhondt, D., Wuyts, N. and Inze, D. 2013. Cell to whole-plant phenotyping: The best is yet to come. *Trends Plant Sci.*, 18: 428–39.

Empig, L. T. and Fehr, W. R. 1971. Evaluation of methods for generation advance in bulk hybrid soybean populations. *Crop Sci.*, 11: 51–4.

Famoso, A. N., Clark, R. T, Shaff, J. E., Craft, E., McCouch, S. R. and Kochian, L. V. 2010. Development of a novel aluminum tolerance phenotyping platform used for comparisons of cereal aluminum tolerance and investigations into rice aluminum tolerance mechanisms. *Plant Physiol.*, 153: 1678–91.

Faris, J. D. 2014. Wheat domestication: Key to agricultural evolutions past and future. In R. Tuberosa, A. Graner and E. Frison (Eds), *Genomics of Plant Genetic Resources, vol. 1 Managing, sequensing and mining genetic resources.* Springer: Netherlands, pp. 439–64.

Fick, G. and Qualset, C. 1976. Seedling emergence, coleoptile length, and plant height relationships in crosses of dwarf and standard-height wheats. *Euphytica*, 25: 679–84.

Fischer, R. A. and Byerlee, D. B. 1991. Trends of wheat production in the warmer areas: Major issues and economic considerations. *In* Wheat for the Non-traditional Warm Areas. Proceedings of Conf., Iguazu, Brazil, 29 Jul.–3 Aug. 1990. CIMMYT: Mexico, DF, pp. 3–27.

Gianola, D. 2013. Priors in whole-genome regression: The bayesian alphabet returns. *Genetics*, 194: 573–96. doi: 10.1534/genetics.113.151753.

Gianola, D., de los Campos, G., Hill, W. G., Manfredi, E. and Fernando, R. 2009. Additive genetic variability and the Bayesian alphabet. *Genetics*, 183: 347–63. doi: 10.1534/genetics.109.103952.

Gillham, N. W. 2001. Sir Francis Galton and the birth of eugenics. *Ann. Rev. Genet.*, 35: 83–101.

Goulden, C. H. 1939. Problems in plant selection. In R. C. Burnett (Ed.), Int. Genet. Congr. Proc. 7th (Edinburgh) Cambridge University Press, pp. 132–3.

Goulden, C. H. 1941. Problems in plant selection. *In* Proc 7th International Genetics Congress. Cambridge Univ. Press: Cambridge, UK, pp. 132–3.

Grafius, J. E. 1965. *A Geometry of Plant Breeding.* Michigan State University. Agri. Exp. Sta. Res. Bulletin No. 7, Michigan, USA, p. 59.

Gupta, P. K., Balyan, H. S., Mir, R. R., Kumar, J., Kumar, A., Kumar, S., Jaiswal, V., Tyagi, S. and Kumari, S. 2011. QTL analysis, association mapping and marker-assisted selection for some quality traits in bread wheat- An overview of the work done at CCS University, Meerut. *J. Wheat Res.*, 3(2): 1–11.

Gupta, P. K., Landridge, P. and Mir, R. R. 2010. Marker-assisted wheat breeding: Present status and future possibilities. *Mol. Breed.*, 26: 145–61.

Gupta, P. K., Mir, R. R., Mohan, A. and Kumar, J. 2008. Wheat genomics: Present status and future prospects. *Int. J. Plant Genom.*, 2008, Article ID 896451, 36 pages, 10.1155/2008/896451.

Gupta, P., Kulwal, P. and Jaiswal, V. 2014. Association mapping in crop plants: Opportunities and challenges. *Adv. Genet.*, 85C: 109–47. doi: 10.1016/B978-0-12-800271-1.00002-0.

Harlan, H. V. and Pope, M. N. 1922. The use and value of backcrosses in small grain breeding. *J. Hered.*, 13: 319–22.

Harlan, H. V., Martini, M. I. and Stevens, H. 1940. *A Study of Methods in Barley Breeding.* USDA Technical Bulletin 720, U.S. Gov. Print. Office, Washington D.C., pp. 720–940.

Harrington, J. B. 1937. The mass-pedigree method in hybridization improvement of cereals. *J. Am. Soc. Agron.*, 29: 379–84.

Heslot, N., Akdemir, D., Sorrells, M. E. and Jannink, J. L. 2014. Integrating environmental covariates and crop modeling into the genomic selection framework to predict genotype by environment interactions. *Theor. Appl. Genet.*, 127: 463–80. doi: 10.1007/s00122-013-2231-5. pmid:24264761.

Hobbs, P. R. 2001. Tillage and crop establishment in South Asian rice-wheat systems: Present and future options. In: Kataki, P. K. ed. The Rice-Wheat Cropping System of South Asia: Efficient Production Management. *J. Crop Prod.*, 4(1): 1–23.

Houle, D., Govindaraju, D. R. and Omholt, S. 2010. Phenomics: The next challenge. *Nat. Rev. Genet.* 11: 855–66.

Izadi-Darbandi, A. and Yazdi-Samadi, B. 2012. Marker-assisted selection of high molecular weight glutenin alleles related to bread-making quality in Iranian common wheat (*Triticumaestivum* L.). *J. Genet.*, 91(2): 193–8.

Jain, S. K. 1961. Studies on the breeding of self-pollinating cereals. The composite cross bulk population method. *Euphytica*, 10: 315–24.

Jarquin, D., Crossa, J., Lacaze, X., Cheyron, P. D., Daucourt, J. and et al. 2014. A reaction norm model for 572 genomic selection using high-dimensional genomic and environmental data. *Theor. Appl. Genet.*, 127: 595–607.

Jenkins, J. A. 1966. The origin of cultivated wheat. *Can. J. Genet. Cytol.*, 8: 220–32.

Jennings, P. R. and Aquino, R. C. 1968. Studies on competition in Rice. III: The mechanism among phenotypes. *Evolution*, 22: 529–42.

Jenson, N. F. 1988. *Plant Breeding Methodology*. John Wiley & Sons: New York.

Johnson, H. W. and Bernard, R. L. 1962. Soybean genetics and breeding. *Adv. Agron.* 14: 149–221.

Jonas, E. and de Koning, D. J. 2013. Does genomic selection have a future in plant breeding? *Trends Biotechnol.*, 31(9): 497–504.

Joshi, A. K., Azab, M., Mosaad, M., Moselhy, M., Osmanzai, M., Gelalcha, S., Bedada, G., Bhatta, M. R., Hakim, A., Malaker, P. K., Haque, M. E., Tiwari, T. P., Majid, A., Jalal Kamali, M. R., Bishaw, Z., Singh, R. P., Payne, T. and Braun, H. J. 2011. Delivering rust resistant wheat to farmers: A step towards increased food security. *Euphytica*, 179: 187–96.

Joshi, A. K., Chand, R., Arun, B., Singh, R. P. and Ortiz, R. 2007b. Breeding crops for reduced-tillage management in the intensive, rice-wheat systems of south Asia. *Euphytica*, 153: 135–51.

Joshi, A. K., Chand, R., Chandola, V. K., Prasad, L. C., Arun, B., Tripathi, R. and Ortiz-Ferrara, G. 2005. Approaches to germplasm dissemination and adoption – reaching farmers in the eastern Gangetic Plains. *In* Proceedings of 7th Intl. Wheat Conference, 27 November–2 December 2005, Mar del Plata, Argentina.

Joshi, A. K., Mishra, B., Chatrath, R., Ortiz Ferrara, G. and Singh, R. P. 2007a. Wheat improvement in India: Present status, emerging challenges and future prospects. *Euphytica*, 157: 431–46.

Kohn, S. 2000. An update of the U.S. baking industry. *Cereal Foods World*, 45: 94–7.

Kottearachchi, N. S., Uchino, N., Kato, K. and Miura, H. 2006. Increased grain dormancy in white-grained wheat by introgression of pre-harvest sprouting tolerance QTLs. *Euphytica*, 152: 421–8.

Kuchel, H., Fox, R., Reinheimer, J., Mosionek, L., Willey, N., Bariana, H. and Jefferies, S. 2007. The successful applicationof a marker-assisted wheat breeding strategy. *Mol. Breed.*, 20: 295–308.

Kumar, U., Joshi, A. K., Kumar, S., Chand, R. and Roder, M. S. 2009. Mapping of resistance to spot blotch disease caused by Bipolaris sorokiniana in spring wheat. *Theor. Appl. Genet.*, 118: 783–92.

Kumar, U., Joshi, A. K., Kumar, S., Chand, R. and Roder, M. S. 2010. Quantitative trait loci for resistance to spot blotch caused by *Bipolaris sorokiniana* in wheat (*T.aestivum* L.) lines 'Ning 8201' and 'Chirya 3'. *Mol. Breed.*, 26: 477–91.

Lillemo, M., Joshi, A. K., Prasad, R., Chand, R. and Singh, R. P. 2013. QTL for spot blotch resistance in bread wheat line Saar co-locate to the biotrophic disease resistance loci Lr34 and Lr46. *Theor. Appl. Genet.*, 126: 711–19.

Liu, J., Liu, D., Tao, W., Li, W., Wang, S., Chen, P., Cheng, S. and Gao, D. 2000. Molecular marker-facilitated pyramiding of different genes for powdery mildew resistance in wheat. *Plant Breed.*, 119: 21–4.

Lobet, G. 2013. http://www.plant-image-analysis.org/ (accessed September 2013).

Massman, J. M., Jung, H.-J. G. and Bernardo, R. 2013. Genome wide selection versus marker-assisted recurrent selection to improve grain yield and stover-quality traits for cellulosic ethanol in maize. *Crop Sci.*, 53: 1–9.

Mendal, G. 1866. Versucheü ber pflanzen-hybriden. *Verhandlungen der naturfoschung Vereins* 4:3–47.

Meuwissen, T. H., Hayes, B. J. and Goddard, M. E. 2001. Prediction of total genetic value using genome-wide dense marker maps. *Genetics* 157: 1819–29.

Miedaner, T. F., Wilde, B. S., Buerstmayr, H., Korzun, V. and Ebmeyer, E. 2006. Stacking quantitative trait loci (QTL) for Fusarium head blight resistance from non adapted sources in an European elite spring wheat background and assessing their effects on deoxynivalenol (DON) content and disease severity. *Theor. Appl. Genet.*, 112: 562–9.

Mishra, V. K., Gupta, P. K., Arun, B., Chand, R., Vashistha, N. K., Vishwakarma, M. K., Yadav, P. S. and Joshi, A. K. 2015. Introgression of a gene for high grain protein content (*Gpc-B1*) into two leading cultivars of wheat in Eastern Gangetic Plains of India through marker assisted backcross breeding. *J. Plant Breed. Crop Sci.*, 7(8): 292–300.

Montague, D. 1998. Global grain perspectives. *World Grain*, September 1998, pp. 13, 14, 16, 18.

Mujeeb-Kazi, A., Rosas, V. and Roldan, S. 1996. Conservation of the genetic variation of *Triticumtauschii* (Coss.) Schmalh. (Aegilopssquarrosaauct. Non L.) in synthetic hexaploid wheats (T. Turgidum L.× T. tauschii; 2n = 6x = 42, AABBDD) and its potential utilization for wheat improvement. *Genet. Resour. Crop. Evol.* 43: 129–34.

Murray, J. 1970. *The First European Agriculture, A Study of the Osteological and Botanical Evidence Until 2000 B.C.* Edinburg University Press: Edinbergh.

Newman, L. H. 1912. *Plant Breeding in Scandinavia.* Canadian Seed Grower's Association: Ottawa.

Nocente, F., Gazza, L. and Pasquini, M. 2007. Evaluation of leaf rustresistance genes *Lr1, Lr9, Lr24, Lr47* and their introgressioninto common wheat cultivars by marker-assistedselection. *Euphytica*, 155: 329–36.

Paliwal, R., Röder, M. S., Kumar, U., Srivastava, J. and Joshi, A. K. 2012. QTL mapping of terminal heat tolerance in hexaploid wheat (T. aestivum L.). *Theor. Appl. Genet.* 125: 561–75. 10.1007/s00122-012-1853-3.

Pandit, D. B., Mandal, M. S. N., Hakim, M. A., Barma, N. C. D., Tiwari, T. P. and Joshi, A. K. 2011. Farmers' preference and informal seed dissemination of first Ug99 tolerant wheat variety in Bangladesh. *Czech J. Genet. Plant Breed.*, 47(Special Issue): S160–64.

Paulsen, G. M. 1987. Wheat stand establishment. In Heyne, E. G. (Ed.), *Wheat and Wheat Improvement*, 2 Edition. American Soc. Agron., Agronomy Monograph No. 13, Madison, WI, pp. 383–9.

Pereyra-Irujo, G. A., Gasco, E. D., Peirone, L. S. and Aguirrezábal, L. A. N. 2012. GlyPh: A low-cost platform for phenotyping plant growth and water use. *Funct. Plant Biol.*, 39: 905–13.

Poland, J., Endelman, J., Dawson, J., Rutkoski, J., Wu, S. and et al. 2012a. Genomic selection in wheat breeding using genotyping-by-sequencing. *Plant Gen.* 5: 103–13.

Poland, J. A., Brown, P. J., Sorrells, M. E. and Jannink, J.-L. 2012b. Development of high-density genetic maps for barley and wheat using a novel wwo-enzyme genotyping-by-sequencing approach. *PloS ONE* 7, e32253.

Qualset, C. O. 1975. Sampling germplasm in a centre of diversity: An example of disease resistance in Ethiopian barley. In O. H. Frankel and J. G. Hawkes (Eds), *Crop Genetic Resources for Today and Tomorrow.* Cambridge University Press: Cambridge, England, pp. 81–96.

Rajaram, S., van Ginkel, M. and Fischer, R. A. 1994. CIMMYT's wheat breeding mega environments (ME). *In* Proceedings of the 8th International Wheat Genetics Symposium, Jul. 19–24 1993. Beijing, China.

Redden, R. J. and Jensen, N. F. 1974. Mass selection and mating systems in cereals. *Crop Sci.*, 14: 345–50.

Renfrew, J. A. 1973. *Palaeoethnobotany. The Prehistoric Food Plants of the Near East and Europe.* Methuen, London.

Reynolds, M. P., Copland, S., Trethowan, R. M. and Skovmand, B. 2001. Physiological approaches to stress breeding for wheat. In book of abstracts, Plant Breeding Sustaining the Future, 16th Eucarpia Congress, Edinburgh International Conference Centre, 10–14 September 2001.

Richards, R. A. 1992. The effect of dwarfing genes in spring wheat in dry environments. II. Growth, water use, and water-use efficiency. *Aust. J. Agr. Res.*, 43: 529–39.

Richards, R. A., Condon, A. G. and Rebetzke, G. J. 2001. Traits to improve yield in dry environments. In M. P. Reynolds, J. I. Ortiz-Monasterio and A. McNab (Eds), *Application of Physiology in Wheat Breeding*. CIMMYT: Mexico, D.F., pp. 88–100.

Ruts, T., Matsubara, S. and Walter, A. 2013. Synchronous high-resolution phenotyping of leaf and root growth in *Nicotiana tabacum* over 24-h periods with GROWMAP-plant. *Plant Methods*, 9: 2.

Sadat, S., Saeid, K. A., Bihamta, M. R., Torabi, S., Salekdeh, S. G. H. and Ayeneh, G. A. L. 2013. Marker assisted selection for heat tolerance in bread wheat. *World App. Sci. J.*, 21(8): 1181–9.

Sharma, D. 1989. Breeding procedures for self-pollinated crops: Concepts and methods. In V. L. Chopra (Ed.), *Plant Breeding*. Oxford & IBH: New Delhi, pp. 21–74.

Shebeski, L. H. 1967. Wheat breeding. In K. F. Nielsen (Ed.), Proceedings of Can. Centennial Wheat Symp. Saskatoon, Saskatchewan, pp. 249–72.

Silvertown, J. W. and Doust, J. L. 1993. *Introduction to Plant Population Biology*. Blackwell Scientific Publications: Oxford, p. 210.

Singh, R., Datta, D., Singh, S. P. and Tiwari, R. 2004. Marker-assisted selection for leaf rust resistance genes *Lr*19 and *Lr*24 in wheat (*Triticumaestivum* L.). *J. Appl. Genet.*, 45(4): 399–403.

Singh, R. P. and Huerta-Espino, J. 2004. The use of 'single-backcross, selected-bulk' breeding approach for transferring minor genes based rust resistance into adapted cultivars. In C. K. Black, J. F. Panozzo and G. J. Rebetzke (Eds), Proceedings of the 54th Australian cereal chemistry conference and 11th wheat breeders assembly, Canberra, Australia, 21–24 September 2004. Publishers Cereal Chemistry Division, Royal Australian Chemical Institute, Victoria, Australia, pp. 48–51.

Singh, R. P., Hodson, D. P., Huerta-Espino, J., Jin, Y., Njau, P., Wanyera, R. and et al. 2008. Will stem rust destroy the world's wheat crop? *Adv. Agron.*, 98: 271–309.

Singh, R. P., Rajaram, S., Miranda, A., Huerto-Espino, J. and Autrique, E. 1998. Comparison of two crossing and four selection schemes for yield, yield traits, and slow rusting resistance to leaf rust in wheat. *Euphytica*, 100: 25–43.

Somers, D. J., Thomas, J., DePauw, R., Fox, S., Humphreys, G. and Fedak, G. 2005. Assembling complex genotypes to resist Fusarium in wheat (*Triticumaestivum* L.). *Theor. Appl. Genet.*, 111: 1623–31.

Spalding, E. and Miller, N. 2013. Image analysis is driving a renaissance in growth measurement. *Curr. Opin. Plant Biol.*, 16: 100–4.

Srinivasa, J., Arun, B., Mishra, V. K., Chand, R., Sharma, D., Bhardwaj, S. C. and Joshi, A. K. 2014. Accessing the spelt gene pool to develop well-adapted bread wheat lines with increased grain zinc and iron. *Crop Sci.* 54: 1–11.

Stoskopf, N. C., Tomes, D. T. and Christie, B. R. 1993. *Plant Breeding. Theory and Practice*. West View Press, Inc.: Boulder, USA.

Suneson, C. A. and Stevens, H. 1953. *Studies With Bulked Hybrid Population of Barley*. U.S. Department Agric. Tech. Bulletin No. 1067, p. 14.

Suneson, C. A. and Wiebe, G. A. 1962. A 'Paul Bunyan' plant breeding enterprise with barley. *Crop Sci.*, 2: 347–8.

Tisne, S., Serrand, Y., Bach, L., Gilbault, E., Ben Ameur, R., Balasse, H. and et al. 2013. Phenoscope: An automated large-scale phenotyping platform offering high spatial homogeneity. *Plant J.*, 74, 534–44. doi:10.1111/tpj.12131.

Tiwari, C., Wallwork, H., Kumar, U., Dhari, R., Arun, B., Mishra, V. K., Reynolds M. and Joshi, A. K. 2013. Molecular mapping of heat tolerance traits in spring wheat (*Triticum aestivum* L.) under hot humid environment of eastern Gangetic plains of India. *Field Crop Res.*, 154: 201–10.

Trethowan, R. M. and Reynolds, M. 2005. Drought resistance: Genetic approaches for improving productivity under stress. *In* Proceedings of the 7th international wheat conference, 27 November–2 December 2005, Mar del Plata, Argentina.

Trethowan, R. M., Reynolds, M., Sayre, K. and Ortiz-Monasterio, I. 2005. Adapting wheat cultivars to resource conserving farming practices and human nutritional needs. *Ann. Appl. Biol.*, 146, 405–13.

Trethowan, R. M., Singh, R. P., Herta-Espino, J., Crossa, J. and Van Ginkel, M. 2001. Coleoptile length variation of near-isogenicRht lines of modern CIMMYT bread and durum wheats. *Field Crops Res.*, 70: 167–76.

Tucker, D. M., Griffey, C. A., Liu, S. and Saghai-Maroof, M. A. 2006. Potential for effective marker-assisted selection of three quantitative trait loci conferring adult plant resistance to powdery mildew in elite wheat breeding populations. *Plant Breed.*, 125: 430–6.

Tyagi, S., Mir, R. R., Kaur, H., Chhuneja, P., Ramesh, B., Balyan, H. S. and Gupta, P. K. 2014. Marker-assisted pyramiding of seven QTL/genes for seven different traits in common wheat (*Triticum aestivum* L.). *Mol. Breed.*, 34(1):167–75.

Vasistha, N. K., Arun, B., Mishra, V. K., Chand, R., Srinivasa, J., Yadav, P. S. and Joshi, A. K. 2015. Enhancing spot blotch resistance in wheat by marker-aided backcross breeding. *Euphytica*, 207:119–33, doi: 10.1007/s10681-015-1548-3.

Vishwakarma, M. K., Mishra, V. K., Gupta, P. K., Yadav, P. S., Kumar, H. and Joshi, A. K. 2014. Introgression of the high grain protein gene *Gpc-B1* in an elite wheat variety of Indo-Gangetic Plains through marker assisted backcross breeding. *Curr. Plant Biol.*, 1: 60–7. www.sciencedirect.com/science/article/pii/S2214662814000115.

Walker, J. T. 1969. Selection and quantitative characters in field crops. *Biol. Rev.*, 44: 207–43.

Weber, J. L. and Broman, K. W. 2001. 7 genotyping for human whole genome scans: Past, present, and future. *Adv. Genet.*, 42:77–96.

Whan, B. R. 1976. The association between coleoptile length and culm length in semidwarf and standard wheats. *J. Aust. Inst. Agr. Sci.*, 42: 194–6.

White J. W., Andrade-Sanchez P., Gore M. A., Bronson K. F., Coffelt T. A., Conley M. M., Feldmann K. A., French A. N., Heun J. T, Hunsaker D. J., Jenks M. A., Kimball B. A., Roth R. L., Strand R. J., Thorp K. R., Wall G. W. and Wang G. 2012. Field-based phenomics for plant genetics research. *Field Crop Res.*, 133: 101–12.

Wilde, F., Korzun, V., Ebmeyer, E., Geiger, H. H. and Miedaner, T. 2007. Comparison of phenotypic and marker-based selection for Fusarium head blight resistance and DON content in spring wheat. *Mol. Breed.*, 19: 357–70.

Witcombe, J. R., Joshi, A., Joshi K. D. and Sthapit, B. R. 1996. Farmer participatpry crop improvement. I. Varietal selection and breeding methods and their impact on biodiversity. *Exp Agr.*, 32: 445–60.

Witcombe, J. R., Joshi, K. D., Rana, R. B. and Virk, D. S. 2001. Increasing genetic diversity by participatory varietal selection in high potential production systems in Nepal and India. *Euphytica*, 122: 575–88.

Würschum Tobias, R. and Jochen C. 2013. Thomas Kraft, Janssen Geert and Zhao Yusheng. 2013. Genomic selection in sugar beet breeding populations. *BMC Genetics* 14: 85. 10.1186/1471-2156-14-85.

Yadav, P. S., Mishra, V. K., Arun, B., Chand, R. and Joshi, A. K. 2015. Validation of molecular markers for stem rust resistance and identification of suitable wheat germplasm targeting Eastern Gangetic Plains of India. *Indian J. Genet.*, 75(4): 518–21.

Yang, Y., Chai, R. and He, Y. 2014. Early detection of rice blast (pyricularia) at seedling stage in nipponbare rice variety using near-infrared hyper-spectral image. *Afr. J. Biotechnol.*, 11: 6809–17.

Zhou, W. C., Kolb, F. L., Bai, G. H., Domier, L. L., Boze, L. K. and Smith, N. J. 2003. Validation of a major QTL for scab resistance with SSR markers and use of marker-assisted selection in wheat. *Plant Breed.*, 122: 40–6.

Chapter 2

Establishment and root development in wheat crops

Peter J. Gregory and Christina K. Clarke, University of Reading, UK

1 Introduction
2 Root development and growth during establishment
3 Crop establishment practices
4 Case study: the effects of embryo size and seed mass on early vigour
5 Research contributions to improved wheat establishment and production
6 Future trends in research
7 Where to look for further information
8 Acknowledgements
9 References

1 Introduction

The early growth stages of wheat are important foundations for the crop's establishment in the field. These include the development and growth of the root system, the production of leaves and tillering, all of which have significant effects on the eventual yield. Emergence is the first visible sign of establishment and comprises the progressive stages of germination (appearance of a root radicle), subcrown internode elongation and coleoptile elongation (Jame and Cutforth, 2004). Germination is stimulated by the process of imbibition, which raises the water content of the seed to about 40% of grain dry weight (Bouaziz and Bruckler, 1989) and allows dissolved gibberellic acid to activate the aleurone layer to secrete amylase. In the endosperm, amylase hydrolyses starch into sugars, which permits respiration, and the radicle then emerges and grows downwards to anchor the seed in the soil (Blake et al., 2003). Following radicle emergence, the coleoptile elongates from the second or third node to form a subcrown internode between the coleoptile node and the node bearing the first leaf. This internode can grow to 8 cm in length and positions the crown 2–3 cm below the soil surface ensuring that the coleoptile and the tip of the first leaf emerge through the soil surface (Blake et al., 2003; Jame and Cutforth, 2004).

Temperature and moisture are the major factors influencing the processes of germination and emergence. The optimum growth temperature for wheat germination is 25°C, but wheat is grown throughout the world due to its ability to grow at temperatures from 3–4°C

http://dx.doi.org/10.19103/AS.2016.0004.13

to 30–32°C and in precipitation ranging from 250 to 1750 mm (Curtis et al., 2002). This variety of environments means that the ideal conditions for establishment and the factors influencing it differ between locations. Wheat can be sown in the spring or autumn depending on the seasonal characteristics of an area.

In this review, we have used the term 'establishment' to encompass development and growth prior to the elongation of the main stem (growth stage 31; Zadoks et al., 1974). We describe how the seedling root system establishes and then outline how four of the major agronomic decisions (time and depth of sowing, seeding rate and soil moisture conditions) affect the initial growth of the plant. Finally, we suggest research required to enhance future wheat production.

2 Root development and growth during establishment

2.1 Seminal and nodal root axes

Germination commences with the emergence of a root axis (root radicle or taproot) rapidly followed by the appearance of other axes developed from the scutellar and epiblast nodes of the embryonic hypocotyl. These seminal roots are typically 5–7 in number, are all visible shortly before the second leaf emerges and persist throughout the crop's life (Manske and Vlek, 2002). Nodal roots (also referred to as basal, shoot-borne or crown roots) are subsequently produced in pairs from primordia, which develop after germination at the base of leaves. In seedlings sown deeper than 5 cm or so, a pair of nodal axes may also

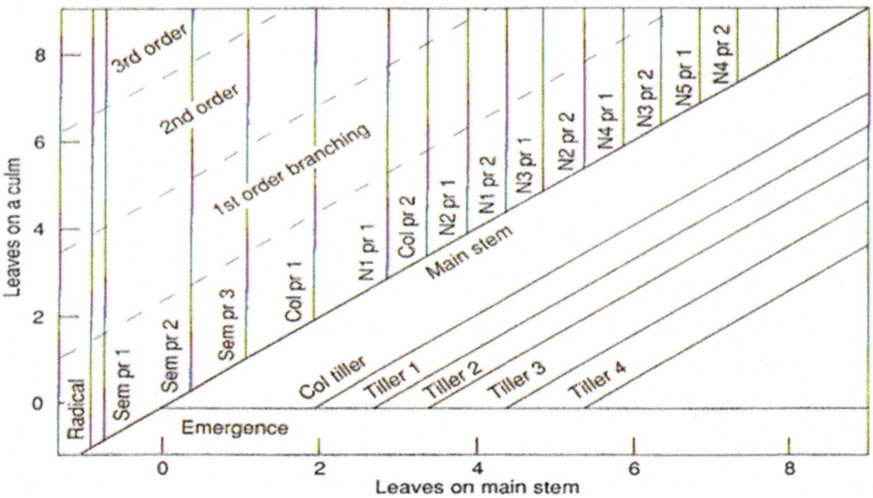

Figure 1 Diagrammatic representation of the coordinated development of wheat leaves, root axes and tillers. The appearance of tillers, root axes and root axis branching relative to the appearance of leaves on the main stem is shown: seminal (sem), coleoptile (col), leaf node (n) and pair (pr).

Source: Reproduced from Gregory (2006) with permission from John Wiley & Sons; based on original work by Klepper et al. (1994).

emerge from coleoptile nodes. The nodal axes appear closely packed on the crown, which is typically 1–2 cm below the soil surface, and continue to develop until stem elongation moves the higher nodes above the soil surface. Typically, the first four nodes of the main stem remain in the crown below the soil surface throughout the crop's life with nodes 6 and 7 generally found above the soil surface; node 5 may either be within the crown or slightly elevated. Branching of axes commences towards the base to form first-order laterals and, depending on temperature and moisture conditions in the soil, typically occurs within a few days. First-order laterals, in turn, branch to produce second-order laterals and so on, although it is rare to find any higher-order branches in plants grown as crops.

Figure 1 summarizes the integrated development of root axes and leaves for the main stem and tillers (Klepper et al., 1984). The figure shows that a wheat plant with five leaves on the main stem (typical of a plant shortly before stem extension commences) could be expected to have all the seminal axes and a pair of coleoptile axes with first-order laterals together with pairs of nodal axes at nodes 1–3; there could also be a coleoptile tiller and three tillers on the main stem. Generally, the number of nodal roots and leaves is linearly related with only slight genetic variation, although unfavourable soil physical conditions may cause partial or complete failure of the nodal root system (Gregory et al., 1978; Manske and Vlek, 1982). Seed size can, however, have a big effect on the number of seminal axes produced (Richards and Passioura, 1981).

2.2 Elongation and growth of the root system

The rate of extension of wheat roots depends on environmental factors, especially temperature and the availability of assimilates. During establishment in autumn and winter, root extension rate often averages only 3–6 mm day^{-1}, but increases to 15–20 mm day^{-1} when temperatures are warmer in spring (Gregory, 2006); in warmer environments, rates of up to 40 mm day^{-1} may occur. For spring-sown wheat, the absolute rates of extension are greater than those of winter-sown crops because of warmer temperatures and greater incidence of radiation. For example, Thorup-Kristensen et al. (2009) determined average rates of downward extension of the rooting depth of about 7 mm day^{-1} for winter wheat and 13 mm day^{-1} for spring wheat, but both crops had the same rate when temperature differences were included (1.3 mm °C day^{-1}). Elongation behind the root tip enables the root to penetrate downwards, eventually reaching depths of about 2 m for winter wheat in deep soils during grain filling. During establishment, though, the depth of rooting is typically $<$0.5 m.

As with shoot growth, the accumulation of dry matter (and length) occurs at an exponential rate during the establishment phase and becomes linear thereafter. Root mass and length are not, though, uniformly distributed with depth, and in most soils, the highest percentage of roots is found in the upper soil layers, especially the cultivated layer. In the United Kingdom, winter wheat crops at maturity had $>$50% of their root length in the top 20 cm of soil with $<$8% below 80–100 cm (Barraclough and Leigh, 1984). Similarly, Entz et al. (1992) found 60–80% of the total root length of two winter and two spring wheat cultivars in the upper 50 cm of soil, with the maximum rooting depth ranging between 110 and 130 cm. In homogenous soils, a simple exponential model of reduced distribution with depth frequently describes root distribution (Gerwitz and Page, 1974; Jackson et al., 1996), a pattern that is evident even during the establishment phase (Gregory et al., 1978).

2.3 Root system architecture

Root architecture defines the spatial configuration of a root system in the soil and describes aspects of the shape of a root system. Besides the axes and their branches, the growth angle of the seminal roots is an important aspect of root architecture and has a significant association with the acquisition efficiency of crops because it affects the depth of rooting (Manske and Vlek, 2002). Several studies have indicated that this root trait shows considerable genotypic variation and may be associated with drought tolerance. For example, in a study of 27 Australian and 3 International Maize and Wheat Improvement Center (CIMMYT, Mexico) wheat genotypes, Manschadi et al. (2008) found that the growth angle of the first pair of seminal axes varied significantly, ranging from 36 to 56 degrees, although the angle of the second pair of axes did not differ significantly between these wheat genotypes with an average of 78.4 degrees, elongating almost horizontally. However, they found that the more drought-tolerant genotypes, adapted to deep clay soils in northern Australia, had a more compact root system with smaller root growth angles of the first pair of seminal axes that resulted in more roots occupying the soil directly under the crop. de Dorlodot et al. (2007) also stated that more drought-tolerant crop species have these more vertically inclined root systems.

2.4 Roots, resource acquisition and yield

During the establishment phase, root length is typically <1 cm cm^{-3} soil at all depths increasing to >5 cm cm^{-3} soil in the topsoil during stem elongation; for example, Hoad et al. (2004) found that root length density of two different winter wheat varieties ranged from 3 to 6 cm cm^{-3} soil in the plough layer at anthesis, but <1 cm cm^{-3} below 40 cm. A root length density of about 1 cm cm^{-3} has been found in modelling studies to be sufficient to extract water and nitrogen efficiently, although greater lengths are required for immobile nutrients such as phosphate (van Noordwijk and de Willigen, 1987; King et al., 2003). During establishment, then, root length densities are rarely sufficiently large to extract all of the water and mobile nutrients present.

Because the availability of nutrients and water vary spatially and temporally, there is no unique relation between the length (or mass) of a root system and final yield. Hoad et al. (2004) reviewed literature for the United Kingdom wheat crops and found that a total root length of more than 15 km m^{-2} was associated with higher-yielding wheat crops (9 t ha^{-1}) because of a relationship with ear number. The optimum root length was about 30–35 m per ear with values above this producing no further increase in yield; values <30 m per ear were associated with lower-yielding wheat crops (Hoad et al., 2004).

3 Crop establishment practices

3.1 Time of sowing

The optimum time to sow a wheat crop is dependent on the cultivar, the crop rotation and the weather and soil conditions. For winter wheat, early sowing gives the crop the best chance to establish itself before harsh weather conditions dominate. In northern mid-latitudes where summer temperatures are cooler, the cropping season is long, spanning from October to August. Yields are high in these areas due to the prolonged amount of time for the crop to intercept light and accumulate dry matter, with the sowing date mostly being determined by the timing of the previous crop harvest (Gooding and Davies, 1997). The earlier the sowing date, the greater is the percentage of seeds resulting in established plants, with an average establishment of 70% in September, but only about 50% in mid-November, in the United Kingdom (HGCA, 2008). Although early establishment of the crop can improve over-winter

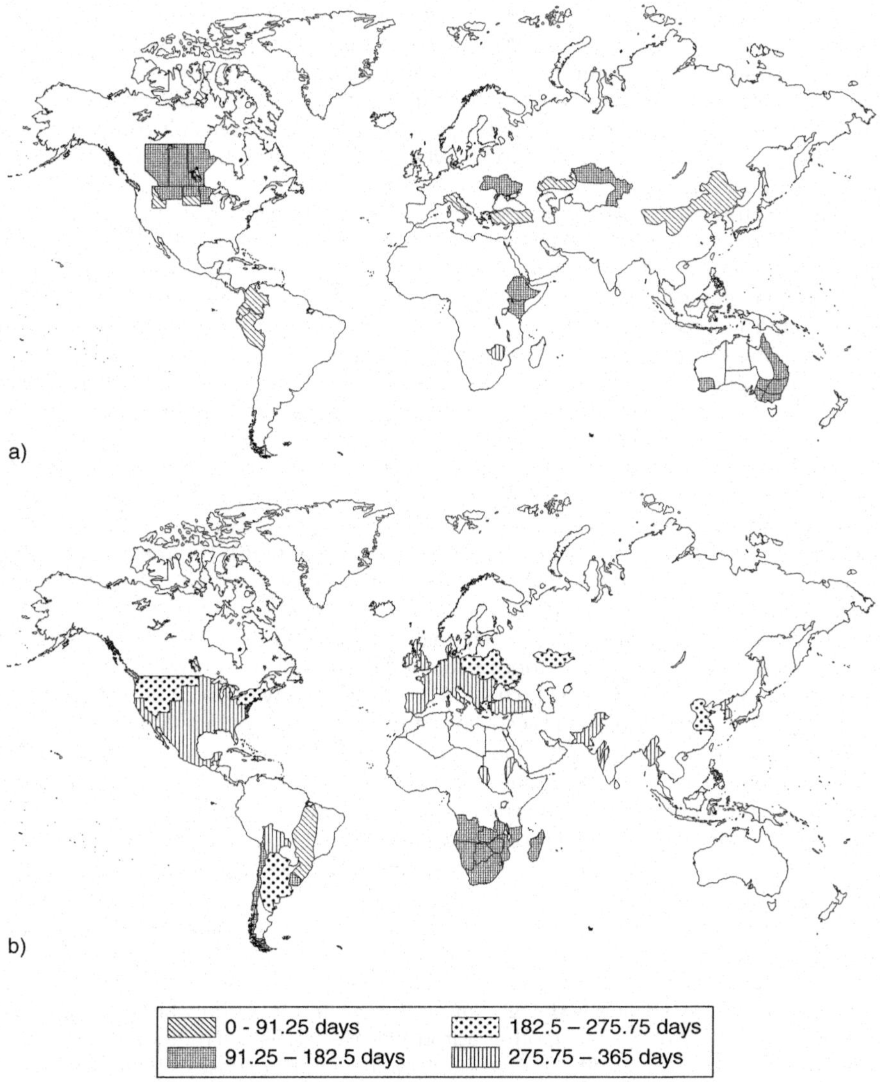

Figure 2 Maps of typical crop planting dates for (a) spring wheat and (b) winter wheat.
Source: Redrawn from Sacks et al. (2010) with permission from John Wiley & Sons.

survival and canopy formation, it also increases exposure to pests and diseases in the autumn season and competition from weeds, which germinate after drilling (Gooding and Davies, 1997; Sacks et al., 2010). Consequently, an early sowing date may result in additional production cost for crop protection chemicals to achieve the benefits of early establishment. These negative impacts can be reduced through choosing an appropriate variety, as some winter wheat varieties are well suited to early drilling through characteristics such as fast growth rate and tillering, which help suppress weeds (Open Field, 2016).

In more extreme winters such as those experienced in Western Canada, sowing date is determined by soil temperature. Wheat plants over-winter in the seedling stage but must establish and develop the root crown before intense cold sets in and the ground freezes. Early sowing can also cause more exposure to pests and diseases such as root rot. Drilling into the previous crop stubble is advantageous because it protects the plant from low soil temperatures by trapping snowfall and insulating the seedling (University of Saskatchewan, 2016).

In cool climates, spring wheat is sown in April/May when there is no risk of frost and temperatures are about 8–14°C. Early sowing is desirable to ensure sensitive wheat growth stages do not coincide with heat stress (Sacks et al., 2010). In much warmer climates such as Australia, rainfall determines when sowing of spring wheat can take place, but it is important to sow in time to ensure that anthesis and grain filling occur during the cool and moist conditions of the winter and spring seasons (Gooding and Davies, 1997; Sacks et al., 2010). Figure 2 shows planting dates for both spring and winter wheat around the world, with the main separation occurring between the northern and southern hemispheres (Sacks et al., 2010). Delayed sowing can have significant impacts on final yield with reductions as large as 140–330 kg/ha/week in Australia (Rebetzke et al., 2007). The growing season is much shorter in these warmer environments, for example, only 90–100 days in Sudan (Ageeb, 1994).

In the rainfed environments of West Asia and North Africa, sowing time has a substantial effect on final crop yield. Farmers tend to delay sowing until 80–100 mm of precipitation has fallen to ensure that the top 20 cm of soil is sufficiently moist for crop establishment. The risk of waiting until later in the season to sow spring wheat is that the shorter season reduces potential yields. Moreover, heavy rain can occur in late December and January, thereby preventing sowing altogether because the soil will be too wet. Delayed sowing can reduce yields by as much as 97 kg ha^{-1} of biomass per day of delay, and above average yields of 2.5 t ha^{-1} were achieved by sowing after just 33 mm of rain (Mahdi et al., 1998).

In contrast, in the Indo-Gangetic Plain, weather conditions permit a double cropping system that usually consists of irrigated spring wheat varieties grown over the winter, followed by rainfed or irrigated rice. The time of sowing of the spring wheat is constrained by the harvest time of the rice crop (Fischer et al., 2014). Delayed sowing of wheat is common as rice is prioritized because it produces more income. This significantly reduces potential yield due to excessive rainfall causing black point and premature sprouting of the grain (Gooding and Davies, 1997). Late sowing also results in reduced establishment and lower tiller production, together with grain filling in the hotter months of March/April when temperatures can reach 39°C (Sattar et al., 2010). Together, these factors result in 0.8–1% yield loss for each day sowing is delayed after 15 November (Aggarwal et al., 1994; Ortiz-Monasterio et al., 1994). However, other agronomic decisions can mitigate these negative effects, including choosing cultivars that grow well in shorter seasons (Sattar et al., 2010) and direct drilling of wheat into the rice crop residues which will speed up sowing and result in the residues assisting weed suppression and water and fuel conservation (Fischer et al., 2014).

3.2 Seed rate

Farmers modify the seeding rate to achieve a desired population depending on the weight of the grain sown and experience of the expected establishment under specific circumstances. As the expected establishment rate declines, the seed rate is increased; this occurs when the quality of the seedbed is poor, seed vigour is low or the seed is broadcast or sown resulting in poor seed-to-soil contact (Gooding and Davies, 1997; Curtis et al., 2002). The mass of the seed also affects seeding rate as larger grains are generally more vigorous and germinate faster. Naylor (1993) compared large and small wheat seeds and found that both sizes germinated well under optimum conditions of 18°C, but small grains were negatively affected by temperature extremes to a greater extent than larger grains and had lower germination under 10°C.

Seed quality is also important for crop establishment. Ideal samples are of an even grade with plump grains free of any diseased or shrivelled seed and do not include crop residues or stones (Curtis et al., 2002). Optimum seed rates give the highest yield for the least amount of seed and, after a certain threshold, increasing seed rate does not increase yield. Optimum seed rate varies from 100 to 120 kg ha^{-1} of seed in the United Kingdom, but farmers in many parts of the world, including Australia, the United States and Canada, commonly use rates <100 kg ha^{-1}. Although a low seed rate reduces input costs, it results in greater reliance on the establishment of the seeds sown as unexpected disease, pests or harsh weather may adversely affect establishment and hence the yield. Thin stands may reduce the efficiency of the crop in accessing light and nutrient resources (Gooding and Davies, 1997), but at very low seed rates, plants will compensate by increased production and survival of tillers. Tiller production may continue until the main stems flower or beyond so that the maturation of tillers differs substantially within one plant, causing difficulties in harvesting because of uneven ripening.

High seed rates increase competition between plants and alter the properties of the stand. As competition increases, percentage germination is reduced by 2.6% for every 100 extra seeds sown per m^2. This results in 67.8% establishment at 100 seeds m^{-2}, reducing to 60% establishment at 400 seeds m^{-2} (Blake et al., 2003). Higher seed rates give greater plant densities and ensure complete light interception due to overlapping canopies. However, neighbouring plants competing for light may result in stands with taller stems of a reduced diameter weakening crop anchorage (Gooding and Davies, 1997; Curtis et al., 2002). Yield benefits are small at higher seed rates because the plant produces fewer crown roots and tillers; there are fewer grains per ear, but more tillers survive to reach the reproductive growth stages (HGCA, 2008). Although dense crop stands can improve weed suppression through increased competition, it is also easier for fungal diseases such as black point, Septoria and powdery mildew to spread. Higher seeding rates are also not suitable for dry environments due to increased rates of water use (Gooding and Davies, 1997).

3.3 Depth of sowing

Optimum sowing depth varies from about 25–100 mm depending on the location and the characteristics of the soil (Gooding and Davies, 1997). Soil moisture is a key factor determining sowing depth, together with the ability of the coleoptile to penetrate the soil surface. The deeper a seed is sown, the more stable will be the environment because it is less likely to be eaten by birds, moisture contents will be higher and the seed will be better protected from surface drying or frost. In temperate conditions, the optimum sowing depth for cereals is about 30–50 mm with 40 mm being advised in the United Kingdom (Gooding and Davies, 1997; HGCA, 2008). The optimum depth changes depending on the time of year as spring wheat is typically sown in drier conditions than winter wheat and, therefore, may benefit from deeper sowing. Deeply sown seeds have slow emergence and poor tillering capabilities. The main reason to sow deeply is to ensure a timely sowing in dry environments where there is no irrigation; soil moisture stored at depth can ensure that the seed germinates and the crop establishes (Rebetzke et al., 2007).

Sowing deeply may hinder crop establishment due to slower emergence which can delay crop development (Gooding and Davies, 1997). The likelihood of abnormal seedlings increases with sowing depth as, although the coleoptile can extend to 70 mm, sowing deeper than this will cause the first emerging leaf to grow through the soil surface unprotected. While the first leaf is morphologically adapted to be better at growing through the soil than later emerging leaves, it can still be damaged or deflected, preventing emergence (Blake et al., 2003). Such seedlings also typically have reduced leaf area index (LAI) due to thinner but longer leaves, and reduced tillering resulting in a reduced biomass (Mahdi et al., 1998; Rebetzke et al., 2007). Shallow sowing depths are common in cold environments such as Canadian wheat-growing areas, where the crop is directly drilled into the previous crop's stubble. Loeppky et al. (1989) found that shallow sowing depths of 10–25 mm at the beginning of the growing season was advantageous to establishment, resulting in yield improvements of 11% in four of the six field trials undertaken. This was a result of faster development of the seedling, allowing it to escape winter damage and produce more tillers and heads per m^2. Sowing more deeply by as little as 17 mm delayed plant emergence and resulted in deeper crown placement (Loeppky et al., 1989).

In dry environments, deep sowing into good soil moisture conditions can assist in optimizing sowing date (facilitating a longer season) and hence prevent more sensitive wheat growth stages coinciding with extreme weather conditions. For example, Mahdi et al. (1998) sowed seeds at 30, 60, 90 and 120 mm depths in rainfed northern Syria. Seed sown at 60 mm had more tillers and 61 and 90% higher LAI, compared to sowing at 120 mm in the two seasons studied. Emergence and establishment were poor for seeds sown at 30 mm due to the dry soil forming a crust at the surface extending to 20–30 mm (Mahdi et al., 1998). In a trial with a single winter wheat variety in the United Kingdom, 100% establishment was obtained at a sowing depth of 68 mm, but this was reduced to 89% for sowing depths of 23 and 53 mm and to only 6% when sown at 143 mm (Kirby, 1993). Leaf length decreased with sowing depth, resulting in a 12-fold difference in the length of leaf 9 between sowing depths of 8 and 98 mm. The rate of crop development was affected by sowing depth with seeds sown at 38 mm or less reaching floret or spikelet stage 176 days after sowing; the proportion of plants at this growth stage decreased as sowing depth increased (Kirby, 1993).

Coleoptile length varies with genotype, so choosing or breeding a wheat cultivar with a longer coleoptile can improve emergence and crop establishment when sown deeply, thereby offsetting yield reductions caused by delayed sowing times waiting for precipitation. A longer coleoptile, resulting in faster emergence and better establishment, may also improve the water-use efficiency of the plant as greater leaf cover reduces soil surface evaporation (Lopezcastaneda and Richards, 1994). Semi-dwarf wheats containing Rht-B1b and Rht-D1b are insensitive to gibberellin and have short coleoptiles due to reduced cell elongation, but those containing Rht8 remain sensitive to gibberellin and might provide semi-dwarf breeding material with improved coleoptile length for deeper sowing (Rebetzke et al., 2007). A study of wheat cultivars containing Rht8, Rht-B1b and Rht-D1b sown at 50, 80 and 110 mm found that although seedling numbers were reduced by an average of 65% at 110 mm compared to 50 mm, the Rht8 and other tall varieties produced longer coleoptiles as well as longer subcrown internodes and therefore had shallower crowns (Rebetzke et al., 2007). Coleoptile length in the Rht8 and tall wheats were 118 and 117 mm, respectively, compared to 97 mm for the Rht-B1b and Rht-D1b semi-dwarfs. Consequently, the Rht8 and tall varieties had greater emergence and seedling vigour when sown at 110 mm than at 50 mm, together with smaller reductions in spike number and grain yield over 2 years (18 and 10% for Rht8, 15 and 7% for tall wheat and 39 and 16% for Rht-B1b and Rht-D1b (Rebetzke et al., 2007)).

3.4 Wetness and dryness

Once sowing has occurred, seeds in the top 50–100 mm of soil can be subjected to wetting and drying cycles (Downs and Cavers, 2000), with either beneficial or adverse effects on the process of germination, depending on the length of the cycles. Wetting and drying cycles are particularly common in semi-arid environments when rainfall is sporadic and water in the topsoil can evaporate quickly. Wetting of the seed does not disrupt germination, but the timing of desiccation, once the process of germination has become active, can cause embryo damage and seed death. Andoh and Kobata (2001) state that seed germination can be maintained if desiccation occurs at imbibition or during biochemical activation of the endosperm, but once germination has begun, desiccation will cause seed death.

Alternatively, wetting and drying cycles of the seed can be used advantageously as a pre-sowing seed treatment to improve germination; this is termed 'seed hardening'. The hardening treatment can improve the stability of seed emergence and establishment in varying soil water conditions to prevent harmful effects of seed dehydration in drying soil (Andoh and Kobata, 2001). Seed of a Norin 61 variety imbibed to achieve a seed water content of 64% for 24 h before re-drying, increased germination and seedling emergence in both irrigated and non-irrigated conditions (Andoh and Kobata, 2001). Germination was accelerated and plant height and seminal root length increased, compared to seed imbibed for 12, 36 and 48 h. This hardening treatment also increased seminal root length at 54 days after sowing in field conditions (Andoh and Kobata, 2001). Changes in properties and proteins in the seed coating facilitate accelerated germination following the wetting and re-drying process (Berrie and Drennan, 1971).

4 Case study: the effects of embryo size and seed mass on early vigour

The seed weight of a wheat cultivar has been shown in several studies to be associated with embryo size which, in turn, influences early vigour and production of leaf biomass during crop establishment (Botwright et al., 2002; Richards and Lukacs, 2002; Spielmeyer et al., 2007). For example, wheat cultivars backcrossed with a high-vigour donor had 10% greater leaf area than the control cultivars (Botwright et al., 2002). Larger grain mass was linearly related to embryo size (Fig. 3) with a 23 mg increase in grain mass (from 24 to 47 mg) increasing embryo size by 25%. The larger seeds produced 37% more leaf area at 50 days after sowing and a 12% increase in yield in years with sufficient rainfall due to

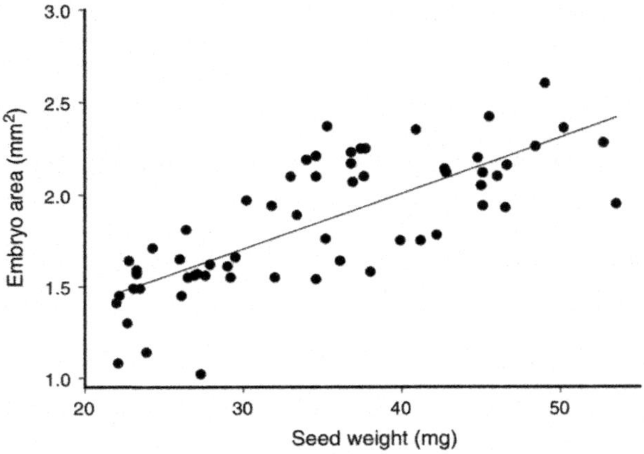

Figure 3 Relationship of embryo size and seed weight of a wheat cultivar, $r^2 = 0.57$.
Source: Reproduced from Botwright et al. (2002) with permission from CSIRO Publishing.

this improved early biomass accumulation (Botwright et al., 2002). Richards and Lukacs (2002) also found a linear increase in embryo size with seed weight and investigated the effect of seed weight on early vigour traits. A 10 mg increase in grain weight of wheat cultivars sourced in Australia and CIMMYT reduced emergence time by 4% but increased plant dry weight by 21%, tiller number by 12% and leaf area by 15% (Richards and Lukacs, 2002).

Seed weight also influences the size of the coleoptile and coleoptile tiller (Rebetzke et al., 2004). Greater seed mass increased the diameter of the coleoptile tiller which, in turn, is associated with reduced time to emergence and increased shoot strength (as shown by Andrews et al., 1997 in grass species). Embryo size was also associated with the frequency of coleoptile tiller formation and with leaf area in plants resulting from a cross of a cultivar with a high frequency of coleoptile tiller production with one with a lower frequency of production (Rebetzke et al., 2008). Coleoptile tiller leaf area also had a significant effect on the average leaf area of wheat seedlings (Liang and Richards, 1994; Richards and Lukacs, 2002; Rebetzke et al., 2008). Although coleoptile tillers are generally smaller than other tillers, their contribution to leaf area and biomass accumulation during wheat establishment is significant and contributes to early vigour (Rebetzke et al., 2008). For example, the presence of a large coleoptile tiller increased plant seedling leaf area and biomass by 24–30%, compared to plants with either a small or no coleoptile tiller (Liang and Richards, 1994). Increased vigour was also demonstrated by an increased rate of leaf expansion with lines having a large coleoptile tiller having a rate 25–35% higher, compared to lines with small or no coleoptile tillers (Liang and Richards, 1994). In this study, the coleoptile tiller was associated with the total leaf area of the plant but did not affect the growth of other tillers; other early vigour traits such as mainstem leaf number, specific leaf area and tiller number were not associated with leaf area (Liang and Richards, 1994).

The genetic control of embryo size in wheat and its influence on early vigour was investigated in more detail in three different doubled haploid populations of Sunco/Tasman; Cranbrook/Halberd and Vigour18/Chuan-Mai 18 (Moore and Rebetzke, 2015).

The populations differed slightly for vigour traits correlating with embryo size, with the largest genetic correlations in the Sunco/Tasman population being between embryo length and specific area; embryo area and total dry weight and embryo width and leaf width. Within the Cranbrook/Halberd population, the strongest correlation with embryo size was with the combined leaf area of the first and second leaves, in addition to the coleoptile tiller length being associated with embryo length and width. Coleoptile tiller length and total dry weight had the greatest association with embryo size in the Vigour18/Chuan-Mai 18 population, with leaf area and leaf width also being associated to a lesser extent (Moore and Rebetzke, 2015). Genetic effects had only a small influence on the phenotypic variance within the populations but co-locating quantitative trait loci (QTL) for embryo size and early vigour were found between populations, possibly signifying genes causing a pleiotropic effect. Additionally, QTL for embryo size and leaf width were found on the Rht-B1b and Rht-D1b chromosomes 4B and 4D. Heritabilities were highest for embryo size (associated with embryo width and length) and leaf area traits but lower for dry weight and coleoptile tiller length, common throughout the double haploid populations (Moore and Rebetzke, 2015). In Australia, it is more appropriate to breed for early vigour through a larger embryo than a greater seed size because the hot and dry end to many wheat-growing seasons makes it difficult to reliably recover large seeds (Rebetzke and Richards, 1999).

Domestication of the wheat crop has favoured the selection of an increased endosperm size. This is evident in the greater grain size but similar embryo weights of cultivated durum wheat (*Triticum turgidum*), compared to its direct relative, wild emmer (*Triticum dicoccoides*; Golan et al., 2015). This is most likely due to the contribution of the endosperm to the milling yield of wheat with the values of the endosperm, germ (embryo and scutellum) and outer layers (pericarp, testa and aleurone) of large kernels stated as 83.5, 2.5 and 14% respectively (Posner and Hibbs, 2011). Subsequently, breeding for a larger embryo may have been a trade-off with endosperm size and the flour yield of the grain. However, the endosperm also provides nourishment to the embryo during seed development and germination and therefore is an important component of embryo growth and function (Golan et al., 2015).

5 Research contributions to improved wheat establishment and production

Early vigour constitutes the increased growth of a wheat plant at the beginning of its life cycle; it can reduce emergence time and increase plant biomass due to increased interception of light, water and other resources and thereby accelerate growth. It is characterized by multiple traits which include high specific leaf area (SLA; leaf area to leaf weight ratio), greater root length and dry weight, a large embryo, a longer coleoptile and the presence of a large coleoptile tiller (Richards and Lukacs, 2002; Liao et al., 2004). The benefit of early vigour has been widely reported in many wheat-growing regions, including Mediterranean environments, such as the wheat-growing areas of Western and Southeastern Australia, the temperate climates of Europe and the diverse conditions experienced in the United States and Canada. Almost universally, rapid establishment is viewed as beneficial both because of the direct boost to crop growth and the indirect benefits provided by reduced evaporation from the soil surface and improved weed suppression and the consequent reduction in herbicide use (Richards and Lukacs, 2002; Spielmeyer et al., 2007).

Additional breeding efforts to improve the establishment and production of wheat in different environments have included exploitation of the genetic diversity of wild relatives of wheat. Introduction of novel genes from diploid and tetraploid ancestors into modern-day cultivars has produced synthetic hexaploid wheats with superior traits that increase yields, compared to local check cultivars (Ogbonnaya et al., 2007; van Ginkel and Ogbonnaya, 2007). This introduced diversity has particularly improved abiotic stress tolerance through traits such as water-use efficiency (WUE), earliness to flowering and increased rooting in the subsoil (Lopes and Reynolds, 2011). Another successful breeding example that exploited the genetic diversity of wheat landraces was the reduction of xylem vessel diameter to conserve soil water for critical growth stages. Backcrossing a Turkish landrace wheat into two commercial Australian varieties in a breeding programme reduced xylem vessel diameter from 65 to <55 μm. In a dry environment, this reduction improved yields between 3 and 11% due to the conservation of water (Richards and Passioura, 1989).

6 Future trends in research

Undoubtedly, research into readily observable traits such as embryo size, longer coleoptiles and high specific leaf area will continue to provide new germplasm for incorporation into breeding programmes around the world. However, increasing attention is also being paid to traits that are belowground, which might be used to improve early growth, the acquisition of resources (especially water, nitrogen and phosphorus) and grain yield.

Studying root systems is more difficult than assessing shoots, but a range of techniques is now emerging to speed up the phenotyping of roots including seedling screens in controlled conditions and novel techniques in soil (Gregory et al., 2009; Das et al., 2015). There is some concern that current wheat crops have been bred in high input/output agricultural systems and that some traits of value in more diverse systems may have been lost (Meyer et al., 2012; White et al., 2013). Moreover, trialling varieties in ploughed soils may obviate the need for root traits to overcome soil mechanical constraints that occur in minimally tilled soils – a practice gaining a more prominent place in crop production systems. These observations, together with a call for more sustainable intensification of crop production, have led to a focus on reducing inputs while simultaneously increasing or maintaining crop yield, together with a greater appreciation of the social consequences of production systems (Godfray et al., 2010; Andres and Bhullar, 2016).

A greater emphasis on matching roots to their soil environment has led to the development of ideas of root ideotypes (Fig. 4). Plants that have root systems that can forage within the topsoil have been suggested for those areas in which P is the major limitation to crop growth because it is immobile and usually concentrated in the topsoil. In contrast, in many regions water is distributed throughout the soil profile and nitrogen is often leached out of the topsoil into the subsoil. In such environments, the ability to grow roots rapidly to depth to prevent resources moving beyond the rooting front is likely to be advantageous. Several rooting traits might contribute to this ideotype, including thick unbranched seminal roots, a combination of relatively unbranched seminal roots with nodal roots with prolific lateral branching and abundant root hairs, and modifications

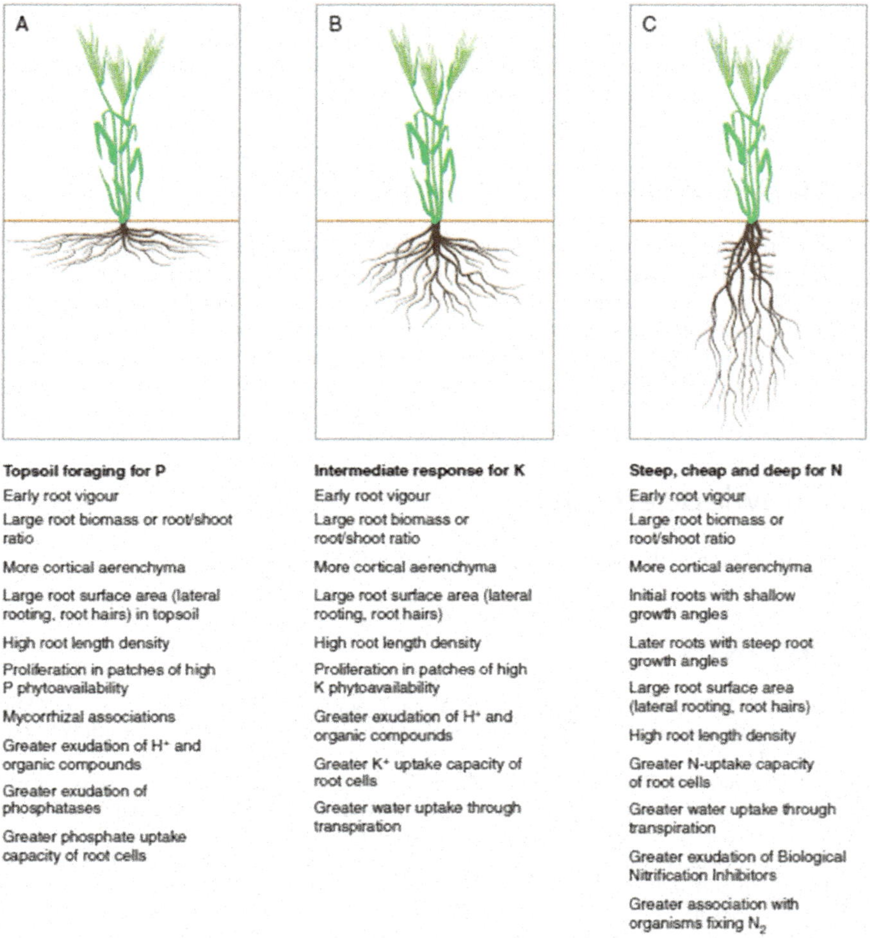

Figure 4 Root ideotypes for efficient acquisition of (A) phosphorus, (B) potassium and (C) nitrogen. Source: Reproduced from White et al. (2013) with permission from Oxford University Press.

to root structure including more aerenchyma and high capacity for nitrate uptake by root epidermal cells (Lynch, 2013; White et al., 2013).

Opportunities also exist to explore genetic modification of wheat root systems including the introgression of root traits from wild relatives. For example, Hendriks et al. (2016) examined the effects of introducing a tillering inhibition (*tin*) gene on the performance of wheat in rainfed environments of Eastern Australia. When grown in the field, lines containing *tin* had a negligible effect on seedling growth, but significantly reduced tiller number, LAI and spike number at anthesis. Root biomass, root:shoot ratio at early stem elongation and root depth at maturity were all increased in the *tin*-containing lines. These changes produced slower rates of soil water use in the *tin*-containing lines, greater stomatal conductance, cooler canopy temperatures and an extended duration of green leaf area which, together, resulted in increased harvest index and grain yield (Hendriks et al., 2016). Such studies suggest that a considerable potential exists for beneficially modifying root systems, especially in the establishment phase.

7 Where to look for further information

General information about the establishment of wheat crops can be found in the textbook on the wheat crop by Gooding and Davies (1997). A good insight into the importance of the coleoptile and its role in the establishment of wheat is given in the paper by Rebetzke et al. (2007). General information about the growth and activity of cereal roots can be found in Gregory (2006), with specific information about the role of roots in extracting water from the subsoil and its consequences for grain yield provided by Kirkegaard et al. (2007).

8 Acknowledgements

CKC is grateful to AHDB Cereals and Oilseeds for financial support.

9 References

Ageeb, O. A. A., 1994. *Agronomic Aspects of Wheat Production in Sudan. Wheat in Heat Stressed Environments: Irrigated, Dry Areas and Rice-Wheat Farming Systems*. CIMMYT, Mexico, pp. 67–74.

Aggarwal, P. K., Kalra, N., Singh, A. K. and Sinha, S. K., 1994. Analyzing the limitations set by climatic factors, genotype, water and nitrogen availability on productivity of wheat. 1. The model description, parametrization and validation. *Field Crop Res.* 38, 73–91.

Andoh, H. and Kobata, T., 2001. Effect of seed hardening, wetting and redrying before sowing, on germination and seedling emergence of a Japanese wheat variety Norin 61 in desiccated soil. *Plant Prod. Sci.* 4, 50–5.

Andrews, M., Douglas, A., Jones, A. V., Milburn, C. E., Porter, D. and McKenzie, B. A., 1997. Emergence of temperate pasture grasses from different sowing depths: importance of seed weight, coleoptile plus mesocotyl length and shoot strength. *Ann. Appl. Biol.* 130, 549–60.

Berrie, A. M. M. and Drennan, D. S. H., 1971. Effect of hydration-dehydration on seed germination. *New Phytol.* 70, 135–42.

Blake, J. J., Spink, J. H. and Dyer, C., 2003. Factors affecting cereal establishment and its prediction. *HGCA Res. Rev.* 51, p. 51.

Botwright, T. L., Condon, A. G., Rebetzke, G. J. and Richards, R. A., 2002. Field evaluation of early vigour for genetic improvement of grain yield in wheat. *Aust. J. Agr. Res.* 53, 1137–45.

Bouaziz, A. and Bruckler, L., 1989. Modeling of wheat imbibition and germination as influenced by soil physical-properties. *Soil Sci. Soc. Am. J.* 53, 219–27.

Curtis, B. C., 2002. Wheat in the world. In B. C. Curtis, S. Rajaram and H. Gomez Macpherson (eds), *Bread Wheat. Improvement and Production*. Food and Agriculture Organization of the United Nations, Rome.

de Dorlodot, S., Forster, B., Pages, L., Price, A., Tuberosa, R. and Draye, X., 2007. Root system architecture: opportunities and constraints for genetic improvement of crops. *Trends Plant Sci.* 12, 474–81.

Downs, M. P. and Cavers, P. B., 2000. Effects of wetting and drying on seed germination and seedling emergence of bull thistle, Cirsium vulgare (Savi) Ten. *Can. J. Bot.* 78, 1545–51.

Entz, M. H., Gross, K. G. and Fowler, D. B., 1992. Root-growth and soil-water extraction by Winter and Spring wheat. *Can. J. Plant Sci.* 72, 1109–20.

Fischer, T., Byerlee, D. and Edmeades, G., 2014. Crop yields and global food security will yield increase continue to feed the world? Introduction. *Aciar. Monogr. Ser.* 158, 1–26.

Golan, G., Oksenberg, A. and Peleg, Z., 2015. Genetic evidence for differential selection of grain and embryo weight during wheat evolution under domestication. *J. Exp. Bot.* 66, 5703–11.

Gooding, J. M. and Davies, W. P., 1997. *Wheat Production and Utilization: Systems, Quality, and the Environment.* CAB International, Wallingford, UK.

Gregory, P. J., 2006. *Plant Roots: Growth, Activity and Interaction with Soils.* Blackwell Publishing, Oxford.

HGCA, 2008. *The Wheat Growth Guide.* Agriculture and Horticulture Development Board Second Edition.

Hoad, S. P., Russel, G., Kettlewell, P. S. and Belshaw, M., 2004. Root system management in winter wheat: practices to increase water and nitrogen use. HGCA Project Report No. 351.

Jame, Y. W. and Cutforth, H. W., 2004. Simulating the effects of temperature and seeding depth on germination and emergence of spring wheat. *Agr. Forest Meteorol.* 124, 207–18.

Kirby, E. J. M., 1993. Effect of sowing depth on seedling emergence, growth and development in barley and wheat. *Field Crop Res.* 35, 101–11.

Kirkegaard, J. A., Lilley, J. M., Howe, G. N. and Graham, J. M., 2007. Impact of subsoil water use on wheat yield. *Aust. J. Agric. Res.* 58, 303-15.

Liang, Y. L. and Richards, R. A., 1994. Coleoptile tiller development is associated with fast early Vigor in wheat. *Euphytica* 80, 119–24.

Liao, M. T., Fillery, I. R. P. and Palta, J. A., 2004. Early vigorous growth is a major factor influencing nitrogen uptake in wheat. *Funct. Plant Biol.* 31, 121–9.

Loeppky, H., Lafond, G. P. and Fowler, D. B., 1989. Seeding depth in relation to plant development, winter survival, and yield of no-till winter-wheat. *Agron. J.* 81, 125–9.

Lopes, M. S. and Reynolds, M. P., 2011. Drought adaptive traits and wide adaptation in Elite Lines derived from Resynthesized Hexaploid wheat. *Crop Sci.* 51, 1617–26.

Lopezcastaneda, C. and Richards, R. A., 1994. Variation in temperate cereals in rain-fed environments. 3. water-use and water-use efficiency. *Field Crop Res.* 39, 85–98.

Mahdi, L., Bell, C. J. and Ryan, J., 1998. Establishment and yield of wheat (Triticum turgidum L.) after early sowing at various depths in a semi-arid Mediterranean environment. *Field Crop Res.* 58, 187–96.

Manschadi, A. M., Hammer, G. L., Christopher, J. T. and deVoil, P., 2008. Genotypic variation in seedling root architectural traits and implications for drought adaptation in wheat (Triticum aestivum L.). *Plant Soil* 303, 115–29.

Manske, G. and Vlek, P., 2002. *Root Architecture – Wheat as a Model Plant.* Marcel Dekker, Inc., New York

Moore, C. and Rebetzke, G., 2015. Genomic regions for embryo size and early vigour in multiple wheat (Triticum aestivum L.) populations. *Agronomy-Basel* 5, 152–79.

Ogbonnaya, F. C., Ye, G. Y., Trethowan, R., Dreccer, F., Lush, D., Shepperd, J. and van Ginkel, M., 2007. Yield of synthetic backcross-derived lines in rainfed environments of Australia. *Euphytica* 157, 321–36.

OpenField, 2016. *Seed.* Advice and Supply. Open Field.

Ortiz-Monasterio, J. I., Dhillon, S. S. and Fischer, R. A., 1994. Date of sowing effects on grain-yield and yield components of irrigated spring wheat cultivars and relationships with radiation and temperature in Ludhiana, India. *Field Crop Res.* 37, 169–84.

Posner, E. S. and Hibbs, A. N., 2011. Wheat: the raw material. In E. S. Posner and A. N. Hibbs (eds), *Wheat Flour Milling,* pp. 1–46. AACC International, Inc., St. Paul, MN.

Rebetzke, G. J., Lopez-Castaneda, C., Acuna, T. L. B., Condon, A. G. and Richards, R. A., 2008. Inheritance of coleoptile tiller appearance and size in wheat. *Aust. J. Agr. Res.* 59, 863–73.

Rebetzke, G. J. and Richards, R. A., 1999. Genetic improvement of early vigour in wheat. *Aust. J. Agr. Res.* 50, 291–301.

Rebetzke, G. J., Richards, R. A., Fettell, N. A., Long, M., Condon, A. G., Forrester, R. I. and Botwright, T. L., 2007. Genotypic increases in coleoptile length improves stand establishment, vigour and grain yield of deep-sown wheat. *Field Crop Res.* 100, 10–23.

Richards, R. A. and Lukacs, Z., 2002. Seedling vigour in wheat-sources of variation for genetic and agronomic improvement. *Aust. J. Agr. Res.* 53, 41–50.

Richards, R. A. and Passioura, J. B., 1989. A breeding program to reduce the diameter of the major xylem vessel in the seminal roots of wheat and its effect on grain-yield in rain-fed environments. *Aust. J. Agr. Res.* 40, 943–50.

Sacks, W. J., Deryng, D., Foley, J. A. and Ramankutty, N., 2010. Crop planting dates: an analysis of global patterns. *Global Ecol. Biogeogr.* 19, 607–20.

Sattar, A., Cheema, M. A., Farooq, M., Wahid, M. A., Wahid, A. and Babar, B. H., 2010. Evaluating the performance of wheat cultivars under late sown conditions. *Int. J. Agric. Biol.* 12, 561–5.

Spielmeyer, W., Hyles, J., Joaquim, P., Azanza, F., Bonnett, D., Ellis, M. E., Moore, C. and Richards, R. A., 2007. A QTL on chromosome 6A in bread wheat (Triticum aestivum) is associated with longer coleoptiles, greater seedling vigour and final plant height. *Theor. Appl. Genet.* 115, 59–66.

University of Saskatchewan, 2016. Crop Establishment. Winter Wheat Production Manual. University of Saskatchewan.

van Ginkel, M. and Ogbonnaya, F., 2007. Novel genetic diversity from synthetic wheats in breeding cultivars for changing production conditions. *Field Crop Res.* 104, 86–94.

van Noordwijk, M. and de Willigen, P., 1987. *Roots, Plant Production and Nutrient Use Efficiency.* PhD Thesis, Wageningen Agricultural University.

White, P. J., George, T. S., Gregory, P. J., Bengough, A. G., Hallett, P. D. and McKenzie, B. M., 2013. Matching roots to their environment. *Ann. Bot.* 112, 207–22.

Zadoks, J. C., Chang, T. T. and Konzak, C. F., 1974. A decimal code for the growth stages of cereals. *Weed Res.* 14, 415–21.

Chapter 3

Conservation tillage for sustainable wheat intensification: the example of South Asia

Vijesh Krishna, Georg-August University of Göttingen, Germany; Alwin Keil, International Maize and Wheat Improvement Center (CIMMYT), India; Sreejith Aravindakshan, Wageningen University, The Netherlands; and Mukesh Meena, Indian Institute of Soil and Water Conservation, India

1 Introduction
2 Factors affecting the diffusion of CT wheat in South Asia
3 Recent evidence of the agronomic and economic impacts of CT wheat in South Asia
4 Constraints to the diffusion of CT practices in wheat in South Asia
5 Recent developments in CT wheat
6 Concluding remarks
7 Where to look for further information
8 Acknowledgements
9 References

1 Introduction

The wheat production sector of South Asia has registered only a sluggish productivity growth in the last two decades (FAOSTAT, 2015). The rate of growth, which is lower than the wheat yield growth realized in the rest of the world (Shiferaw et al., 2013), is inadequate to keep pace with the population expansion in the region. Along with a rapidly changing climate, degradation of the natural resource base as a result of intensive farming and inappropriate use of inputs is often cited as one of the root causes of the problem (Chauhan et al., 2012; Erenstein and Laxmi, 2008; Sekar and Pal, 2012). To achieve a sustainable yield growth rate in wheat, there has been an increased scientific interest in developing environmentally sustainable agronomic practices that are resource conserving, while ensuring financial benefit to farmers. This paradigm shift is closely associated with an increased awareness in the public policy arena towards the negative environmental externalities associated with the traditional farming practices. Most prominent among such resource-conserving technologies in the cereal systems of South Asia is conservation tillage (CT) in wheat (Derpsch et al., 2010), which is identified also as one of the key

http://dx.doi.org/10.19103/AS.2016.0004.14

agricultural practices that can enhance the adaptive capacity of farming communities and the potential of farming practices to mitigate the challenges of climate change in the tropics (Harvey et al., 2014; Sapkota et al., 2015). In this chapter, we examine the recently emerged evidence on the on-farm impacts of the ensemble of CT technologies in wheat production and the possible constraints hindering its widespread diffusion in the Indo-Gangetic Plains (IGP) of South Asia. We use the umbrella term 'CT wheat' to represent zero-tillage (ZT), reduced-tillage and ridge-tillage practices in wheat production.

These agronomic practices are based on the principles of Conservation Agriculture – that is, minimal soil disturbance, residue retention and permanent soil cover, combined with rational crop rotations (Krishna and Veettil, 2014; Hobbs et al., 2008). They facilitate conservation of the natural resource base and enhance the resource use efficiency of farming systems. The CT practices are often heralded as more sustainable than the traditional tillage alternatives, leading to a paradigm shift in tillage and land preparation options, while helping farmers reduce the cost of production (Erenstein and Laxmi, 2008; Krishna and Veettil, 2014). Here, seeds are placed directly with minimum soil disturbance, retaining crop or cover crop residues from the previous season (Derpsch et al., 2010). By minimizing the soil disturbance and providing residue soil cover, CT practices are found to enhance soil fertility and water-use efficiency, thus helping cereal farmers to sustain crop yields over a longer term (Baker and Saxton, 2007; Singh et al., 2014; Saharawat et al., 2010). CT also allows earlier sowing of wheat by reducing the turnaround time. Due to heat-related yield depressions under late sowing conditions, advancing sowing times is a major mechanism for increasing wheat productivity (Ortiz et al., 2008; Erenstein, 2009). Further, CT with residue retention avoids the widespread practice of burning the rice stubbles, thereby reducing air pollution, especially in the northwestern (NW) IGP (Saunders et al., 2012).

The main pathways through which a change from traditional tillage to CT could affect crop yields and farm income are depicted by Erenstein and Laxmi (2008), as well as Brouder and Gomez-Macpherson (2014). Due to differences in factor abundance and agro-climatic conditions, these pathways are not identical across geographical locations and farming systems of the IGP, leading to significant heterogeneity in the on-farm impacts of CT wheat (Ekboir, 2002; Krishna et al., 2012b). So far, the diffusion of CT wheat is concentrated mainly in the NW IGP, diminishing towards the east (Erenstein and Laxmi, 2008; Erenstein, 2010b). This pattern corresponds to the significant socio-economic disparity characterizing the region. The relevance of agro-climatic and socio-economic heterogeneity of the wheat growing tracts of South Asia is examined in Section 2 of this chapter. In Section 3, we review recent studies on the agronomic and economic impacts of CT wheat, as a clear documentation of farm-level impacts is one of the first steps to address the constraints to the diffusion of a production technology. Although the on-farm impacts of CT practices are well-documented in many other parts of the world (Pittelkow et al., 2015), South Asia's CT wheat system is unique in the sense that it is usually followed by transplanted rice in thoroughly puddled fields, making the CT practice discontinuous across seasons (Erenstein et al., 2008). This reduces the possibility of deriving insights from other cereal production systems with the continuous use of CT.

Despite the projected private as well as public benefits, only a minority of farmers is found to adopt CT technologies in South Asia (Kassam et al., 2014). Using survey data from farmers in the rice–wheat cropping systems of four countries of South Asia, Singh et al. (2012) indicate that only about 16–18% of farm households ever adopted CT wheat. Considerable regional variation is also noted, with the highest adoption rate being observed in the NW IGP and the lowest in the east. Possible reasons for the slow diffusion process of CT wheat are elucidated in Section 4, while advancements in developing locally adapted CT machinery are highlighted in Section 5. Concluding remarks are provided in Section 6.

This chapter is based on both a review of recent literature published after 2005 on CT wheat and the authors' own analysis of a set of recent farm household surveys. The latter encompass a household survey conducted among cereal farmers across the IGP during 2010–11 (Pede et al., 2015), and another three targeted CT adoption surveys that were conducted later in Haryana state of NW India (Krishna and Veettil, 2014), Bihar state of eastern India (Keil et al., 2015) and Bangladesh (Aravindakshan et al., 2015).

2 Factors affecting the diffusion of CT wheat in South Asia

Wheat production in South Asia takes place predominantly in the IGP, which has been a major breadbasket for the entire region from the 1970s (Chauhan et al., 2012). However, after 1990, the rate of yield growth has been diminishing for cereal crops in this region (FAOSTAT, 2015). Barring some states of NW IGP, the productivity level of wheat in the Plains is significantly lower than the global average (Krishna et al., 2012b; Shiferaw et al., 2013). There exists a diverse set of production constraints, including late sowing of wheat and associated terminal heat stress, declining groundwater table, input and energy scarcity, and increased infestation of the weed *Phalaris minor* (Ekboir, 2002). The severity of these constraints shows spatial patterns, causing the agronomic and economic potentials of CT wheat to vary within the region. For example, the importance of cattle production in the farm economy and use of rice residues as livestock feed vary widely across the Plains, affecting the adoption of CT wheat with residue retention (Erenstein et al., 2010). In the NW Plains, both productivity per unit of land and cost of cultivation of wheat are significantly higher than in the eastern states, thereby equalizing the per unit cost of wheat grain production in these two regions (DWR, 2014). This implies that an absolute increase in wheat productivity (and/or cost savings) due to CT adoption is financially more beneficial for farmers in the eastern Plains. It must also be noted that the importance of wheat in the cropping system declines towards the east (rice–rice cropping systems dominate rice–wheat systems (Alam et al., 2015)) – reducing the relevance of CT wheat for farmers' livelihoods from a static perspective.[1] Such differences in cropping intensity and system productivity make it difficult to predict the pattern of diffusion and implications of CT technologies. The analysis is further complicated by the fact that the Plains encompass significant disparity also with respect to the incidence of income poverty and resource endowment.

As a generic principle, agricultural innovations well suited to the relatively capital-abundant, labour-scarce NW IGP might not be socially acceptable for the capital-scarce, densely populated eastern regions (Erenstein and Thorpe, 2011). For example, the tractor (which is a typical labour-saving technology) is more popular in the Indian Punjab (around 10 tractors/sq.km of cultivated land) than in West Bengal (<1 tractor/sq.km). A gradient

1. Although the cropping systems in the eastern IGP are more diverse with rice, wheat, maize, jute, sugarcane and vegetables in the portfolio, the intensity of land use is slightly low compared to the NW region (Krishna et al., 2012b; Anonymous, 2013).

analysis that considered land- and labour scarcity showed that the eastern Plains (especially West Bengal state of India and Bangladesh) have twice the population density of the NW Plains (Erenstein, 2010b; Erenstein et al., 2010). While the high population density is expected to reduce the cost of labour in relative terms as a factor of farm production, increasing out-migration from the eastern IGP has likely altered the scenario by driving up rural wages (Datta and Mishra, 2011; Agasty and Patra, 2014).

Alongside relative factor abundance, the persistence of poverty determines the diffusion of agricultural technologies, especially the resource-conserving ones. The NW states of the IGP have poverty rates between 8–12% of the rural population, while the eastern Indian states and Bangladesh have 20–35% (Planning Commission, 2014). The widespread poverty affects the scope of mechanization and supplementary inputs (e.g. herbicides) required for successful adoption of CT wheat. At the same time, increased risk aversion of resource-poor households and their inability to access information while experimenting with new technologies present additional challenges for agricultural extension. Timely availability of machinery is another constraint in technology diffusion among small and marginal farmers, who often do not possess the CT seed drills, and access the technology via custom-hiring arrangements. During the peak sowing period, the relative scarcity of CT drills may critically limit access to the technology for this stratum of farmers (Tripathi et al., 2013; Krishna et al., 2012a). Based on survey data from the Cereal Systems Initiative for South Asia (CSISA; Pede et al., 2015), Table 1 highlights the basic differences in farming and economic attributes of cereal producers in three different locations across the

Table 1 Heterogeneity of wheat production scenarios within the IGP

	Haryana (NW IGP)	Bihar (eastern IGP)	Bangladesh (eastern IGP)
Average monthly per capita expenditure of household (PPP-adjusted international dollars)	90.49[a] (99.25)	55.03[b] (39.08)	42.76[c] (14.69)
Number of adult (above 14 years of age) household members	5.71[a] (3.48)	7.13[b] (4.73)	4.48[c] (1.81)
Operating landholding size (ha/household)	3.37[a] (3.70)	1.39[b] (1.51)	0.57[c] (0.53)
Land-to-labour ratio (operating landholding size/ number of household members above 14 years)	0.68[a] (0.78)	0.27[b] (0.31)	0.15[c] (0.17)
% sample households owning a four-wheel tractor	43.80[a]	2.20[b]	0.00[c]
Area under wheat (ha)[#]	2.78[a] (3.37)	0.60[b] (0.61)	0.24[c] (0.20)
Wheat yield (tonnes per ha)	4.45[a] (0.65)	2.77[b] (1.06)	2.76[b] (0.67)
Market price for wheat (PPP-adjusted international dollars per tonne)[##]	705.20[a] (9.74)	681.28[b] (89.94)	702.87[a,b] (87.79)
Number of observations	324	324	324

Figures in parentheses give standard deviation of sample mean.
[#] Conditional on cultivation. [##] Conditional on sale in the market.
[a, b, c] Difference in mean values with different superscripts are statistically significant at ≤0.05 level.

Sources: Estimated from CSISA surveys (Pede et al. 2015).

IGP – Haryana in NW India, Bihar in eastern India and Bangladesh. Details of the sampling framework and data collection can be found in Pede et al. (2012).

The size of operational holdings, as well as the wheat area cultivated by the households, varies significantly across the aforementioned locations (Table 1). While the number of adults in the household is highest in Bihar and lowest in Bangladesh, there is a pronounced decline in the land-to-labour ratio – being a more appropriate indicator of labour availability – when moving towards the east. The average land-to-labour ratio in Haryana is more than twice that of Bihar and four times that of Bangladesh. In addition, a considerable share of farmers (44%) possesses a four-wheel tractor in Haryana, which is not the case in the two regions of the eastern Plains. Apart from relative labour scarcity that incentivizes the use of labour-saving technologies, the large differences in tractor ownership can be explained by the gradient in farm sizes and wealth levels between the west and the east. Krishna et al. (2012b) indicate that the reach of public extension services also varies widely across the cereal production systems of the IGP. Wheat yield in Haryana, as a result of better access to inputs, information, machinery and other factors, is found to be 61% higher than in the other two regions. Compared to their counterparts in Haryana, the cereal farmers in Bihar and Bangladesh incur 39% and 52% lower consumption expenditures, respectively, illustrating their lower standard of living. These differences in yield and economic status of the households are expected to play a key role in the potential economic impacts of CT across the Plains. In the following section, we will examine these aspects in greater depth.

3 Recent evidence of the agronomic and economic impacts of CT wheat in South Asia

A number of experimental and on-farm studies have been conducted in the last decade, documenting the impacts of CT practices in the wheat production sector of South Asia. The National Agricultural Research Systems of India and Bangladesh have realized the potential of CT wheat to improve the sustainability of soil and water resources in the IGP, and promotion of these technologies is one of the future research and development (R&D) priorities (DWR, 2011; MoA and FAO, 2010). However, the existing evidence on technology impacts is derived largely from field experiments, conducted mostly in the NW Plains. These researcher-managed field experiments may overstate the benefits from CT wheat as compared to farmer management in the real world, where environmental and soil heterogeneity is greater, and farmers' application of technologies and management practices show wide variation (Aravindakshan et al., 2015; Krishna and Veettil, 2014). According to Baudron et al. (2015), the impacts of Conservation Agriculture practices would be more pronounced in systems where (1) the energy (including labour and draft power) available for crop establishment is limited and/or costly, (2) delayed planting results in a significant yield decline and (3) yield is limited by irrigation water availability. While some forms of energy (e.g. machinery) are scarce in the eastern IGP, some others (e.g. human labour) are relatively cheap as compared to the NW Plains. Although delays in wheat sowing are common and consequences severe in the eastern Plains, in many systems (e.g. cotton–wheat) of the NW IGP similar constraints also exist. Due to these factors, it is not easy to foresee in which region the farm-level impacts of CT would be greater.

The uncertainty around CT-related yield effects is further increased by the complexity of interacting biophysical factors and the diversity of impact pathways (Brouder and Gomez-Macpherson, 2014). However, there is a dearth of literature that compiles the existing studies across these two major agro-ecological zones of South Asia.[2]

Based on an exhaustive review of the literature, we compile recent studies on the impacts of CT wheat in South Asia (Table 2). As indicated earlier, the majority of reviewed studies (19 out of 25) were conducted in the NW Plains, especially in the state of Haryana. About half of them are based on researcher-managed field trials and the other half on household surveys. A comparison of studies in the former group is challenging, as the treatment often combined CT with other agronomic practices. For example, Coventry et al. (2011) and Das et al. (2013) examine the wheat productivity differences across different tillage operations, cropping systems and fertilizer regimes. Although such analyses are valuable from an agronomic perspective, deriving the individual impacts of CT is not a straightforward task. Further, while the grain yields are often reported, information on the cost of production is missing in many of these studies, giving an incomplete picture of the financial benefits of CT wheat. Also, there are only a few studies that examine the stability of the observed impacts over time. Those who address this aspect, like Rawson et al. (2007) and Bhushan et al. (2007), report significant temporal impact variability. Multi-year perspective is essential particularly for the farm-based surveys, as the technology uptake by the farmers is dynamic, and the effects of agronomic practices like CT are subject to seasonal fluctuations (Erenstein, 2010a). Finally, the studies based on farm household datasets have the limitation of not fully addressing the potential selection bias between users and non-users of CT. While some studies controlled for observable characteristics in their impact assessment of CT, panel data studies and/or instrumental regression models would be required to control for potential bias due to unobservable heterogeneity.

Irrespective of these limitations, these studies have demonstrated certain consistent patterns with respect to impacts of CT wheat. First, there exists hardly any evidence for yield penalty associated with the technology. While 12 studies did not find any significant difference in the grain yield, 13 studies reported a significant positive impact. In South Asia, CT is practised mostly in the wheat (i.e. winter) season alone, and the long-term system benefits of the technology might not be attainable. Various factors may compromise the CT-induced yield benefits, such as the relative abundance of different (especially broad-leaved) weed species (Chhokar et al., 2007; Kumar et al., 2013b), uneven seedling emergence due to changes in soil density in some environments (Powlson et al., 2014), difficulties in stubble management (Coventry et al., 2015), scarcity of suitable machinery at sowing time (Krishna et al., 2012a) and farmers' inability to manage new technologies (Rawson et al., 2007). The existing evidence on the factors influencing the differential impacts of CT is rather limited.

Secondly, the technology impact on cost saving is prevalent throughout the study region. In the absence of a yield penalty, this would lead to a significant reduction in the unit cost of wheat production, denoting that CT qualifies as a superior technology alternative from an economic perspective, especially in areas where rural poverty is

2. Brouder and Gomez-Macpherson (2014) conducted a meta-analysis of Conservation Agriculture across crops and regions, from which locally adapted and crop-specific policy measures cannot be derived. While Erenstein and Laxmi (2008) systematically reviewed the literature on CT wheat in India, there were far less studies published at the time of analysis (2002–5), limiting the scope to assess the heterogeneous impacts of CT practices that arise due to agro-ecosystem and socioeconomic diversity.

Table 2 Impacts of CT wheat, documented in the literature

Source	Effect on wheat crop: Grain yield	Effect on wheat crop: Cost of cultivation	Effect on wheat crop: Other effects	Study region in IGP
Ali and Erenstein (2013)	Positive			NW (Pakistan)
Aravindakshan et al. (2015)	Positive		High energy use efficiency	Eastern (Bangladesh)
Aryal et al. (2015)	Not sig.	Negative	Reduced CO_2 eq. emission	NW (Haryana)
Bhushan et al. (2007)*	Not sig.			NW (Uttar Pradesh)
Chhokar et al. (2007)*	Positive	Negative		NW (Haryana)
Coventry et al. (2011)*	Not sig.		Higher protein content and grain hardness	NW (Haryana)
Coventry et al. (2015)	Mixed			NW (Haryana)
Das et al. (2013)*	Not sig.			NW (Delhi)
Das et al. (2014)*	Not sig.	Not sig. (without residue); Positive (with residue)	High water productivity	NW (Delhi)
Devi et al. (2015)*	Mixed		High carbon accumulation and nitrogen uptake	NW (Haryana)
Erenstein (2009) and Erenstein et al. (2008)	Positive	Negative	Positive gross water productivity	NW (Haryana)
Erenstein (2009) and Erenstein et al. (2008)	Not sig.	Negative	Insignificant gross water productivity	NW (Punjab, Pakistan)
Keil et al. (2015)	Positive	Negative	Insignificant technical efficiency gain	Eastern (Bihar)
Krishna et al. (2012)	Positive	Negative		Eastern (West Bengal)
Krishna and Veettil (2014)	Positive	Negative	Small technical efficiency gain	NW (Haryana)
Kumar et al. (2013)*	Positive	Negative	Higher energy use efficiency; Higher net income	NW (Uttar Pradesh)
Laxmi and Mishra (2007)	Not sig.	Not sig.	Savings in use of fuel and machinery	NW (Haryana)
Laxmi and Mishra (2007)	Positive	Not sig.	Savings in use of fuel and machinery	Eastern (Bihar)
Rawson et al. (2007)	Not sig.			Eastern (Bangladesh)
Saharawat et al. (2010)*	Positive		Higher net income	NW (Haryana)
Sapkota et al. (2015)*	Positive	Negative	Higher net income	NW (Haryana)
Sekar et al. (2015)	Positive	Negative	Higher net income, irrigation water saving	NW (Haryana), Eastern (Bihar)
Sidhu et al. (2014)	Positive			NW (Punjab, India)
Singh et al. (2014)*	Not sig.			NW (Haryana)
Tripathi et al. (2013)	Not sig.	Negative		NW (Haryana)

Note: The table includes only the peer-reviewed papers with a test of statistical difference in mean values.
* Researcher-managed field trials.

rampant and financial resource constraints exist. Although the finding of CT-related cost savings is unsurprising – the very first extension efforts to promote CT in South Asia itself was showcasing the technology as a way to reduce the cost of cultivation (Coventry et al., 2015) – the exact magnitude of cost reduction varies widely. Unfortunately, there are not many studies examining the economic impacts associated with CT wheat adoption beyond the farm-level profitability changes. The one exception is Ali and Erenstein (2013) which demonstrated that the adoption of CT wheat could help reduce poverty among the rural households of Pakistan.

Thirdly, many studies have indicated significant environmental impacts of CT. Higher energy use efficiency (Aravindakshan et al., 2015; Kumar et al., 2013a), reduced CO_2 emission from tractor-powered tillage operations, improved carbon sequestration (Aryal et al., 2015; Devi et al., 2015) and higher irrigation water-use efficiency (Das et al., 2014) are noted. Among these effects, the irrigation water saving is particularly important. Rodell et al. (2009) estimated that 13–17 cubic kilometres of groundwater is lost permanently from the aquifers in the NW Plains of India every year. More efficient utilization of water does not only mitigate the decline of the groundwater tables, but also entails reduced electricity consumption for irrigation, which in turn leads to lower carbon emissions and, hence, has wider environmental implications. Similar environmental benefits are also reported across smallholder systems outside South Asia (e.g. Baudron et al., 2014). However, Powlson et al. (2014) indicate that the effect of CT wheat on carbon savings and thereby climate change mitigation is often overstated, depending on the distribution of crop residues in the field.

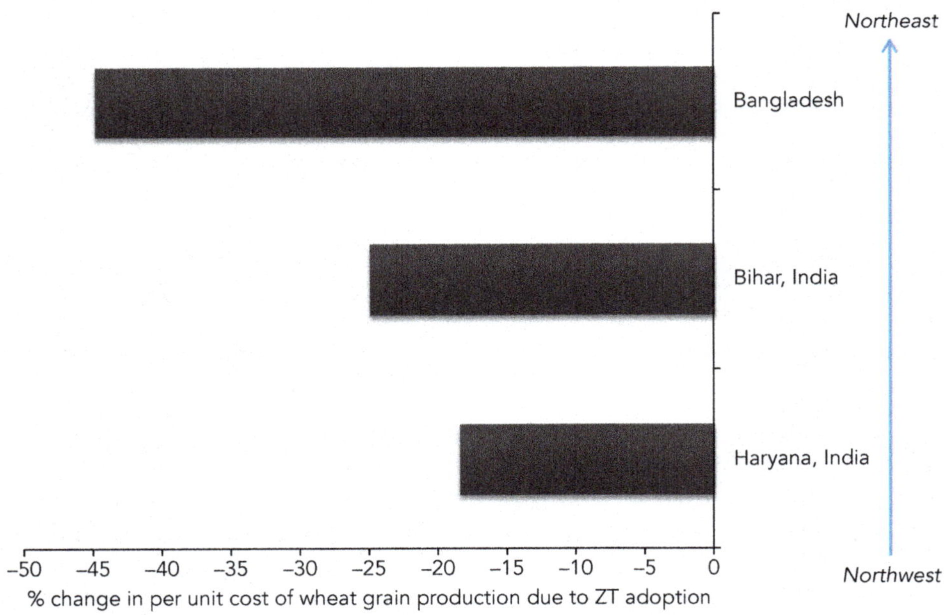

Figure 1 Reduction in per unit cost of wheat production with conservation tillage adoption. Source of data: Authors.

Finally, there has been only limited evidence on the on-farm impacts of CT technology in the eastern IGP, where yield gaps in wheat are particularly large and, at the same time, the potential to increase yields is high due to the relative abundance of (under-developed) water resources (Aggarwal et al., 2004; Anonymous, 2008). Two studies (Krishna et al., 2012a; Rawson et al., 2007) are based on data from researcher-assisted farmer adoption or trials, where the benefits of CT could be overstated. The remaining three studies (Aravindakshan et al., 2015; Keil et al., 2015; Sekar et al., 2015) show a more realistic picture of farmer adoption of CT. From the farm household survey datasets, we estimated the change in per unit cost of wheat production under CT across three locations in the IGP. The percentage savings in per unit cost of wheat production is found lowest (still approximately 15%) in the NW Indian state of Haryana (Fig. 1). Cost reduction of comparable magnitude was noted by other studies conducted in Pakistan Punjab (Ali and Erenstein, 2013) and Haryana (Aryal et al., 2015; Tripathi et al., 2013).[3] The per unit cost savings with CT technology adoption are found to be greater (25–45%) among the wheat growers of the eastern IGP. However, since CT is a relatively new technology in the eastern Plains, livelihood impacts of the technology could not be assessed yet comprehensively. Little is also known about the differential technology impacts and adoption rates across different socio-economic strata. We proceed with a discussion of the existing evidence on the constraints to the widespread and socially inclusive diffusion of CT technology.

4 Constraints to the diffusion of CT practices in wheat in South Asia

In contrast to the widely documented positive agronomic, economic and environmental effects, CT wheat has not yet become widely popular in many parts of the IGP. At present, South Asia's share of the global no-tillage area is only marginal (Derpsch et al., 2010; Kassam et al., 2009, 2014).[4] Coventry et al. (2015) describe the slow process of diffusion of CT practices as particularly puzzling: even after many years of extension programmes and farmer exposure to the technology, the adoption rate remains low and patchy in most districts of the IGP. This concern is shared by others (Farooq and Siddique (2015), for example). We now focus our discussion on the challenges for a widespread diffusion of CT wheat practices, with special attention to the aspect of social inclusiveness, and the economic relevance of developing locally adapted technologies that are tailored to suit the resource endowment of local farmers. While there are numerous studies analysing the diffusion of CT practices all over the world, most of them refer to the United States and Europe (Schneider et al., 2012). The existing studies in South Asia have indicated likely reasons for the slow diffusion of CT wheat in the region, but tend to lack empirical verification. These largely untested yet plausible reasons can be broadly summarized into three categories.

3. The per unit cost savings due to CT wheat adoption in Pakistan is found to be 15%. This information is not available in Ali and Erenstein (2013), but was obtained through personal communication with Akhter Ali (CIMMYT, Pakistan) on 10 November 2015.
4. A recent estimate by the Food and Agricultural Organization of the United Nations estimated that about 1.5 million hectares (1% of arable land) are under different conservation agricultural practices in India (http://www.fao.org/ag/ca/6c.html). However, the method of estimation and the validity of the estimate are unclear.

Table 3 Adoption and awareness about ZT technology in wheat cultivation

% wheat farmers	Haryana	Bihar	Bangladesh
Survey 1: CSISA (Pede et al. 2015)			
Having information on ZT	95.7	8.3	42.3
Ever adopted ZT	36.1	4.9	1.5
Adopters among informed	37.7	59.0	3.5
Survey 2 (by the authors)			
Having information on ZT	100.0	46.1	80.5
Ever adopted ZT	83.9	27.5*	75.9
Adopters among informed	83.9	59.7	94.3

* Estimated from the village census (others on household surveys). The higher level of CT awareness in Survey 2 is likely caused by the focus of the study on villages with at least some CT adoption.

4.1 Limited information availability and access

Conservation Agriculture practices are knowledge-intensive. For example, ZT wheat adopters might require additional information on chemical weed management (Aryal et al., 2015; Rawson et al., 2007; Ekboir, 2002) since a shift in the weed flora towards broad-leaved plants is often observed with the adoption of ZT (Kumar et al., 2013b). Public extension coverage on Conservation Agriculture is too low in India to facilitate faster diffusion of CT wheat (Coventry et al., 2011; Krishna et al., 2012b). The requirement for more specific information on crop production and protection coupled with ineffective institutions for information provision and technology transfer may result in inadequate and discontinuous adoption of the technology within the farming community. Information access and usage also show significant inter-regional differences, as depicted in Table 3. Here, results from two sets of household surveys are compared. In the first set (Pede et al., 2015), study villages were selected randomly, whereas the second set of surveys (conducted by the authors) was targeted at areas with some extension activities on CT wheat in order to obtain a sufficiently large adopter sub-sample. Significant differences can be observed not only across the IGP, but also within a study location across the surveys. The second set of surveys indicates significantly higher rates of awareness and adoption of CT wheat, reinforcing the notion that the diffusion of CT wheat is patchy even across villages within the same area. In fact, based on census data collected in the research villages for the study by Keil et al. (2015), village-level CT wheat adoption rates within one single district of Bihar were found to range from 1 to 93%. Strengthening institutional linkages between farmers, agricultural extension and the scientific community, using mass media for awareness raising, and endorsing participatory methods of research trials and demonstrations could be one important pillar of a strategy to enhance the diffusion of CT technology.

4.2 Heterogeneous technology impacts

Adoption of Conservation Agriculture practices need not always lead to significant yield benefits due to a number of factors like soil type, prevailing weather and management practices including the handling of crop residues (Brouder and Gomez-Macpherson, 2014; Singh et al., 2014). Impact heterogeneity is a major reason for farmers dis-adopting the CT technology in the NW Plains (Planning Commission, 2014; Erenstein, 2010a). In some instances, the adoption of CT wheat could be associated with lower yield due to varietal incompatibility and unavailability of machineries at the required time (Krishna et al., 2012a). There could also be technical limitations, operational problems and incompatibility with other technologies and crops adopted by farmers for the first-generation ZT drills (Ekboir, 2002). For example, the use of combine harvester for harvesting the preceding rice crop makes ZT wheat difficult due to the presence of excessive crop residue in the field (Coventry et al., 2015). The negative perceptions about ZT prevailing among wheat farmers include specificity of ZT technology to certain soil conditions, especially when the adoption is discontinuous across seasons. While continuous adoption of ZT would result in better soil physical properties and improved root growth in the long-term (Singh et al., 2014), increased sub-surface compaction in heavier soils remains a cause of concern particularly when the CT adoption is discontinuous (Sidhu et al., 2014). Further, CT wheat may not be a profitable option in some systems, where livestock production competes with no-till for the use of crop residues. This could be a major problem in the eastern Plains, because rice straw is a major source of cattle feed.

The overall rather positive on-farm impacts of CT reported may also be attributable to a certain extent to the non-inclusion of covariant factors and selection bias in the econometric models used. Similar to the many other technologies, farmer self-selection might be biasing the estimates on impacts of CT wheat. However, Keil et al. (2015) controlled for a wide range of factors on production, management and farmers' level of risk aversion, and estimated a CT-induced yield gain of about half-tonne per hectare compared to conventional tillage wheat. Their estimate, which is consistent with the findings from field trials in the eastern Plains, is based on farmers' practice without full retention of crop residues, due to the importance of rice straw as cattle feed. The authors conclude that the causal mechanisms for the estimated yield gains are likely associated with soil-related factors, especially the reduction of evaporative losses under deficit irrigation conditions. It is also worth noting that the *observed* yield advantage of CT wheat in their study – when comparing the adopter with the non-adopter sub-sample – was only about 200 kg ha^{-1}. This means that the selection bias worked against the CT adopters, rather than in favour of them. This result is plausible, since smaller farmers (who tend to not use CT thus far, as we elaborate below) often attain higher yields than larger farmers due to more intense input use and crop management on the smaller area which is available for cropping (cf. Ellis, 1993). Nevertheless, carrying out farming systems analysis based on panel data, which could systematically control for selection bias in technology adoption and address dis-adoption, would be desirable to obtain a clearer picture of the potentials and limitations of CT wheat in South Asia.

4.3 Institutional and policy hurdles

At times, government policies regarding input and machinery subsidy for the farming sector counteract the diffusion of Conservation Agriculture in general, and ZT wheat in particular. Government subsidies for farm machinery often cover less-sustainable tillage options, such as rotavators (multi-utility machines that complete multiple operations in a single pass); hence, fostering their adoption in place of ZT drills (Erenstein, 2010b). Furthermore, public provision of production factors, such as highly subsidized electricity for groundwater extraction, creates a disincentive for farmers to adopt water-saving conservation agricultural technologies.

Often the small and marginal farmers are left out of the technology benefits, similar to the traditional and capital-intensive innovations in agriculture. The recent studies have noted that household wealth, land tenure, farmer education, availability of CT seed drills and access to agricultural extension and information are some of the key attributes determining the on-farm adoption of the CT wheat technology (Erenstein and Farooq, 2009; Krishna and Veettil, 2014; Ali and Erenstein, 2013). Only a small share of farmers possesses ZT drills, so that the majority accesses the technology via custom-hire services (Ali and Erenstein, 2013). While the literature suggests that the relationship between farm size and adoption of ZT technologies need not always be positive (Laxmi and Mishra, 2007; Erenstein and Laxmi, 2008; Akter and Gathala, 2014), provision of CT services could be biased against the small and marginal farmers with small and often fragmented plots due to higher transaction costs per unit of land. This problem is severe where there is no institutional support or organizational collaboration at the local level to motivate farmers towards the Conservation Agriculture philosophy (Farooq and Siddique, 2015), particularly in the eastern Plains (Krishna et al., 2012a). Using three adoption survey datasets, we examine the patterns in landholding and wheat area across adopters with and without CT seed drills (Table 4). Farmers who own a CT drill typically cultivate a larger area of wheat; hence, they can recover the investment costs within a shorter period at lower financial risk.

Widespread adoption of the technology hinges on an expansion of the network of CT service providers. Although substantial subsidies are provided by the state governments on seed drills to motivate farmers to adopt conservation agriculture practices (Krishna et al., 2012a; Ekboir, 2002), tractor and drill ownership is no tenable goal for most capital-constrained small and medium-sized farmers. Furthermore, efforts are needed to reduce the observed scale bias in accessing such services, which is likely caused by the fixed costs associated with service provision in dispersed fields. Acknowledging that scale bias may lessen over time as the number of service providers increases (competition may make

Table 4 Mean cultivable area and wheat area with respect to CT adoption

	Haryana			Bihar			Bangladesh		
		Area (ha)			Area (ha)			Area (ha)	
	n	Cultivable	Under wheat	*n*	Cultivable	Under wheat	*n*	Cultivable	Under wheat
Non-adopters and dis-adopters	29	3.96[a] (3.31)	3.79[a] (3.30)	642	1.07[a] (1.34)	0.73[a] (0.94)	82	1.09[a] (1.28)	0.14[a] (0.16)
Adopters, do not own CT machinery	111	3.72[a] (3.22)	3.54[a] (3.20)	313	1.67[b] (1.81)	1.16[b] (1.49)	208	0.64[b] (0.59)	0.11[b] (0.11)
Adopters, own CT machinery	40	11.13[b] (6.22)	10.68[b] (6.26)	45	3.86[c] (3.10)	2.62[c] (2.10)	38	2.82[c] (2.06)	0.40[c] (0.46)
Overall	180	5.41 (5.09)	5.17 (5.03)	1000	1.38 (1.72)	0.95 (1.27)	328	1.00 (1.25)	0.15 (0.21)

[a, b, c] Difference in mean values with different superscripts are statistically significant across the rows (within a study area) at ≤0.01 level.

Standard deviation is shown in parentheses. Non-parametric tests were used to find the statistical significance.

Source: Authors.

them less selective in marketing their services), there are three possible pathways to make the technology socially inclusive. First, business models that include demand aggregation and service coordination through local point persons and/or lead farmers at the village level may help overcome the scale bias by reducing the transaction costs of servicing small farms. Second, farmer groups and cooperatives could jointly invest in CT machinery to mitigate capital constraints. Rawson et al. (2007) suggest that, in Bangladesh, a group of half-dozen farmers purchasing a CT drill could recover their investment cost from the additional profits of one single wheat crop. Farmers' participation in cooperatives and groups is high in many parts of South Asia (Krishna et al., 2012b), which could be harnessed effectively for the diffusion of CT wheat technologies. And, third, more location-specific and scale-appropriate machinery could be developed and deployed, which will be elaborated in the next section.

5 Recent developments in CT wheat

The production constraints faced by farmers in the IGP, as we have already discussed, are both system- and location-specific, and hence the attempts to import technologies developed outside a particular system or location might not be appropriate for all farmers living within it. Such system-specificity was evident in the initial diffusion process of CT wheat in the eastern Plains of India (Krishna et al., 2012a).[5] Given this background of high spatial heterogeneity, in this section we discuss two locally adapted CT technologies developed independently in the IGP. The first one is second-generation CT machinery named the 'Turbo Happy Seeder' (Fig. 2a) that enables ZT seeding under the heavy stubble conditions

(a) (b)

Figure 2 Diverse CT machines gaining popularity in the IGP: (a) Turbo Happy Seeder in Punjab, India, photo by V. Singh (Punjab Agricultural University, Ludhiana) and (b) Power Tiller-Operated Seeder (PTOS) in Bangladesh, photo by M.K. Gathala (CIMMYT, Dhaka).

5. When the same CT seed drills were introduced in the eastern Plains by the International Maize and Wheat Improvement Center (CIMMYT), they were found incompatible with the locally popular 35 horse-power (hp) tractors. This was because the imported ZT drills from the NW states of India came with 11-tines, which required 45 hp tractors to operate. Hence, 9-tine drills had to be introduced in the eastern Plains as a technically viable foundation to the diffusion of CT wheat (Krishna et al., 2012a).

of the NW IGP, where the use of rice straw for animal feed is of minor importance. The second one is the 'Power Tiller-Operated Seeder' (PTOS; Fig. 2b) in Bangladesh, operated by low-hp two-wheel tractor. These technologies represent innovations that are system-specific and locally adapted. Since there is a dearth of literature on these machines, our elaborations in this section are partly based on the 'grey literature'.

In the cropping systems of NW IGP, where rice is grown in June–October followed by wheat in November–April, heavy rice stubbles left in the field as a result of increased adoption of combined harvester are rarely used as cattle feed. Farmers usually undertake multiple tillage operations and residue burning in the traditional tillage systems. To address this problem and to facilitate direct seeding of wheat with surface retention of residues and without any soil disturbance, Happy Seeders were developed in NW IGP through different projects funded by the Australian Centre for International Agricultural Research (ACIAR) during the period 2000–10. The third prototype of Happy Seeders, the Turbo Happy Seeder, is more flexible while reducing the dust and making it easier to see the sowing lines (Saunders et al., 2012). The technical features of the machine are detailed by Jat et al. (2013). These new generation CT machines are suitable for seeding under heavy stubble conditions, where the loose residues interfere with tillage and seeding operations. The use of Happy Seeders is associated with significant cost savings by avoiding traditional tillage and stubble incorporation, and also avoiding the yield penalty due to delayed wheat sowing (Sidhu et al., 2007). The technology has extra benefits in terms of reduced incidence of Karnal bunt (*Tilletia indica*), and less termite and weed infestation (Coventry et al., 2015), and is hence possibly more profitable than the other forms of CT machinery (CSISA, 2013). Despite these benefits, the technology faces considerable adoption constraints. In addition to the conventional barriers to CT wheat diffusion, the proponents of the Happy Seeder face the challenge of low demand for the technology, which stems from subsidization of herbicides, irrigation water and electricity in the NW Plains. These subsidies diminish the incentive to adopt the Happy Seeder whose primary benefit is to reduce the dependence on these inputs. Ironically, at the same time, the government provides purchase subsidies for Happy Seeders to jump-start the demand for the technology. Complementing the subsidy schemes, extension programmes demonstrate to farmers how to implement and maximize the benefits of the technology (Saunders et al., 2012). The use of the Turbo Happy Seeder in NW India was heavily supported by the CSISA, funded by the Bill and Melinda Gates Foundation and the United States Agency for International Development (CSISA, 2013).

Another locally adapted machine for CT wheat is the PTOS, which has become popular in the reduced-tillage regime of the eastern IGP, where late sowing is a major challenge in increasing wheat productivity. The technology was introduced in Bangladesh by CIMMYT in 1995. The seed drill is operated by a two-wheel power tiller with a 12 or 16 hp diesel engine. The implement can be detached and replaced with an alternative, more innovative equipment that offers farmers diverse improved tools for land preparation, seeding, fertilizing and other operations. Their small size and efficiency make PTOS particularly appropriate for smallholder farmers (Krupnik et al., 2013). A recent study, based on a household survey among rice–wheat farmers in Bangladesh indicated that the use of PTOS results in higher energy efficiency compared to the other tillage options (Aravindakshan et al., 2015). Hossain et al. (2015) indicate that PTOS have gained popularity among farmers in Bangladesh over the recent past. Constraints to the diffusion of the technology include the high market price of the seeder, limited private sector participation in PTOS manufacturing and scarcity of trained operators who are needed for an effective use of the machine (Akter and Gathala, 2014).

In addition to these two machinery interventions, there have been additions to the CT ensemble across the IGP, especially in the eastern region. Recently, in Bangladesh, experiments have been started with wheat being established on raised beds in a ridge-furrow system known as bed planting, which is considered as an appropriate CT practice under conditions of temporary waterlogging stress, which represent a problem frequently encountered in Bangladesh. Another development in the same region is strip tillage that involves making narrow strips where the soil has been opened up by either in-row chisels or PTOS. In this system, soil tillage is accomplished in a single pass, at which time the seed is sown from a mounted seed box, saving a considerable amount of labour and energy. Finally, and more recently, the two-wheel power tiller-based 'Versatile Strip Tillage Planter' and 'Versatile Multi-crop Planter' are being designed, which have the capability for continuous seeding and/or spaced planting with fertilizer meters for drilling in lines (Haque et al., 2011). Although not commercially available yet, these implements are expected to overcome the challenges of precision seeding and fertilizer application in small, fragmented plots of eastern IGP.

Alongside the development of dynamic technology dissemination and monitoring systems, there has been a renewed interest in redefining the principles of CT practices (Vanlauwe et al., 2014). This interest arises partially due to the dis-adoption of CT wheat reported from different parts of the IGP (Planning Commission, 2014; Krishna and Veettil, 2014; Erenstein, 2010a), as elaborated in Section 4. The usual promotion of broad blanket packages of recommended practices for fertilization and plant protection in conjunction with the use of CT equipment has often failed due to its poor adaptation to local circumstances (Tittonell et al., 2012). As such, optimizing fertilizer and nutrient use and scale-appropriate CT equipment to suit local socio-economic and agro-ecological conditions has generated considerable research interest. Combining CT with location-specific fertilizer use through on-farm benchmarking is advocated in South Asian farming systems for higher productivity and energy use efficiency (Aravindakshan et al., 2015). Nevertheless, adding fertilizer use as a fourth principle of CT has been criticized by Sommer et al. (2014), noting that full CT involving the three principles of reduced tillage, mulching and intercropping/crop rotation leads to significantly higher yields than the use of mineral fertilizers alone. Conversely, the on-going debate of redefining the nature of CT can be considered as one of the recent developments, recognizing the system specificity and economic relevance of locally adapted cultivation practices.

6 Concluding remarks

Smallholder cereal production systems of South Asia span a wide and diverse gradient with respect to labour availability and cultivation intensity. Farmers' ability to access human, financial and natural capital varies significantly, which in turn determines the adoption potential of technologies such as CT wheat in the region. The review of the recent literature shows that despite generating favourable farm impacts and positive externalities, CT wheat has not yet become widely popular in South Asia. Inadequate coverage of the CT wheat technology by the national extension network, variation of yield impacts across farming systems and locations, and the fact that tractor and CT drill ownership is no tenable goal for the small and marginal farmers are noted as major diffusion constraints. Therefore, business models for custom-hiring services with institutional mechanisms to reduce the transaction costs of servicing small farms, farmer group- or cooperative-based investments in CT machinery and the development and deployment of more location-specific and scale-appropriate machinery are recommended to enhance the diffusion of the technology. The acceptance of Turbo Happy Seeders in NW India and PTOS in Bangladesh represent relatively recent innovations that are system-specific, locally adapted and successfully adopted. Since the overall on-farm impacts of CT are found to generate significant positive externalities, governments and other developmental organizations in the region may invest to provide an enabling institutional environment for the accelerated spread of CT as an important element of sustainable wheat intensification in South Asia.

7 Where to look for further information

Further information on the latest research findings and activities around conservation tillage in South Asia and related topics can be found on the CSISA (http://csisa.org) and CIMMYT (http://www.cimmyt.org) websites. In particular, the repository of discussion papers and research articles on the CSISA website (http://csisa.org/research-publications) and the CIMMYT Knowledge Center Blog (http://knowledgecenterblog.cimmyt.org/?s=conservation+agriculture+south+asia) are recommended as rich sources of up-to-date information.

8 Acknowledgements

Financial assistance from the United States Agency for International Development (USAID) and the Bill and Melinda Gates Foundation for the project 'Cereal Systems Initiative for South Asia' (CSISA) is gratefully acknowledged.

9 References

Agasty, M. P. and Patra, R. N. (2014), 'Migration, labor supply, wages and agriculture: A case study in rural Odisha', *Developing Country Studies*, Vol. 4, No. 16, pp. 91–109.

Aggarwal, P. K., Joshi, P. K., Ingram, J. and Gupta, R. K. (2004), 'Adapting food systems of the Indo-Gangetic plains to global environmental change: Key information needs to improve policy formulation', *Environmental Science & Policy*, Vol. 7, No. 6, pp. 487–98.

Akter, S. and Gathala, M. K. (2014), 'Adoption of Conservation Agriculture technology in diversified systems and impact on productivity: Evidence from three districts in Bangladesh'. Paper presented at the 88th Annual Conference of the Agricultural Economics Society, AgroParisTech, 9–11 April 2014, Paris, France.

Alam, M. M., Ladha, J. K., Faisal, M. W., Sharma, S., Saha, A., Noor, S. and Rahman, M. A. (2015), 'Improvement of cereal-based cropping systems following the principles of Conservation Agriculture under changing agricultural scenarios in Bangladesh', *Field Crops Research*, Vol. 175, pp. 1–15.

Ali, A. and Erenstein, O. (2013), 'Impact of zero tillage adoption on household welfare in Pakistan', *Journal of Agricultural Technology*, Vol. 9, No. 7, pp 1715–29.

Anonymous (2008), *Bihar's Agricultural Development: Opportunities and Challenges.* A report of the special task force on Bihar, New Delhi.

Anonymous (2013), *Statistical Yearbook of Bangladesh*, 32nd ed., Bangladesh Bureau of Statistics, Government of the People's Republic of Bangladesh, Dhaka.

Aravindakshan, S., Rossi, F. J. and Krupnik, T. J. (2015), 'What does benchmarking of wheat farmers practicing conservation tillage in the eastern Indo-Gangetic Plains tell us about energy use efficiency? An application of slack-based data envelopment analysis', *Energy*, Vol. 90, pp. 483–93.

Aryal, J. P., Sapkota, T. B., Jat, M. L. and Bishnoi, D. K. (2015), 'On farm economic and environmental impact of zero tillage wheat: A case of North-West India', *Experimental Agriculture*, Vol. 51, No. 01, pp. 1–16.

Baker, J. C. and Saxton, K. E. (2007), *The 'what' and 'why' of no-tillage farming*, Chapter 1, in 'No-tillage seeding in conservation agriculture', 2nd ed., C. J. Baker and K. E. Saxton (eds), FAO and CAB International, Rome (Italy), Oxfordshire (UK), Cambridge (USA).

Baudron, F., Jaleta, M., Okitoi, O. and Tegegn, A. (2014), 'Conservation agriculture in African mixed crop-livestock systems. Expanding the niche', *Agriculture, Ecosystems & Environment*, Vol. 187, pp. 171–82.

Baudron, F., Thierfelder, C., Nyagumbo, I. and Gérard, B. (2015), 'Where to target Conservation Agriculture for African smallholders? How to overcome challenges associated with its implementation? Experience from eastern and southern Africa', *Environments*, Vol. 2, No. 3, pp. 338–57.

Bhushan, L., Ladha, J. K., Gupta, R. K., Singh, S., Tirol-Padre, A., Saharawat, Y. S., Gathala, M. and Pathak, H. (2007), 'Saving of water and labor in a rice–wheat system with no-tillage and direct seeding technologies', *Agronomy Journal*, Vol. 99, No. 5, p. 1288.

Brouder, S. M. and Gomez-Macpherson, H. (2014), 'The impact of Conservation Agriculture on smallholder agricultural yields: A scoping review of the evidence', *Agriculture, Ecosystems & Environment*, Vol. 187, pp. 11–32.

Chauhan, B. S., Mahajan, G., Sardana, V., Timsina, J. and Jat, M. L. (2012), 'Productivity and sustainability of the rice–wheat cropping system in the Indo-Gangetic Plains of the Indian subcontinent', *Advances in Agronomy*, Vol. 117, pp. 315–69.

Chhokar, R. S., Sharma, R. K., Jat, G. R., Pundir, A. K. and Gathala, M. K. (2007), 'Effect of tillage and herbicides on weeds and productivity of wheat under rice–wheat growing system', *Crop Protection*, Vol. 26, No. 11, pp. 1689–96.

Coventry, D. R., Poswal, R. S., Yadav, A., Gupta, R. K., Gill, S. C., Chhokar, R. S., Kumar, V., Sharma, R. K., Kumar, A., Mehta, A., Kleemann, S. and Cummins, J. A. (2011), 'Effect of tillage and nutrient management on wheat productivity and quality in Haryana, India', *Field Crops Research*, Vol. 123, No. 3, pp. 234–40.

Coventry, D. R., Poswal, R. S., Yadav, A., Riar, A. S., Zhou, Y., Kumar, A., Chand, R., Chhokar, R. S., Sharma, R. K., Yadav, V. K., Gupta, R. K., Mehta, A. and Cummins, J. A. (2015), 'A comparison of farming practices and performance for wheat production in Haryana, India', *Agricultural Systems*, Vol. 137, pp. 139–53.

CSISA (2013), *Cereal Systems Initiative for South Asia (Phase II): Annual report*, International Maize and Wheat Improvement Center (CIMMYT), Kathmandu, Available at: http://csisa.org/wp-content/uploads/sites/2/2013/10/CSISA-Annual-Report-Nov-2013-final-v.2.pdf (accessed 9 November 2015).

Das, T. K., Bhattacharyya, R., Sharma, A. R., Das, S., Saad, A. A. and Pathak, H. (2013), 'Impacts of conservation agriculture on total soil organic carbon retention potential under an irrigated agro-ecosystem of the western Indo-Gangetic Plains', *European Journal of Agronomy*, Vol. 51, pp. 34–42.

Das, T. K., Bhattacharyya, R., Sudhishri, S., Sharma, A. R., Saharawat, Y. S., Bandyopadhyay, K. K., Sepat, S., Bana, R. S., Aggarwal, P., Sharma, R. K., Bhatia, A., Singh, G., Datta, S. P., Kar, A., Singh, B., Singh, P., Pathak, H., Vyas, A. K. and Jat, M. L. (2014), 'Conservation Agriculture in an irrigated cotton–wheat system of the western Indo-Gangetic Plains: Crop and water productivity and economic profitability', *Field Crops Research*, Vol. 158, pp. 24–33.

Datta, A. and Mishra, S. K. (2011), 'Glimpses of women's lives in rural Bihar: Impact of male migration', *The Indian Journal of Labour Economics*, Vol. 54, No. 3, pp. 457–77.

Derpsch, R., Friedrich, T., Kassam, A. and Li, H. (2010), 'Current status of adoption of no–till farming in the world and some of its main benefits', *International Journal of Agricultural and Biological Engineering*, Vol. 3, No. 1, pp. 1–25.

Devi, P., Aggarwal, A. and Gupta, S. R. (2015), 'Effect of zero tillage on soil carbon storage and nitrogen uptake in rice–wheat systems in Northern India', *American–Eurasian Journal of Agricultural and Environmental Sciences*, Vol. 15, No. 5, pp. 923–31.

DWR (2011), *Vision 2030*, Directorate of Wheat Research, Karnal.

DWR (2014), *Wheat Scenario: A Snippet*, Directorate of Wheat Research, Karnal (Haryana), India, pp. 1–6, Available at: http://www.dwr.res.in/sites/default/files/Publications/wheatsnippetissue%202-2014.pdf.

Ekboir, J. (2002), *Developing No-Tillage Packages For Small-Scale Farmers*, In J. Ekboir (ed.), World Wheat Overview and Outlook. CIMMYT, Mexico, DF, pp. 1–38.

Ellis, F. (1993), *Peasant Economics: Farm Households and Agrarian Development*, 2nd ed., Cambridge University Press, Cambridge, UK.

Erenstein, O. (2009), 'Specification effects in zero tillage survey data in South Asia's rice–wheat systems', *Field Crops Research*, Vol. 111, No. 1–2, pp. 166–72.

Erenstein, O. (2010a), 'Triangulating technology diffusion indicators: Zero tillage wheat in South Asia's irrigated plains', *Experimental Agriculture*, Vol. 46, No. 03, pp. 293–308.

Erenstein, O. (2010b), 'Village surveys for technology uptake monitoring: Case of tillage dynamics in the Trans-Gangetic Plains', *Experimental Agriculture*, Vol. 46, No. 03, pp. 277–92.

Erenstein, O. and Farooq, U. (2009), 'A survey of factors associated with the adoption of zero tillage wheat in the irrigated plains of South Asia', *Experimental Agriculture*, Vol. 45, No. 02, p. 133.

Erenstein, O., Farooq, U., Malik, R. K. and Sharif, M. (2008), 'On-farm impacts of zero tillage wheat in South Asia's rice–wheat systems', *Field Crops Research*, Vol. 105, No. 3, pp. 240–52.

Erenstein, O., Hellin, J. and Chandna, P. (2010), 'Poverty mapping based on livelihood assets: A meso-level application in the Indo-Gangetic Plains, India', *Applied Geography*, Vol. 30, No. 1, pp. 112–25.

Erenstein, O. and Laxmi, V. (2008), 'Zero tillage impacts in India's rice–wheat systems: A review', *Soil and Tillage Research*, Vol. 100, No. 1–2, pp. 1–14.

Erenstein, O. and Thorpe, W. (2011), 'Livelihoods and agro-ecological gradients: A meso-level analysis in the Indo-Gangetic Plains, India', *Agricultural Systems*, Vol. 104, No. 1, pp. 42–53.

FAOSTAT (2015), *Food and Agriculture Organization of the United Nations, FAOSTAT database ('Crops')*, Rome, Available at: http://faostat3.fao.org/home/E (accessed 1 November 2015).

Farooq, M. and Siddique, K. H. M. (2015), *Conservation Agriculture: Concepts, Brief History, and Impacts on Agricultural Systems*, Springer International Publishing, Heidelberg.

Haque, M. E., Bell, R. W., Islam, A., Sayre, K. and Hossain, M. M. (eds) (2011), *Versatile Multi-Crop Planter For Two-Wheel Tractors: An Innovative Option For Smallholders.* Paper presented at the World Congress on Conservation Agriculture, 26–29 September, Brisbane, Australia.

Harvey, C. A., Chacón, M., Donatti, C. I., Garen, E., Hannah, L., Andrade, A., Bede, L., Brown, D., Calle, A., Chará, J., Clement, C., Gray, E., Hoang, M. H., Minang, P., Rodríguez, A. M., Seeberg-Elverfeldt, C., Semroc, B., Shames, S., Smukler, S., Somarriba, E., Torquebiau, E., van Etten, J. and Wollenberg, E. (2014), 'Climate-smart landscapes: Opportunities and challenges for integrating adaptation and mitigation in tropical agriculture', *Conservation Letters*, Vol. 7, No. 2, pp. 77–90.

Hobbs, P. R., Sayre, K. and Gupta, R. K. (2008), 'The role of conservation agriculture in sustainable agriculture', *Philosophical Transactions of the Royal Society of London. Series B, Biological Sciences*, Vol. 363, No. 1491, pp. 543–55.

Hossain, M. I., Sarker, M. and Haque, M. A. (2015), 'Status of conservation agriculture based tillage technology for crop production in Bangladesh', *Bangladesh Journal of Agricultural Research*, Vol. 40, No. 2, pp. 235–48.

Jat, M. L., Kapil, Kamboj, B. R., Sidhu, H. S., Singh, M., Bana, A., Bishnoi, D., Gathala, M., Saharawat, Y. S., Kumar, V., Kumar, A., Jat, H. S., Jat, R. K., Kumar, V., Sharma, R. K., Singh, R., Sapkota, T. B., Malik, R. K. and Gupta, R. K. (2013), *Operational Manual for Turbo Happy Seeder- Technology for Managing Crop Residues with Environmental Stewardship*, International Maize and Wheat Improvement Center (CIMMYT); Indian Council of Agricultural Research (ICAR), New Delhi.

Kassam, A., Derpsch, R. and Friedrich, T. (2014), 'Global achievements in soil and water conservation: The case of conservation agriculture', *International Soil and Water Conservation Research*, Vol. 2, No. 1, pp. 5–13.

Kassam, A., Friedrich, T., Shaxson, F. and Pretty, J. (2009), 'The spread of conservation agriculture: Justification, sustainability and uptake', *International Journal of Agricultural Sustainability*, Vol. 7, No. 4, pp. 292–320.

Keil, A., D'souza, A. and McDonald, A. J. (2015), 'Zero-tillage as a pathway for sustainable wheat intensification in the Eastern Indo-Gangetic Plains: Does it work in farmers' fields?', *Food Security*, Vol. 7, No. 5, pp. 983–1001.

Krishna, V. V., Aravindakshan, S., Chowdhury, A. and Rudra, B. (2012a), *Farmer Access and Differential Impacts of Zero Tillage Technology in the Subsistence Wheat Farming Systems of West Bengal, India, Socio-Economics Programme Working Paper*, Vol. 7, Mexico DF.

Krishna, V. V., Mehrotra, M. B., Teufel, N. and Bishnoi, D. K. (2012b), *Characterizing the Cereal Systems and Identifying the Potential of Conservation Agriculture in South Asia, Socio-Economics Programme Working Paper*, Vol. 5, Mexico DF.

Krishna, V. V. and Veettil, P. C. (2014), 'Productivity and efficiency impacts of conservation tillage in northwest Indo-Gangetic Plains', *Agricultural Systems*, Vol. 127, pp. 126–38.

Krupnik, T. J., Santos Valle, S., McDonald, A. J., Justice, S., Hossain, I. and Gathala, M. K. (2013), *Made in Bangladesh: Scale-Appropriate Machinery for Agricultural Resource Conservation*, Mexico DF.

Kumar, V., Saharawat, Y. S., Gathala, M. K., Jat, A. S., Singh, S. K., Chaudhary, N. and Jat, M. L. (2013a), 'Effect of different tillage and seeding methods on energy use efficiency and productivity of wheat in the Indo-Gangetic Plains', *Field Crops Research*, Vol. 142, pp. 1–8.

Kumar, V., Singh, S., Chhokar, R. S., Malik, R. K., Brainard, D. C. and Ladha, J. K. (2013b), 'Weed management strategies to reduce herbicide use in zero-till rice–wheat cropping systems of the Indo-Gangetic Plains', *Weed Technology*, Vol. 27, No. 1, pp. 241–54.

Laxmi, V. and Mishra, V. (2007), 'Factors affecting the adoption of resource conservation technology: Case of zero tillage in rice-wheat farming systems', *Indian Journal of Agricultural Economics*, Vol. 62, No. 1, pp. 126–38.

MoA and FAO (2010), *Towards a Food Secure Bangladesh: National Medium Term Priority Framework of Bangladesh (2010–2015)*.

Ortiz, R., Sayre, K. D., Govaerts, B., Gupta, R. K., Subbarao, G. V., Ban, T., Hodson, D., Dixon, J. M., Iván Ortiz-Monasterio, J. and Reynolds, M. (2008), 'Climate change: Can wheat beat the heat?', *Agriculture, Ecosystems & Environment*, Vol. 126, No. 1–2, pp. 46–58.

Pede, V. O., Krishna, V., Teufel, N., Ward, P. S., Yamano, T., Sammadar, A., Spielman, D. J., Mittal, S., Paris, T., Singh, D., Singh, V., Ghimire, S. and Mehrotra, M. (2015), *Cereal Systems Initiative for South Asia (CSISA) Baseline Household Survey 2010–2011*, International Rice Research Institute (IRRI), International Maize and Wheat Improvement Center (CIMMYT), International Food Policy Research Institute (IFPRI), International Livestock Research Institute (ILRI), New Delhi, Available at: http://csisa.org/resources/csisa-phase-i-baseline-data/=QB2015104204.

Pede, V. O., Ward, P. S., Spielman David, J. and Paris, T. (2012), *Summary of the Agro-ecological and Socioeconomic Context for the Cereal Systems Initiative for South Asia (CSISA)*, International Rice Research Institute (IRRI), International Maize and Wheat Improvement Center (CIMMYT), International Food Policy Research Institute (IFPRI), International Livestock Research Institute (ILRI), New Delhi.

Pittelkow, C. M., Linquist, B. A., Lundy, M. E., Liang, X., van Groenigen, K. J., Lee, J., van Gestel, N., Six, J., Venterea, R. T. and van Kessel, C. (2015), 'When does no-till yield more? A global meta-analysis', *Field Crops Research*, Vol. 183, pp. 156–68.

Planning Commission (2014), *Report of the Expert Group to Review the Methodology for Measurement of Poverty*, New Delhi.

Powlson, D. S., Stirling, C. M., Jat, M. L., Gerard, B. G., Palm, C. A., Sanchez, P. A. and Cassman, K. G. (2014), 'Limited potential of no-till agriculture for climate change mitigation', *Nature Climate Change*, Vol. 4, No. 8, pp. 678–83.

Rawson, H. M., Gomez-Macpherson, H., Hossain, A. B. S., Saifuzzaman, M., Rashid, H., Sufian, M. A., Samad, M. A., Sarker, A. Z., Ahmed, F., Talukder, Z. I., Rahman, M., Siddique, M. M. A. B., Hossain, I. and Amin, M. (2007), 'On-farm wheat trials in Bangladesh: A study to reduce perceived constraints to yield in traditional wheat areas and southern lands that remain fallow during the dry season', *Experimental Agriculture*, Vol. 43, No. 01, pp. 21–40.

Rodell, M., Velicigna, I. and Famiglietti, J. S. (2009), 'Satellite-based estimates of groundwater depletion in India', *Nature*, Vol. 460, pp. 999–1002.

Saharawat, Y. S., Singh, B., Malik, R. K., Ladha, J. K., Gathala, M., Jat, M. L. and Kumar, V. (2010), 'Evaluation of alternative tillage and crop establishment methods in a rice–wheat rotation in North Western IGP', *Field Crops Research*, Vol. 116, No. 3, pp. 260–7.

Sapkota, T. B., Jat, M. L., Aryal, J. P., Jat, R. K. and Khatri-Chhetri, A. (2015), 'Climate change adaptation, greenhouse gas mitigation and economic profitability of Conservation Agriculture: Some examples from cereal systems of Indo-Gangetic Plains', *Journal of Integrative Agriculture*, Vol. 14, No. 8, pp. 1524–33.

Saunders, C., Davis, L. and Pearce, D. (2012), *Rice–Wheat Cropping Systems in India and Australia, and Development of the 'Happy Seeder', ACIAR Impact Assessment Series*, Report No. 77, The Australian Centre for International Agricultural Research (ACIAR), Canberra.

Schneider, F., Steiger, D., Ledermann, T., Fry, P. and Rist, S. (2012), 'No-tillage farming: Co-creation of innovation through network building', *Land Degradation & Development*, Vol. 23, No. 3, pp. 242–55.

Sekar, I. and Pal, S. (2012), 'Rice and wheat crop productivity in the Indo-Gangetic Plains of India: Changing pattern of growth and future strategies', *Indian Journal of Agricultural Economics*, Vol. 67, No. 2, pp. 238–52.

Sekar, I., Pal, S. and Kar, A. (2015), 'Economic and environmental impact of resource conserving technology: A study of zero tillage in the Indo-Gangetic Plains of India', *Indian Journal of Soil Conservation*, Vol. 43, No. 2, pp. 187–91.

Shiferaw, B., Smale, M., Braun, H.-J., Duveiller, E., Reynolds, M. and Muricho, G. (2013), 'Past successes and future challenges to the role played by wheat in global food security', *Food Security*, Vol. 5, No. 3, pp. 291–317.

Sidhu, G. S., Bhattacharyya, T., Sarkar, D., Ray, S. K., Chandran, P., Pal, D. K., Mandal, D. K., Prasad, J., Nair, K. M., Sahoo, A. K., Das, T. H., Singh, R. S., Mandal, C., Srivastava, R., Sen, T. K., Chatterji, S., Patil, N. G., Obireddy, G. P., Mahapatra, S. K. and Kumar, K. S. A. (2014), 'Impact of management levels and land-use changes on soil properties in rice-wheat cropping system of the Indo-Gangetic Plains', *Current Science*, Vol. 107, No. 9, pp. 1487–501.

Sidhu, H. S., Manpreet-Singh, Humphreys, E., Yadvinder-Singh, Balwinder-Singh, Dhillon, S. S., Blackwell, J., Bector, V., Malkeet-Singh and Sarbjeet-Singh (2007), 'The Happy Seeder enables direct drilling of wheat into rice stubble', *Australian Journal of Experimental Agriculture*, Vol. 47, No. 7, p. 844.

Singh, A., Phogat, V. K., Dahiya, R. and Batra, S. D. (2014), 'Impact of long-term zero till wheat on soil physical properties and wheat productivity under rice–wheat cropping system', *Soil and Tillage Research*, Vol. 140, pp. 98–105.

Singh, R., Erenstein, O., Saharawat, Y. S., Chaudhary, N. and Jat, M. L. (2012), 'Adoption analysis of resource-conserving technologies in rice (Oryza sativa)–wheat (Triticum aestivum) cropping system of South Asia', *Indian Journal of Agricultural Sciences*, Vol. 82, No. 5, pp. 405–9.

Sommer, R., Thierfelder, C., Tittonell, P., Hove, L., Mureithi, J. and Mkomwa, S. (2014), 'Fertilizer use should not be a fourth principle to define Conservation Agriculture. Response to the opinion paper of Vanlauwe et al. (2014) "A fourth principle is required to define Conservation Agriculture in sub-Saharan Africa: The appropriate use of fertilizer to enhance crop productivity"', *Field Crops Research*, Vol. 169, pp. 145–8.

Tittonell, P., Scopel, E., Andrieu, N., Posthumus, H., Mapfumo, P., Corbeels, M., van Halsema, G. E., Lahmar, R., Lugandu, S., Rakotoarisoa, J., Mtambanengwe, F., Pound, B., Chikowo, R., Naudin, K., Triomphe, B. and Mkomwa, S. (2012), 'Agroecology-based aggradation-Conservation Agriculture (ABACO): Targeting innovations to combat soil degradation and food insecurity in semi-arid Africa', *Field Crops Research*, Vol. 132, pp. 168–74.

Tripathi, R. S., Raju, R. and Thimmappa, K. (2013), 'Impact of zero tillage on economics of wheat production in Haryana', *Agricultural Economics Research Review*, Vol. 26, No. 1, pp. 101–8.

Vanlauwe, B., Wendt, J., Giller, K. E., Corbeels, M., Gerard, B. and Nolte, C. (2014), 'A fourth principle is required to define Conservation Agriculture in sub-Saharan Africa: The appropriate use of fertilizer to enhance crop productivity', *Field Crops Research*, Vol. 155, pp. 10–13.

Chapter 4

Improving water management in winter wheat

Q. Xue, J. Rudd, J. Bell, T. Marek and S. Liu, Texas A&M AgriLife Research and Extension Center at Amarillo, USA

1 Introduction
2 Winter wheat yield
3 Yield determination under water-limited conditions
4 The role of measuring evapotranspiration (ET)
5 Water-use efficiency
6 Wheat yield, evapotranspiration (ET) and water-use efficiency (WUE) relationships
7 Case studies
8 Future trends and conclusion
9 Where to look for further information
10 References

1 Introduction

Wheat (*Triticum aestivum* L. em. Thell) is a major crop in the world and is the number one food grain consumed directly by humans. Wheat is grown in a wide range of environments around the world and has the broadest adaptation of all cereal crop species (Briggle and Curtis, 1987). Therefore, wheat production plays a critical role in the world economy and food security. For 2016/17, the world wheat production is projected to be about 720 million metric tons. The European Union dominates the production with 157 million tons, followed by China (130 million tons), India (88 million tons), Russia (64 million tons), the United States (56 million tons), Canada (28.5 million tons), Pakistan (25.3 million tons), Australia (25 million tons), Ukraine (24 million tons) and Turkey (17.5 million tons) (http://www.statista.com/statistics/237912/global-top-wheat-producing-countries/). Wheat is also widely traded in the world market, with the United States, European Union, Australia, Canada and Russia being the major exporters (Fischer et al., 2014). For example, the United States share of world wheat exports has ranged from about 16% to nearly 40% in last three decades (Fig. 1). Although wheat exports have a declining trend in coming decade, the United States will remain one of the world's leading suppliers of high-quality

http://dx.doi.org/10.19103/AS.2016.0004.15

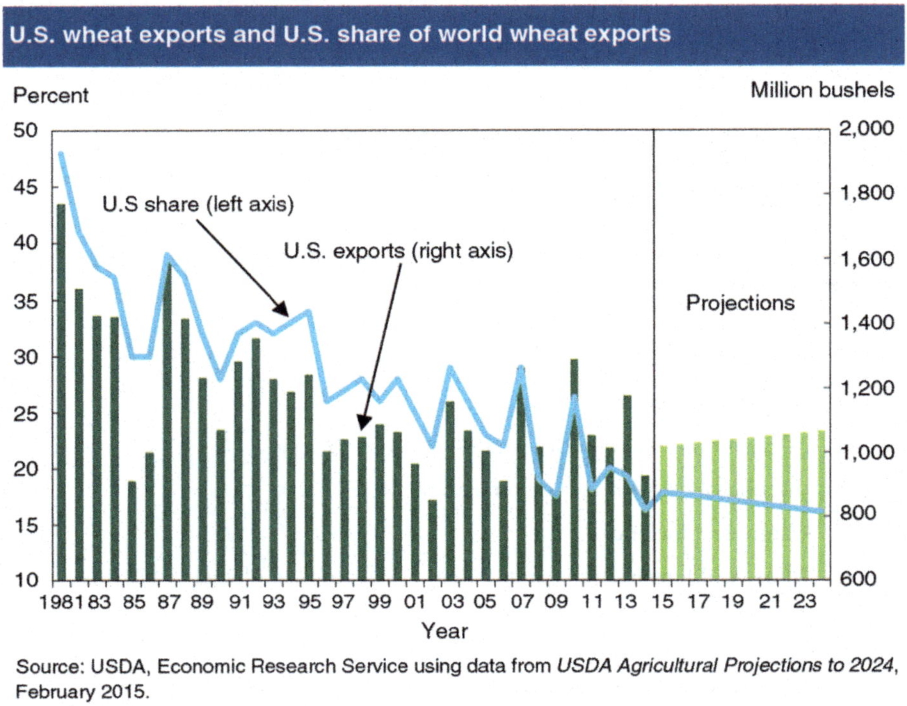

Figure 1 U.S. wheat exports and share of world wheat exports (historic data and projections to 2024).

wheat (USDA-Economic Research Service, 2015) (http://www.ers.usda.gov/data-products/chart-gallery/detail.aspx?chartId=52238).

Globally, winter wheat is mainly grown in Eurasia, North China Plain (NCP), Anatolia area of Iran and the United States. Some areas have high precipitation (>500 mm) such as Europe and eastern United States but other areas such as NCP require irrigation for high yield in winter wheat (Fischer et al., 2014). Nevertheless, drought stress can significantly reduce winter wheat yields, even under high precipitation environment like United Kingdom (Foulkes et al., 2002). In the United States, except the eastern states, winter wheat is mainly produced in the Great Plains and Pacific Northwest states, with a low to moderate precipitation (200–500 mm). The impact of drought stress on winter wheat production in these areas is more significant than others. The Great Plains of North America extend from the northern part of the state of Coahuila, Mexico, through west Texas and northward through the United States to the prairie provinces of Alberta, Manitoba and Saskatchewan of Canada (Fig. 2). In the United States, central and southern Great Plains (SGP) are major winter wheat production areas. In 2015, winter wheat production from the states of Colorado, Kansas, New Mexico, Oklahoma and Texas was 665 million bushels (18 million metric tons) and accounted for 42% of US total winter wheat production (Fig. 3; NASS, 2016).

The SGP includes the Texas High Plains, the Oklahoma panhandle, part of eastern New Mexico, southwestern Kansas and southeastern Colorado. Winter wheat is widely

Figure 2 Great Plains of North America.

grown under dryland (rainfed), full irrigation and deficit irrigation production systems and produced for both grain and winter cattle forage in this region (Musick and Dusek, 1980; Musick et al., 1994; Howell et al., 1995). The region has a semi-arid climate with annual precipitation ranging from 380 mm in the southwest to 580 mm in the northeast and averages about 480 mm. Growing season precipitation for wheat production averages about 250 mm (Musick et al. 1994). The seasonal evapotranspiration (ET) for winter wheat growth ranges from 700 to 950 mm under full irrigation conditions (Musick and Porter, 1990; Musick et al., 1994; Howell et al. 1995, 2007). Therefore, the seasonal precipitation for winter wheat can only meet one-third of the ET required for maximum grain yield.

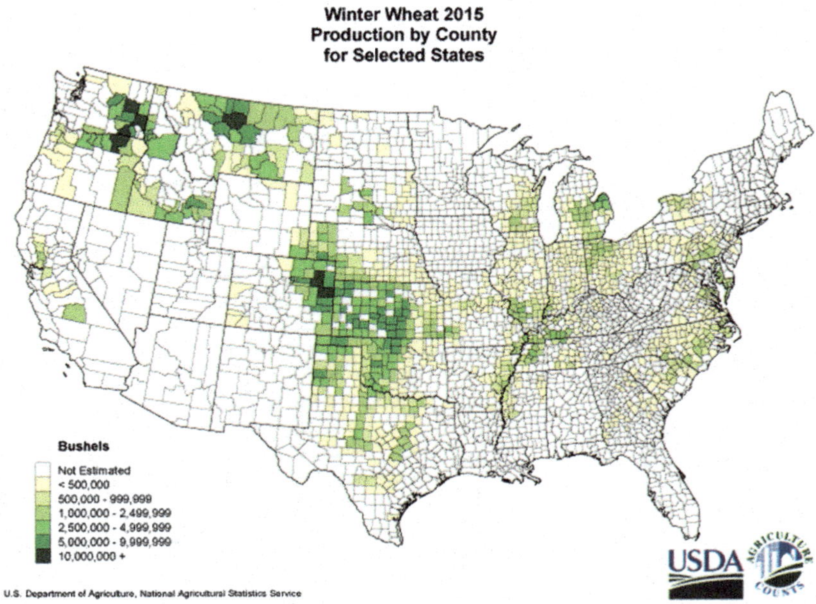

Figure 3 Winter wheat production by county in the United States in 2015 (NASS, 2016). https://www.nass.usda.gov/Charts_and_Maps/Crops_County/ww-pr.php.

As a result, wheat yield and water-use efficiency (WUE) are primarily limited by soil water deficits from late spring to early summer (Musick et al., 1994; Howell et al., 1997). Under dryland conditions, wheat production is largely determined by the amount and effective use of soil water storage and seasonal precipitation. In general, dryland wheat yields are much lower than irrigated wheat in the region due to the drought stress (Musick et al., 1994; Jones and Popham, 1997).

In the U.S. SGP, high wheat yields have been dependent on irrigation. The irrigation water resource is mainly from the Ogallala Aquifer (Fig. 4), and development of irrigation in this region significantly increased during the 1950s. The Ogallala Aquifer is essentially a closed system with minimal recharge capacity, and the dramatic increase in water extraction for crop irrigation resulted in a significant decline in the water table; some areas have experienced up to 50% reduction in predevelopment saturated thickness. Irrigated land area has decreased from a peak of 2.4 million ha in 1974 to 1.9 million ha in 2000 (Colaizzi et al., 2008).

Similar to many other agricultural regions in the world, the SGP faces many challenges in crop production. First, the growing world population continuously requires more production of food, forage and fibre, particularly the major food crops such as wheat and corn. Second, the declining water table in the Ogallala Aquifer and increasing pumping costs will inevitably reduce irrigation levels (Musick et al. 1994; Stone and Schlegel, 2006; Colaizzi et al., 2008). Third, the possibility of increasing frequency and severity of drought stress as well as other abiotic and biotic stresses under changing climate will likely reduce crop yields more frequently. For example, the wheat yield and economic losses from the

Figure 4 Ogallala aquifer.

historic 2011 drought in Texas resulted in $5.2 billion agricultural losses and the losses from wheat alone were $243 million. The last two drought years in 2006 and 2009 resulted in $4.1 and $3.6 billion economic losses in Texas agriculture, respectively (Fannin, 2011).

NCP produces about 69% of wheat in the whole country (Mo et al., 2009). The average annual rainfall in the region ranges from 500 to 650 mm. However, the majority of precipitation occurs during summer months (July–September). Precipitation during the wheat growing season ranges from 100 to 180 mm, or approximately 25–40% of crop water requirement over the growing season (Zhang et al., 1999). Therefore, drought is also an important factor affecting wheat yield without irrigation. Similar to the U.S. SGP, irrigation from groundwater sources is also mainly used in winter wheat production. However, groundwater levels have also decreased considerably in the NCP due to irrigation (Zhang et al., 2003).

Since water is the most important factor affecting crop production, development of crop management practices to conserve water, optimize water use and improve crop WUE becomes essential, particularly under changing climate conditions. Although increasing yield is an ultimate goal to any crop production system, maximizing WUE is of particular importance under water-limited conditions since lower WUE sometime indicates poor management and inefficient use of available soil water (Musick et al., 1994; Howell, 2001; Passioura and Angus, 2010; Xue et al., 2012).

In this chapter, we reviewed the research progress in winter wheat water management and WUE. The major data sources were from long-term field experiments in the U.S. SGP due to the long history of winter wheat research. We also include some research results from NCP of China as irrigation studies have been conducted extensively in the past two decades. We initially discuss the yield, ET, WUE and their relationships. Then, the best management practices will be discussed based on several case studies, ranging from soil and water conservation to genetic improvement of drought tolerance to deficit irrigation practice.

2 Winter wheat yield

Winter grain yield is determined by genetic, environmental and management factors. In the U.S. SGP, water is the major limiting factor for wheat yield. The long-term dryland wheat yields at Bushland, Texas (1984–93), were mostly in range of 1–2 Mg ha^{-1} (Jones and Popham, 1997). Musick and Porter (1990) reported that farm level dryland yield averaged 1.1 Mg ha^{-1} from 1968 to 1986 in the SGP. However, yields in dryland treatment from irrigation studies were frequently over 3 Mg ha^{-1} (Musick et al., 1994; Xue et al., 2006). Stone and Schlegel (2006) summarized long-term winter wheat yield in Tribune, Kansas (1973–2004), and reported an average yield of 2.5 Mg ha^{-1}. Graybosch and Peterson (2010) analysed winter wheat yield in the Great Plains based on the yield data from USDA-coordinated winter wheat regional performance nurseries from 1959 to 2008. They showed that dryland wheat yield ranged from 1.8 to 2.9 Mg ha^{-1}, with an average of 2.3 Mg ha^{-1} in the SGP (Graybosch and Peterson, 2010). This average yield is close to the average yield from Stone and Schlegel (2006). Genetic improvement in wheat yield was evident over the years based on the long-term yield data, with 1.3% yield increase per year up to the middle of 1990s (Graybosch and Peterson, 2010). Our recent studies also demonstrated a significant genetic gain in dryland wheat yield at Bushland, Texas. For example, in a two-year field study, relatively newer cultivar TAM 111 (released in 2003) yielded 2.8 Mg ha^{-1} but a relatively older cultivar, TAM 105 (released in 1979), only yielded 2.3 Mg ha^{-1} in 2010 (Xue et al., 2014).

In the SGP, irrigated wheat yields ranged from 3.0 to 7.7 Mg ha^{-1}, depending on irrigation timing and frequency, and environmental conditions (Xue et al., 2012). In general, yield increased as irrigation amount and frequency increased (Schneider and Howell, 2001; Xue et al., 2006). Under full irrigation, yield genetic gains are more significant as compared to dryland. At Bushland, Texas, long-term irrigated studies showed that the fully irrigated yield in 1980s was in the range of 4–5 Mg ha^{-1} (Musick and Dusek, 1980) but increased to about 7.3 Mg ha^{-1} in the late 1990s (Schneider and Howell, 2001). The high yield of 7.7 Mg ha^{-1} was also found in producer's research and demonstration fields (AgriPartner, 2007). The highest irrigated yield in Tribune, Kansas, was about 6.0 Mg ha^{-1}, which was likely under limited irrigation conditions (Stone and Schegel, 2006). In NCP, winter wheat is mainly grown under irrigation and yields have increased over the past four decades. For example, Zhang et al. (2010) showed that cultivars released in 2000s yielded about 7.0 Mg ha^{-1} as compared to yield of 5.0 Mg ha^{-1} in 1970s. The more recent field studies indicated that irrigated yield can be over 8.0 Mg ha^{-1} (Wang et al., 2016).

3 Yield determination under water-limited conditions

For yield determination under water-limited conditions, several conceptual models have been introduced by different authors (Passioura, 1977; Howell, 2001; Richards et al., 2002; Blum, 2009; Passioura and Angus, 2010; Stewart and Peterson, 2015). For wheat, Passioura (1977) introduced a framework for identifying the important components for yield under water-limited conditions. Grain yield is determined by three components:

$$\text{Yield} = ET \times WUE_{bm} \times HI \quad (1)$$

where ET is the seasonal evapotranspiration, WUE_{bm} is the water-use efficiency for biomass production, and HI is the harvest index, that is, the fraction of biomass partitioning to grains (Passioura, 1977). Since these three components are likely to be largely independent of each other, then an improvement in any one of them should result in an increase in yield. This framework has been proved to be very useful for identifying management strategies under water-limited conditions (Richards, et al., 2002).

Stewart and Peterson (2015) introduced another form of yield determination as

$$\text{Yield} = ET \times T/ET \times 1/TR \times HI \quad (2)$$

where T/ET is the ratio of plant transpiration and evapotranspiration and TR is the transpiration ratio (the amount of water required to produce a unit of biomass). In equation (1), WUE_{bm} is related to transpiration efficiency (TE), transpiration (T) and soil evaporation (Es) (Richards et al., 2002):

$$WUE_{bm} = TE/(1 + Es/T) \quad (3)$$

In equation (2), 1/TR is actually transpiration efficiency (TE).

Based on the above equations, increasing ET and HI and minimizing Es will increase yield under water-limited conditions. In general, TE is a relatively conservative parameter due to the low variability in the basic biochemical efficiency of photosynthesis (Bodner et al., 2015).

In equations (1) and (2), the product of ET x WUE_{bm} or ET x T/ET x 1/TR is biomass at maturity (BM). Therefore, the yield determination equation also can be simply rewritten as follows:

$$\text{Yield} = BM \times HI \quad (4)$$

Improving biomass production, HI or both will lead to higher yield (Blum, 2009). In the southern Great Plains, an early analysis indicated that wheat improvement from 1950s to 1980s was mainly contributed by increased biomass. The potential HI was relatively stable in the semi-dwarf cultivars (Howell, 1990). In recent studies, we also found that higher yield in newer cultivars was a result of higher biomass under dryland conditions

(Xue et al., 2014). The HI has increased by genetic improvement as evidenced in our new study. However, HI is generally low due to drought or heat stress at reproductive stages in semi-arid environments (Schneider and Howell, 2001).

4 The role of measuring evapotranspiration (ET)

ET has been widely used in determining water requirements, irrigation scheduling and assessing crop growth and yield response to water deficit. In the U.S. SGP, seasonal ET for winter wheat has been measured in irrigation studies since 1960s. Musick and Porter (1990) summarized ET requirement in wheat up to late 1980s. Xue et al. (2012) updated the information to early 2000s in the region. Under full irrigation conditions, seasonal ET for winter wheat averaged 710 mm in Bushland, Texas (Musick and Porter, 1990). Xue et al. (2006) reported a similar seasonal ET (691 mm) in a field study in 1993. However, Howell et al. (1995) showed that the seasonal ET under full irrigation was up to 957 mm in a 3-year lysimeter study. Schneider and Howell (2001) showed that seasonal ET at 100% ET requirement averaged 825 mm under centre pivot system. In the NCP, seasonal ET of winter wheat is generally in the range of 400–500 mm, much lower than the U.S. SGP (Hao et al., 2014).

5 Water-use efficiency

In most field studies, WUE is simply defined as the ratio of grain yield and seasonal ET and expressed as kg m^{-3}. Another common unit of WUE is kg ha^{-1} mm^{-1}. Since WUE is the ratio of yield and ET, either higher yield or lower ET can result in higher WUE. However, increasing yield is an ultimate goal for crop production. As such, a higher WUE value should reflect an increased yield, not reduced yield. Economic benefits from increased WUE under water-limited conditions are usually achieved only if yield is maximized for the available water (Sinclair and Muchow, 2001). Frequently, a lower WUE value could be the result of poor management or adverse environmental conditions such as water stress (Musick et al., 1994). In the U.S. SGP, WUE ranged from 0 to 0.8 kg m^{-3}, with an average of 0.4 kg m^{-3} for dryland wheat based on long-term field studies at Bushland, Texas (Jones and Popham, 1997). The major reason for such low WUE values is because of drought stress in dryland wheat in the SGP (Musick and Porter, 1990). In contrast, WUE in irrigated wheat was much higher than that in dryland and ranged from 0.5 to 1.2 kg m^{-3} (Musick et al., 1994; Xue et al., 2006). Therefore, irrigation management plays an important role to increase wheat yield and WUE in the SGP. Since wheat is a relatively drought tolerant crop as compared to corn, deficit irrigation can significantly increase WUE (Musick et al., 1994). In the NCP, WUE in winter wheat is generally higher than that in the U.S. SGP. Hao et al. (2014) summarized the winter wheat irrigation studies in NCP and showed that WUE ranged from 0.9 to 2.3 kg m^{-3}, with an average about 1.6 kg m^{-3}. The main reason for the higher WUE in NCP is high yield (up to 9 Mg ha^{-1}) but relatively low seasonal ET (generally less than 600 mm).

Over the years, WUE has been improved through breeding efforts as evidenced in both the SGP and the NCP. In SGP, Musick and Porter (1990) showed that wheat WUE increased from 0.44 kg m^{-3} with the cultivar Concho grown in 1950s to 0.54 kg m^{-3} with the cultivar Tascosa in the late 1960s and to 0.94 kg m^{-3} for the semi-dwarf cultivar in the early 1980s. Schneider and Howell (2001) reported a higher WUE of 1.14 kg m^{-3} in late 1990s. In the NCP, Zhang et al. (2010) demonstrated wheat WUE increased from 1.0 to 1.2 kg m^{-3} for cultivars from the early 1970s to 1.4–1.5 kg m^{-3} for recently released cultivars in Northern China Plain.

6 Wheat yield, evapotranspiration (ET) and water-use efficiency (WUE) relationships

The wheat yield-ET relationship has been reported in different studies (Musick et al., 1994; Schneider and Howell, 2001; Stone and Schlegel, 2006; Xue et al., 2012; Hao et al., 2014). Musick et al. (1994) summarized the yield-ET relationship based on long-term data in dryland and irrigated plots from 1958 to 1992 at Bushland, Texas. They showed a linear relationship between grain yield and seasonal ET pooling dryland and irrigated data together. The linear regression between yield and ET resulted in a slope of 1.22 kg grain yield per m^3 of seasonal ET (kg m^{-3}) and a threshold of 206 mm ET (Musick et al., 1994). Xue et al. (2012) analysed the yield-ET relationship again by updating the data to late 1990s. Similarly, there was a significant linear relationship between yield and ET, and the regression resulted in a slope of 1.06 kg m^{-3} and a threshold of 164 mm ET ($Y = 0.0106X–1.7393$, $R^2 = 0.75$, $P < 0.001$) (Fig. 5). The new data showed a slightly lower slope but a lower ET threshold as compared to Musick et al. (1994). Stone and Schlegel (2006) summarized dryland wheat data from 1974 to 2004 and showed a linear relationship between wheat yield and ET ($Y = 0.01X–1.838$, $R^2 = 0.64$, $P < 0.0001$). The slope was 1.0 kg

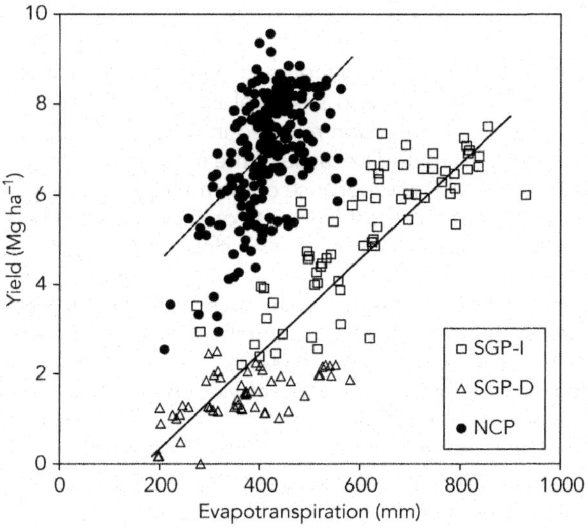

Figure 5 The linear relationship between yield and evapotranspiration (ET) in the U.S. SGP and NCP. I–irirgated; D–dryland. Regression equations: SHP-$Y = 0.0106X–1.7393$, $R^2 = 0.75$, $P < 0.001$; NCP-$Y = 0.0117X + 2.1945$, $R^2 = 0.31$, $P < 0.001$. Sources: Xue et al. (2012); Hao et al. (2014).

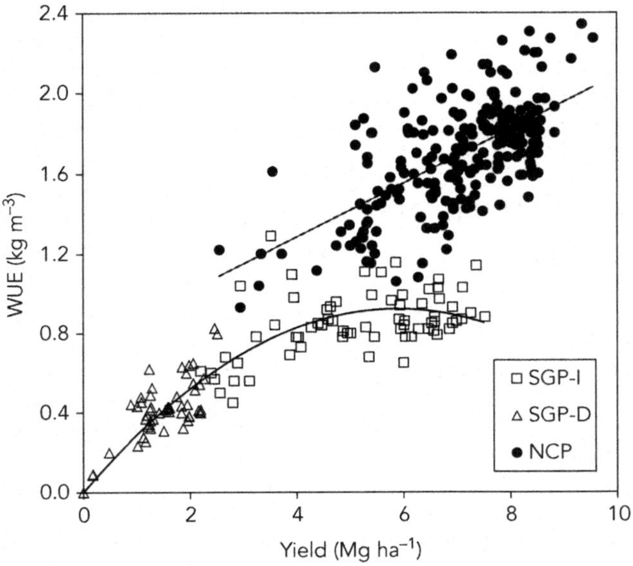

Figure 6 The relationship between WUE and yield in the U.S. SGP and NCP. I–irrigated; D–dryland. Regression equations: SGP-$Y = -0.0267X^2+0.3138X$, $R^2=0.79$, $P < 0.001$; NCP-$Y = 0.1338X + 0.7482$, $R^2 = 0.40$, $P < 0.001$). (Sources: see Fig. 4).

m^{-3} and the ET threshold was 183 mm, which were close to the results from Bushland in Fig. 5. The WUE based on the above linear regression analysis was about 1.0 kg m^{-3} in the SGP region.

Hao et al. (2014) summarized field studies in multiple years in NCP. There was also a linear relationship between yield and seasonal ET ($Y = 0.0117X + 2.1945$, $R^2 = 0.31$, $P < 0.001$) (Fig. 5). The linear regression resulted in a slope of 1.17 kg m^{-3}. However, the ET threshold could not be estimated because of the more scattered data and low R^2 value. The more scattered data in NCP was because of the multi-locations while data in SGP were mainly from one or two locations (Bushland, TX, or Tribune, KS). Nevertheless, the linear regression of yield and ET in NCP had a greater slope (1.17 kg m^{-3}) in NCP than that in SHP (1.06 kg m^{-3}).

Figure 6 showed the relationship between WUE and wheat yield in both SGP and NCP. The combined data showed a clear linear relationship between WUE and yield ($Y = 0.2086X + 0.1175$, $R^2 = 0.75$, $P < 0.0001$). However, the WUE-yield relationship was a quadratic function when the full range of yield was considered for SGP. The WUE increased linearly when yield increased up to 4–5 Mg ha^{-1}. When yield increased further, WUE maximized and even tended to decrease. In NCP, the WUE increased linearly as yield increased. Fig. 6 indicates that higher WUE generally can be achieved with higher yields. However, curvilinear relationship between WUE and yield in the SHP showed that WUE might not be the highest when yield was in the high range. The ET demand could be as high as 12 mm per day and was frequently over 60 mm per week in irrigated wheat due to high winds and associated high vapour pressure deficit (Howell et al., 1995) (Fig. 7). In the NCP, daily ET was generally less than 10 mm (Liu et al., 2002).

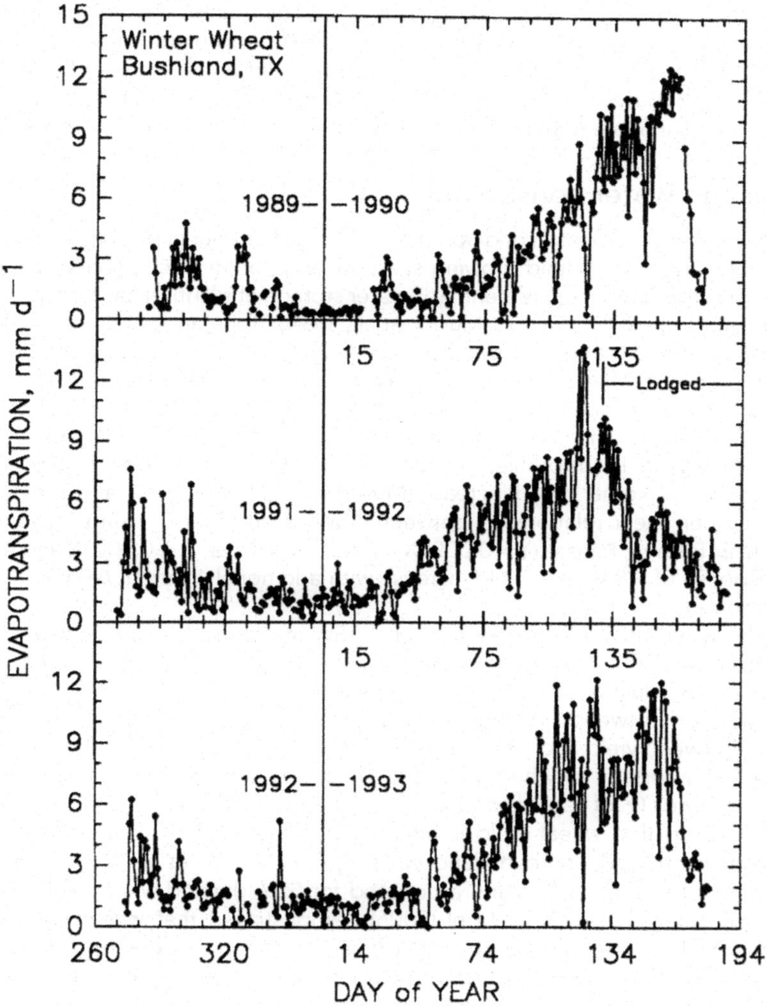

Figure 7 Daily evapotranspiration during winter wheat growing season under fully irrigated conditions in Bushland, Texas (Howell et al., 1995, used with permission).

7 Case studies

The above analysis of yield, ET, WUE and their relationships demonstrates that there are large variations in yield and WUE at different ET levels for winter wheat production. As such, development of management practices is important to improve yield and WUE under water-limited conditions. Generally, there are two ways to improve crop performance: breeding and management practice. Improving wheat yield and WUE through breeding has been a major focus in semi-arid environments (Richards et al., 2002, 2010; Reynolds et al., 2005).

Management practices are equally important as breeding under water-limited conditions (Passioura and Angus, 2010; Richards et al., 2010). Improved crop management is responsible for a large portion of increased productivity under water-limited conditions (Anderson, 2010). Based on the long-term field studies in the SGP and NCP, we provided four case studies that are important for wheat water management in the two regions.

7.1 Soil and water conservation

Under semi-arid environmental conditions in the SGP, dryland wheat production is dependent on both soil water storage and growing season precipitation since precipitation only can meet part of seasonal ET requirement. Soil water storage at planting has long been a major focus in dryland wheat production (Musick et al., 1994; Stone and Schlegel, 2006). Musick et al. (1994) analysed the relationship between wheat yield and available soil water (ASW) at planting (1.8 m profile) based on 34 years of dryland wheat data in Bushland, Texas. There was a linear relationship between yield and ASW at planting ($Y = 0.0157X - 0.94$, $R^2 = 0.34$, $P < 0.001$). The linear regression resulted in a yield response of 1.57 kg m^{-3} to soil water storage in 1.8 m profile. In another word, increasing 1 mm of ASW at planting would lead to 15.7 kg ha^{-1} yield increase (Musick et al., 1994). Stone and Schlegel (2006) showed another linear relationship between ASW at emergence using 30-year dryland wheat data in Tribune, Kansas ($Y = 0.0098X + 0.828$, $R^2 = 0.32$, $P < 0.0001$), and resulted in a yield response of 9.8 kg ha^{-1} yield increase with additional 1 mm of ASW at emergence in 1.8 m profile.

The linear relationships between yield and ASW at planting or emergence emphasize the importance of preseason soil water storage and conservation of precipitation for dryland wheat production. For the linear regressions from Musick et al. (1994), the slope of the regression between yield and ASW at planting was much higher than that of the regression between yield and seasonal ET (1.57 vs. 1.22 kg m^{-3}). This suggests that ASW at planting is important for wheat plants to effectively use precipitation during growing season. The lower R^2 of regression between yield and ASW at planting or emergence (<0.40) indicated that wheat yield is affected by other factors. Seasonal precipitation is also significantly related to dryland wheat yield. Musick et al. (1994) showed that precipitation from October to June accounted for 55% dryland wheat yield variation in Bushland, Texas. Stone and Schelgel (2006) demonstrated that wheat yield increased 8.3 kg ha^{-1} with additional 1 mm of seasonal precipitation. Nevertheless, high yield in dryland wheat only can be achieved with both high ASW at planting and seasonal precipitation. In our recent dryland wheat studies, wheat in 2010–11 season was largely related to soil water storage and yield averaged only 0.65 Mg ha^{-1} because there was only 38 mm of seasonal precipitation (versus 250 mm in a regular season) (Xue et al., 2014). The dryland wheat in 2011–12 was mainly dependent on seasonal precipitation because of low soil water storage due to historic drought in 2011. As a result, wheat yield averaged 0.89 Mg ha^{-1} (Ajayi et al., 2016). In the 2009–10 season, dryland yield average 2.5 Mg ha^{-1} because of more ASW and seasonal precipitation as compared to other two seasons (Xue et al., 2014).

Soil water storage is determined by the amount of precipitation as well as the efficiency of precipitation storage during the fallow period. In the SGP, management practices for increasing soil water storage have focused on crop rotation, tillage systems and residue management (Musick et al., 1994; Jones and Popham, 1997; Stone and Schlegel, 2006; Stewart et al., 2010). Dryland wheat may be grown under different rotation systems such as continuous wheat (CW), wheat-fallow (WF) and wheat-sorghum-fallow (WSF). The tillage systems include conventional tillage and no-till with wheat residue mulch. Jones and Popham (1997) investigated the effects of rotation and tillage systems on dryland wheat and sorghum production and WUE in a 10-year study (1984–93) at Bushland, Texas. Among the different rotation systems, ASW at planting was always lower in CW system than in WF and WSF systems. The 10-year average ASW amounts at planting were 156, 212 and 205 mm for CW, WF and WSF, respectively. For the tillage systems, no-till significantly increased ASW at planting (199 mm) as compared to stubble mulch (183 mm). The precipitation storage efficiency (PSE) was generally low in Bushland, Texas, for wheat. Among the various treatments, WSF generally had higher PSE (about 17%) than WF (11%) and CW under stubble mulch (14%). However, no-till with wheat residue mulch in CW resulted in a higher PSE (20%) than WSF, indicating no-till is an important practice for improving PSE under the Great Plains environment (Jones and Popham, 1997). The above long-term study clearly demonstrated that WSF system under no-till provided significant benefits for improving wheat yield and WUE, as a result of increasing ASW at planting and PSE. On average, annual grain production was 1.8 Mg ha^{-1} for WSF (wheat and sorghum) but only 1.0 Mg ha^{-1} for CW (wheat only) (Jones and Popham, 1997). Norwood (1994) compared conventional and no-till tillage under WF and WSF systems. Although they did not find differences in wheat WUE between conventional tillage and no-till, less evaporation and runoff in no-till system promoted water moving to deeper soil profile in the WF and WSF systems. Stone and Schlegel (2006) sorted wheat yield and WUE data by tillage and showed that no-till significantly increased dryland wheat yield and WUE. The wheat yield response to ET was greater with no-till (1.38 kg m^{-3}) than with conventional tillage (0.86 kg m^{-3}). Stone and Schlegel (2006) also summarized the additional water increase in the soil profile during fallow period as a result of no-till, comparing with conventional tillage in the Great Plains. The soil water increase due to no-till ranged from 15 mm in SGP to 87 mm in central Great Plains, depending on rotation systems.

7.2 Irrigation and irrigation scheduling

Irrigation remains the most effective way to sustain high crop productivity. In the southern Great Plains, irrigation technology has changed significantly in last four decades, from furrow irrigation in early years (1950s–1970s) to current central pivot sprinkler and subsurface drip irrigation systems. The history and trends of irrigation research and development in the region have been reviewed in different eras from 1990s to recent time (Musick et al., 1990; Colaizzi et al., 2008; Evett et al., 2014). The details of irrigation history, economic impact, and research and development trends can be found in these three reviews. Based on these review papers, there are some general consensus for irrigation in the SGP. First, irrigation supplies from Ogallala Aquifer are declining continuously. Second, irrigation efficiency has been increased significantly from 1950s to current time. Third, the future challenge will be how to more efficiently use reduced amount of irrigation in centre pivot system. However, sustaining irrigation is crucial to sustainable crop production in the region because irrigation can significantly increase WUE as compared to dryland (Evett et al., 2014).

In order to increase the efficiency of the irrigation application, irrigation should be scheduled using measurements of crop ET and/or soil moisture depletion. Advanced irrigation scheduling to enhance the WUE includes management of both water and soils. Consequently, irrigation scheduling is largely related to ET when striving to increase the seasonal crop WUE. ET for a specific crop is normally calculated using a crop-specific coefficient (Kc) and the reference ET (ET_o). ET_o is the calculated ET from a reference crop (turf grass or alfalfa) using weather station data including temperature, relative humidity, precipitation, wind speed and solar radiation. Irrigation is also scheduled according to changes in soil moisture. In the past, using ET was not practiced for by many irrigators but ET today can be readily calculated from meteorological data directly from local weather stations. State and commercial service providers in many regions of winter production make available daily ET estimates as part of their outreach programs or consulting service. Recently, estimates of daily and forecasted weekly reference ET are now being made available by the U.S. National Weather Service. International plans for these same estimates are being considering for global locations and applications in the near future. Soil moisture sensors provide irrigators the knowledge of soil moisture fluxes in the root zone, which provides information of when to initiate and terminate irrigation and ability to create a seasonal soil water balance.

The irrigation system of choice must be able to efficiently deliver and apply the amount of water needed to meet the crop water requirement. Knowledge of an irrigation system's efficiency is critical to ensure the crop is uniformly irrigated at the targeted level. Efficiency describes fraction of water stored in the root zone compared to the volume of water delivered by the system, which accounts for drainage below the root zone, evaporation and the uniformity of the application. Irrigation systems used to irrigate wheat include pressurized systems (sprinkler and subsurface drip irrigation) and surface irrigation (furrow and flooded borders), but according to the 2013 Nation Agricultural Statistics Service 2013 survey, 97% of irrigation is applied with a pressurized system of which are predominately sprinkler. For subsurface drip irrigation systems, on expansive soils, very deep cracks develop as wheat matures and dries down. Most producers have discovered that rodents invade the cracks and damage the drip tape resulting in the need for significant repairs. Sprinkler systems include centre pivot, linear move, wheel roll and travelling gun sprinklers. System efficiencies vary between systems; subsurface drip systems generally operate at efficiencies greater than 90% and typically can be in the 95 to 98% range, but the application efficiencies of surface and sprinkler systems can vary greatly. Sprinkler application efficiencies vary generally from 70% to 95%, depending on sprinkler nozzle design, height of the applicator or nozzles, temperature and wind conditions. Surface systems are generally the lowest efficiency system. The drip and center pivot systems offer the greatest opportunity to implement advanced irrigation scheduling management in winter wheat production.

7.3 Deficit irrigation

In the U.S. SGP, with the declining irrigation water supplies from Ogallala Aquifer and increasing energy costs, application of less irrigation water than the plants require for high yield will be the primary practice in the future for irrigated wheat production. Deficit irrigation has been studied and practiced for over three decades and shown to be a viable management practice for improving yield and WUE (Eck, 1988; Musick, et al., 1994; Schneider and Howell, 2001; Xue, et al., 2003, 2006). Deficit irrigation is defined as the application of less water than it is required for full ET and maximum yield (English et al., 1990; Musick, et al. 1994). It is called by a variety of other names such as partial irrigation, regulated deficit irrigation, ET deficit irrigation and limited irrigation (English et al., 1990; Kang et al., 2000; Fereres and Soriano, 2007). The overall goal of deficit irrigation is to increase WUE, either by reducing irrigation frequency or by eliminating the least productive irrigations. Management of deficit irrigation is very different from full irrigation management. Rather than working to minimize crop water stress, plants must be allowed to certain levels of water stress (English et al., 1990; Zhang et al., 1998).

Compared to dryland wheat, deficit irrigation significantly increased grain yield and WUE (Xue et al., 2012). Table 1 showed the responses of wheat yield, WUE, ET, irrigation water-use efficiency (IWUE), and harvest index (HI) to different irrigation levels (from deficit to full irrigation) in two growing seasons under a centre pivot irrigation system. Irrigation application significantly increased wheat yield and WUE as compared to dryland treatment. For example in 1997–98 season, the dryland yield was only 2.5 Mg ha^{-1} and WUE only 0.57 kg m^{-3}. However, yields in irrigated treatments ranged from 4.5 to 7.2 Mg ha^{-1} and WUE from 0.85 to 1.06 kg m^{-3}. Among the irrigation treatments, deficit irrigation (50% ET requirement) resulted in highest WUE and IWUE. For 50% ET treatment, yield was 86–95% of full irrigation (100% ET) (Schneider and Howell, 2001). In another field study, Xue et al. (2006) showed that deficit irrigation of 100 mm at booting stage increased yield by 46% and WUE by 23% as compared to dryland treatment. Compared to full irrigation of 400 mm, a deficit irrigation of 220 mm at jointing and anthesis achieved 84% of the yield at full irrigation and resulted in 45% irrigation water savings. Recently, we used long-term weather data and a simulation model further confirmed these results (Attia et al., 2016). In the NCP, Zhang et al. (1998) showed that wheat yield at one-irrigation (80 mm) was 84–100% of the yield at four-irrigation (320 mm). As a result, WUE was increased to 1.55 kg m^{-3} at one-irrigation treatment from 1.22 kg m^{-3} at four-irrigation treatment

Table 1 Irrigation amounts, yield, evapotranspiration (ET), water-use efficiency (WUE), irrigation water-use efficiency (IWUE) and harvest index (HI) under different irrigation levels in two growing seasons under sprinkler irrigation system (Schneider and Howell, 2001, used with permission)

Crop	Irrigation		Yield	ET	WUE	IWUE	HI
Season	% full ET	mm	Mg ha^{-1}	mm	kg m^{-3}	kg m^{-3}	
1997–98	0%	0	2.46d	432	0.57c	–	0.16d
	25%	109	4.48c	527	0.85b	1.93a	0.25c
	50%	213	6.79b	638	1.06a	2.13a	0.29a
	75%	319	6.43b	745	0.86b	1.30b	0.27b
	100%	419	7.17 a	824	0.87b	1.19b	0.29a
1998–99	0%	0	4.08e	557	0.73b	–	0.28c
	25%	73	4.93d	622	0.80a	1.36a	0.29b
	50%	144	5.85c	697	0.84a	1.26a	0.30ab
	75%	221	6.30b	782	0.81a	1.01b	0.31a
	100%	297	6.79a	825	0.82a	0.88b	0.32a

Growing season precipitation: 212 mm in 1997–98; 334 mm in 1998–99.

(Zhang et al., 1998). More recent studies confirmed the above results in the NCP (Qiu et al., 2008; Zhang et al., 2011).

Many field studies have been conducted in the SGP and NCP to elucidate the mechanisms for increased yield and WUE under deficit irrigation conditions. First, maintaining high ASW at planting and allowing soil drying at an early stage are important for successful practice of deficit irrigation. Xue et al. (2003) showed that mild soil water stress in the spring promoted root growth and development. For example, root length density along the soil profile was higher in dryland than in irrigated treatment at booting. With the deep root system, irrigation can be delayed until booting stage or anthesis if only one-irrigation is allowed (Xue et al., 2003). Similarly, Zhang et al. (1998) and Zhang et al. (2011) also reported that mild water stress at early stage led to a relatively deeper root system in NCP environment.

Second, irrigation must be applied at critical stages for wheat yield determinations when only limited irrigation is allowed. Critical growth stages for irrigating winter wheat generally occur from early spring growth to early grain development in the SHP. When only one-irrigation allowed, irrigation between jointing and anthesis resulted in same yield and WUE. However, irrigation at grain filling did not increase yield and WUE (Xue et al., 2003, 2006). Irrigation at critical stage also reduced the severity of soil water stress and allowed plants to use more water during grain filling (Zhang et al., 1998; Xue et al., 2003, 2006). Xue et al. (2003, 2006) demonstrated that one-irrigation at booting stage maintained high soil water content and significantly reduced water stress at anthesis. As a result, plants in one-irrigation at booting stage had higher photosynthetic rate and stomatal conductance than dryland and one-irrigation at jointing (Xue et al., 2006). Similar results were also found in the NCP (Zhang et al., 1998, 2011).

Third, increased yield and WUE under deficit irrigation is related to increased HI. HI is determined during grain filling by both current photosynthesis and remobilization of pre-anthesis carbon reserve from stems. For the maintenance of current photosynthesis to meet the carbohydrate supply, higher photosynthesis rate and longer green leaf area duration are advantageous under drought conditions. Zhang et al. (2011) showed that deficit irrigation changed wheat canopy structure and increased the proportion of non-leaf organs for photosynthesis. As a result, canopy photosynthetic capacity was improved during grain filling and grain yield was increased under deficit irrigation. When the photosynthesis during grain filling is reduced by drought stress, remobilization of carbon reserves can be important to grain filling. The contribution of remobilized carbon reserves to grain yield in wheat varied from 5% to 90%, depending on the environmental conditions (Asseng and van Herwaarden, 2003; Xue et al., 2006). The increased HI under appropriate deficit irrigation was due to increase in both current photosynthesis and the remobilization of pre-anthesis carbon reserves (Xue et al., 2006).

7.4 Genetic improvement of drought tolerance and WUE

Since drought stress is inevitable during wheat growing season under water-limited conditions, particularly in the U.S. SGP, improving wheat drought tolerance through breeding continuously remains an important part of overall crop improvement (Xue et al., 2012, 2014). Wheat yield has been increasing through breeding in the SGP. For example, regional cultivar trials in the Texas High Plains showed that newly released cultivars such as TAM 111 (released in 2003) and TAM 112 (released in 2005) consistently had higher yields than a historic check, TAM W-101 (released in 1969) under both dryland and irrigated conditions. On average, the yield benefits with newly released cultivars ranged from 15% to 35% over check cultivars (Bean, 2010, 2011; Xue et al., 2014). The yield benefits from two newly released cultivars, TAM 111 and TAM 112, were particularly beneficial in the SGP. TAM 111 and TAM 112 were top two cultivars planted in the Texas High Plains and western Kansas acreage in 2011 and 2012 (NASS, 2011, 2012a, b). Despite the genetic improvement in yield, yield is still the primary trait used in cultivar selection for drought tolerance. Therefore, better understanding physiological mechanisms of drought tolerance and identification of plant traits are important for developing cultivars with improved drought tolerance and WUE.

In last few years, we have been conducting both field and greenhouse experiments to understand the physiological basis of yield determination, drought tolerance and WUE in wheat genotypes. We have mainly focused on cultivars and genotypes developed by Texas A&M AgriLife Research. In a 2-year field study conducted in Bushland, Texas, Xue et al. (2014) investigated the physiological basis of yield determination and WUE in 10 genotypes under dryland and irrigated conditions. The results showed that the newer cultivars such as TAM 111 and TAM 112 had higher yield, biomass, WUE and WUE_{bm} than a relatively older cultivar (TAM 105) under drought. Moreover, TAM 111 and TAM 112 were more drought tolerant than TAM 105 and another drought susceptible line (TX86A5606). The WUE or WUE_{bm} was determined by yield or biomass as genotypic differences in ET were not significant. Biomass at anthesis significantly contributed to increased yield under drought. For dryland wheat, remobilization of stem carbon reserves also contributed to yield. Foulkes et al. (2002) showed that winter wheat cultivars with more drought tolerance had higher HI and more carbon remobilization under drought in the United Kingdom. Our further studies demonstrated that higher yield in drought tolerant cultivars under dryland was related to cooler canopy temperature and more effective use of soil water in deeper profile. In addition, the two drought tolerant cultivars (TAM 111 and TAM 112) had different mechanisms to respond drought. TAM 111 performed well with a relatively high soil moisture conditions at early stage. However, TAM 112 can sustain longer and severe drought conditions (Pradhan et al., 2014a,b). In the NCP of China, Zhang et al. (2010) showed that newer cultivars not only had higher yield but also required less irrigation water, indicating that newer cultivars are more drought tolerant than older cultivars.

Reddy et al. (2014) further conducted greenhouse study to understand the differences in the physiological and transcriptomic responses of TAM 111 and TAM 112 to drought stress during grain filling. Whole-plant data indicated that TAM 112 used more water, produced more biomass and grain yield under water stress compared to TAM 111. Leaf-level data at the grain filling stage indicated that TAM 112 had elevated abscisic acid (ABA) content and reduced stomatal conductance and photosynthesis as compared to TAM 111. Sustained drought stress during the grain filling also resulted in greater flag leaf transcriptome changes in TAM 112 than TAM 111. The two cultivars had very different transcripts associated with photosynthesis, carbohydrate metabolism, phytohormone metabolism and other dehydration responses under drought stress.

Based on these studies, some important physiological traits (e.g. biomass at anthesis, canopy temperature, carbon reserve remobilization and ABA) associated with drought tolerance and WUE can be identified under water-limited conditions in the SGP. Further research is needed to evaluate these traits in large breeding populations and identify QTLs and molecular markers. Because the effective use of soil water is important for ET and yield in the SGP, a better understanding root traits will also be needed in the near future.

8 Future trends and conclusion

Although progress has been made to improve wheat yield and WUE in last few decades in the U.S. SGP, the current WUE level is still low (<1.2 kg m^{-3}) as compared to other winter wheat production areas such as NCP (about 2.0 kg m^{-3}). Therefore, developing better management strategies to improve yield and WUE is still a challenge for agricultural scientists, particularly under changing climate scenario. For dryland wheat production in the SGP, management practices (e.g. no-till, residue mulching and rotation) have resulted in significant gain in soil water storage during fallow periods over the last four to five decades. Future research must address how to efficiently use soil water. Since drought stress is inevitable during wheat growing season, improving drought tolerance through breeding will be an increasingly important part of overall crop improvement. A better understanding of crop response to drought stress and identification of plant traits will lead to development of improved germplasm and cultivars in the region. Over the years, there have been some progresses for better understanding drought tolerance and WUE in winter wheat under drought in the SGP. However, further studies are still needed to identify more traits related to improved drought tolerance and WUE. Currently, the major challenge is high-throughput field phenotyping as more and more genotypes are readily available by genotyping and genomic technologies.

Declining irrigation water continues to challenge irrigated wheat production in the SGP. Although several management factors (improved cultivars, fertilization and pests control, etc.) contributed to wheat yield and WUE improvement, irrigation has played a vital role to increase wheat yield and WUE in the region. Irrigated wheat yields can be 2–4 times higher than dryland yields. Therefore, irrigation will be an important management practice for a long time. As irrigation water becomes more limiting, deficit irrigation will be the primary practice in the future for irrigated wheat production. This is also true in NCP. Currently, irrigation frequency has reduced to 1–2 irrigations per season as compared to historically 4 irrigations during winter wheat growing season. Emerging precision application tools and irrigation systems control technologies currently being developed will enhance and compliment the understanding of practices and genetics as well yield and WUE in winter wheat production.

9 Where to look for further information

More information regarding water management and drought tolerance in wheat can be found by searching the key words in scholar.google.com.

The general crop water management topics also can be found in the websites of the departments of agronomy or soil & crop sciences in many universities such as Texas A&M University, Kansas State University, University of Nebraska and University of California, to name a few.

Popular journals: *Crop Science, Agronomy Journal, Field Crops Research, Functional Plant Biology, European Journal of Agronomy, Agricultural and Water Management, Irrigation Science, Journal of Experimental Botany, Advances in Agronomy* and so on.

Websites: plantstress.com.

Conferences: American Society of Agronomy, Crop Science Society of America, Soil Science Society of America International Annual meeting; American Society of Plant Biologists Annual Meeting; Inter-Drought Conference; American Society of Agricultural and Biosystems Engineering Annual Meeting.

10 References

AgriPartners (2007). Irrigation and Cropping Demonstrations. http://amarillo.tamu.edu/amarillo-center-programs/agripartners/.

Ajayi, S., Reddy, S. K., Gowda, P. H., Xue, Q., Rudd, J. C., Pradhan, G., Liu, S., Stewart, B. A. Biradar, C. and Jessup, K. E. (2016). Spectral reflectance models for characterizing winter wheat genotypes. *Journal of Crop Improvement* 30: 176–95. DOI:10.1080/15427528.2016.1138421.

Anderson, W. K. (2010). Closing the gap between actual and potential yield of rainfed wheat. The impacts of environment, management and cultivar. *Field Crops Research* 116: 14–22.

Attia, A., Rajan, N., Xue, Q., Ibrahim, A. and Hays, D. (2015). Application of DSSAT-CERES-Wheat model to simulate winter wheat response to irrigation management in the Texas High. *Agricultural Water management* 165: 50–60.

Asseng, S. and van Herwaarden, A. F. (2003). Analysis of the benefits to yield from assimilates stored prior to grain filling in a range of environments. *Plant Soil* 256: 217–29.

Bean, B. W. (2010). 2010 wheat variety trials conducted in the Texas and New Mexico High Plains. http://amarillo.tamu.edu/files/2010/11/Wheat-Variety-Descriptions-2010-7_26.pdf.

Bean, B. W. (2011). 2011 wheat variety trials conducted in the Texas and New Mexico High Plains. http://amarillo.tamu.edu/files/2010/11/Wheat-Variety-Trial-Summary-Final2011.pdf.

Blum, A. (2009). Effective use of water (EUW) and not water-use efficiency (WUE) is the target of crop yield improvement under drought stress. *Field Crop Research* 112:119–23.

Bodner, G., Nakhforoosh, A. and Kaul, H. (2015). Management of crop water under drought: a review. *Agronomy for Sustainable Development* 35: 401–42.

Bringle, R. W. and Curtis, B. C. (1987). Wheat worldwide. In Hey, E. G. (ed.), *Wheat and Wheat Improvement*, Second Edition, pp. 1–32. ASA-CSSA-SSSA, Madison, WI.

Colaizzi, P. D., Gowda, P. H., Marek, T. H. and Porter, D. O. (2008). Irrigation in the Texas High Plains: A brief history and potential reductions in demand. *Journal of Irrigation and Drainage Engineering* 58. doi:10.1002/ird.418.

Eck, H. V. (1988). Winter wheat response to nitrogen and irrigation. *Agronomy Journal* 80: 902–8.

English, M. J, Musick, J. T. and Murty, V. V. N. (1990). Deficit irrigation. In *Management of Farm Irrigation Systems*. American Society of Agricultural Engineers, St. Joseph, MI, pp. 631–63.

English, M. J. (1990). Deficit irrigation: Observations in the Columbia Basin. *Journal of Irrigation and Drainage Engineering* 116: 413–26.

Evett, S. R., Colaizzi, P. D., Susan, A. O'Shaughnessy, S. A., Lamm, F. R., Trout, T. J. and Kranz, W. L. (2014). The future of irrigation on the U.S. Great Plains. *Proceedings of the 26th Annual Central Plains Irrigation Conference*, Burlington, CO., 25–26 February 2014. Available from CPIA, 760, N. Thompson, Colby, Kansas, USA.

Fanin, B. (2011). Texas agricultural drought losses reach record $5.2 billion. Texas AgriLife News, 17 August 2011. http://agrilife.org/today/2011/08/17/texas-agricultural-drought-losses-reach-record-5-2-billion/.

Fereres, E., and Soriano, M. A. (2007). Deficit irrigation for reducing agricultural water use. *Journal of Experimental Botany* 58 (2): 147–59.

Fischer, R. A., Byerlee, D. and Edmeades, G. O. (2014). Crop yields and global food security: will yield increase continue to feed the world? *Canberra: Australian Centre for International Agricultural Research*. http://aciar.gov.au/publication/mn158

Foulkes, M. J., Scott, R. K. and Sylvester-Bradley, R. (2002). The ability of wheat cultivars to withstand drought in UK conditions: formation of grain yield. *Journal of Agricultural Science* 138: 153–69.

Graybosch, R. A. and Peterson, C. J. (2010). Genetic improvement in winter wheat yields in the Great Plains of North America, 1959–2008. *Crop Science* 50: 1882–90.

Hao, B., Xue, Q., Zhang, Y. H., Stewart, B. A. and Wang, Z. M. (2014). Deficit irrigation in winter wheat-U. S. Southern High Plains and North China Plain. *Journal of Arid Land Studies* 24–1: 129–32.

Howell, T. A. (2001). Enhancing water use efficiency in irrigated agriculture. *Agronomy Journal* 93: 281–9.

Howell, T. A., Tolk, J. A., Evett, S. R., Copeland, K. S. and Dusek, D. A. (2007). Evapotranspiration of deficit irrigated sorghum and winter wheat. *Proceedings USCID 4th International Conference*, pp. 223–39.

Howell, T. A., Steiner, J. L., Schneider, A. D. and Evett, S. R. (1995). Evapotranspiration of irrigated winter wheat: Southern High Plains. *Transactions of the ASAE* 38: 745–59.

Howell, T. A. (1990). Grain, dry matter yield relationships for winter wheat and grain sorghum – Southern High Plains. *Agronomy Journal* 82: 914–18.

Howell, T. A., Steiner, J. L., Schneider, A. D., Evett, S. R. and Tolk, J. A. (1997). Seasonal and maximum daily evapotranspiration of irrigated winter wheat, sorghum, and corn Southern High Plains. *Transactions of the ASAE* 40: 623–34.

Jones, O. R. and Popham, T. W. (1997). Cropping and tillage systems for dryland grain production in the Southern High Plains. *Agronomy Journal* 89: 222–32.

Kang, S., Zhang, L., Liang, Y., Hu, X., Cai, H. and Gu, B. (2002). Effects of limited irrigation on yield and water use efficiency of winter wheat in the Loess Plateau of China. *Agricultural Water Management* 55: 203–16.

Liu, C., Zhang, X. and Zhang, Y. (2002). Determination of daily evaporation and evapotranspiration of winter wheat and maize by large-scale weighing lysimeter and micro-lysimeter. *Agricultural and Forest Meteorology* 111: 109–20.

Mo, X., Liu, S., Lin, Z. and Guo, R. (2009). Regional crop yield, water consumption and water use efficiency and their responses to climate change in the North China Plain. *Agriculture, Ecosystems and Environment* 134: 67–78.

Musick, J. T and Porter, K. B. (1990). Wheat. In Stewart, B. A. and Nielson, D. R. (eds), *Irrigation of Agricultural Crops*, Agron. Monogr. 30. Madison: ASA-CSSA-SSSA, pp. 597–638.

Musick, J. T., and Dusek, D. A. (1980). Planting date and water deficit effects on development and yield of irrigated winter wheat. *Agronomy Journal* 72: 45–52.

Musick, J. T., Pringle, F. B., Harman, W. L. and Stewart, B. A. (1990). Long-term irrigation trends–Texas High Plains. *Applied Engineering In Agriculture* 6: 717–24.

Musick, J. T., Jones, O. R., Stewart, B. A. and Dusek, D. A. (1994). Water–yield relationships for irrigated and dryland wheat in the U.S. Southern Plains. *Agronomy Journal* 86: 980–6.

NASS. (2011). 2010 Texas Wheat Variety Survey Results. http://www.nass.usda.gov/Statistics_by_State/Texas/Publications/Crop_Reports/Wheat/twheat_var.htm.

NASS. (2016). Winter Wheat: Production Acreage by County https://www.nass.usda.gov/Charts_and_Maps/Crops_County/ww-pr.

NASS. (2012a). Texas 2012 Wheat Variety Results. http://www.nass.usda.gov/Statistics_by_State/Texas/Publications/tx_wheat_varieties.pdf.

NASS. (2012b). Wheat Varieties. http://www.nass.usda.gov/Statistics_by_State/Kansas/Publications/Crops/Whtvar/whtvar12.pdf.

Norwood, C. (1994). Profile water distribution and grain yield as affected by cropping system and tillage. *Agronomy Journal* 86: 558–63.

Passioura, J. B. and Angus, J. F. (2010). Improving productivity of crops in water-limited environments. *Advances in Agronomy* 106: 37–75.

Passioura, J. B. (1977). Grain yield, harvest index and water use of wheat. *Journal of the Australian Institute of Agricultural Science* 43: 117–20.

Pradhan, G., Xue, Q., Liu, S., Rudd, J. C. and Jessup, K. E. (2014a). Effective use of soil water contributed to high yield in wheat in the U.S. Southern High Plains. *Journal of Arid Land Studies* 24–1: 153–6.

Pradhan, G., Xue, Q., Liu, S., Rudd, J. C., Jessup, K. E. and Mahan, J. R. (2014b). Cooler canopy temperature contributed to higher yield in new drought tolerant cultivars. *Crop Science* 54: 2275–84.

Qiu, G., Wang, L., He, X., Zhang, X., Chen, S., Chen, J. and Yang, Y. (2008). Water use efficiency and evapotranspiration of winter wheat and its response to irrigation regime in the north China plain. *Agricultural and Forest Meteorology* 148: 1848–59.

Reddy, S. K., Liu, S., Rudd, J. C., Xue, Q., Payton, P., Finlayson, S. A., Mahan, J., Akhunova, A., Holalu, S. V. and Lu, N. (2014). Physiology and transcriptomics of water-deficit stress responses in wheat cultivars TAM 111 and TAM 112. *Journal of Plant Physiology* 171: 1289–98.

Reynolds, M. P., Mujeeb-Kazi, A. and Sawkins, M. (2005). Prospects for utilising plant-adaptive mechanisms to improve wheat and other crops in drought- and salinity-prone environments. *Annals of Applied Biology* 146: 239–59.

Richards, R. A., Rebetzke, G. J., Condon, A. G. and van Herwaarden, A. F. (2002). Breeding opportunities for increasing the efficiency of water use and crop yield in temperate cereals. *Crop Science* 42: 111–21.

Richards, R. A., Rebetzke, G. J., Watt, M., Condon, A. G., Spielmeyer, W. and Dolferus, R. (2010). Breeding for improved water productivity in temperate cereals: phenotyping, quantitative traits loci, markers and the selection environment. *Functional Plant Biology* 37: 85–97.

Schneider, A. D. and Howell, T. A. (2001). Scheduling deficit irrigation with data from an evapotranspiration network. *Transaction of the ASAE* 44: 1617–23.

Sinclair, T. R. and Muchow, R. C. (2001). System analysis of plant traits to increase grain yield on limited water supplies. *Agronomy Journal* 93: 263–70.

Stewart, B. A. and Peterson, G. A. (2015). Managing green water in dryland agriculture. *Agronomy Journal* 107: 1544–53.

Stewart, B. A., Baumhardt, R. L. and Evett, S. R. (2010). Major advances of soil and water conservation in the U.S. Southern Great Plains. Zoebeck, T. M., Schillinger, W. F.,editors. Soil and Water Conservation Advances in the United States. Special Publication 60. Madison, WI: Soil Science Society of America, Inc., pp. 103–29.

Stone, L. R. and Schlegel, A. J. (2006). Yield–water supply relationships of grain sorghum and winter wheat. *Agronomy Journal* 98: 1359–66.

USDA-ERS. (2015). U.S. share of world wheat exports continues to decline. http://www.ers.usda.gov/data-products/chart-gallery/detail.aspx?chartId=52238.

Wang, B., Zhang, Y., Hao, B., Xu, X., Zhao, Z., Wang, Z. and Xue, Q. (2016). Grain yield and water use efficiency in extremely-late sown winter wheat cultivars under two irrigation regimes in the north china plain. *PLoS ONE* 11(4): e0153695. doi:10.1371/journal.pone.0153695.

Wang, F. Z.e, Sayre, K., Li, S., Si, J., Feng, B. and Kong, L. (2009). Wheat cropping systems and technologies in China. *Field Crops Research* 111: 181–8.

Xue, Q., Liu, W. and Stewart, B. A. (2012). Improving wheat yield and water use efficiency under semi-arid environment – The US Southern Great Plains and China's Loess Plateau. In R. Lal and B. A. Stewart (eds), *Advances in Soil Science: Soil Water and Agronomic Productivity*. Taylor & Francis.

Xue, Q., Zhu, Z., Musick, J. T., Stewart, B. A. and Dusek, D. A. (2003). Root growth and water uptake in winter wheat under deficit irrigation. *Plant Soil* 257: 151–61.

Xue, Q., Zhu, Z., Musick, J. T., Stewart, B. A. and Dusek, D. A. (2006). Physiological mechanisms contributing to the increased water-use efficiency in winter wheat under deficit irrigation. *Journal of Plant Physiology* 163: 154–64.

Zhang, H., Wang, X., You, M. and Liu, C. (1999). Water – yield relations and water-use efficiency of winter wheat in the North China Plain. *Irrigation Science* 19: 37–45.

Zhang, X., Pei, D. and Hu, C. (2003). Conserving groundwater for irrigation in the North China Plain. *Irrigation Science* 21: 159–66.

Zhang, Y., Zhang, Y., Wang, Z. and Wang, Z. J. (2011). Characteristics of canopy structure and contributions of non-leaf organs to yield in winter wheat under different irrigated conditions. *Field Crops Research*123: 187–95.

Zhang, J. H., Sui, X. Z., Li, B., Su, B. L., Li, J. M. and Zhou, D. X. (1998). An improved water-use efficiency for winter wheat grown under reduced irrigation. *Field Crops Research* 59: 91–8.

Zhang, X., Chen, S., Sun, H., Wang, Y. and Shao, L. (2010). Water use efficiency and associated traits in winter wheat cultivars in the North China Plain. *Agriculture Water Management* 97: 1117–25.

Chapter 5

Post-harvest wheat losses in Africa: an Ethiopian case study

Tadesse Dessalegn, Tesfaye Solomon, Tesfaye Gebre Kristos, Abiy Solomon, Shure Seboka and Yazie Chane, Ethiopian Institute of Agricultural Research, Ethiopia; Bhadriraju Subramanyam and Kamala A. Roberts, Kansas State University, USA; and Fetien Abay and Rizana Mahroof, South Carolina State University, USA

1 Introduction
2 Methods
3 Demographics
4 Crop production information
5 Post-harvest issues
6 Preventing post-harvest losses
7 Information flow and training requirements
8 Gender and food security issues
9 Conclusion
10 Acknowledgements
11 Where to look for further information
12 References

1 Introduction

Wheat is one of the major cereals and a staple crop in Africa. Demand for wheat is increasing in sub-Saharan Africa because of income growth, urbanization and dietary diversification (Jayne et al., 2010a; Negassa et al., 2013). Sub-Saharan countries and Africa as a whole produce only about 30% and 40% of their domestic requirements, respectively (Negassa et al., 2013). This results in heavy dependence on imports which makes the region highly vulnerable to global market and supply shocks. In Ethiopia in 2013 wheat was cultivated on 1.7 million ha, with an annual production of 4.04 million tons (Fig. 1) (FAOSTAT, 2015). This represented the largest area of cultivation and highest overall production in sub-Saharan Africa (Table 1). Wheat is grown by more than 4.7 million smallholder farmers in Ethiopia. It is the third most important crop in terms of production, after teff (a grain crop) and maize, and the fourth most important crop in area coverage, after teff, maize and

http://dx.doi.org/10.19103/AS.2016.0004.18

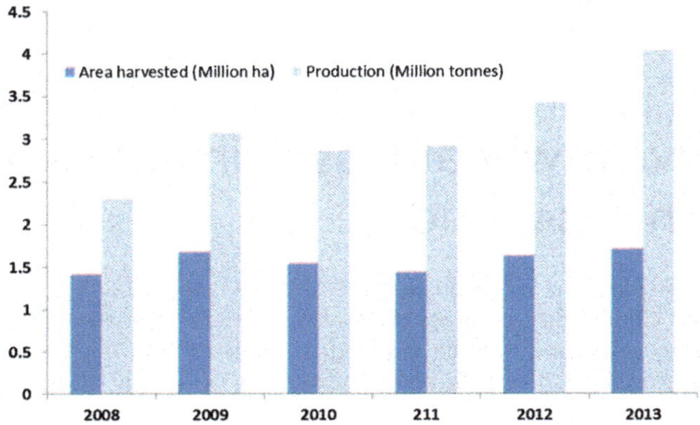

Figure 1 Area and production of wheat in Ethiopia during 2008–2013. (FAOSTAT, http://faostat.fao.org/site/567/DesktopDefault.aspx?PageID=567#ancor, accessed February 2015). The new website is http://faostat3.fao.org.

Table 1 Area (ha), production (tonnes) and rank of wheat in selected countries in sub-Saharan Africa (FAOSTAT, accessed February 2015)

	2011		2012		2013		
Countries	Area	Production	Area	Production	Area	Production	Rank
Angola	3650	4000	3400	4000	3420	3958	
Burundi	11500	9787	9434	4196	8828	6423	
Ethiopia	1437485	2916334	1627647	3434706	1706324	4039113	1
Kenya	131509	268482	148703	441754	160000	485846	3
Malawi	1216	1850	1295	1956	1269	1784	
Mali	9844	33842	10349	40071	6900	27430	
Nigeria	99000	165000	90000	100000	80000	80000	
South Africa	604700	2005000	511000	1915000	500000	1760000	2
Uganda	13000	23000	14000	20000	14200	20000	
Zambia	37631	237332	37209	253522	41810	273584	
Zimbabwe	12000	23000	11000	20000	10000	25000	

sorghum (CSA, 2013). Wheat production is steadily increasing despite serious constraints such as recurrent epidemics of diseases such as rusts and *Septoria* (Singh et al., 2008; Teferi and Gebreslassie, 2015).

Ethiopia's wheat farmers are the greatest producer of wheat in sub-Saharan Africa (Table 1), yet Ethiopia is not self-sufficient in its wheat production and imports an average of more than 1 million tons per annum (Index Mundi, 2016). Once a net exporter of wheat, Ethiopia is now a net importer of wheat due to many factors including war and crop failures (Hailu, 1991), as well as population growth outpacing increases in production (Reuben et al.,

2005). Increasing yield is frequently cited as an important issue for increasing food security (Adugna et al., 1991; Hailu, 1991; van Keulen and Hengsdijk, 2005; Bekele et al., 2009; Jayne et al., 2010b).

Wheat in Ethiopia is predominantly grown in the midland to highland regions of the country, at elevations of 1900–2700 m, in temperatures ranging from 6 to 23°C, with annual rainfall of 35–180 mm (Hailu, 1991; White et al., 2001). Since wheat is mainly a rainfed crop, wheat growers utilize the long rainy season (Meher). This is the main growing season which starts in June and lasts until the rains begin to decrease in October, with the wheat harvest starting in December (Hailu, 1991; White et al., 2001). The soil types found in these areas are clay-rich Vertisols, less clay-rich Nitisols, black soils, red soils and brown soils, which have their own input needs (Asnakew et al., 1991; White et al., 2001).

Increasing domestic production to meet local demands is not enough to make Ethiopia self-sufficient in wheat production. A more meaningful means of improving food security in sub-Saharan Africa would be to improve post-harvest protection practices aimed at protecting the harvested wheat throughout the wheat value chain (Negassa, et al., 2013; World Bank, 2011). Understanding where improvements can be made, and how best to make them, will optimize efforts in these areas. However, but there are gaps in available data (Affognon et al., 2015). These gaps need to be filled if a programme to improve food security in Ethiopia is to be as successful as a recent post-harvest loss eradication programme in Uganda by World Food Programme (Costa, 2015).

Post-harvest losses can be quantitative (e.g. reduced volume or weight of grain), qualitative (e.g. loss of nutritional or processing quality, including contamination with aflatoxins) and economic (e.g. reduced value or access to some markets). According to estimates provided by the African Postharvest Losses Information System (APHLIS), physical grain losses range from 10 to 20% (www.aphlis.net). It has been estimated that overall post-harvest grain losses for sub-Saharan Africa could be as high as US$4 billion/ year, around 15% of total production (World Bank, 2011). This compares to the findings reported for post-harvest losses of wheat in Ethiopia discussed later in this chapter. These suggest high average losses of around 17%, with losses ranging from 14 to as high as 23%.

Post-harvest loss can occur at any stage along the post-harvest value chain. Understanding the circumstances around harvest and post-harvest operations for a given crop will help reduce post-harvest losses and improve the income of farm households. Improvements in post-harvest management practices will help avoid both quantitative and qualitative losses and maintain the quality of the grain for various end uses. It is necessary to develop effective strategies for the post-harvest value chain (field to market) that avoid deterioration in wheat grain quantity and quality, satisfy market demands, and improve the income and food security of smallholder farmers. Reduction in wheat post-harvest losses will also help to offset the costs of importing wheat.

In order to better understand post-harvest issues of Ethiopian wheat farmers, a questionnaire-based survey was conducted in 2014 to understand farmer perceptions about wheat production, potential sources of post-harvest losses from harvesting to marketing and post-harvest loss mitigation measures used by farmers, as well as post-harvest protection information needs. The survey was part of a project funded for a five-year period (2014–2018) by the United States Agency for International Development. This project resulted in setting up of the Feed the Future Innovation Lab for the Reduction of Post-Harvest Loss at Kansas State University, Manhattan, Kansas. The Ethiopian part of the project was also supported by funds from the ADM Institute for Prevention of Postharvest

Loss, University of Illinois, Urbana-Champaign, Illinois. Our work builds on previous surveys of Ethiopian grain farmers by Blum and Bekele (2001), Abebe and Bekele (2003), Pender and Gebremedhin (2006), and Gebremedhin and Hoekstra (2007).

2 Methods

The study was conducted in November 2014. It surveyed wheat growers from four regions of Ethiopia: the Oromia, Amhara, and Tigray regions and the Southern Nations Nationalities and Peoples (SNNP) region. The top 100 wheat growing zones and districts are located in these four regions (Warner et al., 2015). A 13-page survey questionnaire with 65 questions was used to collect primary data from wheat farmers in the study areas. The first and second stages involved a selection of zones and districts that are representative of wheat-based farming systems. Fourteen villages were selected for the study. Carefully selected enumerators pre-tested the questionnaire and later collected data on input use, outputs, post-harvest losses, and socio-economic and farm characteristics. The 200 respondents were from highland (21%), midland (73%) and lowland (6%) ecological zones. The interviews were conducted using structured and semi-structured formats. The estimation of post-harvest losses was carried out using SPSS software (SPSS, 2009) and the post-harvest loss online calculator developed by APHLIS (Hodges et al., 2011, 2014).

3 Demographics

The survey asked numerous questions including gender of household head, number of household members, number of children, size of farm and distance to nearest market.

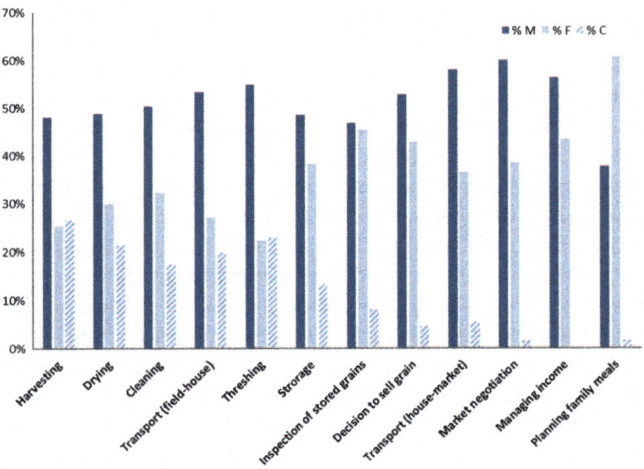

Figure 2 Post-harvest and household division of labour (percentage) by males (solid dark blue), females (solid light blue) and children (diagonal mid-blue).

Participation in post-harvest activities was also included. The age of the household's head, family size and wheat farm size showed variation across the regions studied. The average age of the surveyed wheat farmers was 43 years, but the ages ranged from 20 to 87 years (Table 2). The average farm household had a total of six members.

Farmers had an average of 1.42 ha of land for wheat cultivation during 2014–15. The minimum and the maximum land owned for wheat by the sampled households were 0.13 ha in Amhara and 8 ha in Oromia. Farmers in Oromia, on average, allotted more land for wheat cultivation (2.17 ha) compared to farmers in SNNP (1.1 ha), Amhara (1 ha) and Tigray (0.56 ha). The average distance from home to the nearest market centre was 7 km, with a minimum of less than 1 km, for all regions. The maximum distance reported was 45 km in Amhara. Generally, the wheat farmers in SNNP region travelled shorter distances to market than the other regions. Respondents who reported 0 km distance to the nearest market may have traders coming to the farm gate or village, as is the case for maize growers in Malawi, Zambia and Kenya (Jayne et al., 2010a).

Of the 200 respondents, 92% of households were headed by males and only 8% by females. Men play a major role in wheat harvesting, drying, cleaning, transportation from the field to the farm, threshing, storage, the decision to sell the grain, transportation from

Table 2 Socio-economic characteristics of sampled wheat farmers

Region	Variables	Mean	*N*	Minimum	Maximum
Tigray	Age of the household head	44.2	30	28	68
	Family size	6.8	30	2	10
	Proximity to the nearest market (Km)	7.7	23	0	30
	Wheat farm size (ha)	0.56	30	0.25	2.9
Amhara	Age of the household head	47.2	66	24	78
	Family size	6	66	2	13
	Proximity to the nearest market (Km)	7.1	63	0	45
	Wheat farm size (ha)	1	66	0.13	3
Oromia	Age of the household head	40.7	80	20	87
	Family size	7.2	79	1	21
	Proximity to the nearest market (Km)	7.0	73	0	20
	Wheat farm size (ha)	2.17	79	0.25	8
SNNP	Age of the household head	34.7	15	25	58
	Family size	7.6	15	5	11
	Proximity to the nearest market (Km)	5.7	13	0	15
	Wheat farm size (ha)	1.1	15	0.5	2
Total	Age of the household head	43	191	20	87
	Family size	6.7	190	1	21
	Proximity to the nearest market (Km)	7.0	172	0	45
	Wheat farm size (ha)	1.42	190	0.13	8

homes to the market, market negotiations and managing income (Fig. 2). Both men and women play an equal role in inspecting stored wheat, and women play a major role in planning family meals. Children play a slightly bigger role in harvesting and threshing than women, but their role is minimal in all other post-harvest operations.

4 Crop production information

The following is an overview on farm production by wheat growers in the regions covered by our survey. Crop production is the main source of income for wheat producing farmers. About 82% of respondents had livestock which provided secondary source of income. Twenty-five different varieties of bread and durum wheat were grown in the studied zones across the four regions (data not shown), but there were five main varieties grown across the four study regions (Table 3). Varietal variation of wheat produced across the studied regions reflects subtle differences in local climates, soil types, and grower's needs and priorities (White et al., 2001; Pender and Gebremedhin, 2006; Negassa et al., 2013).

The main selection criteria of wheat varieties from our survey by farmers were seed price, seed availability, productivity, followed by length of crop cycle and stress resistance (Table 4). There was no distinction between insect resistance in the field or during storage. Grain colour was considered the most important criterion for choosing grain for consumption by 91% of respondents, closely followed by kernel size (Danielsen et al., 2014).

Respondents to the survey practised mixed farming, with bread wheat, maize, teff, faba bean, potato, pepper and finger millet being the major crops grown in the 2013/2014 cropping period (Fig. 3). Devereux et al. (2008) noted that even if families can produce a surplus on their land, they will sell most of their crops in order to purchase more calorific and nutritious foods than those they grow. The available literature does not indicate that this practice has changed since the 2006–2008 global food crisis, suggesting that increasing costs of food, from the main harvest period, are outweighed by the benefits of a varied diet, which helps in reducing malnutrition and undernutrition, especially in children below five years of age (Ali et al., 2012; Nguyen et al., 2013; Jones et al., 2014; Ford and Stein, 2016). Mean productivity of the different zones was highest at Arsi, Oromia region (4.1 ton ha^{-1}), followed by Hadia, SNNP region (3.9 ton ha^{-1}), and Semen Shewa, Amhara region (3.8 ton ha^{-1}). The lowest productivity was recorded at western Arsi, Oromia region (2.3 ton ha^{-1}), and eastern Tigray, Tigray region (2.1 ton ha^{-1}).

Table 3 Main wheat varieties, by region

Region	Variety	Number of respondents (%)
Amhara	Kekeba	78 (62)
Oromia	Digalu and Kubsa	114 (25 and 25)
Tigray	Danda'a	14 (57)
SNNP	Digalu	21 (43)

Table 4 Criteria used to select wheat varieties

	Criteria	*N*	Percentage
Farmers	Cost of seed	182	91
	Crop demonstration/research data	14	7
	Disease resistance	92	46
	Drought resistance	110	55
	End use quality	66	33
	Growing period	116	58
	Insect resistance	112	56
	Lodging[1] resistance	75	37.5
	Only variety known	62	31
	Resistance to water logging	94	47
	Salt tolerance	72	36
	Yield per hectare	139	69.5
Cooks	Grain colour	172	91
	Grain not damaged or crushed	153	83
	Grain not damaged by insects	166	66
	Grain not stained	155	85
	Kernel size	161	87

[1]In machinery

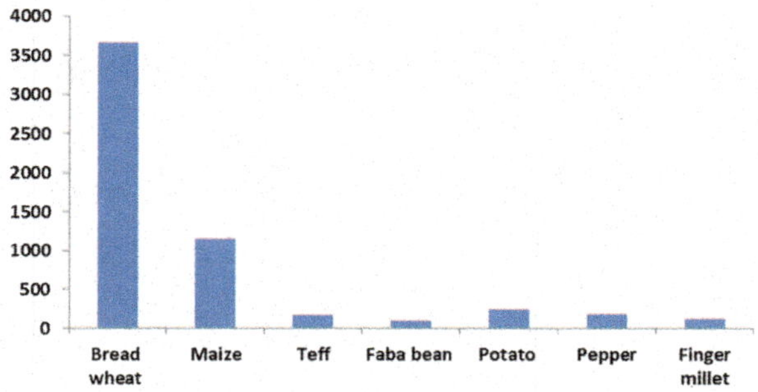

Figure 3 Crop production of surveyed farmers 2013/2014 (tons).

5 Post-harvest issues

The post-harvest period begins when wheat is harvested and ends with sales or consumption of grains. The 'how' and 'where' losses occur will be discussed in this section, followed by 'who does what' as well as current methods of managing post-harvest losses. The losses (kg ha^{-1}) were estimated across wheat zones in different regions (Table 5) based on the

Table 5 Post-harvest loss estimates (kg ha^{-1}) in different zones in surveyed regions

	Tigray		Amhara			Oromia			SNNP	
PH stages	SE Tigray	Eastern Tigray	West Gojam	East Gojam	Semen Shewa	Arsi	West Arsi	Bale	Hadiya	Mean
Harvesting	230.2	142.2	237.0	203.1	257.3	277.6	155.7	223.4	264.0	221.2
Threshing	119.0	73.5	122.5	105.0	133.0	143.5	80.5	115.5	136.5	114.3
Cleaning	72.1	44.5	74.2	63.6	80.6	86.9	48.8	70.0	82.7	69.3
Packaging/bagging	5.4	3.4	5.6	4.8	6.1	6.6	3.7	5.3	6.2	5.2
Transportation (farm to storage)	38.1	23.5	39.2	33.6	42.6	45.9	25.8	37.0	43.7	36.6
Farm storage	90.8	56.1	93.5	80.1	101.5	109.5	61.4	88.1	104.1	87.2
Transportation (storage to market)	6.8	4.2	7.0	6.0	7.6	8.2	4.6	6.6	7.8	6.5
Market storage	3.4	2.1	3.5	3.0	3.8	4.1	2.3	3.3	3.9	3.3
Milling/crushing/grinding	14.6	9.0	15.1	12.9	16.3	17.6	9.9	14.2	16.8	14.0
Total	579.7	358.1	596.8	511.5	647.9	699.1	392.2	562.7	665.0	557.0

overall yield potential. Mean losses across zones and regions were highest at harvesting (221 kg ha^{-1}), threshing (114 kg ha^{-1}) and during farm storage (87.2 kg ha^{-1}).

When wheat grain is handled at harvest and in the post-harvest chain, post-harvest losses occur from a number of different causes (Table 6). High grain moisture (11.8% of respondents), insects (11.6%) and rodents (11.1%) in storage were reported as major causes for post-harvest loss, although severity varied from region to region. The Oromia, Amhara and Tigray regions experienced the most post-harvest losses due to moisture or insects and rodents in storage.

The percentage post-harvest loss of wheat grain at different stages in post-harvest operations was estimated using SPSS, and using the post-harvest loss calculator for two

Table 6 Different causes of wheat post-harvest losses by region

Variable	Count (*N*) Percentage (%)	Region Tigray	 Amhara	 Oromia	 SNNPR	Total
Broken kernel	N	20	10	21	3	54
	%	74.1	37	41.2	30	47
Harvesting method	N	18	15	39	10	82
	%	64.3	48.4	63.9	71.4	61.2
Insect in the field	N	20	21	52	5	98
	%	71.4	58.3	76.5	50	69
Insect in storage	N	25	50	51	2	128
	%	86.2	90.9	81	25	82.6
Moisture	N	25	45	51	9	130
	%	83.3	84.9	81	81.8	82.8
Mould in the field	N	20	9	12	2	43
	%	71.4	45	40	25	50
Mould in storage	N	19	8	19	2	48
	%	76	42.1	52.8	40	56.5
Rodents in the field	N	26	24	39	7	96
	%	89.7	63.2	67.2	87.5	72.2
Rodents in storage	N	27	39	50	7	123
	%	90	84.8	80.6	70	83.1
Spillage	N	17	2	27	4	50
	%	70.8	12.5	56.3	36.4	50.5
Threshing method	N	17	19	40	7	83
	%	65.4	54.3	64.5	53.8	61
Transportation	N	1	11	27	5	44
	%	12.5	55	56.3	45.5	50.6

Table 7 Estimated wheat post-harvest losses at different stages for all surveyed regions

			PHL calculator estimates at two scenarios[2]	
Harvest and PH stages	Wheat losses (%)[1]	N	No rain at harvesting	Rain at harvest
Harvesting	6.8	183	6.8	16.3
Threshing	3.5	178	3.5	3.5
Cleaning	2.1	175	–	–
Packaging/bagging	0.2	168	–	–
Transportation (farm to storage)	1.1	165	1.2	1.2
Farm storage	2.7	180	2.7	2.7
Transportation (storage to market)	0.2	165	1	1
Market storage	0.1	166	2.7	2.7
Milling/grinding	0.4	172	–	–
Total	17.1		14	23

[1] SPSS calculated.
[2] APHLIS online calculated.

Table 8 Post-harvest loss estimates for wheat in Ethiopian Birr (ETB) at harvest in the absence (NR) and presence (WR) of rain for the four regions in 2014. US$1 = 22 ETB

Region	Production (tonnes)	Average price (ETB/ tonne)	Estimated value (ETB)	Average weight loss (%)	Value of weight loss (NR) (ETB)	Average weight loss (%)	Value of weight loss (WR) (ETB)
Tigray	379.5	615.9	233,730	0.14	32,722	0.23	53,758
Amhara	2,060	615.9	1,268,795	0.14	177,631	0.23	279,134
Oromia	4,900	615.9	3,017,861	0.14	422,500	0.23	663,929
SNNPR	588	615.9	362,143	0.14	50,700	0.23	79,671

different scenarios (Table 7). Total post-harvest losses during various stages of handling was estimated to be high at 17.1%, ranging from 14 to 23%, depending on levels of rainfall. In all of the estimates, losses were highest at harvesting (6.8–16.3%), followed by threshing (3.5%) farm and market storage (2.7%). According to respondents, loss at harvesting was mainly due to shattering of the standing grains, if there was no rain at harvest. Excessive rain during harvest spoils the grains due to moulds because farmers leave the harvested stalks in the fields to dry in the sun.

The economic losses due to post-harvest weight loss for the Tigray, Amhara, Oromia and SNNP regions was estimated using APHLIS post-harvest loss calculator (Table 8). The monetary losses that resulted from grain weight loss were Ethiopian Birs or ETB 32722, ETB 177631, ETB 422500 and ETB 50700 for Tigray, Amhara, Oromia and South regions, respectively, when no rain occurred during harvesting. If rain occurs during harvesting, losses increase significantly.

Figure 4 Two gotera, traditional storage units.

Harvesting methods for the sampled wheat farming households were: traditional manual techniques in the Tigray region; 95% traditional in the Amhara region with 5% using combine harvesters; and, in the Oromia region, 80% using combine harvesters and 20% traditional methods. There was no data on harvesting methods in SNNP. The majority of surveyed farmers used oxen for threshing, with some losses due to oxen eating grains. The oxen also urinate and defecate while trampling the grain during threshing. Grain losses during cleaning was marginal, accounted for about 2%.

At harvest farmers determine grain moisture by biting the kernel. Farmers expressed their interest in monitoring moisture control at storage using a moisture meter and dryer if the price of the moisture meter was not a limiting factor. If each woreda (district) or kebele (peasant association) had cooperatively owned dryers, wheat farmers who belonged to these organizations would be able to achieve consistent levels of dryness in their grain. For smallholder farmers, who are more likely to dry their wheat in the sun, harvest timing was important so that the threshed wheat could dry sufficiently. Before wheat can be stored, it needs to be dried to prevent moulds and mycotoxin production and reduce insect damage. Wheat growers in Ethiopia predominantly rely on solar/sun drying. However, there are cabinet dryers used for drying fruits and vegetables and these are currently being modified and optimized to dry grains.

Harvested grain should be stored properly to prevent damage by insects and moulds during storage. Several species of stored product insects and moulds have been reported from wheat and other grains in Ethiopia. Abraham et al. (2008) has provided a comprehensive review of insect pests and moulds associated with stored grains in Ethiopia. Traditional storage units, such as gota, gotera and underground pits (Fig. 4) are suboptimal storage facilities (Blum and Bekele, 2001; Tefera et al., 2011; Costa, 2014), because these structures can be easily infested by insects and rodents, and mould damage is common in improperly dried grains and in grains stored in underground pits because of seepage of moisture. The most popular storage methods were fertilizer bags, gotera (traditional storage structures made from plant materials), jute bags, polypropylene bags and warehouses (Fig. 5). Most farmers reported inspecting their grain once a month using visual and olfactory senses (e.g. to check colour or smell). Respondents stored grain for between 3 to 12 months for different purposes, including consumption, later sale, for seed or for storing surpluses after a big harvest.

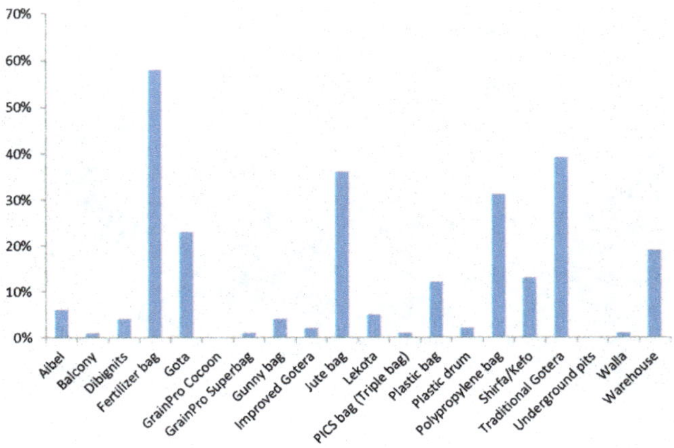

Figure 5 Methods of storing grain (percentages based on 200 farmers).

The storage methods used by respondents vary in their effectiveness and affordability. As previously mentioned, insects play a major role in deterioration of stored wheat grains and post-harvest losses and control measures for these pests need to be devised to minimize post-harvest losses. Gotera are better than nothing, but are not optimal as they are not hermetic and grains are still affected by insect pests and moulds (Costa, 2014). Hermetic forms of storage, e.g. Purdue Improved Crop Storage (PICS) bags, are only effective when used properly. Without training hermetic bags are not properly used (Costa, 2015) and uptake of PICS and similar technologies is limited (Ibro et al., 2014). Once training is provided, post-harvest losses decrease dramatically (Costa, 2014, 2015; Moussa et al., 2014).

Farmers who store their wheat use different storage control methods in order to minimize and avoid losses during storage (Table 9). Among the different kinds of storage

Table 9 Methods used to control storage losses

Method	Percentage usage
Traditional herbs	10
Mix with teff	2
Mix with ashes	2
Mix with sawdust	2
Filter cake	1
Actellic dust	27
Malathion dust	28
Fumigant (phosphine)	35
Fungicides	6
Smoking	6
Drying	79

Table 10 Factors affecting selection of control methods

Factor for selection	Respondents (%)
Traditional practice or custom	79
Ease of use	53
Locally available materials	39
Effectiveness of method	30
Affordable price	28
Prior positive results	17
Received training on	10
Others	7

control methods, 79% of the surveyed farmers use drying, as previously mentioned. Fumigation was the second most popular method for controlling losses from insects. Farmers' preference to select a given control method depended on different factors. These included traditional practices, ease of use, locally availability of the material, control effectiveness and affordability (Table 10).

Lack of training in the use of specific technologies is not always the main issue compared to access to technologies. Costa (2015) found the uptake of metal silos was influenced by farmers' ability to get the silo to their farm. If farmers had to travel to collect a silo, they would be less likely to do so, but a silo delivered to their farm was utilized. There may be similar issues with distance and transportability among respondents to our survey as the average distance to economic centres was 7 km. Farmers do not all have the means of transporting large objects such as a family-size metal silo.

Not all farmers store their wheat; many sell their wheat right after harvest because of an urgent need for cash (Abebe and Bekele, 2003). Cash is needed to pay loans and taxes, even if prices for grains right after harvest are lower than prices in 3–6 months after harvest (Abebe and Bekele, 2003). Farmers allocate cash from grain sales to a diverse range of household needs. Different expenditures indicated by respondents were for clothing, children's education, food, celebrating holidays and savings (Fig. 6). Money for business investment is scarce, indicating the income is insufficient to start such activities. Ethiopian farmers have access to credit systems, but this often involves poorer farmers using credit for food or agricultural inputs, while the poorest cannot access credit systems at all because they lack sufficient collateral (Uraguchi, 2010; World Bank, 2011; Tadesse, 2014; World Bank Group, 2016).

Those farmers that do sell wheat either have to travel to the nearest market or use traders who come directly to the farm gate (Jayne et al., 2010a). Jayne et al. (2010a) found maize growers in Zambia, Kenya and Malawi receive 60–90% of the retail value for their grain from grain traders, indicating a need for greater training in marketing and negotiation among grain growers. The probability that a grain grower owns a mobile phone has been found to increase with increasing distance from markets (Tadesse and Bahiigwa, 2015). Various circumstances may prevent farmers from selling their grain. These range from reluctance to sell due to poor or fluctuating market prices, home consumption, retaining seed for the next season and the need for emergency grain when other food sources were insufficient. The long-term effect of multiple poor harvests is annual reductions in

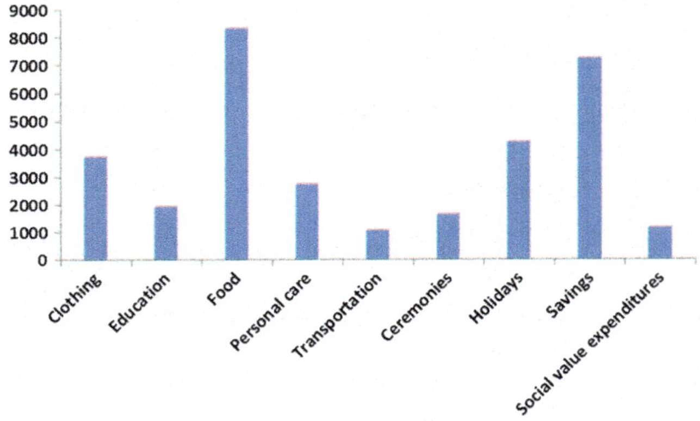

Figure 6 Annual household budget allocations (ETB) (US$1 = 22 ETB).

income, with corresponding reductions in spending or selling off assets (Devereux et al., 2008). Information on where reductions might occur was not collected in our survey. Uraguchi (2010) found Ethiopian and Bangladeshi households affected by the 2007–2008 food price hikes allocated more time looking for cheaper food, reduced the quality and variety of foods consumed, while 16% of Ethiopian households removed children from school.

Respondents said the price of wheat grain varied from ETB 400–900 during the season – the highest prices were noted after storage and reached ETB 1000–1500 at some marketplaces. This level of price variation indicates wheat growers would benefit greatly from either improved on-farm storage practices or warehouse storage systems (Abebe and Bekele, 2003; Jayne et al., 2010b). Farmers who owned cell phones were more likely to sell their grain to traders than cooperatives (Tadesse and Bahiigwa, 2015).

Budget allocations for household and farm were considered separately. For the year of the survey, the averages for fertilizer and herbicide expenditure were ETB 2523 and ETB 1683, respectively (*N* = 191 farmers). Widely used varieties of wheat often require extra inputs to reach their production potential, mostly fertilizers (Adugna et al., 1991; Negassa et al., 2013) which are not always affordable or easily accessible (unless they are organic fertilizers, which can conflict with other needs – animal feed, cooking fuel, etc.) (Assefa, 2005; Tadesse, 2014).

At the time of our survey, respondents said grain unfit for human consumption was mostly fed to animals. Blum and Bekele (2001) found some farmers could not afford to waste any grain and ate damaged grain themselves. The effects on their health depend on the type of damage to the grain. Insect damage reduces the nutritional quality of the grain, while grains affected by moulds such as aflatoxins can make consumers very ill (Wagacha and Muthomi, 2008). Illness from eating damaged wheat was reported by 3% of the respondents to our survey, but they did not specify which household members or how many had been ill. Improving on-farm/in-house storage conditions for wheat will not only reduce post-harvest losses. It will reduce, if not eliminate, the frequency or need to consume damaged or contaminated grains, increasing food and nutritional security (Ali et al., 2012; Nguyen et al., 2013; Jones et al., 2014; Ford and Stein, 2016).

6 Preventing post-harvest losses

It is clear that post-harvest losses are an issue of national food security for Ethiopia. Reducing post-harvest losses across wheat growing regions requires awareness of the issues, plus a variety of inputs and strategies, which are discussed below.

7 Information flow and training requirements

As far as current trends in provision of information and training to farmer training are concerned, respondents were unsatisfied with existing levels of support. They had a long list of needs. These included more information about weather conditions (particularly during the harvest period), better moisture measurement, better storage techniques (including using pesticides during storage) and better market information. Farmers expressed their eagerness to receive training in the future in these areas. When it came to training on post-harvest losses, more than half of the respondents (58%) claimed they had not received training or information in any form on post-harvest loss prevention.

This indicates that the awareness of the factors attributing to post-harvest loss was low among respondents, limiting their ability to take measures to reduce losses. Increasing farmer training through large meetings or farmer field schools can increase participants' food security (Larsen and Lilleør, 2014), especially if there is coverage of what most interests farmers. Increasing education and training in post-harvest loss reduction, and the research behind it, will require continued funding from local and international bodies (Kitinoja et al., 2011; World Bank, 2011). However, it is not just a case of providing more training opportunities or training in local languages. More targeted training is required in the areas specified by the respondents, including training targeted towards female farmers. Rugumamu (2009) found post-harvest losses experienced by maize farmers were a function of gender roles and technologies employed.

Ethiopia's wheat farmers, and all farmers across sub-Saharan Africa, would greatly benefit from reliable seasonal forecasts (Coe and Stern, 2011). Information on when rains are likely to start will help them decide when to harvest and how to dry their wheat, as the timing of the harvest affects the nutritional qualities of the grain. Respondents in Kassie et al. (2013) received most of their weather forecast information from village meetings but found the process unreliable.

Almost all wheat farmers were engaged or involved in one or more forms of farmers' organizations, that is, general cooperatives, unions, development associations, research or crop production groups, saving association, woman and youth associations. The different organizations have different purposes and services for the community. Farmers believed that the most trusted crop production and stored grain management information was received from development agents. Other organizations such as offices of agriculture at different levels, research centres, projects such as the East Africa Agricultural Productivity Project, neighbouring farmers, farmers training centres, seed enterprises and farmers' unions were also perceived as reliable sources of information about crop production. Ethiopia's long history of government-run extension programmes (Adugna et al., 1991) places government researchers, extension officers and farmers in good stead for further capacity building in post-harvest loss prevention (World Bank, 2011).

Table 11 Preferred methods for receiving information ($N = 200$)

	N	%
Cell phones	68	34
Demonstration trials	124	62
Fellow farmers	136	68
Input suppliers	89	44.5
Internet	52	26
Large meetings	176	88
One-on-one delivery	126	63
Printed materials	81	40.5
Radio programmes	127	63.5
Religious leaders	73	36.5
Television programmes	79	39.5

The five most preferred and used methods of receiving new information by farmers were large meetings, fellow farmers, radio programmes, one-on-one delivery and demonstration trials (Table 11). Grain traders also rely on radio programmes for supply and demand information, national policy changes and external market grain prices (Gabre-Madhin and Amha, 2005). Some educated farmers have used internet sources to gain information (Gebremedhin and Hoekstra, 2007; Bekele et al., 2009; World Bank, 2011). Tedesse and Bahiigwa (2015) observed that farmer age and education level were correlated with cell phone ownership and use, indicating that as younger, more educated farmers take over family farms, extension organizations might consider distributing some information via cell phones.

8 Gender and food security issues

It has been suggested that 'Capacity development is critical to achieving lasting change' (Costa, 2014). Our survey included 16 female lead households and even though one third of them were members of local farmer's associations, their involvement is likely to be limited. Taboos around what women can and cannot do inhibit female farmers from ploughing their own fields, with or without oxen, and their inclusion in agricultural extension programmes is rare (Pender and Gebremedhin, 2006). Increasing membership to women farmer associations and women only extension programmes will go some way to improving food security for female lead households.

Improvements for female lead households, and poorer households in general, are linked to pressures to sell grain immediately after harvest. As previously mentioned, price of food increases over the course of the year, being lowest just after harvest and steadily increasing until the next harvest (Abebe and Bekele, 2003; Devereux, 2008). Selling grain immediately may solve the issue of on-farm or household storage but then replace it with issues of food insecurity during the lean months leading up to the next harvest (Devereux,

2008; World Bank, 2011). For farmers who do produce surplus grains, but are unable to store it, an alternative to selling it immediately would be utilizing local warehouse receipt systems (Abebe and Bekele, 2003). Where they exist, warehouse receipt systems allow grain farmers to store surplus grain at a centralized warehouse until they need their grain for sales or consumption. Using warehouse receipt systems has the potential to even out grain prices across the year, levelling out grain prices, improving livelihoods and food security on average and providing farmers access to these systems (Abebe and Bekele, 2003; World Bank, 2011).

Metal silos are highly effective at storing grains, beans and cowpeas (Costa, 2014), but the high upfront price is prohibitive for smaller producers without subsidies or loan systems (Tefera et al., 2011). The success of metal silos in Kenya, Uganda and Burkina Faso is encouraging, even if initial uptake is slower in sub-Saharan Africa than it was in Latin America (World Bank, 2011). It is noticeable that female farmers in Latin America have experienced increased status and self-esteem with the introduction of metal silos, as they are in charge of the silo's contents (SDC, 2008). Access to credit, extension programmes and membership of farmer associations increased uptake of new maize storage in Mozambique (Cunguara and Darnhofer, 2011). However, female farmers in Kenya were found to be constrained in their ability to take up new technologies such as improved storage due to poor access to credit (Ndiritu et al., 2014).

An initial assessment of the use of PICS bags found them to be cost-effective for maize producers in Kenya, Ghana, Tanzania, Malawi and Mozambique, especially if used for more than one season (Jones et al., 2011). Ibro et al. (2014) found that female cowpea farmers in Niger, Nigeria and Bukina Faso were more likely to use PICS if they received training and the bags were locally available. If PICS bags are made available to Ethiopian wheat farmers and proper training is provided, they may be a cost-effective and viable means of wheat storage for female and poorer wheat growers. It is important to note, in this respect, that technology targeting one aspect can create new issues in another area because increased efficiencies in one area might increase the workload further along the post-harvest chain, placing some household members at a disadvantage (Beuchelt and Badstue, 2013)

9 Conclusion

Food security in Ethiopia is an increasingly important issue and reducing post-harvest losses is vital to increasing the nation's food security. Saying that post-harvest losses need to be reduced is easy enough, but actually reducing the losses and improving food security is proving to be a challenge. It is not impossible, but there is also no 'one-size-fits-all' solution.

10 Acknowledgements

The authors thank the following people and organizations for their assistance and support: USAID Feed the Future Innovation Lab for the Reduction of Post-Harvest Losses; Kansas State University; Ethiopian partner organizations (each by name); names of enumerators.

11 Where to look for further information

FAO (2015), Regional strategic framework: reducing food losses and waste in the Near East and North Africa region. Food and Agriculture Organization of the United Nations, Cairo.

Kiaya, V. (2014), Post-harvest losses and strategies to reduce them. Technical paper on Post-harvest losses. Action Contre la Faim (ACF) International.

Rosegrant, M. W., Magalhaes, E., Valmonte-Santos, R. A. and Mason-D'Croz, D. (2015), Returns to investment in reducing postharvest food losses and increasing agricultural productivity growth: post-2015 consensus. Working Paper. Copenhagen Consensus Center.

USDA (2015), Record 2014/2015 Wheat Harvest Expected for Ethiopia. 23 January 2015. http://pecad.fas.usda.gov/highlights/2015/01/Ethiopia/Index.htm (accessed 28 April 2016).

12 References

Abebe, H. G. and Bekele, H. (2003), Farmers' post-harvest grain management choices under liquidity constraints and impending risks: implications for achieving food security objectives in Ethiopia, http://www.eap.gov.et/sites/default/files/post%20harves%20grain%20mgt.%20Abebe.%20H.pdf (accessed 29 February 2016).

Adugna, H., Workneh, N. and Bisrat, R. (1991), Technology transfer for wheat production in Ethiopia. In Hailu, G.-M., Tanner, D. G. and Mengistu, H. (Eds), *Wheat Research in Ethiopia: A Historical Perspective*. IAR/CIMMYT: Addis Ababa, Ethiopia, pp. 277–99.

Affognon, H., Mutungi, C., Sanginga, P. and Borgemeister, C. (2015), Unpacking postharvest losses in sub-Saharan Africa: a meta-analysis, *World Dev.*, 66, 49–68.

Ali, D., Rawat, R., Subandoro, A. and Menon, P. (2012), Infant and young child feeding (IYCF) practices in Ethiopia and Zambia and their association with child nutrition: analysis of demographic and health survey data, *Afr. J. Food Agr. Nutr. Dev.*, 12(2), 5895–914.

Asnakew, W., Tekaligne, M., Mengesha, B. and Tefera, A. (1991), Soil fertility management studies on wheat in Ethiopia. In Hailu, G.-M., Tanner, D. G. and Mengistu, H. (Eds), *Wheat Research in Ethiopia: A Historical Perspective*. IAR/CIMMYT: Addis Ababa, Ethiopia, pp. 137–72.

Assefa, A. (2005), Soil nutrient flows and integrated nutrient management in the northern Ethiopian Highlands. In Pender, J., Ruben, R., Jabbar, M. and Gabre-Madhin, E. (Eds), *EDRI-ESSP Policy Working Paper No. 1: Policies for Improved Land Management and Agricultural Market Development in the Ethiopian Highlands*. Workshop Summary Paper, Ethiopian Development Research Institute, August 2005.

Bekele, A., Viljoen, M. F., Ayele, G. and Ali, S. (2009), Effect of farm size on efficiency of wheat production in Moretna-Jirru district in central Ethiopia, *Indian J. Agr. Econ.*, 64(1), 133–43.

Beuchelt, T. D. and Badstue, L. (2013), Gender, nutrition- and climate-smart food production: opportunities and trade-offs, *Food Sec.*, 5, 709–21.

Blum, A. and Bekele, A. (2001), Ethiopian farmers' practices of on-farm grain storage, *Trop. Sci.*, 41, 23–30.

Central Statistical Agency (CSA) (2014), Area and production of major crops private peasant holdings, meher season, May 2014. Addis Ababa.

Coe, R. and Stern, R. D. (2011), Assessing and addressing climate-induced risk in sub-Saharan rainfed agriculture: lessons learned, *Exp. Agr.*, 47(2), 395–410.

Costa, S. J. (2014), *Reducing Food Losses in Sub-Saharan Africa (Improving Post-Harvest Management and Storage Technologies of Smallholder Farmers): An 'Action Research' Evaluation Trial from Uganda and Burkina Fuso*, UN World Food Programme: Kampala, Uganda.

Costa, S. J. (2015), *Taking it to Scale: Post-Harvest Loss Eradication in Uganda 2014–2015*, UN World Food Programme: Kampala, Uganda.

Cunguara, B. and Darnhofer, I. (2011), Assessing the impact of improved agricultural technologies on household income in rural Mozambique, *Food Policy*, 36(3), 378–90.

Danielsen, K., Wong, F., Hunter, C., Kamminga, E., Mukhopadhyay, M., Mudege, N. N. and Badstue, L. (2014), *Wheat Gender Audit 2013 Summary Report*, CRP Wheat, http://maize.org/wp-content/uploads/sites/5/2014/11/FINAL-Gender-Audit-Summary-report-MAIZE-191114.pdf

Devereux, S., Vaitla, B. and Hauenstein Swan, S. (2008), *Seasons of Hunger: Fighting Cycles of Starvation Among the World's Rural Poor*, Pluto Press: London, GBR.

FAOSTAT (2015), *Area and Production of Wheat*, http://faostat.fao.org/site/567/DesktopDefault.aspx?PageID=567#ancor. Updated: 04 August 2014.

Ford, N. D. and Stein, A. D. (2016), Risk factors affecting child cognitive development: a summary of nutrition, environment, and maternal-child interaction indicators for sub-Saharan Africa, *J. Dev. Orig. Health Dis.*, 7(2), 197–217.

Hailu, G.-M. (1991), Wheat production and research in Ethiopia. In Hailu, G.-M., Tanner, D. G. and Mengistu, H. (Eds), *Wheat Research in Ethiopia: A Historical Perspective*, IAR/CIMMYT: Addis Ababa, Ethiopia, pp. 1–15.

Hodges, R. J., Buzby, J. C. and Bennett, B. (2011), Postharvest losses and waste in developed and less developed countries: opportunities to improve resource use, *J. Agr. Sci.*, 149, 37–45.

Hodges, R. J., Bernard, M. and Rembold, F. (2014), APHLIS – Postharvest Cereal Losses in sub-Saharan Africa, their Estimation, Assessment and Reduction, Joint Research Centre, European Union: Luxembourg.

Ibro, G., Sorgho, M. C., Idris, A. A., Moussa, B., Baributsa, D. and Loweberg-DeBoer, J. (2014), Adoption of cowpea hermetic storage by women, *J. Stored Prod. Res.*, 58, 87–96.

Index Mundi (2016), *Ethiopia Wheat Imports by Year*. http://www.indexmundi.com/agriculture/?country=et&commodity=wheat&graph=imports. (accessed February 2016).

Jayne, T. S., Mason, N., Myers, R., Ferris, J., Mather, D., Beaver, M., Lenski, N., Chapoto, A. and Boughton, D. (2010a), Patterns and trends in food staples markets in Eastern and Southern Africa: toward the identification of priority investments and strategies for developing markets and promoting smallholder productivity growth. MSU International Development Working Paper, No. 104.

Jayne, T. S., Mather, D. and Mghenyi, E. (2010b), Principal challenges confronting smallholder agriculture in sub-Saharan Africa, *World Dev.*, 38(10), 1384–98.

Jones, M., Alexander, C. and Lowenberg-DeBoer, J. (2011), An initial investigation of the potential for hermetic Purdue improved crop storage (PICS) bags to improve incomes for maize producers in sub-Saharan Africa. Working Paper #11–3, Department of Agricultural Economics, Purdue University, West Lafayette, Indiana.

Jones, A. D., Ickes, S. B., Smith, L. E., Mbuya, M. N. N., Chasekwa, B., Heidkamp, R. A., Menon, P., Zongrone, A. A. and Stoltzfus, R. J. (2014), World Health Organization infant and young child feeding indicators and their associations with child anthropometry: a synthesis of recent findings, *Matern. Child Nutr.*, 10, 1–17.

Kassie, B. T., Hengsdijk, H., Roetter, R., Kahiluoto, H., Asseng, S. and van Ittersum, M. (2013), Adapting to climate variability and change: experiences from cereal-based farming in the Central Rift and Kobo Valleys, Ethiopia, *Environ. Manag.*, 52, 1115–31.

Kitinoja, L., Saran, S., Roy, S. K. and Kader, A. A. (2011), Postharvest technology for developing countries: challenges and opportunities in research, outreach and advocacy, *J. Sci. Food Agr.*, 91, 597–603.

Larsen, A. F. and Lilleor, H. B. (2014), Beyond the field: the impact of farmer field schools on food security and poverty alleviation, *World Dev.*, 64, 843–59.

Moussa, B., Abdoulaye, T., Coulibaly, O., Baributsa, D. and Lowenberg-DeBoer, J. (2014), Adoption of on-farm hermetic storage for cowpea in West and Central Africa in 2012, *J. Stored Prod. Res.*, 58, 77–86.

Ndiritu, S. W., Kassie, M. and Shiferaw, B. (2014), Are there systematic gender differences in the adoption of sustainable agricultural intensification practices? Evidence from Kenya, *Food Policy*, 49, 117–27.

Negassa, A., Shiferaw, B., Koo, J., Sonder, K., Smale, M., Braun, H. J., Gbegbelegbe, S., Guo, Z., Hodson, D., Wood, S., Payne, T. and Abeyo, B. (2013), *The Potential for Wheat Production in Africa: Analysis of Biophysical Suitability and Economic Profitability*, CIMMYT: Mexico, D. F.

Nguyen, P. H., Avula, R., Ruel, M. T., Saha, K. K., Ali, D., Tran, L. M., Frongillo, E. A., Menon, P. and Rawat, R. (2013), Maternal and child dietary diversity are associated in Bangladesh, Vietnam, and Ethiopia, *J. Nutr.*, 143, 1176–83.

Pender, J. and Gebremedhin, B. (2006), Land management, crop production, and household income in the highlands of Tigray, Northern Ethiopia: an econometric analysis. In Pender, E. et al. (Eds), *Strategies for Sustainable Land Management in the East African Highlands*, International Food Policy Research Institute, Washington DC, USA, pp. 107–39.

Reuben, R., Kruseman, G. and Tesfay, G. (2005), Policies and technologies for sustainable land management in the Ethiopian Highlands. In Pender, J., Ruben, R., Jabbar, M., Gabre-Madhin, E. (Eds), *EDRI-ESSP Policy Working Paper No. 1: Policies for improved land management and agricultural market development in the Ethiopian Highlands.* Workshop Summary Paper, Ethiopian Development Research Institute, August 2005.

Rugumamu, C. P. (2009), Assessment of post-harvest technologies and gender relations in maize loss reduction in Pangawe village Eastern Tanzania, *Tanz. J. Sci.*, 35, 67–75.

Singh, R. P., Hodson, D. P., Huerta-Espino, J., Jin, Y., Njau, P., Wanyera, R., Herrera-Foessel, S. A. and Ward, R. W. (2008), Will stem rust destroy the world's wheat crop? *Adv. Agron.*, 98, 271–309. https://doi.org/10.1016/s0065-2113(08)00205-8

Tadesse, M. (2014), Fertilizer adoption, credit access, and safety nets in rural Ethiopia, *Agr. Finance Rev.*, 74(3), 290–310.

Tadesse, G. and Bahiigwa, G. (2015), Mobile phones and farmers' marketing decisions in Ethiopia, *World Dev.*, 68, 296–307.

Tefera, T., Kanampiu, F., de Groote, H., Hellin, J., Mugo, S., Kimenju, S., Beyene, Y., Boddupalli, P. M., Shiferaw, B. and Banziger, M. (2011), The metal silo: an effective grain storage technology for reducing post-harvest insect and pathogen losses in maize while improving smallholder farmers' food security in developing countries, *Crop Prot.*, 30, 240–5.

Teferi, T. A. and Gebreslassie, Z. S. (2015), Occurrence and intensity of wheat *Septoria tritici* blotch and host response in Tigray, Ethiopia, *Crop Prot.*, 68, 67–71.

Uraguchi, A. B. (2010), Food price hikes, food security, and gender equality: assessing the roles and vulnerability of women in households of Bangladesh and Ethiopia, *Gend. Dev.*, 18(3), 491–501.

van Keulen, H. and Hengsdijk, H. (2005), Agroecological analyses: production potential for Tigray Region. In Pender, J., Ruben, R., Jabbar, M. and Gabre-Madhin, E. (Eds), EDRI-ESSP Policy Working Paper No. 1: Policies for Improved Land Management and Agricultural Market Development in the Ethiopian Highlands. Workshop Summary Paper, Ethiopian Development Research Institute, August 2005.

Wagacha, J. M. and Muthomi, J. W. (2008), Mycotoxin problem in Africa: current status, implications to food safety and health and possible management strategies, *Int. J. Food Microbiol.*, 124, 1–12.

Warner, J., Stehulak, T. and Kasa, L. (2015), Woreda-Level crop production rankings in Ethiopia: A pooled data approach, International Food Policy Research Institute: Addis Ababa, Ethiopia. http://ebrary.ifpri.org/cdm/ref/collection/p15738coll2/id/130090 (accessed 9 May 2016).

White, J., Tanner, D. G. and Corbett, J. D. (2001), *An Agro-Climatological Characterization of Bread Wheat Production Areas in Ethiopia.* NRG-GIS Series 1, CIMMYT: Mexico, D. F.

World Bank (2011), Missing food: The case of post-harvest grain losses in sub-Saharan Africa, The World Bank.

World Bank Group (2016), Global Economic Prospects, January 2016: Spillovers amid Weak Growth, World Bank: Washington, DC.

Part 2

Wheat crop management

Chapter 6

Wheat crop modelling to improve yields

J. R. Guarin and S. Asseng, University of Florida, USA

1 Introduction

2 Extrapolation from crop models in time and space

3 Experiments for testing crop models

4 Adapting crop models for climate change impact

5 Natural resource management

6 Managing seasonal variability

7 Decision support for farmers

8 Future trends

9 Conclusion

10 Where to look for further information

11 Acknowledgements

12 References

1 Introduction

The global population is continuously growing, and agricultural practices will have to be adapted to match the rapidly increasing demand for food to ensure sustainability and survival (Godfray et al., 2010). This chapter discusses the use of crop modelling as a tool to explore adaptation options. Crop modelling is 'a formal way to present mathematical algorithms that represent quantitative knowledge about how a crop grows in interaction with its environment' (Asseng et al., 2014, p. 102). Crop models use dynamic simulations based on the principles of crop physiology, soil science and agro-meteorology (Loomis et al., 1979). Models often simulate crop development dynamics (phenology), biomass accumulation, yield, water use and nutrient uptake using daily solar radiation, maximum and minimum temperatures, rainfall, initial soil conditions, soil characteristics, cultivar characteristics and crop management.

Crop modelling originated in the 1960s when scientists first developed mathematical relationships between solar radiation and biomass growth (De Wit, 1965; Monteith, 1965b; Duncan et al., 1967). In the early 1970s, initial applications of crop models began in agricultural research (Hesketh et al., 1971), eventually leading to the development of ARC-WHEAT1 (Porter, 1984) and CERES-Wheat (Ritchie, 1985) in the 1980s. Also in the

http://dx.doi.org/10.19103/AS.2016.0004.27

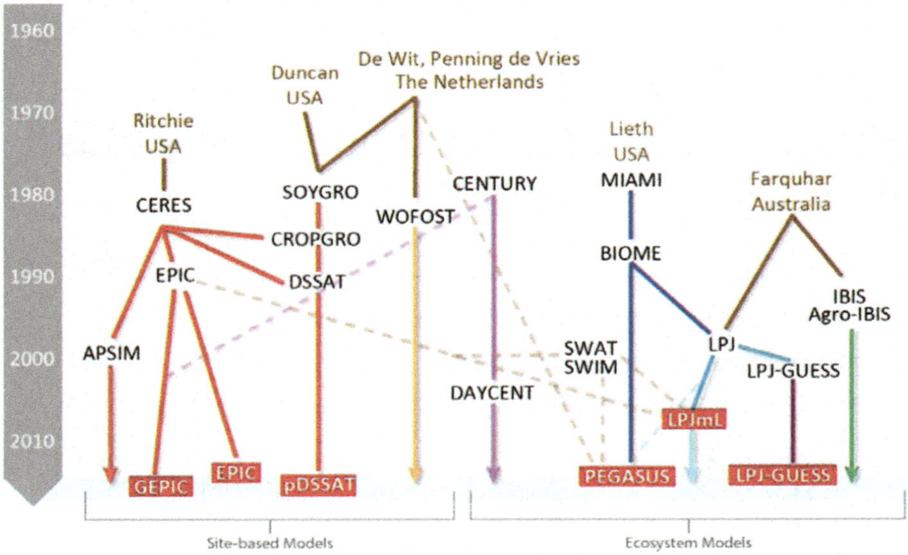

Figure 1 Crop model genealogy for site-based and ecosystem models (including wheat models).
Source: Modified after Rosenzweig et al. (2014).

1980s, Farquhar et al. (1980) developed a detailed photosynthesis model still used in many crop models today. In the 1990s, researchers began incorporating climate change impacts into crop models to assess the effects of atmospheric carbon dioxide (CO_2) (Rosenzweig and Parry, 1994). At that time, models also started merging into crop modelling platforms. Crop modelling platforms combine models of different crops with software that facilitates evaluation and application of crop models for different purposes (Jones et al., 2003). Crop modelling platforms include the Environmental Policy Integrated Climate (EPIC) model (Kiniry et al., 1995), Decision Support System for Agrotechnology Transfer (DSSAT) (Jones et al., 2003) (http://dssat.net/), Agricultural Production Systems Simulator (APSIM) (Keating et al., 2003) (http://www.apsim.info/Wiki/), CropSyst (Stockle et al., 2003) (http://www.bsyse.wsu.edu/CS_Suite/CropSyst/), Simulateur multidisciplinaire pour les Cultures Standard (STICS) (Brisson et al., 2003) (http://www7.avignon.inra.fr/agroclim_stics_eng/) and 'Wageningen crop models' (van Ittersum et al., 2003).

Over the past decade, fundamental and technological advancements made in both single-crop models and crop modelling platforms have led to numerous integrations between existing models and the development of diverse new models with new approaches (Fig. 1). Table 1 lists most of the wheat crop models currently used in research and applications.

While crop models have become a valuable, integrated part of many agricultural research activities, new challenges have arisen with this development. These new challenges include the following: 1) improving models to simulate the impacts of climate extremes (e.g. heat, frost, droughts, CO_2), 2) simulating multiple climate impact interactions (e.g. differing effects of elevated CO_2 on crops experiencing high or low temperatures), 3) incorporating

Table 1 Available wheat crop models (33)

Model (version)	Reference	Documentation
AFRCWHEAT2	(Porter, 1993)	Request from jrp@plen.ku.dk
APSIM-E	(Wang et al., 2002; Keating et al., 2003; Chen et al., 2010)	http://www.apsim.info
APSIM-Nwheat (V.1.55)*	(Asseng et al., 1998b; Keating et al., 2003; Asseng et al., 2004)	http://www.apsim.info
APSIM-Wheat (V.7.4)	(Keating et al., 2003)	http://www.apsim.info
AQUACROP (V.4.0)	(Steduto et al., 2009; Vanuytrecht et al., 2014)	http://www.fao.org/nr/water/aquacrop.html
CropSyst (V.3.04.08)	(Stockle et al., 2003)	http://modeling.bsyse.wsu.edu/CS_Suite_4/CropSyst/index.html
DAISY (V.5.19)	(Hansen et al., 1991; Hansen et al., 2012)	https://code.google.com/p/daisy-model/
DSSAT-CERES (V.4.0.1.0)	(Ritchie, 1985; Hoogenboom and White, 2003; Jones et al., 2003)	http://www.dssat.net
DSSAT-CROPSIM (V4.5.1.013)	(Hunt and Pararajasingham, 1995; Jones et al., 2003)	http://www.dssat.net
EPIC-Wheat (V1102)	(Williams et al., 1989; Kiniry et al., 1995)	http://epicapex.brc.tamus.edu/
Expert-N (V3.0.10) - CERES (V2.0)	(Ritchie et al., 1987; Stenger et al., 1999; Priesack et al., 2006; Biernath et al., 2011)	http://www.helmholtz-muenchen.de/en/iboe/expertn/
Expert-N (V3.0.10) - GECROS (V1.0)	(Stenger et al., 1999; Biernath et al., 2011)	http://www.helmholtz-muenchen.de/en/iboe/expertn/
Expert-N (V3.0.10) - SPASS (2.0)	(Stenger et al., 1999; Wang and Engel, 2000; Yin and van Laar, 2005; Priesack et al., 2006; Biernath et al., 2011)	http://www.helmholtz-muenchen.de/en/iboe/expertn/
Expert-N (V3.0.10) - SUCROS (V2)	(Goudriaan and Van Laar, 1994; Stenger et al., 1999; Priesack et al., 2006; Biernath et al., 2011)	http://www.helmholtz-muenchen.de/en/iboe/expertn/
FASSET (V.2.0)	(Olesen et al., 2002; Berntsen et al., 2003)	http://www.fasset.dk
GLAM (V.2)	(Challinor et al., 2004; Li et al., 2010)	http://www.see.leeds.ac.uk/research/icas/climate-impacts-group/research/glam/
HERMES (V.4.26)	(Kersebaum, 2007; Kersebaum, 2011)	http://www.zalf.de/en/forschung/institute/lsa/forschung/oekomod/hermes
INFOCROP (V.1)	(Aggarwal et al., 2006)	http://www.iari.res.in
LINTUL4 (V.6)	(Spitters and Schapendonk, 1990; Shibu et al., 2010)	http://models.pps.wur.nl/models

(*Continued*)

Table 1 (Continued)

Model (version)	Reference	Documentation
LINTUL5 (V.5)	(Spitters and Schapendonk, 1990; Shibu et al., 2010)	http://models.pps.wur.nl/models
LOBELL	(Gourdji et al., 2013)	Request from dlobell@stanford.edu
LPJmL (V3.2)	(Gerten et al., 2004; Bondeau et al., 2007; Müller et al., 2007; Rost et al., 2008; Fader et al., 2010; Beringer et al., 2011)	http://www.pik-potsdam.de/research/projects/lpjweb
MCWLA-Wheat (V.2.0)	(Tao et al., 2009a; Tao et al., 2009b; Tao and Zhang, 2010; Tao and Zhang, 2013)	Request from taofl@igsnrr.ac.cn
MONICA (V.1.0)	(Nendel et al., 2011)	http://monica.agrosystem-models.com
OLEARY (V.7)	(O'Leary et al., 1985; O'Leary and Connor, 1996a; O'Leary and Connor, 1996b; Latta and O'Leary, 2003)	Request from gjoleary@yahoo.com
SALUS (V.1.0)	(Senthilkumar et al., 2009; Basso et al., 2010)	http://salusmodel.glg.msu.edu
SIMPLACE (V.1)	(Angulo et al., 2013)	Request from frank.ewert@uni-bonn.de
SIRIUS (V2010)	(Jamieson et al., 1998a; Jamieson and Semenov, 2000; Lawless et al., 2005; Semenov and Shewry, 2011)	http://www.rothamsted.ac.uk/mas-models/sirius.php
SiriusQuality (V.2.0)	(Martre et al., 2006; Ferrise et al., 2010; He et al., 2010)	http://www1.clermont.inra.fr/siriusquality/
SSM-Wheat	(Amir and Sinclair, 1991; Soltani et al., 2013)	Request from marco.bindi@unifi.it
STICS (V.1.1)	(Brisson et al., 1998; Brisson et al., 2003)	http://www.avignon.inra.fr/agroclim_stics_eng/
WHEATGROW	(Cao and Moss, 1997; Yan et al., 2001; Cao et al., 2002; Li et al., 2002; Hu et al., 2004; Pan et al., 2006; Pan et al., 2007)	Request from yanzhu@njau.edu.cn
WOFOST (V.7.1)	(Boogaard and Kroes, 1998)	http://www.wofost.wur.nl

*Also in DSSAT modelling platform.
Source: Modified after Asseng et al. (2015).

quality aspects of grain yield (e.g. grain protein content and composition) and nutrition (e.g. Zn, Fe), 4) considering externalities (e.g. leaching of nitrate and pesticides) and soil constraints (e.g. acidity, salinity, access water), 5) incorporating pest and diseases into crop models and 6) linking crop models to genetics. These challenges can only be met with better understanding of these processes incorporated into existing crop models, which will require availability and testing from high-quality field experiments that include detailed, dynamic sampling (measurements over time).

2 Extrapolation from crop models in time and space

One major advantage of crop models is their potential to extrapolate data from limited field experimentation. The ability to extrapolate data beyond field testing allows for system studies across a wide range of specified environmental conditions, well beyond any possible field experiment, with consistent inputs. Crop model simulations depend on model design and input methodology resulting in various time and spatial outputs; thus, it is crucial to understand the interactions and area between the crop and environment being modelled.

2.1 Treatments

A powerful characteristic of crop models is the ability to simulate agronomic treatments, similar to treatments in field experiments. Treatment simulations allow a wider range of variables over longer periods (e.g. many decades) than field experimentation, making crop models a valuable tool for testing treatments in variable growing environments. Figure 2a illustrates a treatment study in which total wheat biomass output was simulated using two different nitrogen (N) fertilizer treatments over a complete growing season. These simulations allow for a dynamic comparison of total wheat biomass growth under two different regimes of N fertilizer supply without the need for continuous field measurements. Figure 2b shows a model simulation of grain yields over 70 years for three different soils; this data would have been difficult to collect with consistent inputs over numerous years in the field. Crop models have the ability to simulate multiple treatments across many seasons, which saves users time and effort in understanding cropping systems and can help them make future decisions.

2.2 Time

Crop simulation models are used to predict how a crop will develop over a certain period (e.g. months, weeks or days). For example, Asseng et al. (2013b) demonstrated how crop productivity has changed in a non-agricultural region in Argentina from 1937 to 2005. The area studied was assumed unsuitable for cropping due to lack of rainfall in the past (Fig. 2b). However, the study found that additional rainfall due to recent climate change has increased rainfall-limited yield potential to a level that could make this region suitable for wheat cropping. Even though no field experiments took place in this non-agricultural region, the crop simulation led to a new understanding that was previously not available.

2.3 Field

Crop models simulate components on varying spatial scales using a conceptual framework based on hierarchy theory (Ewert et al., 2009). Hierarchy theory partitions crop model applications into a nested approach in which each level shares similar time and spatial scales (Ewert et al., 2011); the smallest level is usually a single field or farm. Figure 3 illustrates how a model was used to simulate wheat yields on a 70-ha field with multi-year rainfall records and multi-soil characteristics in a semi-arid environment in Australia. This

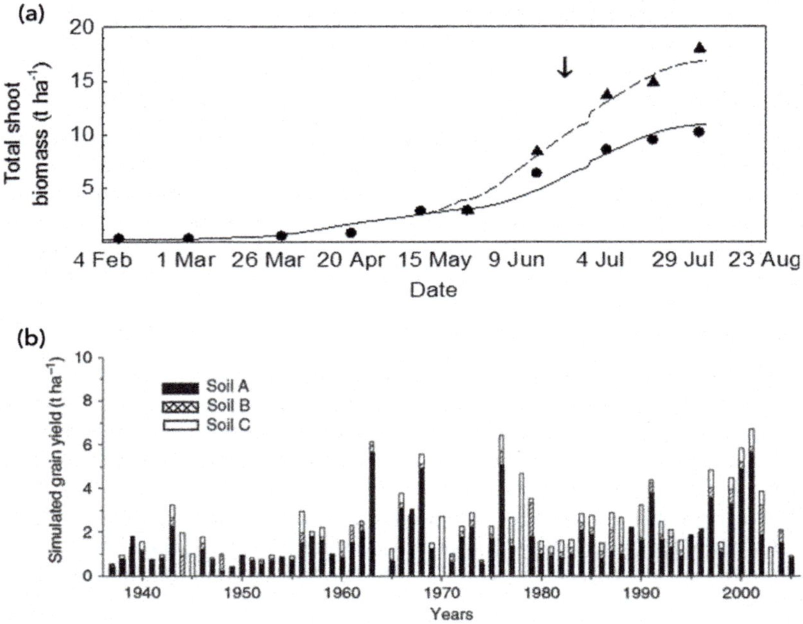

Figure 2 (a) Effects of N fertilizer on wheat total shoot biomass. Simulated (lines) and observed (symbols) for N1, (circle —) and N3 (triangle - - -) fertilizer treatments at the Eest, Netherlands, 1983. The arrow indicates the date of observed anthesis. Observed data after Groot and Verberne (1991). Source: Asseng et al. (2000). (b) Simulated wheat grain yields to explore the wheat production potential of a current non-agricultural region using APSIM-Nwheat. Simulations are from 1937 to 2005 for a location west of the current wheat-growing region of Argentina, which received increasing rainfall over the past 70 years, for three soils (A: fine sand, B: coarse sand and C: loam) using a current modern cultivar and a summer fallow. The crops were sown every year on June 5, with a 100 kg N ha^{-1} fertilizer application. Simulations were reset each year; simulated yield variation and trend are because of rainfall variability and trend. Source: Modified after Asseng et al. (2013b).

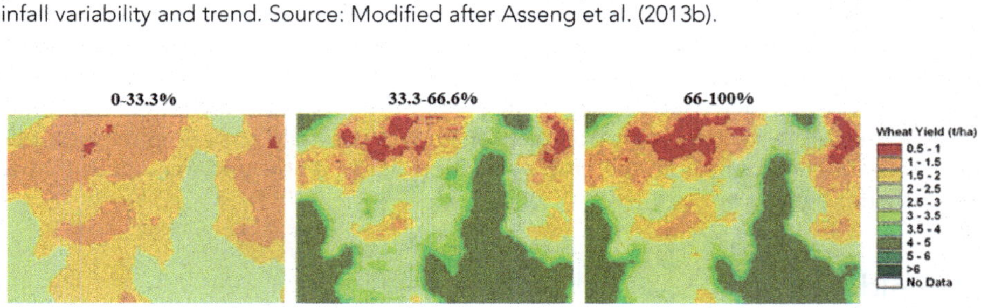

Figure 3 Simulated wheat yields for initially dry soil conditions with 60 kg N/ha fertilizer for average of 0–33.3% percentile driest (260 mm) May–October rainfall years, for average of 33.3–66.6% percentile (370 mm) May–October rainfall years and for average of 66–100% percentile (470 mm) May–October rainfall years for a field near Three Springs, Western Australia. Map is 1000 m wide by 700 m, top of the map is north.

Source: Modified after Wong and Asseng (2006).

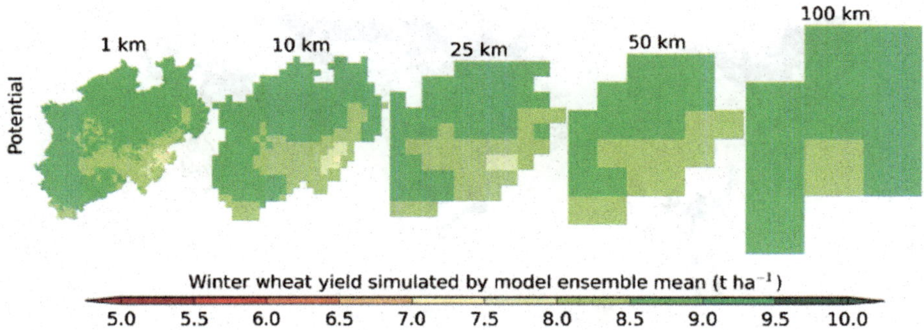

Figure 4 Simulated potential yield of winter wheat (ensemble mean, average over all 13 models from 1982 to 2011) at 5 resolutions (1, 10, 25, 50 and 100 km) for the state of North Rhine-Westphalia, Germany.

Source: Zhao et al. (2015).

example shows multiple treatment simulations on a local scale to describe the spatial yield response across a field during many years of growing seasons.

2.4 Region

To simulate a larger area than a field, such as a region or country, many simulated points can be aggregated to a larger scale (Ewert et al., 2011). Figure 4 shows an example of a regional model application. Zhao et al. (2015) used the ensemble mean of several crop models to simulate potential yield of winter wheat across 30 years (1982–2011) in a temperate region of Germany. Simulations were conducted at different spatial resolutions (1, 10, 25, 50 and 100 km) to illustrate the change in results with different resolution. Higher resolution resulted in greater simulated detail and accuracy, but required more detailed input information, which is often not available and requires more computer processing power.

2.5 Global

The global scale is the largest aggregation level that crop models can be used to simulate. Figure 5 shows the ability of a crop model to project future wheat yields on a global scale. For example, Rosegrant et al. (2014) used wheat crop simulations to estimate relative grain yield improvements when introducing drought-tolerant varieties compared to current varieties, taking into consideration climate change until 2050. The study found an increase in global wheat yield with regional differences when introducing drought-tolerant crop characteristics. This conclusion would have been impossible to determine at a global level using field experimentation.

The spatial scale in crop model applications depends on the study area (e.g. local, regional or global). However, it is crucial to understand the uncertainties introduced when aggregating input data to upscale crop models from fields to larger areas (Zhao et al., 2015) (further discussed in Section 4.2).

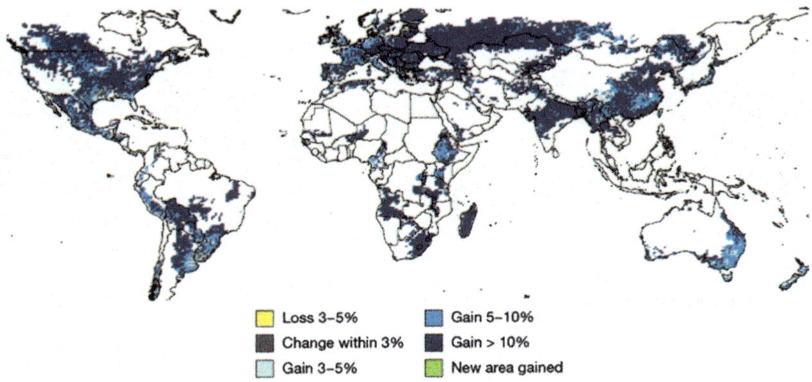

Figure 5 Maximum impact benefit map for wheat yield, MIROC A1B climate scenario, in 2050. Map shows the percentage improvement in yields when switching from the normal variety to the drought-tolerant variety for a climate projection of 2050. MIROC = Model for Interdisciplinary Research on Climate. A1B = Greenhouse gas emissions scenario that assumes fast economic growth, a population that peaks mid-century, and the development of new and efficient technologies, along with a balanced use of energy sources.

Source: Rosegrant et al. (2014). Reproduced with permission from the International Food Policy Research Institute.

3 Experiments for testing crop models

Simulation outputs should always be compared to field experimentations when possible to evaluate and improve crop models (Boote et al., 1996; Ziska and Bunce, 2007). Simulation models are tested against measured data under diverse growing conditions to validate crop models (Asseng et al., 1998b; Jamieson et al., 1998b; Probert et al., 1998; Lv et al., 2013). Data sets used for model calibration and validation consist of data describing initial soil conditions; crop-specific management; daily weather; phenology of a crop; yields; yield components and nutrient contents from intermediate harvests; and measurements of energy, water and CO_2 fluxes (Kersebaum et al., 2015). To validate a model's grain yield predictability, simulated growth dynamics are compared to observed growth dynamics for a wide range of crop management and growing environments and quantified with statistical performance parameters (e.g. root mean squared deviation of simulated to observed data). The following subsections describe high-quality field experiments used for model testing.

3.1 Nitrogen application rate experiment in the Netherlands

Groot and Verberne (1991) collected data from a series of winter wheat field experiments from several sites, seasons and N fertilization regimes in the temperate maritime climate of the Netherlands. The experiments comprised three different N treatments at each site with 2–3 weekly observations that measured soil mineral N content, soil water content, groundwater table, dry matter production and distribution, N uptake, N distribution and root length density. This experimental data has been widely used for testing simulation models for crop growth and development, soil N dynamics and N uptake (Asseng et al., 2000; Olesen et al., 2000; Wang and Engel, 2002).

One of the crop model testing studies used a crop model to simulate above- and below-ground growth, grain yield, soil water, soil mineral N and N uptake of wheat at the same sites in the Netherlands to determine different N rate and timing treatments (Fig. 2a). The model was tested and calibrated using the previous data collected and then validated using additional field measurements at the same sites. Using the additional field measurements for model validation, the model simulations of dry weight grain yields, grain protein concentrations, shoot growth, phenology, kernels m^{-2}, specific grain weight, grain N, soil mineral N and N uptake were found acceptable based on the statistical analysis performed (coefficient of determination [r^2], root mean squared deviation), thus proving that the model could reproduce detailed field measurements across different growing conditions.

3.2 Water stress experiment in New Zealand

In an experimental study at Lincoln, Canterbury, New Zealand, Jamieson et al. (1995) determined the drought response of wheat using a series of experiments in a mobile rain shelter. Wheat was sown in the rain shelter with 180 kg N/ha applied over the growing period. Eleven experimental treatments subjected the crops to drought of varying duration at different crop developmental stages to determine the drought influence on wheat growth and yield. The study found that wheat yield reduction with drought was associated mainly with reduced grain number. The data collected from this experiment has been used in crop modelling experiments to compare various wheat models, evaluate accuracy of model outputs and reproduce crop growth affected by different levels of water stress through simulations of total biomass and grain yields (Jamieson et al., 1998b; Asseng et al., 2004).

3.3 Free-air carbon dioxide enrichment experiment in Arizona

The free-air carbon dioxide enrichment (FACE) experiment conducted at Maricopa, Arizona, included several growing seasons to determine the interactive effects of elevated atmospheric CO_2 concentrations with limited water supply and limited N on a spring wheat cultivar. The experiment showed an increase of wheat yields due to the elevated CO_2 concentrations, but that the level of N and water availability of crops influenced the overall response to CO_2 levels (Kimball et al., 2002). Several crop models have been tested with CO_2 data from the Maricopa FACE experiment to simulate the phenology, growth and grain yield of wheat under future climatic conditions (Grant et al., 1995; Kartschall et al., 1995; Tubiello et al., 1999; Jamieson et al., 2000; Grossman-Clarke et al., 2001; Asseng et al., 2004; Ko et al., 2010). Using the Maricopa FACE experiment data, most crop models were able to simulate crop growth and development under elevated CO_2 concentrations with acceptable accuracy (based on statistical analyses), indicating that crop models can simulate the impact of increased CO_2 on crop growth in high-input and irrigated cropping systems.

3.4 FACE experiment in Australia

Three years of measured wheat growth and water use data was collected from the Australian Grains FACE (AGFACE) experiment (Mollah et al., 2009) in the semi-arid environment of southern Australia. Wheat was grown at elevated atmospheric CO_2 concentrations with varying water and N fertilization regimes to observe the effects on wheat growth, yield and water use. Several crop model studies have used the AGFACE data for comparison of simulated results under elevated CO_2 concentrations (Nuttall et al., 2012; O'Leary et al.,

2015). One study tested the performance of six wheat crop models against the AGFACE data and found that the crop simulation performance in low-input, rain-fed cropping systems for all six models was similar to the experimental measurements of accumulated biomass, yield and water use response (O'Leary et al., 2015).

3.5 Hot Serial Cereal experiment in Arizona

The Hot Serial Cereal (HSC) experiment was a well-irrigated and fertilized experiment in Maricopa, Arizona, with time-of-sowing and artificial heating treatments under field conditions using a single spring wheat cultivar (Wall et al., 2011; Ottman et al., 2012). The experiment used infrared radiation heaters for artificial heat treatments and infrared thermometers to control the wheat canopy temperature to determine the effects of future temperature rise on wheat growth and yield. The study found that increases in temperature will decrease wheat yield for late plantings and shift optimum planting windows to earlier dates depending on location.

A crop modelling study used data from the HSC experiment to simulate yield in 30 different wheat crop models in a 'blind experiment' (individual replicates used in this simulation were previously not publicly available; modellers only had access to phenology and yield information of a single treatment in the normal temperature range) to determine the effects of increasing temperature on simulated wheat yields (Asseng et al., 2015a). Figure 6 illustrates how the models simulated grain yield and that many simulations became more variable at higher temperatures (>22°C). The model ensemble median simulation was more accurate in estimating crop temperature impacts on yield than any single model. The findings highlight a key issue in the predictability of some existing crop models and provide a reminder that further process understanding and model development for certain processes are still needed.

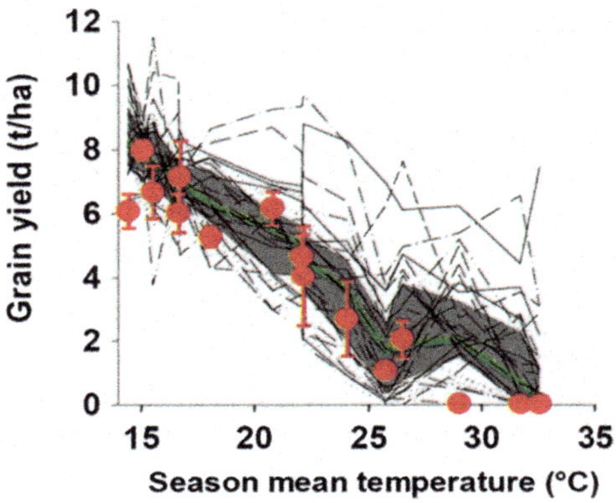

Figure 6 Observations and multi-model simulations of grain yield at different mean seasonal temperatures for the Hot Serial Cereal experiment, Maricopa, Arizona with *Triticum aestivum* L. cultivar Yecora Rojo with time-of-sowing and infrared heat treatments. Observed values ± 1 standard deviation (s.d.) are shown as symbols with 30 simulated values shown by thin lines. Multi-model ensemble median (thick line) and interval between the 25th and 75th percentiles (shaded grey) based on the 30 models are shown as well.

Source: Modified after Asseng et al. (2015a).

4 Adapting crop models for climate change impact

Atmospheric CO_2 has increased from a pre-industrial concentration of 280 ppm (parts per million) to 403 ppm in 2016 (Dlugokencky and Tans, 2016), while global temperature has increased by 0.90°C from 1880 to 2015 (NOAA National Centers for Environmental Information, 2016). These trends suggest a significant change in climate, and further changes are inevitable regardless of efforts to reduce global emissions of greenhouse gases (IPCC, 2013).

The relationship between climate change and agricultural systems is complex, but crop models are often used to understand the impact of climate change and assist in the development of adaptation strategies (Ewert et al., 2015). The model's ability to project the impact of future climate scenarios enables scientists, farmers and policymakers to prepare improved crop management and breeding strategies.

4.1 Approach

In order to account for climate change in simulations, crop models are paired with general circulation models (i.e. global climate models, GCMs). GCMs are three-dimensional mathematical representations that generate the processes of interactions between the atmosphere, land surface, oceans and sea ice to simulate climate scenarios (Kang et al., 2009). GCMs have course spatial resolutions (typically between 1 and 5 degrees in latitude or longitude, or between 100 and 500 km depending on location), but many times a finer (sub-grid-scale) resolution is needed for regional decision-making and impact assessment. Such finer resolutions can be achieved through climate downscaling, which is explained below.

The first step in creating a climate scenario with a GCM is to determine a baseline period. This serves as the reference period from which the modelled future change in climate is calculated. Thereafter, in order to achieve a finer resolution, several methods of downscaling can be used to apply coarse-scale climate change projections to a high-resolution observed climate baseline (Wilby et al., 2004). One of the most common and simple downscaling methods applied is the 'change factor method' (i.e. delta change factor method), which calculates the difference between the GCM future values (e.g. mean GCM temperatures) and the baseline period values (e.g. mean baseline temperatures), and adds this difference to the observed reference values of the baseline period (Anandhi et al., 2011). Another method of obtaining finer spatial resolutions is using statistical downscaling. Statistical downscaling develops quantitative relationships between large-scale atmospheric variables (predictors) and local surface variables (predictands), often making the predictand a function of the predictor (Wilby et al., 2004). Statistical relationships based on linear regressions are developed between large-scale predictors and local predictands and are used to better project possible changes in climate extremes.

GCMs have large uncertainties from the modelled processes within GCMs themselves, initial conditions and future greenhouse gas emissions, and downscaling GCM climate scenarios to local scales (e.g. fields) for crop models adds additional uncertainty (Wilby et al., 2004; Asseng et al., 2013a). The reliability of downscaling is critical for the results of crop impact; therefore, the uncertainties and limits in downscaling and GCMs must be considered when studying climate change impacts by using several emission scenarios and GCMs. The use of several emission scenarios and GCMs is achieved through the use of multiple crop models in model intercomparison programmes (MIPs) for multi-model climate impact assessments; one example is the Agricultural Model Intercomparison and Improvement Project (AgMIP; www.agmip.org). AgMIP is a major international collaborative effort to combine the climate, crop and economic modelling communities with state-of-the-art information technology to produce improved climate impact projections, crop models and economic models for the agricultural sector (Rosenzweig et al., 2013). A new AgMIP initiative is coordinating detailed crop model simulations on a global network for climate change impact assessments following a set of common protocols of emission scenarios and GCMs (Ruane et al., 2014).

4.2 Scales

Scaling refers to the extrapolation of information from one scale to another scale in space, time or both (Li and Wu, 2006). In crop modelling, scaling is needed when data are generated at a small scale (e.g. field level), but large-scale (e.g. regional level) issues need to be addressed (e.g. climate change adaptation scenarios). An ideal scaling method would be a high-resolution simulation for all points in the region being studied, giving the true values at each point. Unfortunately, this is usually not possible owing to the vast number of data points required. An ideal feasible scaling method is a resolution fine enough to provide accurate spatial variability for the purpose of the study, while also saving efforts in data collection and model runtime (Zhao et al., 2015).

Crop models may use several approaches to scale outputs, such as data extrapolation, data interpolation, data aggregation or stratified sampling (Ewert et al., 2011). Data extrapolation uses one or a few locations within an area to represent the entire area of interest, whereas data interpolation derives data using limited known data combined with physical information to obtain some of the prevailing heterogeneity within the area of interest (van Bussel et al., 2016). Data aggregation divides the area being studied into smaller zones and takes the aggregate value in each zone. Stratified sampling divides the area into zones (or strata) with similar conditions based on variables (e.g. temperature seasonality, potential evapotranspiration seasonality, growing degree days with base temperature of 0°C, aridity index) and then selects a random sampling location within each zone to represent that zone as a whole (van Bussel et al., 2016). The selection of scaling approach is ultimately up to the modeller's discretion, although previous studies have found that using interpolated weather data can introduce considerable errors into the model (Van Wart et al., 2013).

4.3 Rainfall

Climate change is likely to alter rainfall amounts, patterns and distributions differently across the globe. Total annual rainfall will have the tendency to increase at higher latitudes and near the equator, while subtropical rainfall is likely to decrease and become more variable (Giorgi and Bi, 2005). Alterations in rainfall affect wheat development and growth because the water from rainfall infiltrates the soil and is absorbed by crop roots. For modelling purposes, the daily amount of water in the soil depends on the initial amount of water in the soil along with any additional precipitation, irrigation and subtractions, including evaporation, transpiration, drainage and run-off that occur throughout the day.

Most models have a soil–water sub-module coupled with climate and crop growth modules to determine the availability of water for the crop. This sub-module calculates the soil–water balance throughout the day using rainfall data, equations for potential evapotranspiration (Penman, 1948; Monteith, 1965a; Priestley and Taylor, 1972) and uptake rate constraints (Jamieson et al., 2010).

In crop models, water stress is usually simulated using a stress index based on a water supply–demand rate ratio, with a maximum value of one (Ritchie, 1985; Porter, 1993), or a fraction of available soil water in a soil (Amir and Sinclair, 1991; Keating et al., 2001). The stress index is then incorporated into the crop growth module to reduce leaf area expansion, photosynthetic rate or radiation-use efficiency (RUE), or to accelerate leaf senescence (Hu et al., 2004; Liu et al., 2010; Andarzian et al., 2011).

4.4 Temperature

Crop models use temperature as one of the main drivers for development and growth. Simulating the progress of crop development is usually based on a concept of thermal time (Cao and Moss, 1997; Yin and Struik, 2010) or physiological development days (Cao and Moss, 1997; Wang and Engel, 1998). Thermal time is a summation of cumulative differences between daily mean temperature and a specified base temperature, with units of degree days. Total physiological development days are the minimum number of days for development under optimal temperature (Wang and Engel, 1998).

The use of thermal time allowed Wang and Engel (1998) to develop the temperature response function, which simulates wheat development rate and vernalization regarding changes in temperature. Vernalization refers to a low-temperature accumulation requirement of wheat before the beginning of spikelet (axillary bud) formation (Ritchie, 1991). Many wheat crop models account for vernalization and photoperiod impact (crop sensitivity to daily light/dark periods).

The temperature effect on wheat growth rate and biomass accumulation is simulated using a temperature response curve that modifies the wheat photosynthesis rate or RUE. Modelling the impacts of temperature change is challenging and studies have shown that certain models underestimate high-temperature impact on crop senescence (Lobell et al., 2012). Some models include a heat stress component to accelerate crop senescence under high temperatures (Keating et al., 2001; Asseng et al., 2011).

Changes in temperature also affect the vapour pressure deficit (VPD), which influences crop–water stress. Increased VPD means lower relative humidity and a greater possibility of a crop losing water to the atmosphere, leading to an increased possibility of water stress. When modelling VPD and water stress, it is important to understand the relationship between minimum and maximum temperature changes (Easterling et al., 1997; Roderick and Farquhar, 2002; Wessolek and Asseng, 2006). Temperature increases are a major driving factor in drought stress because VPD increases non-linearly with higher air temperatures (Lobell et al., 2015). Also, many models ignore the impact of diurnal temperature range on grain yield (Lobell, 2007).

Historical weather data shows that climate change is not only altering temperature and rainfall patterns but also increasing the frequency and severity of extreme events (Coumou and Rahmstorf, 2012; Orlowsky and Seneviratne, 2012). Brief periods of extreme hot or cold temperatures in wheat can be devastating to grain yield depending on the developmental stage (Ferris et al., 1998), so models must incorporate effects of extreme temperature stress to quantify future impacts and to develop adaptation strategies (Liu et al., 2014). Many crop models account for extreme temperature stress by setting critical temperature thresholds of heat (or frost) stress at certain developmental stages that increase

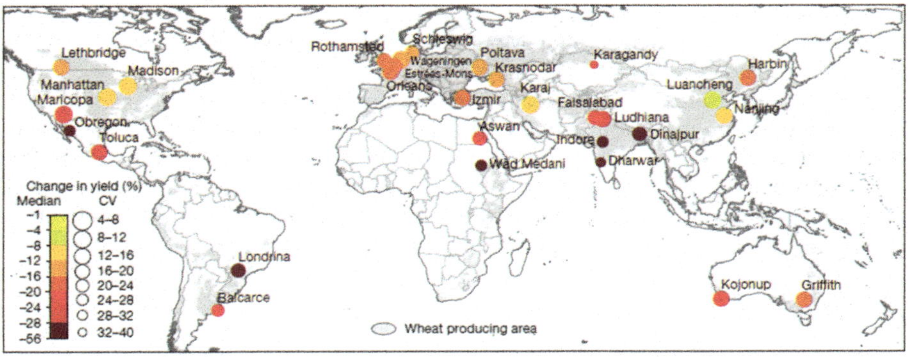

Figure 7 Simulated global wheat grain yield change with higher temperatures. Relative median grain yield for +4°C temperature increase imposed on the 1981–2010 period for the 30-model ensemble using region-specific cultivars. Simulation model uncertainty was calculated as the coefficient of variation (CV%) across 30 models and plotted as circle size. The larger the circle, the less the uncertainty.

Source: Asseng et al. (2015a).

senescence or reduce biomass when temperatures remain above (or below) the thresholds for extended periods of time (Asseng et al., 2011; Moriondo et al., 2011; Eitzinger et al., 2013; Barlow et al., 2015). Several crop models also consider the temperature effects on yield quality, such as wheat grain protein content (Asseng and Milroy, 2006) and wheat grain protein components (Martre et al., 2006).

Many crop models are used to predict crop development while accounting for temperature increases. Figure 7 illustrates a study by Asseng et al. (2015a) that modelled global wheat production while accounting for rising temperature. The study found that climate warming is already slowing yield gains at many wheat-producing areas and that global wheat production is estimated to decrease by 6% for each °C of further global temperature increase.

4.5 Other impact factors

Other factors, including elevated CO_2, reduced solar radiation and increased ground-level ozone, can impact wheat crop growth and development.

Elevated CO_2 affects crop growth in two major ways, simultaneously. First, the intercellular CO_2 concentration increases, which causes increased net photosynthetic rates and, in turn, increased crop RUE. Second, stomatal conductance decreases, which leads to reduced transpiration and increased transpiration efficiency (TE) (Farquhar et al., 1978). Some crop models account for these effects by scaling RUE, using a calculated ratio of light-limited photosynthetic response at elevated CO_2 concentration compared to the parameterized (baseline) CO_2 concentration (Goudriaan and Vanlaar, 1978; Reyenga et al., 1999). TE, or total biomass per unit water transpired, is usually scaled by a specified factor that increases linearly as the CO_2 concentration increases, illustrating the effects of elevated CO_2 on the ratio between the internal CO_2 concentration of the crop and the external ambient CO_2 concentration (Gifford and Morison, 1993; Duursma et al., 2013). Some crop models use a constant multiplier for daily total crop production in elevated CO_2 (Tubiello et al., 2007).

The main way crop models account for changes in solar radiation is through simulation of light interception and RUE. Some models simulate RUE by calculating the daily growth rate from photosynthesis equations, integrated over a layered canopy at sub-intervals, while other models use an aggregated RUE approach (Monteith, 1977). In addition, some models account for the impact of diffuse radiation (incoming solar radiation that has been scattered by molecules and particles in the atmosphere) with a diffuse fraction in radiation (Yang et al., 2013).

Ozone (O_3) is a form of oxygen that is both beneficial and harmful depending on its location in the atmosphere. O_3 in the stratosphere is beneficial because it absorbs solar ultraviolet radiation (UV-B); however, it is a strong oxidizer, making it a harmful pollutant and potentially destructive force to crops when it occurs in the lower troposphere (ground level) (Reich and Amundson, 1985). In crops such as wheat, increased O_3 in the lower troposphere can reduce the activity and amount of Rubisco, an enzyme crucial for photosynthesis, leading to reduced photosynthesis, reduced yield and accelerated leaf senescence (Lehnherr et al., 1987; Grandjean and Fuhrer, 1989; Giorgi and Bi, 2005).

To account for the effects of O_3 in crop modelling, Ewert et al. (1999) modified the AFRCWHEAT model (Porter, 1993) and the LINTUL model (Spitters and Schapendonk, 1990) by assuming that O_3 damage is caused by O_3 uptake. AFRCWHEAT was modified so that O_3 reduces the light saturation rate of leaf photosynthesis and induces a short-term response, which is reversible depending on leaf age. LINTUL was modified so that O_3 decreases Rubisco concentration and increases the costs of detoxification and repair processes. When modelling O_3, the crop response to O_3 depends on crop development (Ewert et al., 1999).

5 Natural resource management

Traditional agricultural management focuses solely on increasing crop productivity (e.g. growth and yield). At present, focus has shifted towards natural resource management (NRM), or agricultural management that implements a more sustainable use of soil, rainfall, irrigation and nutrients to increase agricultural productivity. Focus on NRM is expanding due to soil degradation, water shortages, limited cropping area and climate change. Crop models allow farmers to develop agricultural practices and strategies that increase crop production while considering environmental impacts and natural resource availability. (Section 7 describes several decision support systems available to farmers.)

5.1 Sowing

The main decisions made when planting a wheat crop are choosing the cultivar, the planting intensity and the sowing date. Determining the maximum economic return of these decisions is generally difficult to quantify because sowing dates vary depending on location and climate. Crop model simulations can generate potential yields under different sowing strategies to maximize yield and economic return.

One challenge when sowing crops is that climate change affects seasonal temperatures and rainfall, altering ideal sowing dates. Several studies have used crop models to account for climate change and identify optimal sowing dates and yield benefits of earlier sowing (Asseng et al., 2008; Cammarano et al., 2012; Zheng et al., 2012; Andarzian et al., 2015).

5.2 Cultivars and traits

A crop phenotype with specific traits or characteristics is the observable expression of crop genes, including its structure, development, physiological properties, phenology or behaviour resulting from the interaction of the crop genotype (genetic structure) and the environment. Understanding wheat genotypes and phenotypes can assist to breed alternative wheat cultivars for improved agricultural purposes (e.g. drought-resistant cultivars) (Chenu et al., 2013). Crop models simulate wheat growth and development using genotype, environment and management effects (GxExM) (Jones et al., 2003; Keating et al., 2003; Chenu et al., 2011), while phenological stages are simulated in response to day length (photoperiod), temperature (thermal time and vernalization) and the interactions of these inputs (Zheng et al., 2013). Crop models help determine the effects of breeding new varieties and/or adding new cultivars to a location, such as targeted breeding for adaptation to climate change (Ludwig and Asseng, 2010). Studies in the North China Plain using crop models have found that adding new winter wheat cultivars could prolong the crop growth period through increased thermal time required for development; this could reduce the negative warming effects from climate change in this region (Liu et al., 2010; Li et al., 2016).

Wheat crop models can provide a reasonable approach for testing new wheat ideotypes; these are a set of cultivar parameters that mimic the genotype for an 'ideal wheat type'. Crop models can test ideotypes so they are optimized for specified locations and future adaptation needs, while using less time and resources than field experiments (Sylvester-Bradley et al., 2012; Zheng et al., 2012). Figure 8 illustrates the study by Semenov et al. (2014) that used a crop model to simulate 100-year mean grain yields for possible wheat ideotypes under a future climate scenario at 10 European locations compared to current wheat yields. The study identified a combination of specific traits with a large potential to increase yields (more than double) in temperate environments.

Crop models can estimate the effect of single and combined traits on growth and yield and the degree of additivity or cancelling compensation of the traits (Boote et al., 2013). Models can test traits of a crop grown in various *in silico* environments to observe if the traits will have positive or negative effects on crop growth, yield or protein content (Martre et al., 2015a).

Modelling crop phenology requires algorithms with cultivar-specific parameters, but parameter estimation for large numbers of genotypes can be time-consuming and costly (Bertin et al., 2010), limiting the rate at which new genotypes are incorporated into models. A study by Zheng et al. (2013) attempted to link parameters with genetics by modifying a wheat model to incorporate gene effects into the estimation of wheat heading time; this resulted in a gene-based model that showed new longer-season varieties may flower later when sown early in a season, with a potential to increase yield. The ability of crop models to simulate genotype and phenological interactions allows breeders and farmers to choose the ideal crop cultivar for their specific location and needs.

5.3 Nitrogen fertilization

N fertilization is crucial for maximizing agricultural production, but crop nutrient needs vary over time and space, which makes it a challenge to optimize application rate and timing. For example, varying levels of N may be needed in different areas of a field (Pierce and Nowak, 1999). Another challenge is that N fertilization can lead to nitrate leaching in the soil and groundwater under excessive rainfall or irrigation. Crop models can simulate the long-term effects of N on crop growth and yield and assess the risk

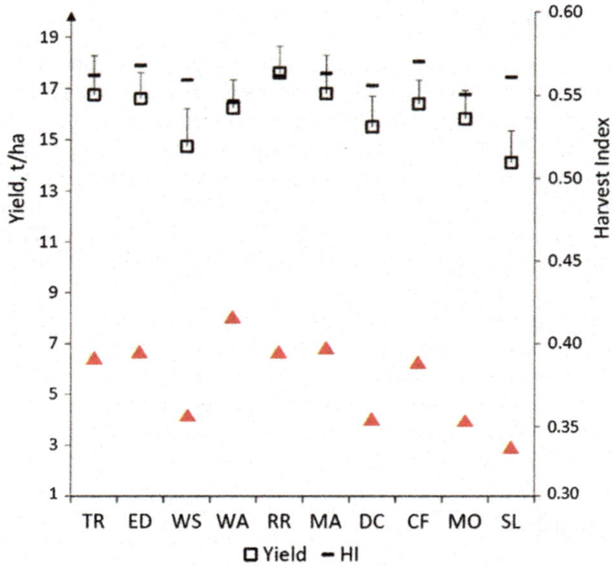

Figure 8 Simulated wheat yields with new traits for a future climate. Squares show grain yields with 95-percentiles (presented as top error bars) and dashes show harvest index as simulated by Sirius using future wheat ideotypes for the HadCM3(A1B) climate scenario at 10 European locations (from right to left: Tylstrup, Denmark; Edinburgh, United Kingdom; Warsaw, Poland; Wageningen, Netherlands; Rothamsted, United Kingdom; Mannheim, Germany; Debrecen, Hungary; Clermont-Ferrand, France; Montagnano, Italy; and Seville, Spain). Triangles show grain yield averages for each country from 2011 to 2013 (FAOSTAT, 2015) with assumed 11% moisture correction. HadCM3 = Global climate model used. A1B = Greenhouse gas emissions scenario that assumes fast economic growth, a population that peaks mid-century, and the development of new and efficient technologies, along with a balanced use of energy sources.

Source: Modified after Semenov et al. (2014).

associated with a specific N fertilizer rate, based on the integrations of the crop–soil modules (Basso et al., 2012). Simulation models have been used to identify optimal N fertilizer application rates for different management zones within a spatially variable field using parameterized management zones (Basso et al., 2011). Simulation models have also been used to determine amounts of nitrate leaching for various soils and rainfall conditions (Asseng et al., 1998a). Crop model simulations help determine zone-specific N fertilization strategies to result in improved yields and reduced nitrate leaching in the field.

Model simulations have shown that N-use efficiency of wheat varies with rainfall season, soil type and initial soil water conditions (Asseng et al., 2001a; Basso et al., 2013); this information would have been difficult to determine from field experiments (see Section 6 for case study).

Wheat grain quality also depends on N fertilization; for example, N applied during the wheat stem elongation period can increase yield and protein content in some environments (Colecchia et al., 2013). Many crop model simulations have been used in studies to assess the effects of N management on wheat growth and grain protein accumulation (Asseng et al., 2002; Martre et al., 2003; Martre et al., 2006).

5.4 Irrigation and drainage

Rainfall is a scarce resource in many parts of the world, and it can limit crop growth and yield. Some agricultural regions use irrigation to compensate for limited rainfall. Crop models can be used to manage crop irrigation because most models consider water balance modules that simulate the complex crop–water balance processes (e.g. rainfall, irrigation, surface drainage, run-off, soil evaporation, transpiration and root uptake) (Singh et al., 2008; Steduto et al., 2009). Heng et al. (2007) used a crop model to determine that wheat grain yields in the arid and semi-arid rain-fed regions of West Asia and North Africa could be improved by adjusting N management to soil type and by applying small amounts of supplemental irrigation.

Drainage below the potential maximum of the root zone occurs when rainfall or irrigation exceeds the soil storage capacity and can cause dry land salinity in some regions (Rengasamy and Olsson, 1993). Drainage is difficult to measure in the field and varies with rainfall and soils. Crop models have been used to estimate the drainage potential in various soils and fields (Asseng et al., 2001b), providing a faster and easier alternative to field experimentation (Verburg et al., 2012).

5.5 Spatial variability

Spatial variability refers to the spatial variations in growing conditions due to variations in weather, soil, nutrient supply, water availability and crop management within a specified area that impact crop production. Spatial variations are often difficult to quantify because they can exhibit values that differ across even a small scale within a field. Crop models can estimate spatial variability based on input variables and parameterization but understanding the scale and resolution desired for the area of interest is important. As Section 5.3 discussed, combining crop models with specified spatial field information can assist to develop improved management practices (e.g. N fertilizer strategies) (Basso et al., 2011, 2013). Crop models have also been coupled with regional soil and weather variability data to develop excess water management strategies to limit dry land salinity in a region (Pracilio et al., 2003).

An effective method of estimating spatial variability is combining crop models with agricultural remote sensing data (e.g. information about crops collected from aerial sensor technology). Combining crop models with remote sensing data can assist to calibrate and test models with detailed spatial and temporal data (Trombetta et al., 2016). Methods used to integrate crop models with remote sensing data include driving variable methods, assimilation, re-initialization and re-calibration (Moulin et al., 1998). A driving variable method creates a new variable or updates a variable (e.g. Leaf Area Index) in a crop model derived from remote sensing data. Assimilation creates new parameters in the model derived from remote sensing observations. Re-initialization adjusts model initial conditions, and re-calibration adjusts model parameters with remote sensing observations. Applying these methods to incorporate remote sensing data into crop models helps to improve management of spatial variability.

6 Managing seasonal variability

Wheat growth in rain-fed agriculture depends on seasonal rainfall, which is highly variable and contributes to the variability in grain yields (Stephens and Lyons, 1998; Iizumi et al.,

2013; Wang et al., 2015). Predicting the type of season (e.g. wet or dry) before it begins can allow producers to adjust management practices to save input costs in low-yielding seasons or increase inputs to achieve higher potential yields in wet, high-yielding seasons (Moeller et al., 2008; Asseng et al., 2012).

Crop modelling enables exploration of the benefits from seasonal forecasts to manage seasonal variability to improve efficiency and profitability. Models can use long-term historical weather data to simulate a wide range of management practices to optimize management strategies for season types based on seasonal forecasts (Asseng et al., 2014). One study combined long-term simulation models with seasonal forecasts and estimated that adjusting crop management (e.g. N fertilization) to southern oscillation index phases showed that it could increase profits by up to 20% (Hammer et al., 1996). Also, pairing a crop model with El Niño Southern Oscillation (ENSO)-based forecasts and adjusting crop management to ENSO phases could increase wheat yields and gross margins in Paraguay (Ramirez-Rodrigues et al., 2014).

Figure 9 illustrates a study that used a crop model coupled with seasonal forecasts generated from the Predictive Ocean-Atmosphere Model for Australia (POAMA) to

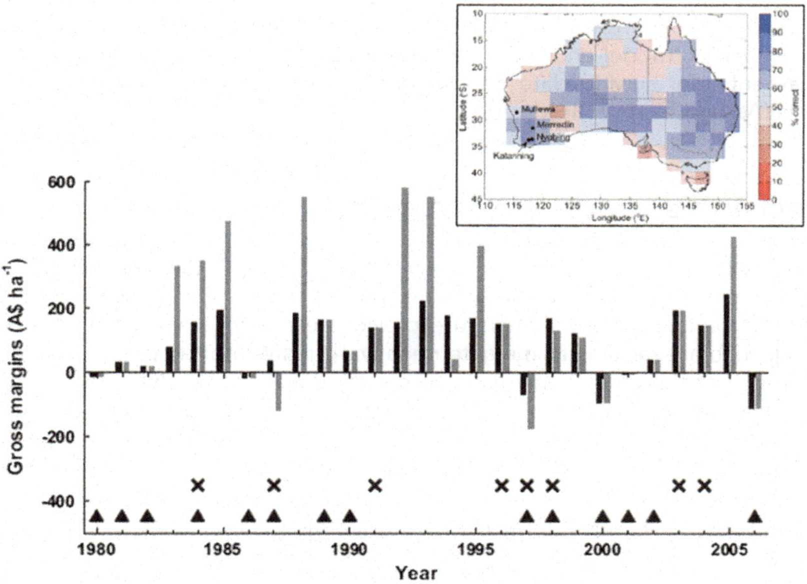

Figure 9 Simulated gross margins (A$ ha^{-1}) of wheat crops using a two-category seasonal rainfall (May–October) POAMA forecast for N fertilizer management (grey bars) compared to not using a seasonal forecast (black bars). Years with a wrong POAMA seasonal rainfall forecasts are indicated with a cross and years with observed below median seasonal rainfall forecasts are indicated with a triangle in the lower part of the figure. Simulations are from APSIM-Nwheat for an initially dry (at start of season) clay soil at Nyabing in the Western Australian wheat belt. Wheat price at A$200 Mg^{-1}, N fertilizer cost at A$1 kg^{-1} N and a profit target of A$2 per each 1A$ invested in N fertilizer. **Inlet:** Percent (%) correct two-category (above and below median) seasonal (May–October) rainfall forecast across Australia using POAMA. Study site Nyabing is indicated on the map.

Source: Modified after Asseng et al. (2012).

determine optimal N fertilizer applications for wheat production in Australia. The study found that using seasonal rainfall forecasts to predict rainfall season types can assist in adjusting N fertilizer applications to produce yields closer to maximum achievable yields. The crop management based on the seasonal forecast resulted in up to A$50 ha^{-1} of additional gross margins from improved crop management. The inlet of Figure 9 displays the varying skill of the seasonal forecast model across Australia, with statistically significant forecasting skill for locations south of the Western Australian wheat belt, but less statistically significant skill towards the central and northern wheat belt.

Seasonal forecasting systems are not perfect and do occasionally fail. Figure 9 shows how forecasting a wrong season type results in a lost opportunity by under-fertilizing a crop when the achievable yield was high, or by over-fertilizing a crop when the actual achievable yield was low. However, Fig. 9 also shows that the benefits from correct seasonal forecasts can outweigh the negative impact of wrong forecasts, resulting in an overall positive benefit. Also, the better water-holding soils showed a higher value in applying a forecast than the poorer water-holding soils as less water was lost through drainage. The value of a seasonal forecast was much higher with dry initial soil water conditions than with initially wet soil conditions (Asseng et al., 2012). Recently, the value of short-term weather forecasts of up to ten days has been evaluated with a crop simulation model and it indicated large potential benefits from adjusting crop management (e.g. sowing date and N fertilizer and fungicide applications) to such forecasts (Asseng et al., 2016).

7 Decision support for farmers

A major key to adapting food production for the future is to improve the link between agricultural practices and climate information, especially for smallholder farmers in developing countries. Many farmers use previous experience and weather patterns to adapt agricultural practices to mitigate negative climate impacts (e.g. heat, drought or frost stress), but the increasingly variable climate causes more difficulty for farmers to manage seasonal climate risks. One way to manage these risks is through the use of in-season decision support systems that incorporate crop models, such as Yield Prophet (www.yieldprophet.com.au), Crop Prophet (https://www.cropprophet.com) or AgroClimate (www.agroclimate.org).

Yield Prophet is a web interface and mobile application built around the APSIM crop modelling system that produces a comprehensive crop report based on simulated growing conditions using farm location, cultivar details, soil data, hundred-year weather data and seasonal climate forecasts (Hochman et al., 2009). This crop report gives the latest seasonal weather forecasts and estimated impacts on crop yields, including updates on growth stage, yield potential, available soil water and N, water and N stress indices, and risks of exposure to heat or frost stress in critical crop developmental stages. Other generated reports allow users to examine 'what-if' scenarios, such as profitability from various N fertilizer or irrigation treatments, as well as best time of sowing for soil water and N management. Yield Prophet can also be used to investigate impacts of climate change scenarios on current management practices and impacts of adaptation options. Crop Prophet is a similar web interface crop forecasting system that predicts end-of-season yield of several crops in the United States using high-resolution weather and satellite data combined with field survey information. It also helps farmers determine profitability and best management practices for various treatment scenarios.

AgroClimate is a web interface and mobile application decision support system that helps farmers in the south-eastern United States make informed decisions on managing county-level climate risks (Fraisse et al., 2006). It compares seasonal forecasts, short-term forecasts and historical climate patterns in specific regions to current conditions and El Niño/La Niña events to determine the anticipated impacts of climate patterns on agricultural commodities. AgroClimate is used to link climate and crop development to determine the possibility and impact of crop disease, hurricanes, wildfires, drought, and extreme temperatures on crop yield and livestock.

Crop models can also be combined with breeder support systems to help farmers and breeders determine ideal cultivars and traits for specific environments or conditions. The Genomic Open-source Breeding Informatics Initiative (http://cbsugobii05.tc.cornell.edu/wordpress/) is developing databases and software tools to maintain and organize large quantities of genomic data that will allow breeders to use genotypic information to help make crop variety selections more efficient and reduce breeding lines with poor breeding value for staple crops. The Integrated Breeding Platform (https://www.integratedbreeding.net/) is another initiative that has developed the Breeding Management System (BMS), a suite of interconnected software to help breeders in phenotyping and genotyping activities, so that the development of new crop varieties can match the increasing global demand for food. The use of crop models and breeding support systems can help smallholder farmers and breeders prepare for future climatic risks and environmental scenarios. Managing the risks associated with a variable climate is necessary to successfully adapt agriculture to future scenarios and can reduce the cycle of poverty, vulnerability and dependence on external assistance brought about by climate-related disasters.

8 Future trends

Crop simulation models are increasingly used to provide comprehensive assessments of various agricultural, climatic and environmental scenarios. The expanding ability of crop models to assess agricultural, climate change and environmental impacts on various scales (e.g. points-regions-global) has advanced agricultural science and opened new areas of interdisciplinary research. This expansion of research has raised additional questions and issues that could not be studied previously (e.g. the issue of food security on a global scale). Crop models will continue to provide a key role in addressing novel questions and issues through the use of larger and more detailed data sets combined with model improvements.

Model improvements are needed for interactions of climate factors with other environmental factors, extreme events and crop feedbacks (e.g. source–sink relationships) (Martre et al., 2003; Triboi et al., 2006; Asseng et al., 2015b). A challenge for most crop models is the lack of sensitivity to short-term severe stresses of one to two days related to extreme cold or warm temperatures that can affect yield-determining steps (Lobell et al., 2009). Figure 6 exemplified how multi-model grain yield simulations diverged with higher temperatures, which emphasized the need for model improvement in response to high temperatures. It is also important to consider single or infrequent extreme events such as hailstorms, floods or wind gusts. The increasing frequency of extreme events due to climate change adds an additional modelling challenge because predicting the timing of extreme events is poor (IPCC, 2013). Improving climate model prediction of extreme events and their effects on crop growth is necessary for determining future climate change impacts. Continued research and field experimentation with climate factors (e.g. temperature treatment factors) (Ottman et al., 2012) at different crop developmental stages is essential for improving crop model simulations.

Crop models can assist agricultural decision-making as natural resources (e.g. water, nutrients and land) become increasingly limited. Most crop models consider N uptake mechanisms, but only a few crop models consider other soil nutrients (e.g. potassium and phosphorus), which must be accounted for since nutrient limitation can affect crop growth, especially under elevated CO_2 (Wolf, 1996; Ziska and Bunce, 2007). Additionally, alterations in soil factors due to climate change may cause nutrients to become more limiting in future crop growth. Understanding soil changes with climate change and considering these in crop models will be important (Boote et al., 2013).

Climate change is likely to affect yield quality (e.g. grain protein) (Kimball et al., 2001), which is often poorly simulated. Improvement requires a better understanding and integration of the physiology of yield quality into crop models (Martre et al., 2006). Other factors affecting yield quality and overall crop health are pests and diseases. Crop models can be used to simulate effects of pests and diseases on crops by linking crop models with disease models (Bregaglio and Donatelli, 2015). However, modelling the occurrence, movement and dynamics of diseases in relation to crops and their dynamic interactions are still a challenge.

Crop models can also be extended into the crop genetic underlying structure to improve the understanding of climate-sensitive growth processes. Crop models can simulate crop traits and linking genotype-to-phenotype relationships is a novel area, but additional research is needed due to the complexity of understanding and integrating genotype by phenotype (GxP) interactions into crop models (Boote et al., 2013). Improving GxP prediction requires understanding and algorithms that represent the underlying genetic structure to generate a phenotype of a crop based on simulated dynamics. However, the added rigour to improve GxP predictions should not increase model complexity so that parameterization becomes problematic (Hammer et al., 2010).

The use of multi-model ensembles can improve accuracy and reduce uncertainty of model results (Martre et al., 2015b), but only projections are improved and not the individual models themselves. The use of multiple crop models in MIPs for multi-model climate impact assessments is a promising method for projecting future crop productivity and for comparing results between modelling groups (Challinor et al., 2013). AgMIP is a major international collaborative effort to combine interdisciplinary modelling communities with the goal of improving climate impact projections, crop models and economic models for agricultural advancement (Rosenzweig et al., 2013). A critical area for improvement considered in AgMIP is more intensive testing of crop models with experimental data of increased temperature and elevated CO_2 (Boote et al., 2013).

Crop models are successfully applied in agricultural research, decision-making and policy support, but future applications in agronomy, breeding, climate change and NRM require consideration of new system factors and improved model accuracy.

9 Conclusion

Crop simulation models are tools to help scientists, farmers and policymakers develop strategies for increasing agricultural productivity and sustainability in various climatic and environmental scenarios. Crop models use mathematical algorithms that represent quantitative knowledge of crop growth and environmental interactions to create dynamic simulations based on the principles of crop physiology, soil science and agro-meteorology. Models use daily solar radiation, maximum and minimum temperatures, rainfall, initial soil conditions, soil characteristics, cultivar characteristics and crop management inputs to simulate crop development dynamics (phenology), biomass accumulation, yield, water use and nutrient uptake.

Crop models can extrapolate data beyond field experimentation, allowing for system studies across a wide range of specified environmental conditions, well beyond any possible field experiment, with consistent inputs. Crop models can simulate various treatments of many variables over longer periods than field experimentation for any location at various spatial scales.

Simulation outputs are tested against measured data under diverse growing conditions to evaluate crop models. Data sets used for model testing consist of data describing initial soil conditions, crop-specific management, daily weather, phenology of the crop, yields, yield components and nutrient contents from intermediate harvests and measurements of energy, water and CO_2 fluxes.

Crop models are used to assess the impact of climate change and assist in the development of adaptation strategies. The relationship between climate change and agricultural systems is complex, but the model's ability to project the impact of future climate scenarios enables scientists, farmers and policymakers to improve crop management and breeding strategies. To account for climate change in simulations, crop models are paired with GCMs that generate the processes of interactions between the atmosphere, land surface, oceans and sea ice to simulate climate scenarios. GCMs have a range of course and fine spatial resolutions so it is important to select the resolution that provides accurate spatial variability needed for the regional decision-making and impact assessment being studied.

Crop models can be used to assess the climatic impacts on crops based on projected rainfall, temperature (including extreme events), elevated CO_2, solar dimming and increased ground-level ozone. Crop model simulations can assist to determine sustainable approaches to maximum economic return of various sowing dates, effects and adaptations of breeding new cultivar varieties at specified locations as well as optimal N fertilization and irrigation rates while managing spatial variability. The ability of crop models to simulate the effects of climate and NRM allows farmers to develop agricultural practices and strategies that increase crop production with consideration to environmental impacts and natural resource availability.

Crop modelling enables researchers to explore the benefits of using seasonal forecasts to manage seasonal variability to improve efficiency and profitability. Crop models can use long-term historical weather data to simulate a wide range of management practices to optimize management strategies for season types based on seasonal forecasts. Predicting the type of season (e.g. wet or dry) in rain-fed agriculture before the season begins can enable producers to adjust management practices to save input costs in low-yielding seasons or increase inputs to achieve higher potential yields in wet, high-yielding seasons.

10 Where to look for further information

For understanding general model approaches, the textbook by Wallach et al. (2014) is recommended. Both Boote et al. (2013) and Asseng et al. (2015b) give an overview of crop modelling approaches and applications (including other crops than wheat).

Basso et al. (2016) provides a comprehensive crop model review that includes one of the widest used wheat models, CERES-Wheat.

For a comprehensive summary of modelling approaches for simulating elevated CO_2 effects, see Tubiello and Ewert (2002) and Yin (2013).

Additional information on modelling grain protein percentages can be found in a special issue by Martre et al. (2006).

For information on current international research and projects in the crop modelling community, visit the AgMIP website (www.agmip.org).

11 Acknowledgements

We acknowledge the support provided by the CGIAR Research Program on Climate Change, Agriculture and Food Security and the CGIAR Research Program on Wheat.

12 References

Aggarwal, P., Banerjee, B., Daryaei, M., Bhatia, A., Bala, A., Rani, S., Chander, S., Pathak, H. and Kalra, N., 2006. InfoCrop: A dynamic simulation model for the assessment of crop yields, losses due to pests, and environmental impact of agro-ecosystems in tropical environments. II. *Performance of the model. Agricultural Systems* 89, 47–67.

Amir, J. and Sinclair, T.R., 1991. A model of water limitation on spring wheat growth and yield. *Field Crops Research* 28, 59–69.

Anandhi, A., Frei, A., Pierson, D.C., Schneiderman, E.M., Zion, M.S., Lounsbury, D. and Matonse, A.H., 2011. Examination of change factor methodologies for climate change impact assessment. *Water Resources Research* 47, 10.

Andarzian, B., Bannayan, M., Steduto, P., Mazraeh, H., Barati, M.E., Barati, M.A. and Rahnama, A., 2011. Validation and testing of the AquaCrop model under full and deficit irrigated wheat production in Iran. *Agricultural Water Management* 100, 1–8.

Andarzian, B., Hoogenboom, G., Bannayan, M. and Shirali, M., 2015. Determining optimum sowing date of wheat using CSM-CERES-Wheat model. *Journal of the Saudi Society of Agricultural Sciences* 14, 189–99.

Angulo, C., Rötter, R., Lock, R., Enders, A., Fronzek, S. and Ewert, F., 2013. Implication of crop model calibration strategies for assessing regional impacts of climate change in Europe. *Agricultural and Forest Meteorology* 170, 32–46.

Asseng, S., Bar-Tal, A., Bowden, J.W., Keating, B.A., Van Herwaarden, A., Palta, J.A., Huth, N.I. and Probert, M.E., 2002. Simulation of grain protein content with APSIM-Nwheat. *European Journal of Agronomy* 16, 25–42.

Asseng, S., Dunin, F.X., Fillery, I.R.P., Tennant, D. and Keating, B.A., 2001a. Potential deep drainage under wheat crops in a Mediterranean climate. II. Management opportunities to control drainage. *Australian Journal of Agricultural Research* 52, 57–66.

Asseng, S., Ewert, F., Martre, P., Rotter, R.P., Lobell, D.B., Cammarano, D., Kimball, B.A., Ottman, M.J., Wall, G.W., White, J.W., Reynolds, M.P., Alderman, P.D., Prasad, P.V.V., Aggarwal, P.K.,

Anothai, J., Basso, B., Biernath, C., Challinor, A.J., De Sanctis, G., Doltra, J., Fereres, E., Garcia-Vile, M., Gayler, S., Hoogenboom, G., Hunt, L.A., Izaurralde, R.C., Jabloun, M., Jones, C.D., Kersebaum, K.C., Koehler, A.K., Muller, C., Kumar, S.N., Nendel, C., O'Leary, G., Olesen, J.E., Palosuo, T., Priesack, E., Rezaei, E.E., Ruane, A.C., Semenov, M.A., Shcherbak, I., Stockle, C., Stratonovitch, P., Streck, T., Supit, I., Tao, F., Thorburn, P.J., Waha, K., Wang, E., Wallach, D., Wolf, I., Zhao, Z. and Zhu, Y., 2015a. Rising temperatures reduce global wheat production. *Nature Climate* Change 5, 143–7.

Asseng, S., Ewert, F., Rosenzweig, C., Jones, J.W., Hatfield, J.L., Ruane, A.C., Boote, K.J., Thorburn, P.J., Rotter, R.P., Cammarano, D., Brisson, N., Basso, B., Martre, P., Aggarwal, P.K., Angulo, C., Bertuzzi, P., Biernath, C., Challinor, A.J., Doltra, J., Gayler, S., Goldberg, R., Grant, R., Heng, L., Hooker, J., Hunt, L.A., Ingwersen, J., Izaurralde, R.C., Kersebaum, K.C., Mueller, C., Kumar, S.N., Nendel, C., O'Leary, G., Olesen, J.E., Osborne, T.M., Palosuo, T., Priesack, E., Ripoche, D., Semenov, M.A., Shcherbak, I., Steduto, P., Stoeckle, C., Stratonovitch, P., Streck, T., Supit, I., Tao, F., Travasso, M., Waha, K., Wallach, D., White, J.W., Williams, J.R. and Wolf, J., 2013a. Uncertainty in simulating wheat yields under climate change. *Nature Climate Change* 3, 827–32.

Asseng, S., Fillery, I.R.P., Anderson, G.C., Dolling, P.J., Dunin, F.X. and Keating, B.A., 1998a. Use of the APSIM wheat model to predict yield, drainage, and NO3'(-) leaching for a deep sand. *Australian Journal of Agricultural Research* 49, 363–77.

Asseng, S., Fillery, I.R.P., Dunin, F.X., Keating, B.A. and Meinke, H., 2001b. Potential deep drainage under wheat crops in a Mediterranean climate. I. Temporal and spatial variability. *Australian Journal of Agricultural Research* 52, 45–56.

Asseng, S., Foster, I. and Turner, N.C., 2011. The impact of temperature variability on wheat yields. *Global Change Biology* 17, 997–1012.

Asseng, S., Jamieson, P.D., Kimball, B., Pinter, P., Sayre, K., Bowden, J.W. and Howden, S.M., 2004. Simulated wheat growth affected by rising temperature, increased water deficit and elevated atmospheric CO_2. *Field Crops Research* 85, 85–102.

Asseng, S., Keating, B.A., Fillery, I.R.P., Gregory, P.J., Bowden, J.W., Turner, N.C., Palta, J.A. and Abrecht, D.G., 1998b. Performance of the APSIM-wheat model in Western Australia. *Field Crops Research* 57, 163–79.

Asseng, S., McIntosh, P.C., Thomas, G., Ebert, E.E. and Khimashia, N., 2016. Is a 10-day rainfall forecast of value in dry-land wheat cropping? *Agricultural and Forest Meteorology* 216, 170–6.

Asseng, S., McIntosh, P.C., Wang, G.M. and Khimashia, N., 2012. Optimal N fertiliser management based on a seasonal forecast. *European Journal of Agronomy* 38, 66–73.

Asseng, S. and Milroy, S.P., 2006. Simulation of environmental and genetic effects on grain protein concentration in wheat. *European Journal of Agronomy* 25, 119–28.

Asseng, S., Milroy, S.P. and Poole, M.L., 2008. Systems analysis of wheat production on low water-holding soils in a Mediterranean-type environment I. Yield potential and quality. *Field Crops Research* 105, 97–106.

Asseng, S., Travasso, M.I., Ludwig, F. and Magrin, G.O., 2013b. Has climate change opened new opportunities for wheat cropping in Argentina? *Climatic Change* 117, 181–96.

Asseng, S., van Keulen, H. and Stol, W., 2000. Performance and application of the APSIM Nwheat model in the Netherlands. *European Journal of Agronomy* 12, 37–54.

Asseng, S., Zhu, Y., Basso, B., Wilson, T. and Cammarano, D., 2014. Simulation Modeling: Applications in Cropping Systems. In N. K. Van Alfen (ed.), *Encyclopedia of Agriculture and Food Systems*. Academic Press, Oxford, pp. 102–12.

Asseng, S., Zhu, Y., Wang, E. and Zhang, W., 2015b. Chapter 20: Crop modeling for climate change impact and adaptation. In V. Sadras, and D. Calderini (eds.), *Crop Physiology*. Academic Press, San Diego, pp. 505–46.

Barlow, K.M., Christy, B.P., O'Leary, G.J., Riffkin, P.A. and Nuttall, J.G., 2015. Simulating the impact of extreme heat and frost events on wheat crop production: A review. *Field Crops Research* 171, 109–19.

Basso, B., Cammarano, D., Fiorentino, C. and Ritchie, J.T., 2013. Wheat yield response to spatially variable nitrogen fertilizer in Mediterranean environment. *European Journal of Agronomy* 51, 65–70.

Basso, B., Cammarano, D., Troccoli, A., Chen, D. and Ritchie, J., 2010. Long-term wheat response to nitrogen in a rainfed Mediterranean environment: Field data and simulation analysis. *European Journal of Agronomy* 33, 132–8.

Basso, B., Liu, L. and Ritchie, J.T., 2016. *A Comprehensive Review of the CERES-Wheat, -Maize and -Rice Models' Performances*. Advances in Agronomy. Elsevier, pp. 27–132.

Basso, B., Ritchie, J.T., Cammarano, D. and Sartori, L., 2011. A strategic and tactical management approach to select optimal N fertilizer rates for wheat in a spatially variable field. *European Journal of Agronomy* 35, 215–22.

Basso, B., Sartori, L., Cammarano, D., Fiorentino, C., Grace, P.R., Fountas, S. and Sorensen, C.A., 2012. Environmental and economic evaluation of N fertilizer rates in a maize crop in Italy: A spatial and temporal analysis using crop models. *Biosystems Engineering* 113, 103–11.

Beringer, T., Lucht, W. and Schaphoff, S., 2011. Bioenergy production potential of global biomass plantations under environmental and agricultural constraints. *Global Change Biology Bioenergy* 3, 299–312.

Berntsen, J., Petersen, B., Jacobsen, B., Olesen, J. and Hutchings, N., 2003. Evaluating nitrogen taxation scenarios using the dynamic whole farm simulation model FASSET. *Agricultural Systems* 76, 817–39.

Bertin, N., Martre, P., Genard, M., Quilot, B. and Salon, C., 2010. Under what circumstances can process-based simulation models link genotype to phenotype for complex traits? Case-study of fruit and grain quality traits. *Journal of Experimental Botany* 61, 955–67.

Biernath, C., Gayler, S., Bittner, S., Klein, C., Hogy, P., Fangmeier, A. and Priesack, E., 2011. Evaluating the ability of four crop models to predict different environmental impacts on spring wheat grown in open-top chambers. *European Journal of Agronomy* 35, 71–82.

Bondeau, A., Smith, P.C., Zaehle, S., Schaphoff, S., Lucht, W., Cramer, W. and Gerten, D., 2007. Modelling the role of agriculture for the 20th century global terrestrial carbon balance. *Global Change Biology* 13, 679–706.

Boogaard, H. and Kroes, J., 1998. Leaching of nitrogen and phosphorus from rural areas to surface waters in the Netherlands. *Nutrient Cycling in Agroecosystems* 50, 321–4.

Boote, K.J., Jones, J.W. and Pickering, N.B., 1996. Potential uses and limitations of crop models. *Agronomy Journal* 88, 704–16.

Boote, K.J., Jones, J.W., White, J.W., Asseng, S. and Lizaso, J.I., 2013. Putting mechanisms into crop production models. *Plant Cell and Environment* 36, 1658–72.

Bregaglio, S. and Donatelli, M., 2015. A set of software components for the simulation of plant airborne diseases. *Environmental Modelling & Software* 72, 426–44.

Brisson, N., Gary, C., Justes, E., Roche, R., Mary, B., Ripoche, D., Zimmer, D., Sierra, J., Bertuzzi, P., Burger, P., Bussiere, F., Cabidoche, Y.M., Cellier, P., Debaeke, P., Gaudillere, J.P., Henault, C., Maraux, F., Seguin, B. and Sinoquet, H., 2003. An overview of the crop model STICS. *European Journal of Agronomy* 18, 309–32.

Brisson, N., Mary, B., Ripoche, D., Jeuffroy, M.H., Ruget, F., Nicoullaud, B., Gate, P., Devienne-Barret, F., Antonioletti, R., Durr, C., Richard, G., Beaudoin, N., Recous, S., Tayot, X., Plenet, D., Cellier, P., Machet, J.M., Meynard, J.M. and Delecolle, R., 1998. STICS: a generic model for the simulation of crops and their water and nitrogen balances. I. Theory and parameterization applied to wheat and corn. *Agronomie* 18, 311–46.

Cammarano, D., Payero, J., Basso, B., Stefanova, L. and Grace, P., 2012. Adapting wheat sowing dates to projected climate change in the Australian subtropics: Analysis of crop water use and yield. *Crop & Pasture Science* 63, 974–86.

Cao, W., Liu, T., Luo, W., Wang, S., Pan, J. and Guo, W., 2002. Simulating organic growth in wheat based on the organ-weight fraction concept. *Plant Production Science* 5, 248–56.

Cao, W., Moss and D.N., 1997. Modelling phasic development in wheat: a conceptual integration of physiological components. *Journal of Agricultural Science* 129, 163–72.

Challinor, A., Wheeler, T., Craufurd, P., Slingo, J. and Grimes, D., 2004. Design and optimisation of a large-area process-based model for annual crops. *Agricultural and Forest Meteorology* 124, 99–120.

Challinor, A.J., Smith, M.S. and Thornton, P., 2013. Use of agro-climate ensembles for quantifying uncertainty and informing adaptation. *Agricultural and Forest Meteorology* 170, 2–7.

Chen, C., Wang, E. and Yu, Q., 2010. Modeling wheat and maize productivity as affected by climate variation and irrigation supply in North China plain. *Agronomy Journal* 102, 1037–49.

Chenu, K., Cooper, M., Hammer, G.L., Mathews, K.L., Dreccer, M.F. and Chapman, S.C., 2011. Environment characterization as an aid to wheat improvement: interpreting genotype-environment interactions by modelling water-deficit patterns in North-Eastern Australia. *Journal of Experimental Botany* 62, 1743–55.

Chenu, K., Deihimfard, R. and Chapman, S.C., 2013. Large-scale characterization of drought pattern: a continent-wide modelling approach applied to the Australian wheatbelt spatial and temporal trends. *New Phytologist* 198, 801–20.

Colecchia, S.A., Basso, B., Cammarano, D., Gallo, A., Mastrangelo, A.M., Pontieri, P., Del Giudice, L., Pignone, D. and De Vita, P., 2013. On the relationship between N management and grain protein content in six durum wheat cultivars in Mediterranean environment. *Journal of Plant Interactions* 8, 271–9.

Coumou, D. and Rahmstorf, S., 2012. A decade of weather extremes. *Nature Climate Change* 2, 491–6.

De Wit, C.T., 1965. Photosynthesis of leaf canopies. Agricultural Research Reports; Centre for Agricultural Publication and Documentation 663, 57.

Dlugokencky, E. and Tans, P., 2016. Trends in Atmospheric Carbon Dioxide. NOAA/ESRL.

Duncan, W.G., Loomis, R.S., Williams, W.A. and Hanau, R., 1967. A model for simulating photosynthesis in plant communities. *Hilgardia* 38, 181–205.

Duursma, R.A., Payton, P., Bange, M.P., Broughton, K.J., Smith, R.A., Medlyn, B.E. and Tissue, D.T., 2013. Near-optimal response of instantaneous transpiration efficiency to vapour pressure deficit, temperature and CO_2 in cotton (Gossypium hirsutum L.). *Agricultural and Forest Meteorology* 168, 168–76.

Easterling, D.R., Horton, B., Jones, P.D., Peterson, T.C., Karl, T.R., Parker, D.E., Salinger, M.J., Razuvayev, V., Plummer, N., Jamason, P. and Folland, C.K., 1997. Maximum and minimum temperature trends for the globe. *Science* 277, 364–7.

Eitzinger, J., Thaler, S., Schmid, E., Strauss, F., Ferrise, R., Moriondo, M., Bindi, M., Palosuo, T., Rotter, R., Kersebaum, K.C., Olesen, J.E., Patil, R.H., Saylan, L., Caldag, B. and Caylak, O., 2013. Sensitivities of crop models to extreme weather conditions during flowering period demonstrated for maize and winter wheat in Austria. *Journal of Agricultural Science* 151, 813–35.

Ewert, F., Rotter, R.P., Bindi, M., Webber, H., Trnka, M., Kersebaum, K.C., Olesen, J.E., van Ittersum, M.K., Janssen, S., Rivington, M., Semenov, M.A., Wallach, D., Porter, J.R., Stewart, D., Verhagen, J., Gaiser, T., Palosuo, T., Tao, F., Nendel, C., Roggero, P.P., Bartosova, L. and Asseng, S., 2015. Crop modelling for integrated assessment of risk to food production from climate change. *Environmental Modelling & Software* 72, 287–303.

Ewert, F., van Ittersum, M.K., Bezlepkina, I., Therond, O., Andersen, E., Belhouchette, H., Bockstaller, C., Brouwer, F., Heckelei, T., Janssen, S., Knapen, R., Kuiper, M., Louhichi, K., Olsson, J.A., Turpin, N., Wery, J., Wien, J.E. and Wolf, J., 2009. A methodology for enhanced flexibility of integrated assessment in agriculture. *Environmental Science & Policy* 12, 546–61.

Ewert, F., van Ittersum, M.K., Heckelei, T., Therond, O., Bezlepkina, I. and Andersen, E., 2011. Scale changes and model linking methods for integrated assessment of agri-environmental systems. *Agriculture Ecosystems & Environment* 142, 6–17.

Ewert, F., van Oijen, M. and Porter, J.R., 1999. Simulation of growth and development processes of spring wheat in response to CO_2 and ozone for different sites and years in Europe using mechanistic crop simulation models. *European Journal of Agronomy* 10, 231–47.

Fader, M., Rost, S., Muller, C., Bondeau, A. and Gerten, D., 2010. Virtual water content of temperate cereals and maize: Present and potential future patterns. *Journal of Hydrology* 384, 218–31.

FAOSTAT, 2015. Food and Agriculture Organization of the United Nations, FAOSTAT database. FAO.

Farquhar, G.D., Caemmerer, S.V. and Berry, J.A., 1980. A biochemical model of photosynthetic CO_2 assimilation in leaves of C3 species. *Planta* 149, 78–90.

Farquhar, G.D., Dubbe, D.R. and Raschke, K., 1978. Gain of the feedback loop involving carbon sioxide and stomata. Theory and measurement. *Plant Physiology* 62, 406–12.

Ferris, R., Ellis, R.H., Wheeler, T.R. and Hadley, P., 1998. Effect of high temperature stress at anthesis on grain yield and biomass of field-grown crops of wheat. *Annals of Botany* 82, 631–9.

Ferrise, R., Triossi, A., Stratonovitch, P., Bindi, M. and Martre, P., 2010. Sowing date and nitrogen fertilisation effects on dry matter and nitrogen dynamics for durum wheat: An experimental and simulation study. *Field Crops Research* 117, 245–57.

Fraisse, C.W., Breuer, N.E., Zierden, D., Bellow, J.G., Paz, J., Cabrera, V.E., Garcia, A.G.Y., Ingram, K.T., Hatch, U., Hoogenboom, G., Jones, J.W. and O'Brien, J.J., 2006. AgClimate: A climate forecast information system for agricultural risk management in the southeastern USA. *Computers and Electronics in Agriculture* 53, 13–27.

Gerten, D., Schaphoff, S., Haberlandt, U., Lucht, W. and Sitch, S., 2004. Terrestrial vegetation and water balance - hydrological evaluation of a dynamic global vegetation model. *Journal of Hydrology* 286, 249–70.

Gifford, R.M. and Morison, J.I.L., 1993. Crop responses to the global increase in atmospheric carbon dioxide concentration. In D. R. Buxton, R. A. Forsberg, B. L. Blad, H. Asay, G. M. Paulsen and R. F. Wilson (eds), *International Crop Science, I. Crop Science Society of America Inc.* {a}, 677 South Segoe Road, Madison, Wisconsin 53711, USA, pp. 325–31.

Giorgi, F. and Bi, X., 2005. Regional changes in surface climate interannual variability for the 21st century from ensembles of global model simulations. *Geophysical Research Letters* 32, 5.

Godfray, H.C.J., Beddington, J.R., Crute, I.R., Haddad, L., Lawrence, D., Muir, J.F., Pretty, J., Robinson, S., Thomas, S.M. and Toulmin, C., 2010. Food security: The challenge of feeding 9 Billion people. *Science* 327, 812–18.

Goudriaan, J. and Van Laar, H.H., 1994. *Modelling Potential Crop Growth Processes. Textbook With Exercises.* Kluwer Academic Publishers, Dordrecht, The Netherlands.

Goudriaan, J. and Vanlaar, H.H., 1978. Calculation of daily totals of the gross CO_2 assimilation of leaf canopies. *Netherlands Journal of Agricultural Science* 26, 373–82.

Gourdji, S.M., Mathews, K.L., Reynolds, M., Crossa, J. and Lobell, D.B., 2013. An assessment of wheat yield sensitivity and breeding gains in hot environments *Proceedings of the Royal Society B-Biological Sciences* 280.

Grandjean, A. and Fuhrer, J., 1989. Growth and leaf senescence in spring wheat (*Triticum-aestivum*) grown at different ozone concentratrions in open-top field chambers. *Physiologia Plantarum* 77, 389–94.

Grant, R.F., Garcia, R.L., Pinter, P.J., Hunsaker, D., Wall, G.W., Kimball, B.A. and LaMorte, R.L., 1995. Interaction between atmospheric CO_2 concentration and water deficit on gas exchange and crop growth: Testing of ecosys with data from the Free Air CO_2 Enrichment (FACE) experiment. *Global Change Biology* 1, 443–54.

Groot, J.J.R. and Verberne, E.L.J., 1991. Response of wheat to nitrogen fertilization, a data set to validate simulation models for nitrogen dynamics in crop and soil. *Fertilizer Research* 27, 349–83.

Grossman-Clarke, S., Pinter, E.J., Kartschall, T., Kimball, B.A., Hunsaker, D.J., Wall, G.W., Garcia, R.L. and LaMorte, R.L., 2001. Modelling a spring wheat crop under elevated CO_2 and drought. *New Phytologist* 150, 315–35.

Hammer, G.L., Holzworth, D.P. and Stone, R., 1996. The value of skill in seasonal climate forecasting to wheat crop management in a region with high climatic variability. *Australian Journal of Agricultural Research* 47, 717–37.

Hammer, G.L., van Oosterom, E., McLean, G., Chapman, S.C., Broad, I., Harland, P. and Muchow, R.C., 2010. Adapting APSIM to model the physiology and genetics of complex adaptive traits in field crops. *Journal of Experimental Botany* 61, 2185–2202.

Hansen, S., Abrahamsen, P., Petersen, C.T. and Styczen, M., 2012. DAISY: model use, calibration, and validation. *Transaction of the ASABE* 55, 1317–35.

Hansen, S., Jensen, H., Nielsen, N. and Svendsen, H., 1991. Simulation of nitrogen dynamics and biomass production in winter-wheat using the Danish simulation model DAISY. *Fertilizer Research* 27, 245–59.

He, J., Stratonovitch, P., Allard, V., Semenov, M.A. and Martre, P., 2010. Global sensitivity analysis of the process-based wheat simulation model siriusquality1 identifies key genotypic parameters and unravels parameters interactions. *Procedia – Social and Behavioral Sciences* 2, 7676–7.

Heng, L.K., Asseng, S., Mejahed, K. and Rusan, M., 2007. Optimizing wheat productivity in two rain-fed environments of the West Asia-North Africa region using a simulation model. *European Journal of Agronomy* 26, 121–9.

Hesketh, J.D., Baker, D.N. and Duncan, W.G., 1971. Simulation of growth and yield in cotton – respiration and carbon balance. *Crop Science* 11, 394–8.

Hochman, Z., van Rees, H., Carberry, P.S., Hunt, J.R., McCown, R.L., Gartmann, A., Holzworth, D., van Rees, S., Dalgliesh, N.P., Long, W., Peake, A.S., Poulton, P.L. and McClelland, T., 2009. Re-inventing model-based decision support with Australian dryland farmers. 4. Yield Prophet (R) helps farmers monitor and manage crops in a variable climate. *Crop & Pasture Science* 60, 1057–70.

Hoogenboom, G. and White, J.W., 2003. Improving physiological assumptions of simulation models by using gene-based approaches. *Agronomy Journal* 95, 82–9.

Hu, J.C., Cao, W.X., Zhang, J.B., Jiang, D. and Feng, J., 2004. Quantifying responses of winter wheat physiological processes to soil water stress for use in growth simulation modeling. *Pedosphere* 14, 509–18.

Hunt, L.A. and Pararajasingham, S., 1995. CROPSIM-wheat – a model describing the growth and development of wheat. *Canadian Journal of Plant Science* 75, 619–32.

Iizumi, T., Sakuma, H., Yokozawa, M., Luo, J.J., Challinor, A.J., Brown, M.E., Sakurai, G. and Yamagata, T., 2013. Prediction of seasonal climate-induced variations in global food production. *Nature Climate Change* 3, 904–8.

IPCC, 2013. Climate Change 2013: The Physical Science Basis. Contribution of Working Group I to the Fifth Assessment Report of the Intergovernmental Panel on Climate Change. In T. F. Stocker, D. Qin, G.-K. Plattner, M. Tignor, S. K. Allen, J. Boschung, A. Nauels, Y. Xia, V. Bex and P. M. Midgley (eds), Cambridge, United Kingdom and New York, NY, USA, p. 1535.

Jamieson, P. and Semenov, M., 2000. Modelling nitrogen uptake and redistribution in wheat. *Field Crops Research* 68, 21–9.

Jamieson, P., Semenov, M., Brooking, I. and Francis, G., 1998a. Sirius: a mechanistic model of wheat response to environmental variation. *European Journal of Agronomy* 8, 161–79.

Jamieson, P.D., Asseng, S., Chapman, S.C., Dreccer, M.F., White, J.W., McMaster, G.S., Porter, J.R. and Semenov, M.A., 2010. Modelling wheat production. In M. van Ginkel A. Bonjean and W. Angus (eds), *The World Wheat Book*. Lavoisier Publishing, Paris, p. 40.

Jamieson, P.D., Berntsen, J., Ewert, F., Kimball, B.A., Olesen, J.E., Pinter, P.J., Porter, J.R. and Semenov, M.A., 2000. Modelling CO_2 effects on wheat with varying nitrogen supplies. *Agriculture Ecosystems & Environment* 82, 27–37.

Jamieson, P.D., Martin, R.J. and Francis, G.S., 1995. Drought influences on grain yield of barley, wheat, and maize. *New Zealand Journal of Crop and Horticultural Science* 23, 55–66.

Jamieson, P.D., Porter, J.R., Goudriaan, J., Ritchie, J.T., van Keulen, H. and Stol, W., 1998b. A comparison of the models AFRCWHEAT2, CERES-wheat, Sirius, SUCROS2 and SWHEAT with measurements from wheat grown under drought. *Field Crops Research* 55, 23–44.

Jones, J.W., Hoogenboom, G., Porter, C.H., Boote, K.J., Batchelor, W.D., Hunt, L.A., Wilkens, P.W., Singh, U., Gijsman, A.J. and Ritchie, J.T., 2003. The DSSAT cropping system model. *European Journal of Agronomy* 18, 235–65.

Kang, Y.H., Khan, S. and Ma, X.Y., 2009. Climate change impacts on crop yield, crop water productivity and food security – A review. *Progress in Natural Science-Materials International* 19, 1665–74.

Kartschall, T., Grossman, S., Pinter, P.J., Garcia, R.L., Kimball, B.A., Wall, G.W., Hunsaker, D.J. and LaMorte, R.L., 1995. A simulation of phenology, growth, carbon dioxide exchange and yields under ambient atmosphere and free-air carbon dioxide enrichment (FACE) Maricopa, Arizona, for wheat. *Journal of Biogeography* 22, 611–22.

Keating, B.A., Carberry, P.S., Hammer, G.L., Probert, M.E., Robertson, M.J., Holzworth, D., Huth, N.I., Hargreaves, J.N.G., Meinke, H., Hochman, Z., McLean, G., Verburg, K., Snow, V., Dimes, J.P., Silburn, M., Wang, E., Brown, S., Bristow, K.L., Asseng, S., Chapman, S., McCown, R.L., Freebairn, D.M. and Smith, C.J., 2003. An overview of APSIM, a model designed for farming systems simulation. *European Journal of Agronomy* 18, 267–88.

Keating, B.A., Meinke, H., Probert, M.E., Huth, N.I. and Hills, I.G., 2001. NWheat: Documentation and performance of a wheat module for APSIM. *Tropical Agriculture Technical Memorandum*, 1–66.

Kersebaum, K., 2007. Modelling nitrogen dynamics in soil-crop systems with HERMES. *Nutrient Cycling in Agroecosystems* 77, 39–52.

Kersebaum, K.C., 2011. Special features of the HERMES model and additional procedures for parameterization, calibration, validation, and applications. In L. R. Ahuja and L. Ma (eds), *Methods of Introducing System Models into Agricultural Research*. Advances in Agricultural Systems Modeling Series 2, Madison (ASA-CSSA-SSSA), pp. 65–94.

Kersebaum, K.C., Boote, K.J., Jorgenson, J.S., Nendel, C., Bindi, M., Fruhauf, C., Gaiser, T., Hoogenboom, G., Kollas, C., Olesen, J.E., Rotter, R.P., Ruget, F., Thorburn, P.J., Trnka, M. and Wegehenkel, M., 2015. Analysis and classification of data sets for calibration and validation of agro-ecosystem models. *Environmental Modelling & Software* 72, 402–17.

Kimball, B.A., Kobayashi, K. and Bindi, M., 2002. Responses of agricultural crops to free-air CO_2 enrichment. *Advances in Agronomy*, 77, 293–368.

Kimball, B.A., Morris, C.F., Pinter, P.J., Wall, G.W., Hunsaker, D.J., Adamsen, F.J., LaMorte, R.L., Leavitt, S.W., Thompson, T.L., Matthias, A.D. and Brooks, T.J., 2001. Elevated CO_2, drought and soil nitrogen effects on wheat grain quality. *New Phytologist* 150, 295–303.

Kiniry, J.R., Major, D.J., Izaurralde, R.C., Williams, J.R., Gassman, P.W., Morrison, M., Bergentine, R. and Zentner, R.P., 1995. EPIC model parameters for cereal, oilseed, and forage crops in the northern great-plains region. *Canadian Journal of Plant Science* 75, 679–88.

Ko, J., Ahuja, L., Kimball, B., Anapalli, S., Ma, L.W., Green, T.R., Ruane, A.C., Wall, G.W., Pinter, P. and Bader, D.A., 2010. Simulation of free air CO_2 enriched wheat growth and interactions with water, nitrogen, and temperature. *Agricultural and Forest Meteorology* 150, 1331–46.

Latta, J. and O'Leary, G., 2003. Long-term comparison of rotation and fallow tillage systems of wheat in Australia. *Field Crops Research* 83, 173–90.

Lawless, C., Semenov, M. and Jamieson, P., 2005. A wheat canopy model linking leaf area and phenology. *European Journal of Agronomy* 22, 19–32.

Lehnherr, B., Grandjean, A., Machler, F. and Fuhrer, J., 1987. The effect of ozone in ambient air on ribulosebisphosphate carboxylase oxygenase activity decreases photosynthesis and grain yield in wheat. *Journal of Plant Physiology* 130, 189–200.

Li, C., Cao, W. and Zhang, Y., 2002. Comprehensive Pattern of Pr mordium Initiation in Shoot Apex of Wheat. *ACTA Botanica Sinica*, 273–8.

Li, H. and Wu, J., 2006. Uncertainty analysis in ecological studies: An overview. In J. Wu, K. B. Jones, H. Li, and O. L. Loucks (eds), *Scaling and Uncertainty Analysis in Ecology: Methods and Applications*. Springer, Po Box 17, 3300 Aa Dordrecht, Netherlands, pp. 45–66.

Li, K.N., Yang, X.G., Tian, H.Q., Pan, S.F., Liu, Z.J. and Lu, S., 2016. Effects of changing climate and cultivar on the phenology and yield of winter wheat in the North China Plain. *International Journal of Biometeorology* 60, 21–32.

Li, S., Wheeler, T., Challinor, A., Lin, E., Xu, Y. and Ju, H., 2010. Simulating the impacts of global warming on wheat in China using a large area crop model. *Acta Meteorologica Sinica* 24, 123–35.

Liu, B., Liu, L.L., Tian, L.Y., Cao, W.X., Zhu, Y. and Asseng, S., 2014. Post-heading heat stress and yield impact in winter wheat of China. *Global Change Biology* 20, 372–81.

Liu, Y.A., Wang, E.L., Yang, X.G. and Wang, J., 2010. Contributions of climatic and crop varietal changes to crop production in the North China Plain, since 1980s. *Global Change Biology* 16, 2287–99.

Lobell, D.B., 2007. Changes in diurnal temperature range and national cereal yields. *Agricultural and Forest Meteorology* 145, 229–38.

Lobell, D.B., Cassman, K.G. and Field, C.B., 2009. Crop yield gaps: Their importance, magnitudes, and causes. *Annual Review of Environment and Resources* 34, 179–204.

Lobell, D.B., Hammer, G.L., Chenu, K., Zheng, B.Y., McLean, G. and Chapman, S.C., 2015. The shifting influence of drought and heat stress for crops in northeast Australia. *Global Change Biology* 21, 4115–27.

Lobell, D.B., Sibley, A. and Ortiz-Monasterio, J.I., 2012. Extreme heat effects on wheat senescence in India. *Nature Climate Change* 2, 186–9.

Loomis, R.S. and Rabbinge, R., Ng, E., 1979. Explanatory models in crop physiology. *Annual Review of Plant Physiology and Plant Molecular Biology* 30, 339–67.

Ludwig, F. and Asseng, S., 2010. Potential benefits of early vigor and changes in phenology in wheat to adapt to warmer and drier climates. *Agricultural Systems* 103, 127–36.

Lv, Z.F., Liu, X.J., Cao, W.X. and Zhu, Y., 2013. Climate change impacts on regional winter wheat production in main wheat production regions of China. *Agricultural and Forest Meteorology* 171, 234–48.

Martre, P., He, J.Q., Le Gouis, J. and Semenov, M.A., 2015a. In silico system analysis of physiological traits determining grain yield and protein concentration for wheat as influenced by climate and crop management. *Journal of Experimental Botany* 66, 3581–98.

Martre, P., Jamieson, P.D., Semenov, M.A., Zyskowski, R.F., Porter, J.R. and Triboi, E., 2006. Modelling protein content and composition in relation to crop nitrogen dynamics for wheat. *European Journal of Agronomy* 25, 138–54.

Martre, P., Porter, J.R., Jamieson, P.D. and Triboi, E., 2003. Modeling grain nitrogen accumulation and protein composition to understand the Sink/Source regulations of nitrogen remobilization for wheat. *Plant Physiology* 133, 1959–67.

Martre, P., Wallach, D., Asseng, S., Ewert, F., Jones, J.W., Rotter, R.P., Boote, K.J., Ruane, A.C., Thorburn, P.J., Cammarano, D., Hatfield, J.L., Rosenzweig, C., Aggarwal, P.K., Angulo, C., Basso, B., Bertuzzi, P., Biernath, C., Brisson, N., Challinor, A.J., Doltra, J., Gayler, S., Goldberg, R., Grant, R.F., Heng, L., Hooker, J., Hunt, L.A., Ingwersen, J., Izaurralde, R.C., Kersebaum, K.C., Mueller, C., Kumar, S.N., Nendel, C., O'Leary, G., Olesen, J.E., Osborne, T.M., Palosuo, T., Priesack, E., Ripoche, D., Semenov, M.A., Shcherbak, I., Steduto, P., Stoeckle, C.O., Stratonovitch, P., Streck, T., Supit, I., Tao, F., Travasso, M., Waha, K., White, J.W. and Wolf, J., 2015b. Multimodel ensembles of wheat growth: many models are better than one. *Global Change Biology* 21, 911–25.

Moeller, C., Smith, I., Asseng, S., Ludwig, F.a nd Telcik, N., 2008. The potential value of seasonal forecasts of rainfall categories – Case studies from the wheatbelt in Western Australia's Mediterranean region. *Agricultural and Forest Meteorology* 148, 606–18.

Mollah, M., Norton, R. and Huzzey, J., 2009. Australian grains free-air carbon dioxide enrichment (AGFACE) facility: design and performance. *Crop & Pasture Science* 60, 697–707.

Monteith, J.L., 1965a. Evaporation and environment. *Symposia of the Society for Experimental Biology* 19, 205–34.

Monteith, J.L., 1965b. Light distribution and photosynthesis in field crops. *Annals of Botany* 29, 17–37.

Monteith, J.L., 1977. Climate and efficiency of crop production in Britain. *Philosophical Transactions of the Royal Society of London Series B-Biological Sciences* 281, 277–94.

Moriondo, M., Giannakopoulos, C. and Bindi, M., 2011. Climate change impact assessment: the role of climate extremes in crop yield simulation. *Climatic Change* 104, 679–701.

Moulin, S., Bondeau, A. and Delecolle, R., 1998. Combining agricultural crop models and satellite observations: from field to regional scales. *International Journal of Remote Sensing* 19, 1021–36.

Müller, C., Eickhout, B., Zaehle, S., Bondeau, A., Cramer, W. and Lucht, W., 2007. Effects of changes in CO_2, climate, and land use on the carbon balance of the land biosphere during the 21st century. *Journal of Geophysical Research-Biogeosciences* 112.

Nendel, C., Berg, M., Kersebaum, K., Mirschel, W., Specka, X., Wegehenkel, M., Wenkel, K. and Wieland, R., 2011. The MONICA model: Testing predictability for crop growth, soil moisture and nitrogen dynamics. *Ecological Modelling* 222, 1614–25.

NOAA National Centers for Environmental Information, 2016. State of the Climate: Global Analysis for Annual 2015. NOAA.

Nuttall, J.G., O'Leary, G.J., Khimashia, N., Asseng, S., Fitzgerald, G. and Norton, R., 2012. 'Haying-off' in wheat is predicted to increase under a future climate in south-eastern Australia. *Crop & Pasture Science* 63, 593–605.

O'Leary, G.J., Christy, B., Nuttall, J., Huth, N., Cammarano, D., Stockle, C., Basso, B., Shcherbak, I., Fitzgerald, G., Luo, Q.Y., Farre-Codina, I., Palta, J. and Asseng, S., 2015. Response of wheat growth, grain yield and water use to elevated CO_2 under a Free-Air CO_2 Enrichment (FACE) experiment and modelling in a semi-arid environment. *Global Change Biology* 21, 2670–86.

O'Leary, G.J. and Connor, D.J., 1996a. A simulation model of the wheat crop in response to water and nitrogen supply .1. Model construction. *Agricultural Systems* 52, 1–29.

O'Leary, G.J. and Connor, D.J., 1996b. A simulation model of the wheat crop in response to water and nitrogen supply .2. Model validation. *Agricultural Systems* 52, 31–55.

O'Leary, G.J., Connor, D.J. and White, D.H., 1985. A simulation-model of the development, growth and yield of the wheat crop. *Agricultural Systems* 17, 1–26.

Olesen, J., Petersen, B., Berntsen, J., Hansen, S., Jamieson, P. and Thomsen, A., 2002. Comparison of methods for simulating effects of nitrogen on green area index and dry matter growth in winter wheat. *Field Crops Research* 74, 131–49.

Olesen, J.E., Jensen, T. and Petersen, J., 2000. Sensitivity of field-scale winter wheat production in Denmark to climate variability and climate change. *Climate Research* 15, 221–38.

Orlowsky, B. and Seneviratne, S.I., 2012. Global changes in extreme events: Regional and seasonal dimension. *Climatic Change* 110, 669–96.

Ottman, M.J., Kimball, B.A., White, J.W. and Wall, G.W., 2012. Wheat growth response to increased temperature from varied planting dates and supplemental infrared heating. *Agronomy Journal* 104, 7–16.

Pan, J., Zhu, Y. and Cao, W., 2007. Modeling plant carbon flow and grain starch accumulation in wheat. *Field Crops Research* 101, 276–84.

Pan, J., Zhu, Y., Jiang, D., Dai, T.B., Li, Y.X. and Cao, W.X., 2006. Modeling plant nitrogen uptake and grain nitrogen accumulation in wheat. *Field Crops Research* 97, 322–36.

Penman, H.L., 1948. Natural evaporation from open water, bare soil and grass. *Proceedings of the Royal Society of London Series a-Mathematical and Physical Sciences* 193, 120–45.

Pierce, F.J. and Nowak, P., 1999. Aspects of precision agriculture. *Advances in Agronomy* 67, 1–85.

Porter, J.R., 1984. A model of canopy development in winter wheat. *Journal of Agricultural Science* 102, 383–92.

Porter, J.R., 1993. AFRCWHEAT2: a model of the growth and development of wheat incorporating responses to water and nitrogen. *European Journal of Agronomy* 2, 69–82.

Pracilio, G., Asseng, S., Cook, S.E., Hodgson, G., Wong, M.T.F., Adams, M.L. and Hatton, T.J., 2003. Estimating spatially variable deep drainage across a central-eastern wheatbelt catchment, Western Australia. *Australian Journal of Agricultural Research* 54, 789–802.

Priesack, E., Gayler, S. and Hartmann, H., 2006. The impact of crop growth sub-model choice on simulated water and nitrogen balances. *Nutrient Cycling in Agroecosystems* 75, 1–13.

Priestley, C.H.B. and Taylor, R.J., 1972. Assessment of surface heat flux and evaporation using large-scale parameters. *Monthly Weather Review* 100, 81–92.

Probert, M.E., Dimes, J.P., Keating, B.A., Dalal, R.C. and Strong, W.M., 1998. APSIM's water and nitrogen modules and simulation of the dynamics of water and nitrogen in fallow systems. *Agricultural Systems* 56, 1–28.

Ramirez-Rodrigues, M.A., Asseng, S., Fraisse, C., Stefanova, L. and Eisenkolbi, A., 2014. Tailoring wheat management to ENSO phases for increased wheat production in Paraguay. *Climate Risk Management* 3, 24–38.

Reich, P.B. and Amundson, R.G., 1985. Ambient levels of ozone reduce net photosynthesis in tree and crop species. *Science* 230, 566–70.

Rengasamy, P. and Olsson, K.A., 1993. Irrigation and sodicity. *Australian Journal of Soil Research* 31, 821–37.

Reyenga, P.J., Howden, S.M., Meinke, H. and McKeon, G.M., 1999. Modelling global change impacts on wheat cropping in south-east Queensland, Australia. *Environmental Modelling & Software* 14, 297–306.

Ritchie, J.T., 1991. Wheat phasic development. Modeling plant and soil systems., 31–54.

Ritchie, J.T., Godwin, D.C. and Otter-Nacke, S., 1985. *CERES-Wheat. A Simulation Model of Wheat Growth and Development*. Texas A & M University Press, College Station, TX.

Ritchie, S., Nguyen, H. and Holaday, A., 1987. Genetic diversity in photosynthesis and water-use efficiency of wheat and wheat relatives. *Journal of Cellular Biochemistry*, 43.

Roderick, M.L. and Farquhar, G.D., 2002. The cause of decreased pan evaporation over the past 50 years. *Science* 298, 1410–11.

Rosegrant, M.W., Koo, J., Cenacchi, N., Ringler, C., Robertson, R., Fisher, M., Cox, C., Garrett, K., Perez, N. and Sabbagh, P., 2014. Food security in a world of natural resource scarcity: *The Role of Agricultural Technologies*. 250.

Rosenzweig, C., Elliott, J., Deryng, D., Ruane, A.C., Muller, C., Arneth, A., Boote, K.J., Folberth, C., Glotter, M., Khabarov, N., Neumann, K., Piontek, F., Pugh, T.A.M., Schmid, E., Stehfest, E., Yang, H. and Jones, J.W., 2014. Assessing agricultural risks of climate change in the 21st century in a global gridded crop model intercomparison. *Proceedings of the National Academy of Sciences of the United States of America* 111, 3268–73.

Rosenzweig, C., Jones, J.W., Hatfield, J.L., Ruane, A.C., Boote, K.J., Thorburne, P., Antle, J.M., Nelson, G.C., Porter, C., Janssen, S., Asseng, S., Basso, B., Ewert, F., Wallach, D., Baigorria, G. and Winter, J.M., 2013. The Agricultural Model Intercomparison and Improvement Project (AgMIP): Protocols and pilot studies. *Agricultural and Forest Meteorology* 170, 166–82.

Rosenzweig, C. and Parry, M.L., 1994. Potential impact of climate change on world food supply. *Nature* 367, 133–8.

Rost, S., Gerten, D., Bondeau, A., Lucht, W., Rohwer, J. and Schaphoff, S., 2008. Agricultural green and blue water consumption and its influence on the global water system. *Water Resources Research* 44.

Ruane, A.C., McDermid, S., Rosenzweig, C., Baigorria, G.A., Jones, J.W., Romero, C.C. and Cecil, L.D., 2014. Carbon-Temperature-Water change analysis for peanut production under climate change: A prototype for the AgMIP Coordinated Climate-Crop Modeling Project (C3MP). *Global Change Biology* 20, 394–407.

Semenov, M.A. and Shewry, P.R., 2011. Modelling predicts that heat stress, not drought, will increase vulnerability of wheat in Europe. *Scientific Reports* 1, 5.

Semenov, M.A., Stratonovitch, P., Alghabari, F. and Gooding, M.J., 2014. Adapting wheat in Europe for climate change. *Journal of Cereal Science* 59, 245–56.

Senthilkumar, S., Basso, B., Kravchenko, A.N. and Robertson, G.P., 2009. Contemporary evidence of soil carbon loss in the US corn belt. *Soil Science Society of America Journal* 73, 2078–86.

Shibu, M., Leffelaar, P., van Keulen, H. and Aggarwal, P., 2010. LINTUL3, a simulation model for nitrogen-limited situations: Application to rice. *European Journal of Agronomy* 32, 255–71.

Singh, A.K., Tripathy, R. and Chopra, U.K., 2008. Evaluation of CERES-Wheat and CropSyst models for water-nitrogen interactions in wheat crop. *Agricultural Water Management* 95, 776–86.

Soltani, A., Maddah, V. and Sinclair, R., 2013. SSM-Wheat: a simulation model for wheat development, growth and yield. *International Journal of Plant Production* 7, 711–40.

Spitters, C.J.T. and Schapendonk, A.H.C.M., 1990. Evaluation of breeding strategies for drought tolerance in potato by means of crop growth simulation. *Plant and Soil* 123, 193–203.

Steduto, P., Hsiao, T., Raes, D. and Fereres, E., 2009. AquaCrop-The FAO crop model to simulate yield response to water: I. concepts and underlying principles. *Agronomy Journal* 101, 426–37.

Stenger, R., Priesack, E., Barkle, G. and Sperr, C., 1999. Expert-N A tool for simulating nitrogen and carbon dynamics in the soil-plant-atmoshpere system. Land Treatment collective proceedings Technical Session, New Zealand.

Stephens, D.J. and Lyons, T.J., 1998. Rainfall-yield relationships across the Australian wheatbelt. *Australian Journal of Agricultural Research* 49, 211–23.

Stockle, C., Donatelli, M. and Nelson, R., 2003. CropSyst, a cropping systems simulation model. *European Journal of Agronomy* 18, 289–307.

Sylvester-Bradley, R., Riffkin, P. and O'Leary, G., 2012. Designing resource-efficient ideotypes for new cropping conditions: Wheat (Triticum aestivum L.) in the High Rainfall Zone of southern Australia. *Field Crops Research* 125, 69–82.

Tao, F., Yokozawa, M. and Zhang, Z., 2009a. Modelling the impacts of weather and climate variability on crop productivity over a large area: A new process-based model development, optimization, and uncertainties analysis. *Agricultural and Forest Meteorology* 149, 831–50.

Tao, F. and Zhang, Z., 2010. Adaptation of maize production to climate change in North China Plain: Quantify the relative contributions of adaptation options. *European Journal of Agronomy* 33, 103–16.

Tao, F. and Zhang, Z., 2013. Climate change, wheat productivity and water use in the North China Plain: A new super-ensemble-based probabilistic projection. *Agricultural and Forest Meteorology* 170, 146–65.

Tao, F., Zhang, Z., Liu, J. and Yokozawa, M., 2009b. Modelling the impacts of weather and climate variability on crop productivity over a large area: A new super-ensemble-based probabilistic projection. *Agricultural and Forest Meteorology* 149, 1266–78.

Triboi, E., Martre, P., Girousse, C., Ravel, C. and Triboi-Blondel, A.M., 2006. Unravelling environmental and genetic relationships between grain yield and nitrogen concentration for wheat. *European Journal of Agronomy* 25, 108–18.

Trombetta, A., Iacobellis, V., Tarantino, E. and Gentile, F., 2016. Calibration of the Aqua Crop model for winter wheat using MODIS LAI images. *Agricultural Water Management* 164, 304–16.

Tubiello, F.N. and Ewert, F., 2002. Simulating the effects of elevated CO_2 on crops: Approaches and applications for climate change. *European Journal of Agronomy* 18, 57–74.

Tubiello, F.N., Rosenzweig, C., Kimball, B.A., Pinter, P.J., Wall, G.W., Hunsaker, D.J., LaMorte, R.L. and Garcia, R.L., 1999. Testing CERES-wheat with free-air carbon dioxide enrichment (FACE) experiment data: CO_2 and water interactions. *Agronomy Journal* 91, 247–55.

Tubiello, F.N., Soussana, J.F. and Howden, S.M., 2007. Crop and pasture response to climate change. *Proceedings of the National Academy of Sciences of the United States of America* 104, 19686–90.

van Bussel, L.G.J., Ewert, F., Zhao, G., Hoffmann, H., Enders, A., Wallach, D., Asseng, S., Baigorria, G.A., Basso, B., Biernath, C., Cammarano, D., Chryssanthacopoulos, J., Constantin, J., Elliott, J., Glotter, M., Heinlein, F., Kersebaum, K.-C., Klein, C., Nendel, C., Priesack, E., Raynal, H., Romero, C.C., Rötter, R.P., Specka, X. and Tao, F., 2016. Spatial sampling of weather data for regional crop yield simulations. *Agricultural and Forest Meteorology* 220, 101–15.

van Ittersum, M.K., Leffelaar, P.A., van Keulen, H., Kropff, M.J., Bastiaans, L. and Goudriaan, J., 2003. On approaches and applications of the Wageningen crop models. *European Journal of Agronomy* 18, 201–34.

Van Wart, J., Grassini, P. and Cassman, K.G., 2013. Impact of derived global weather data on simulated crop yields. *Global Change Biology* 19, 3822–34.

Vanuytrecht, E., Raes, D., Steduto, P., Hsiao, T.C., Heng, L.K., Vila, M.G. and Moreno, P.M., 2014. AquaCrop: FAO's crop water productivity and yield response model. *Environmental Modelling & Software* 62, 351–60.

Verburg, K., Bond, W.J. and Hunt, J.R., 2012. Fallow management in dryland agriculture: Explaining soil water accumulation using a pulse paradigm. *Field Crops Research* 130, 68–79.

Wall, G.W., Kimball, B.A., White, J.W. and Ottman, M.J., 2011. Gas exchange and water relations of spring wheat under full-season infrared warming. *Global Change Biology* 17, 2113–33.

Wallach, D., Makowski, D., Jones, J.W. and Brun, F., 2014. *Working with Dynamic Crop Models. Methods, Tools and Examples for Agriculture and Environment.* Academic Press, Elsevier.

Wang, B., Chen, C., Liu, D.L., Asseng, S., Yu, Q. and Yang, X.H., 2015. Effects of climate trends and variability on wheat yield variability in eastern Australia. *Climate Research* 64, 173–86.

Wang, E. and Engel, T., 2000. SPASS: a generic process-oriented crop model with versatile windows interfaces. *Environmental Modelling & Software* 15, 179–88.

Wang, E., Robertson, M.J., Hammer, G.L., Carberry, P.S., Holzworth, D., Meinke, H., Chapman, S.C., Hargreaves, J.N.G., Huth, N.I. and McLean, G., 2002. Development of a generic crop model template in the cropping system model APSIM. *European Journal of Agronomy* 18, 121–40.

Wang, E.L. and Engel, T., 1998. Simulation of phenological development of wheat crops. *Agricultural Systems* 58, 1–24.

Wang, E.L. and Engel, T., 2002. Simulation of growth, water and nitrogen uptake of a wheat crop using the SPASS model. *Environmental Modelling & Software* 17, 387–402.

Wessolek, G. and Asseng, S., 2006. Trade-off between wheat yield and drainage under current and climate change conditions in northeast Germany. *European Journal of Agronomy* 24, 333–42.

Wilby, R., Charles, S., Zorita, E., Timbal, B., Whetton, P. and Mearns, L., 2004. Guidelines for use of climate scenarios developed from statistical downscaling methods. IPCC, 27.

Williams, J.R., Jones, C.A., Kiniry, J.R. and Spanel, D.A., 1989. The EPIC crop growth-model. *Transactions of the Asae* 32, 497–511.

Wolf, J., 1996. Effects of nutrient supply (NPK) on spring wheat response to elevated atmospheric CO_2. Plant and Soil 185, 113–23.

Wong, M.T.F. and Asseng, S., 2006. Determining the causes of spatial and temporal variability of wheat yields at sub-field scale using a new method of upscaling a crop model. *Plant and Soil* 283, 203–15.

Yan, M., Cao, W. and C. Li, Z.W., 2001. Validation and evaluation of a mechanistic model of phasic and phenological development in wheat. *Chinese Agricultural Science* 1, 77–82.

Yang, X.Y., Asseng, S., Wong, M.T.F., Yu, Q., Li, J. and Liu, E.M., 2013. Quantifying the interactive impacts of global dimming and warming on wheat yield and water use in China. *Agricultural and Forest Meteorology* 182, 342–51.

Yin, X. and van Laar, H.H., 2005. *Crop Systems Dynamics: An Ecophysiological Simulation Model of Genotype-by-Environment Interactions*. Wageningen Academic Publishers, Wageningen, The Netherlands.

Yin, X.Y., 2013. Improving ecophysiological simulation models to predict the impact of elevated atmospheric CO_2 concentration on crop productivity. *Annals of Botany* 112, 465–75.

Yin, X.Y. and Struik, P.C., 2010. Modelling the crop: from system dynamics to systems biology. *Journal of Experimental Botany* 61, 2171–83.

Zhao, G., Hoffmann, H., Bussel, L.G.J.v., Enders, A., Specka, X., Sosa, C., Yeluripati, J., Tao, F., Constantin, J., Raynal, H., Teixeira, E., Grosz, B., Doro, L., Zhao, Z., Nendel, C., Kiese, R., Eckersten, H., Haas, E., Vanuytrecht, E., Wang, E., Kuhnert, M., Trombi, G., Moriondo, M., Bindi, M., Lewan, E. and Bach, M., 2015. Effect of weather data aggregation on regional crop simulation for different crops, production conditions, and response variables. *Climate Research* 65, 141–57.

Zheng, B.Y., Biddulph, B., Li, D.R., Kuchel, H. and Chapman, S., 2013. Quantification of the effects of VRN1 and Ppd-D1 to predict spring wheat (Triticum aestivum) heading time across diverse environments. *Journal of Experimental Botany* 64, 3747–61.

Zheng, B.Y., Chenu, K., Dreccer, M.F. and Chapman, S.C., 2012. Breeding for the future: what are the potential impacts of future frost and heat events on sowing and flowering time requirements for Australian bread wheat (Triticum aestivium) varieties? *Global Change Biology* 18, 2899–2914.

Ziska, L.H. and Bunce, J.A., 2007. Predicting the impact of changing CO_2 on crop yields: Some thoughts on food. *New Phytologist* 175, 607–17.

Chapter 7

Integrated crop management of wheat

Brian L. Beres, Reem Aboukhaddour and Haley Catton, Agriculture and Agri-Food Canada, Lethbridge Research and Development Centre, Canada

1 Introduction
2 Impact of agronomists on agriculture
3 Case study: ICM for wheat stem sawfly (*Cephus cinctus* Norton)
4 Future trends and conclusion
5 Where to look for further information
6 References

1 Introduction

Agronomy is the study of plants, soils and their interactions with each other, as well as with the environment and agro-ecological processes. Agronomy by nature is a cross-disciplinary science that addresses yield stability and exploitable yield gaps through the holistic application of a set of principles that serve to enhance an entire crop production system. These principles are common but specifics of their application can vary with production zone, and whether inputs and potential productivity are high or low. Why is agronomy important? Roughly one-half of crop yield potential is realized through agronomic factors; thus, attainable wheat yield and improved system stability can only be achieved and then exceeded when new genetics are coupled with innovative integrated crop management (ICM) strategies. For example, hybrid maize yields (*Zea mays* L.) improved over conventional maize only when agronomic factors were modified; thus, management contributed to roughly 50% of realized yield potential (Duvick, 2005). Specific to wheat (*Triticum aestivum* L.), the 'Green Revolution' refers to the development of semi-dwarf wheat cultivars; however, yield potential was often not realized and did not become revolutionary until sound agronomic packages were developed (Kirkegaard and Hunt, 2010).

Integrated crop management (ICM) is the practice of using multiple agronomic and pest control methods to maximize yield and net returns and to minimize inputs while maintaining economic and environmental sustainability (Bradley et al., 2002). Thus, ICM is a holistic framework concerned with optimally and sustainably managing abiotic and

http://dx.doi.org/10.19103/AS.2016.0004.28

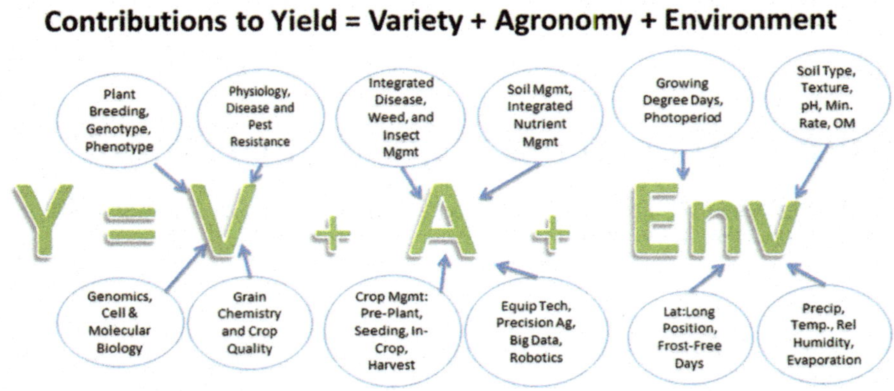

Figure 1 Schematic of the derivation of wheat grain yield through contributions of genetics, agronomy and the environment. 'V' and 'A' are the fixed effects we can control. 'Env' is the proxy random effect for error variance. The Breeder 'Captains' the 'V Team' toward variety release creating the genetic potential for yield, while the Agronomist 'Captains' the 'A Team' toward a holistic system that minimizes the yield gap of genetic potential vs. on-farm yield.

biotic production components to allow actual crop yield to approach the genetic potential yield. This is accomplished by maximizing potential yield through improved crop breeding, mitigating production losses by optimizing nutrients, moisture and agronomic practices, and using integrated pest management to reduce losses from weeds, diseases and insects (Fig. 1). The role of the agronomist and ICM research, therefore, is to develop methods and practices for wheat production that close the gap between genetic potential and realized yield at the farm gate (Fig. 2).

The complexity of factors contributing to yield, summarized in Fig. 1, is evidence that single-factor 'magic bullets' are not likely to be developed that will address or reduce wheat production yield gaps, particularly when used in isolation. Rather, as Kirkegaard et al. (2010, 2014b) have argued, success will be born from synergies among the novel genetic and management strategies currently being pursued through enhanced cross-disciplinary collaborations along the research and development (R&D) continuum so as to capture them at the farm.

They point out two examples of incremental transformation as it is being applied to food production systems in Australia which suggest this is possible – one is focused on improving the water-use efficiency of dryland grain farms (Kirkegaard and Hunt, 2010; Kirkegaard et al., 2014b) and a second focused on the integration of dual-purpose (graze and grain) crops into mixed livestock and grain farms (Dove and Kirkegaard, 2014). In both cases, a range of innovations from different scientific disciplines combined with direct grower collaboration to generate step changes in productivity that have been adopted on-farm. In Canada, there has also been a shift from studying one component of the system to research that combine many individual components (Beres et al., 2010a–c; Beres et al., 2011d). This chapter will examine the principles of agronomy behind these approaches and present case studies where ICM solutions have been employed to achieve optimal wheat production.

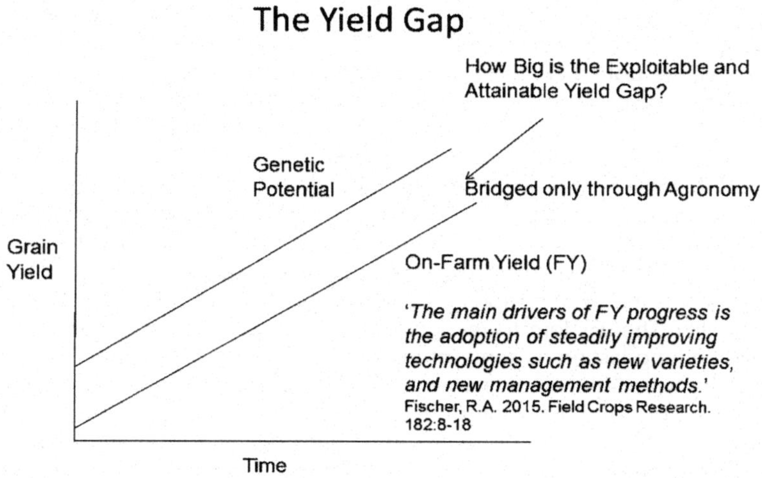

Figure 2 The gap that exists between genetic potential and grain yield achieved at the farm gate.

2 Impact of agronomists on agriculture

Agronomists operate in many spheres of the wheat sector including as scientists in public research institutes and universities, as technologists in private sector seed or crop input industries or as on-farm consultants. An attribute common to all is that agronomists are involved at the production level and, as such, their actions can have immediate and direct impact at the farm gate. This requires skills in problem solving, communication, knowledge transfer and team building. A broad skills set and the ability to assemble quickly a cross-disciplinary team is important for success in the contemporary practice of agronomy. Issues and challenges from wheat production systems can range from biotic or abiotic factors that arise from the soil, the plant or their interaction with the environment. The following are examples of issues and challenges that emerged with wheat production and how ICM was manipulated to address an issue or exploit potential within the cropping system.

2.1 Conservation agriculture

The conversion of the land to crop production expedited settlement of the prairies in the northern and central Great Plains of North America. The common legislative action to promote settlement in Canada and the United States was to offer homesteading opportunities to settlers whereby 160 acres of land was granted if the settler resided on the land for three years, broke at least 15 acres of land in that period, and built a domicile (Fulton and Sonntag, 2010). Unique to Canada's settlement of the prairies was the simultaneous arrival of both the farmer and the scientist as the Dominion of Canada was establishing a network of research stations/farms to facilitate the successful establishment of grain farms. As was the case in North America and other regions around the globe, soil health issues began to emerge soon after native prairie or other lands were converted to crop production. In his review of Canadian prairie agricultural beginnings, Dr Henry Janzen eloquently described how Dominion of Canada scientists collaborated with Rothamsted researchers, Lawes and Gilbert, to monitor the changes to prairie soils (Janzen, 2001). Concern over the potential for soil degradation was expressed early on as researchers recognized that the practice of continuous cropping, aggressive tillage practices, and the export of grain and straw from land (due to stationary threshing) was mining the soil of nutrients and limiting wheat yields. By the turn of the twentieth century, scientists already understood that soil fertility would decline over time as a result of tillage (Fulton, 2010). Fallow was proposed as a practice to recharge soil water and address nitrogen access as tillage on summer fallow would promote the conversion of nitrogen to plant available forms (Brandt, 2010). Moreover, experiments were established to identify alternative cropping systems that would support a wheat-based approach but would address issues of sustainability. The wisdom to recognize the issue early in this era seems remarkable and the vision long-lasting as some of the original rotational experiments established at the turn of the twentieth century remain today. These experiments provided the scientific evidence to support the adoption of methods that would be more sustainable over the monoculture wheat or wheat-fallow systems that dominated the landscape. Dr Janzen noted that these early researchers such as Angus McKay touched on concepts of reduced tillage and direct seeding as alternatives to fall ploughing, which exposes the land to soil erosion effects through the winter. Thus, the principles of cropping diversity, crop–livestock interfaces/mixed farming and even demonstration farms for knowledge transfer were not devised by modern-day scientists, but were developed, studied and employed near the turn of the twentieth century (Janzen, 2001).

While the issue of soil degradation was readily identified, and solutions and innovations developed quite rapidly, the challenges around adoption proved insurmountable. Farmers could not be convinced to integrate these sustainable methods as perception of risk to farm net incomes was deemed too high. This was largely because alternative crops either did not produce enough yield to be economically viable or limited grain yield of wheat that followed in the rotation (Janzen, 2001). Advances to reduce conventional tillage on fallow resulted in the 'Noble Blade', an implement with a wide blade that severed weeds below ground at the roots but left the surface undisturbed to allow the shoot portion to remain anchored in the soil (Mathews, 1945). The one-way disc planter eliminated multiple tillage passes, which facilitated a fallow-wheat-wheat system instead of wheat fallow (Lafond et al., 2011). Incorporation of herbicides in the 1940s and 1950s would reduce the need for tillage further, although uptake was sporadic for concern over the environment and increased soil compaction or disease prevalence without tillage. Thus, monoculture wheat-fallow systems under dryland farming conditions remained relatively static for decades until the late 1970s and early 1980s, with fallow hectares peaking during the period of 1950–80 (Fig. 3).

The transformation of agriculture through the evolution of no-till farming is an example of a problem that was not solved until not only a systems approach was applied that cut across multiple disciplines including machinery engineering and knowledge transfer, but also government policy was established to support the initiative (Lafond et al., 2011). As described by Mr Stewart Brandt (Brandt, 2010): *What evolved was a group of farmers, equipment manufacturers, extension specialists, and research scientists who shared a common belief that only by working together could they identify and resolve the key issues that prevented widespread adoption of conservation agriculture. What bound this group together was the knowledge that failure to address soil degradation would ultimately deplete the soil resource to the point where large areas would no longer be*

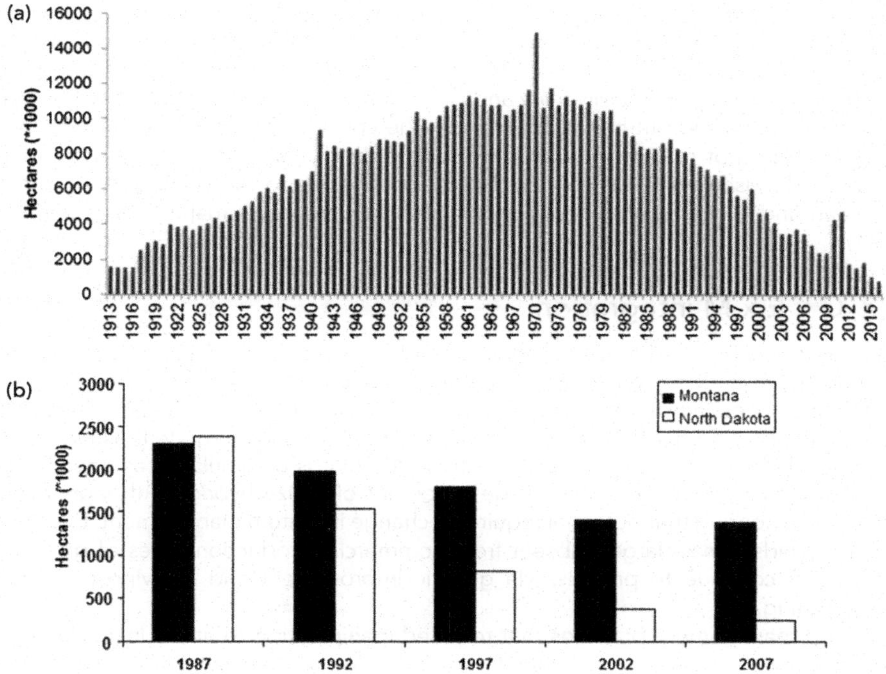

Figure 3 Estimated summer fallow area for the prairie provinces of Canada (a), and for Montana and North Dakota, USA (b). Source: (Anonymous, 2007; Anonymous, 2016a; Beres et al., 2011c).

suitable for crop production. Through a collaborative team effort, they identified and solved problems preventing widespread adoption of conservation tillage. The initial problems involved stubble residue management, seed and fertilizer placement, and weed management. The result was the development of straw choppers for combines that evenly spread residues and seed drills and shank openers capable of placing seed and fertilizer to uniform and shallow depth. Conversion to no-till and direct seeding was embraced globally as other important wheat-producing regions such as Australia (Llewellyn et al., 2012) and Brazil (Bollinger et al., 2005) also widely adopted the practice during the 1980s.

Improved methods for optimum crop management and seeding directly into crop stubble were also coupled with the latest genetics, and rotational diversity proved not only possible but the entire system profitable. The benefits of minimum or no-till systems began to be reported extensively (Janosky et al., 2002; Lindwall and Anderson, 1981), and Derksen et al. (2002) suggested that the conventional wheat-fallow system in the northern Great Plains could be modified with diversified cropping systems. Studies also showed that no-till and direct-seeding practices could increase soil moisture retention, improve residue preservation and reduce soil erosion (Larney et al., 1994; Peterson et al., 1996; Roget et al., 1987). Lower prices for cereal grains, continual improvements in equipment and machinery design, influence of government policy and concern for environmental sustainability of intensive tillage also contributed to the shift to greater cropping diversity including pulses with less fallow (Zentner et al., 2002), which can enhance both soil N and wheat yield (Miller et al., 2006). The result was a significant reduction in the area left fallow each year across the Canadian prairies (Anonymous, 2009) and in Montana and North Dakota, USA (Fig. 3) (Anonymous, 2007). This provided the opportunity for agronomists to revisit the principles of crop diversity and alternative cropping systems that Angus Mackay and others espoused but had not achieved. The conversion to continuous cropping and direct seeding not only resulted in large-scale crop diversification across the prairies of Canada and elsewhere, but improved production levels to the point that infrastructure issues around rail transport of small grains for export are now a major issue (Lafond et al., 2011).

2.2 G x E x M integration

Conservation tillage and direct seeding revolutionized farming systems and reversed the soil health and degradation issues that plagued many regions; however, in isolation it did not necessarily improve crop yields (Giller et al., 2015). Rather, it provided a transformational production platform used by agronomists to integrate the latest genetics (G) into a holistic wheat management (M) system that is manipulated as needed to suit specific agro-ecozones (E). Like the development of maize hybrids with higher genetic potential, unlocking that potential required a change in system management. Even though wheat hybrids remain largely absent from commercial production, rates of gain in yield production continue to progress via genetic improvement and evolving management practices (Fig. 4).

In-crop management practices that required manipulation to utilize increased genetic yield potential in wheat, in a no-till, direct-seeded system included sowing density, integrated weed management and nitrogen management. Earlier recommendations for sowing density in wheat were as low as 75–100 plants m^{-2}. These rates produced acceptable yields and were considered optimum in the context of managing water-limiting factors, or, when water was not an issue, the plasticity effect of plant tillering was relied upon as the mechanism to produce additional seeds. For example, the potential for winter wheat to compensate for lower seeding rates or large reductions in plant density (Holen et al., 2001; Whaley et al., 2000) through increased tillering capacity may explain a reluctance to alter seeding rates. Also, a study conducted in the Netherlands reported that winter wheat grain yield was optimized with a plant density of 100 plants m^{-2} (Darwinkle, 1978). However, newer studies suggested that these rates did not fully exploit the yield potential of winter wheat. Lafond and Gan (1999) observed optimal yields at a seeding rate of 134 kg ha^{-1}. Although their results were reported on a volume basis, which can vary significantly with kernel size, shape and weight, the rate was approximately twice that of earlier recommendations. In the case of winter wheat in Canada, the system of no-till and updated agronomics preceded the genetic component, which resulted in a major decline in acreage in the mid-1980s (Fowler, 1983). New genetics released in the twenty-first century were coupled with the latest in no-till machinery design, and sowing rates were adjusted to the recommended rates from the latest agronomic studies, that is, 450 seeds m^{-2}, which is up to 3x that of earlier practices (Beres et al., 2010a). High-yielding spring wheat cultivars responded similarly to higher seeding rates as both durum (*Triticum turgidum* L.) and common wheat displayed high grain yield, and stand establishment, spike density and grain bulk density were associated with the highest seeding rate of 450 seeds m^{-2} (Beres et al., 2011a). These new recommendations for spring wheat were considered extreme but may be modest considering the environment

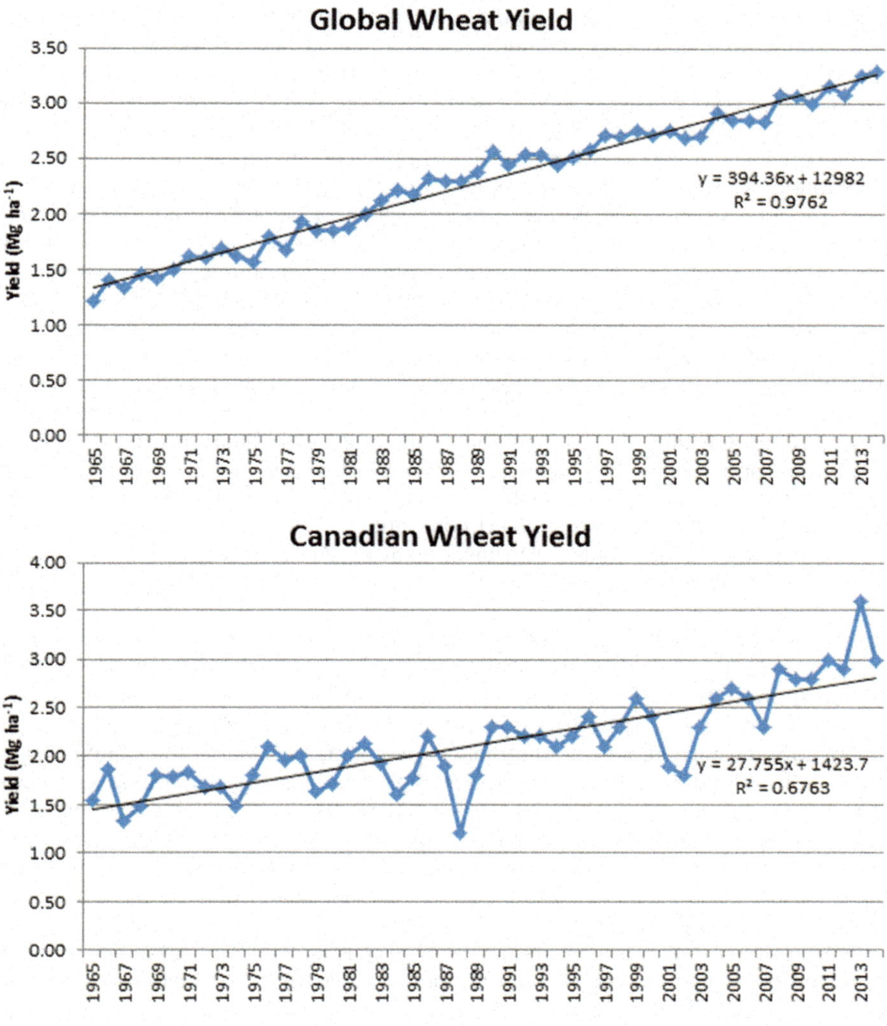

Figure 4 Global and Canadian wheat yield trends over the period 1965–2014. Source: FAO (Anonymous, 2016b); Statistics Canada (Anonymous, 2016c).

was intentionally designed to be less than ideal, that is, direct-seeded wheat-wheat in a semi-arid environment.

Although progress in wheat production was achieved by coupling reduced tillage, genetic improvement and seeding rates, there are additional opportunities to integrate weed management considerations into these systems. Earlier studies of seeding rates in wheat measured responses in relation to grain production or canopy architecture. In other crops, seeding rates were integrated with weed management tactics to further the understanding of the interactions of crop productivity and competitiveness. For example, Harker et al. (2003)

reported significant changes to crop yield and weed biomass when canola (*Brassica napus* L.) seeding rates were integrated with herbicide timing and cultivar selection.

Barley (*Hordeum vulgare* L.) studies have shown that higher seeding rates using cultivars with differing competitive abilities enhanced crop competitiveness against wild oat (*Avena fatua* L.) (Harker et al., 2009; O'Donovan et al., 2000; Watson et al., 2006).

Cultivar selection and weed management practices are equally important tools in an ICM strategy for wheat. Blackshaw (1994) reported a reduction in the competitive ability of shorter cultivars of winter wheat in the presence of downy brome *(Bromus tectorum* L.*)*. Thomas et al. (1994) described the positive influence of plant height for winter wheat cultivars on competitive ability and yield. Cultivar development in most wheat classes has focused on the development of high-yielding, short-stature or semi-dwarf varieties. A benefit is that harvest management and direct seeding of subsequent crops is enhanced with the reduction in straw residue. One drawback of shorter cultivars is their weaker competitive ability with weeds. Higher sowing densities exploited the greater yield potential of shorter varieties while providing improved competitive ability. Moreover, higher sowing densities coupled with alternative herbicide timings further reduce weed competition and reduce the reliance on multiple herbicide applications (Beres et al., 2010b). Although semi-dwarf cultivars generally exhibit weaker competitive ability and require greater herbicide inputs, weed competitive ability varies among short cultivars as some suppress weeds better than taller cultivars (Andrew et al., 2015). Integration of improved competitive ability, high sowing density and narrow row spacing will provide system sustainability over a range of environments.

Sustainability and system stability for wheat production is also influenced by how wheat is sequenced with other crop phases. One of the most reliable methods for determining wheat responses to crop sequence is to conduct long-term rotational experiments. The long-term rotations established at Rothamsted, United Kingdom (Lawes and Gilbert, 1894), influenced the establishment of rotational experiments in Canada, the United States and Europe. Of the rotations that remain today, their value has not diminished as knowledge around soil dynamics, long-term yield trends and net returns and risk continues to evolve. For example, long-term data was used by scientists to calculate the net return risk for wheat-fallow systems where they concluded that fallow was not necessary and reduced phases of fallow-improved net returns with a risk that would only be of concern to very risk-averse growers (Smith et al., 2015a,b; 2016).

Break crops in rotation can also be studied to measure the response of wheat for one or two cycles following the preceding crop (Angus et al., 2015). While the benefits of rotational diversity have long been established, the immediate benefits of a preceding crop on the performance of the subsequent wheat phase or phases is still being investigated. An excellent review by Angus et al. (2015) studied >900 comparisons of wheat following a break crop. They concluded that wheat responds best when preceded by the following crops ranked lowest to highest in terms of positive wheat responses: oats<canola=mustard=flax<field peas=faba beans=chick peas=lentils=lupins. Moreover, two break crops produced 0.2–0.4 t ha^{-1} more wheat yield than a single crop. The variables that influence these findings include root disease severity, soil water availability, and nitrogen, weed suppression and nitrogen and non-nitrogen effects of legumes (Angus et al., 2015). The magnitude of effect or responses may vary by regional specific production systems and seasonal conditions (Kirkegaard et al., 1997).

The shift to conservation agriculture allows for the ability to diversify a wheat-based cropping system with cereal and non-cereal phases, which also provides an opportunity to integrate a livestock phase. Integration of forage crops can generally lead to higher grain crop yields, reduced weed abundance and distribution, improved soil quality, and less reliance on energy costs associated with inorganic fertilizers (Entz et al., 2002). While grazing of the forage phase is an old practice, an alternative is the dual-purpose use of cereals and canola for forage during the vegetative stage, which has been successfully implemented in Australia and elsewhere (Bell et al., 2014). Benefits of this integration provide reduced risk, increased diversity and overall livestock and crop productivity in the range of 25–75% (Bell et al., 2014). Integration of forage and livestock into wheat systems not only is compatible with conservation agriculture but can increase the overall efficiency of input use, providing greater cropping flexibility and profitability (Hunt et al., 2016; Kirkegaard et al., 2014a).

2.3 Challenges facing agronomists

While there is much reverence around the tenets of conservation agriculture, there have been some unintended consequences with its adoption. The unexpected responses from facets of a wheat production system when managed under conservation agriculture serve as a reminder of the complexity underlying cropping systems. Adoption of no-till and the preservation of crop residues have resulted in colder and often wetter soils at planting, which can also delay seedling emergence. Integration of crop protection inputs such as dual fungicide and insecticide seed treatments can help offset weak systems comprised of poor stand establishment and lower yield performance (Beres et al., 2016). However, challenges with cold soils extend further as it was discovered that volatilization of N can indeed occur in cold soils (Engel et al., 2011). Solutions to overcome this challenge involved surface applications of polymer-coated urea (PCU) or N-butyl thiophosphoric triamide (NBPT), which have been reported to reduce ammonia volatilization losses in soil conditions ranging from dry and acidic (Rochette et al., 2009) to cold and wet (Engel et al., 2011). Volatilization also occurs from shallow banding, which is traditionally considered a sustainable practice. No-till systems over time have altered the rhizosphere such that hydrolysis of N and subsequent volatilization are now an issue (Rochette et al., 2009). Integration of enhanced efficiency fertilizers to mitigate N loss and seed treatments to improve plant responses to abiotic and biotic stress not only provide sustainability to the system but lessen system contributions to atmospheric emissions.

Strict adherence to the principles of conservation agriculture – (1) diverse rotations, (2) reduced tillage or no-till systems and (3) preservation of crop residues – is being challenged because all these components do not necessarily lead to sustainable wheat production systems (Giller et al., 2015; Kirkegaard et al., 2014a). Agronomists are faced with sustainability issues in wheat systems that include herbicide-resistant weeds, soil compaction and residue management. Preservation of crop residues and stubble has been so successful in some regions that the thatch layer created from decades of no-till is becoming too deep and interfering with proper seed to soil contact at planting. Strategic tillage, as an example, would violate conservation agriculture principles in the eyes of some of the strongest advocates; however, a more rationalized approach is being proposed. In areas where soil structure is not a concern and where crop–livestock integration is a priority, strategic tillage may need to be considered (Kirkegaard et al., 2014a). Moreover, sustainable intensification of wheat production systems, particularly in smallholder farming systems, may require a modified approach to address the low adoption rates for a pure conservation agriculture approach (Giller et al., 2015).

Innovations in conservation cropping practices that overcome one or more issues can also have unintended negative consequences on soil-borne disease. For example, while the evolution of conservation farming and no-till greatly enhanced soil health and productivity, *F. avenaceum* from cereal residues was greater under reduced-till than conventional-till. However, a bright spot in the findings is that *F. graminearum* inoculum was lowest under zero-till (Fernandez et al., 2009). Thus, strategic tillage may have a place in the management of disease if crop sequencing cannot reduce the presence of the pathogen.

The introduction of glyphosate for weed management greatly facilitated the adoption of no-till. One unintended consequence of this innovation, however, was increased infection by *Fusarium* spp. (Fernandez et al., 2009) resulting largely from increased root colonization of glyphosate-treated weeds by *Fusarium* (Rahe et al., 1990) (Kawate et al., 1997). The problem illustrates the consequences that can arise from over-reliance on a single strategy. Effective weed management in wheat requires integration of multiple chemistries and other cultural practices that reduce the weed seed bank and minimizes opportunities for *Fusarium* to colonize glyphosate-treated weeds.

An ominous challenge ahead for agronomists and the wheat value chain is climate change. While various scenarios exist depending on the climate modelling employed, the regional effects generally remain the same. Most wheat-producing regions will become hotter and drier, including almost all of Africa and Australia (Asseng et al., 2015). Wheat production in these areas will be at risk if systems cannot adapt to environmental changes. All scenarios

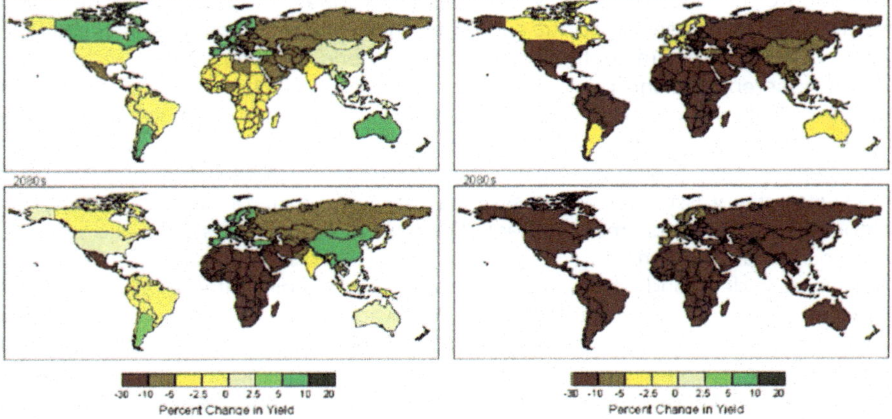

Figure 5 Potential changes (%) in national cereal yields to 2080 compared to 1990, under the HadCM3 SRES A1F1 with and without CO_2. Source: (Parry et al., 2004).

summarized by Parry et al. (2004) project negative mean cereal yields of ~ −16% to −22%. Figure 5 is the most conservative scenario presented by Parry et al. (2004). To compound the problem, current disparities between developing and developed regions would only widen with climate change. Adaptation will be critical and is already underway. Drought tolerance traits tend to be a priority with most private and public breeding programmes targeting arid and semi-arid regions. Collaborations between agronomists and breeders will foster innovations that may allow for successful adaptation strategies. For example, wheat genotypes bred for extra-long coleoptiles have been identified and agronomists in Australia are working with breeders on integration strategies. This trait can be exploited to protect yield when used in very early seeding dates that facilitates asynchrony between sensitive crop phenological phases such as flowering and the hottest, driest periods of the growing season. The longer coleoptile allows deeper planting into moisture that was inaccessible with traditional varieties (Kirkegaard and Hunt, 2010). Thus, the integration of early seeding dates x long coleoptile genetics x crop sequencing that optimize soil water and nutrient reserves may evolve into a successful adaptation strategy (Rebetzke et al., 2005).

Not all regions will suffer declines in yield. The continental regions, and in some cases parts of China/Mongolia and South America, may realize improved cereal production over time and within the context of these climate change scenarios. Cereal production for Canada in particular will likely improve because of climate change (Fig. 5). The only adaptation that may be required could be the adoption of greater use of winter cereals over those with a spring growth habit to capture precipitation events that may occur with less frequency but with greater intensity and at earlier periods in the growing season. Lastly, if the climate change scenarios are correct, a major challenge for wheat in improved areas will extend beyond management of the current wheat system as there will be competition from a wider array of crops to choose from including soybean and C4 plants such as maize.

2.4 Summary

The aim of the previous section was to introduce the reader to the evolution of conservation agriculture and how the transformation of this innovation laid the foundation on which the agronomist built an integrated and sustainable management system for global wheat production. While some may argue that in its purest form conservation agriculture cannot be practised in all wheat-producing regions the reality is that some level of conservation is inevitable in all locations to ensure yield stability and sustainability in the future. The keys to lasting impact of any innovative practices are universal. Agronomists must engage with stakeholders and then build a cross-disciplinary team spanning upstream technologies and breeding, to machinery R&D and then applied research to address specific issues. Political will is also critical at times as many innovations may have to be enshrined in government policy or even legislation before any scale-up can occur.

3 Case study: ICM for wheat stem sawfly (*Cephus cinctus* Norton)

We have previously discussed examples of G x E x M integration in relation to effects on water management, soil health and disease management. The following case study is an example of where the philosophy of holistic integration of agronomic practices was applied systematically to address an insect pest that remains a threat to wheat sustainability in many wheat-producing regions.

The wheat stem sawfly (WSS) is a stem-boring insect that has been a major pest of wheat in the North American Great Plains for more than 110 years. Damage from WSS can result in ≥30% yield loss, and cost US$350 million annually (Beres et al., 2007). This insect presents a challenge for agronomists and sustainable wheat production in regions prone to attack as it cannot effectively be controlled with insecticides. However, there are both pre-crop and in-crop strategies including the expression of resistance in wheat cultivars and the performance of biocontrol insects that may be manipulated with agronomic practices. The biology and integrated management of the WSS system was thoroughly reviewed by Beres et al. (2011b). Here we present a summary of that review and subsequent research that culminated with the development of a decision support system to help growers integrate agronomic components into a holistic management strategy (Beres et al., 2011d).

Adults emerge from the previous year's crop stubble in late spring to early summer and, following mating, the adult female seeks out a suitable host plant to oviposit, usually an adjacent wheat field (Criddle, 1922). Populations can surge exponentially in one season as females can lay up to 50 eggs, depositing one per stem per visit into the elongating stems of wheat plants. Host preference is for a wheat plant in the early reproductive stages, that is, stem elongation, pre-boot. The WSS produces one generation per year and spends its entire egg, larval and pupal stages enclosed within the stems of wheat plants. Only one larva per stem can survive; others tend to be cannibalized by the first larva to emerge. Larvae feed and develop within the stem cavity, damaging stem parenchyma and vascular

Figure 6 Wheat damage caused by wheat stem sawfly *Cephus cinctus* Norton near Foremost, Alberta, Canada, in 2005. Source: B. L. Beres (AAFC).

tissues, reducing photosynthetic activity, leading to the first of two sources of yield loss. When a host plant begins to senesce, the larva will move to the base of the stem, cut a groove around the inside of the stem and plug the stem with frass to create its overwintering chamber. This stem-cutting activity is the second source of yield loss (≥15%), as cut stems topple over, making them difficult to harvest with modern mechanized equipment (Fig. 6). The phenology lends itself to successful survival as the WSS protects itself within the stem of the wheat host for nearly 97% of its life cycle, and only emerges from the stem as free-flying adults in June or July.

There are four points in the WSS life cycle where holistic ICM strategies can be implemented to minimize the impacts of this pest: (1) pre-crop strategies to reduce the local number of adult female WSS available to oviposit, (2) planting and in-crop strategies to reduce the amount of oviposition and to reduce egg or larval survival prior to the cutting stage and (3) harvest management strategies to reduce or prevent damage from cutting and to enhance biocontrol of WSS. We summarize the agronomic factors manipulated to improve the sustainability of wheat production in areas prone to attack.

3.1 Pre-crop strategies to reduce number of adult female WSS

The number of adult females emerging in a local area in any year is a function of the local population size from the previous year, overwintering survival, and immigration of females from surrounding areas. Continuous or wheat-fallow systems in association with dry-weather cycles enhance WSS populations, while wet-weather patterns tend to inhibit reproduction and egg deposition (Wallace and McNeal, 1966). Monoculture wheat production provides the sawfly with an abundance of nearby hosts each spring when the pest emerges from the previous year's infested wheat stubble. Crop rotational diversity would disrupt WSS dynamics but, historically in Canada and even today in some part of the United States, many producers are reluctant to rotate into immune broad leaf crops as continuous wheat provides relatively low economic risk and higher returns compared to other cropping systems in semi-arid regions (Zentner et al., 2006).

Overwintering WSS survival can be disrupted through cultivation or stubble management practices such as burning, or tillage practices such as double-discing and ploughing, but these practices are neither highly effective nor sustainable, and obviously violate conservation principles around preservation of crop residues (Beres et al., 2011b). Newer studies, however, have reported that pre-seed heavy harrow operations and no-till, direct seeding to manage infested stubble reduce WSS populations compared to a wheat-fallow system (Beres et al., 2011c). Higher disturbance drills with 23 cm row spacing would also reduce populations over low-disturbance drills. These practices reduce and disrupt stubble but leave it anchored into the soil, consistent with conservation principles (Beres et al., 2011c).

While cropping diversity and updated pre-crop strategies are effective to reduce WSS populations, the success of the system requires consistent practices from farm to farm. As adult WSS are mobile and can fly, immigration from surrounding fields is a realistic source of ovipositing females. Indeed (Rand et al., 2014), the observed WSS infestation levels correlated positively with the amount of wheat cover within a radius of 5 km. This makes any one producer's levels of WSS infestation partially determined by the management practices of their neighbour's. Landscape-level processes and management of agricultural insects is a growing field of research (Schellhorn et al., 2008; Veres et al., 2013), and will be crucial to determine relevant spatial scales for cooperative pest management between landowners.

3.2 Planting and in-crop strategies to reduce WSS oviposition

Summer mortality of WSS larvae can range from 28 to 84% (Buteler et al., 2015), and is largely the result of two factors: the use of resistant (i.e. solid-stemmed) wheat cultivars and parasitism from natural enemies. Solid-stemmed wheat cultivars were first developed for bread wheat in the 1940s and more recently in winter wheat (Morrill et al., 1992) and durum wheat (Nilsen et al., 2016). Solid-stem cultivars develop pith within the culm of the stems, which as a consequence can crush the deposited eggs or decrease the survivorship of developing larvae, preventing both damage to the plant and perpetuation of the next generation of WSS. However, the use of solid-stem varieties does not come without trade-offs. First, expression of stem solidness in most cultivars is not reliable, and depends on intense sun exposure (Holmes, 1984). Shading or precipitation-related parameters can prevent the expression of resistance (Beres et al., 2017), requiring that solid-stem varieties be seeded at lower densities (≤300 seeds m^{-2}), and supplemented with less N fertilizer (Beres et al., 2012), which may compromise attainable yield (Beres et al., 2011a). However, lower seeding densities required to express solidness may produce smaller but more female-biased WSS populations, which would potentially increase WSS infestations in the following year (Cárcamo et al., 2005, 2016). Some solid-stemmed cultivars also produce lower yields than hollow-stem varieties, particularly in the absence of WSS, as plants partition carbon away from grain filling into stem tissue (Szczepaniec et al., 2015); however, exceptional hexaploid solid-stemmed cultivars do exist (Beres et al., 2007; Beres et al., 2011b), and no yield penalty has been observed in durum (Nilsen et al., 2016). Solid-stemmed wheat cultivars are not available in every market class of wheat. For example, the premium bread wheat class in Canada, Canada Western Red Spring, no longer has a solid-stemmed cultivar option, yet it is the #1 wheat class for export. In this case, growers should adopt sowing densities as high as 450 seeds m^{-2} as this optimizes grain yield and also lowers WSS infestation rates (Beres et al., 2011a).

For spring wheat, delaying seeding by 12–20 days is an option to escape peak adult WSS activity; however, this strategy is not recommended because it will compromise yield and also reduce weed competitive ability (Morrill and Kushnak, 1999). The use of winter wheat may also cause asynchrony between the host and WSS as the crop tends to be advanced beyond the growth phase where it is an attractive host. However, in some regions of the United States, WSS populations have evolved to an earlier emergence period to synchronize with winter wheat (Morrill and Kushnak, 1996). Alternative strategies include trap crops where the perimeter of a field is planted to a solid-stemmed cultivar and the interior sown to a high-performing hollow-stemmed cultivar. This can reduce apparency of target wheat fields to WSS, or focus oviposition into border strips of resistant plants or plants to be destroyed. A cultivar blend of both hollow-stemmed and solid-stemmed wheat across the entire field is also plausible. These strategies can be effective under moderate WSS pressure, but lose efficacy under heavy pressure (Beres et al., 2009). Trap cropping could be improved if pheromone or plant volatiles could be used to attract ovipositing females, but more research is needed to refine such technologies for WSS (Buteler and Weaver, 2012).

3.3 Harvest management strategies to enhance biocontrol of WSS

WSS is attacked by several native parasitoid wasps, *Bracon cephi* (Gahan) and *Bracon lissogaster* Muesebeck (Hymenoptera: Braconidae). These native species serve as biological control agents and can cause 34–60% WSS larval mortality under favourable conditions (Buteler et al., 2015; Cárcamo et al., 2016). The two species have similar life histories and produce two generations per year, although their activity varies regionally. Year-to-year variation is impacted by these biocontrol wasps, and thus WSS population dynamics are impacted by weather; wet conditions tend to delay wheat senescence in later summer and ensure WSS larval hosts are available for the second wasp generation (Holmes et al., 1963). Cropping practices can also influence the populations of these biocontrol agents. Planting later-maturing varieties of wheat can ensure successful completion of the second generation of wasps that emerge in fall (Beres, 2011), and leaving stubble height ≥15 cm conserves most of the parasitoid overwintering habitat (Beres, 2011; Meers, 2005).

Conservation and efficacy of the parasitoid is challenging as efficacy tends to be nullified if completion of a second generation fails to occur. Furthermore, solid-stem varieties and the biocontrol wasps may interact negatively with each other, as larval mortality caused by solid stems can leave fewer hosts available for the parasitoids. Carcamo et al. (2016) reported that using a 1:1 blend of hollow-stemmed and solid-stemmed cultivars can provide some WSS control while augmenting parasitoid populations. Further research is needed to understand and improve the interaction between resistant wheat cultivars and WSS parasitoids.

An artificial neural network model to predict pith expression in solid-stemmed cultivars has been developed (Beres et al., 2017) based on precipitation-related weather data. Producers growing a solid-stemmed cultivar can use the model to determine if any action is warranted based on the cutting damage predicted by the model. If a WSS larval infestation is present and/or substantial cutting is anticipated, producers can swath their fields early prior to stem lodging. Although an added operation, it is very effective at protecting yield by collecting girdled stems safely into a windrow. If swathing is not utilized and the field lodges from WSS cutting, combines will require specialized flex headers equipped with crop lifters, and may have to alter harvest directions to optimize the recovery of toppled stems. This is costly, damages equipment, and is not 100% effective. Furthermore, this action has additional negative consequences as the short-stubble low-cutting heights reduce snow capture, soil health and moisture retention, and biocontrol wasp survival (Beres, 2011).

3.4 A decision support system to manage wheat stem sawfly

The management tactics summarized were combined with pest surveillance and monitoring strategies to provide growers with a knowledge reference key to support wheat management decisions. Risk maps are available and can be reviewed prior to spring sowing (Meers and Barkley, 2015), which allows producers to make informed decisions regarding cultivar selection, wheat field selection and crop phases. In-crop surveillance is recommended to assess site-specific risk and to determine the need for action based on the level of sawfly infestation (Fig. 7). Predicted risk of cutting damage by WSS can be categorized as low, medium or high, based on infested stems observed in the ranges of 0–20%, 20–40% and >40% stems infested, respectively (Fig. 7). Threat levels make prediction about the environment in which the wheat crop will be grown, and a producer can then make additional informed decisions regarding the need for pre-seed harrowing, sowing densities that balance pith expression with yield potential, nitrogen rates for pith expression versus optimizing yield and the need for swathing to collect stems and preserve tall stubble. For example, if the environment

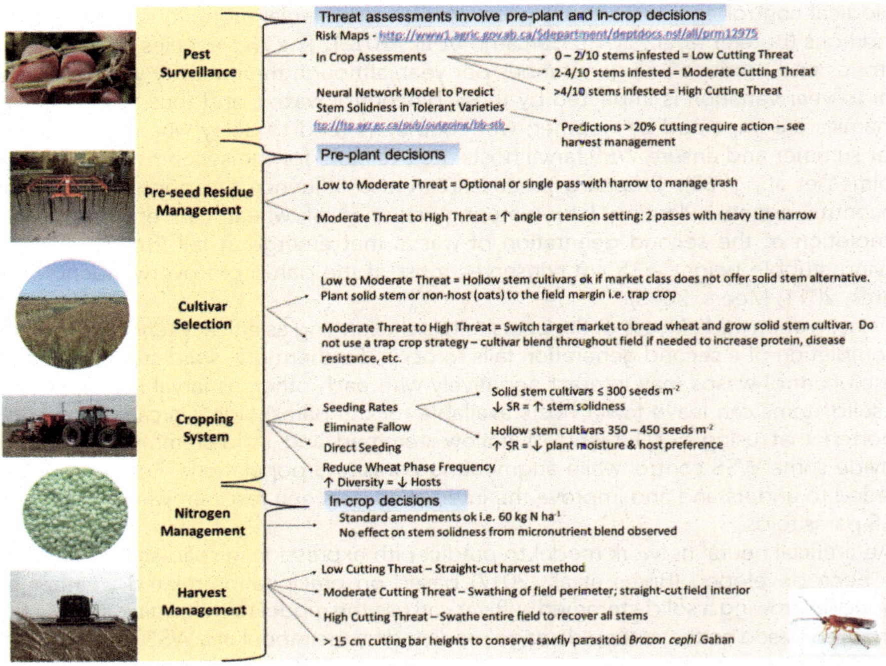

Figure 7 Decision support system schematic for the management of wheat stem sawfly *Cephus cinctus* Norton. Source: (Beres et al., 2011d).

predicts high WSS pressure, a holistic strategy would integrate continuous cropping with appropriate pre-seed residue management/direct-seeding, solid-stemmed cultivars, seeding rates no greater than 300 seeds m^{-2}, ~ 60 kg N ha^{-1} and harvest cutting heights of at least 15 cm (Fig. 7).

3.5 Conclusions

The WSS is an example of an economic pest threat that requires G x E x M integration utilizing input from several cross-cutting disciplines. In this case study, the 'G' was solid-stem germplasm or other cultivars with excellent straw strength to withstand toppling after infestation and cutting; the 'E' is not only an area prone to attack, which is exacerbated in dry cycles of weather, but it is also the wheat stem and the plant's ability to produce pith based on genetics and environment, and the plant's response to management, that reduces infestation. The 'M' was the agronomic package developed to mitigate damage. This case study is also an example of where an integrated system for wheat production in areas prone to attack was developed upon the underlying principles of conservation agriculture. We have also learned that over-reliance on a particular strategy at the expense of another jeopardizes sustainability and yield. The decision support system should not be considered a static management tool – each facet should be reviewed and refined as cropping systems and the interactions with WSS evolve.

4 Future trends and conclusion

As soon as a kernel of wheat is planted in the soil the gap forms between its genetic potential and realized yield – a gap that will only widen with time. The development of traits in the future will no doubt increase genetic potential; however, the impact will be less than desirable if factors contributing to the yield gap are not identified and addressed. Known factors contributing to the yield gap in wheat include water deficits, nitrogen loss and heat stress/shock, but others exist and more will emerge. A common goal globally should be to accurately document wheat yield gaps by country/region using models and but ground-truthing those results to ensure accuracy. Australia, led by Drs J. Kirkegaard and R. Fischer, has laid the ground work for ground-truthing model outputs with a methodology that is easily replicated elsewhere. China has also reported on yield gaps and estimated that they are currently in the range of 62–71% of potential yield (Smith et al., 2016). The following trends identified are not meant to be a complete list of trends but serve as examples of anticipated innovations that could increase the yield potential of wheat.

Biotechnology will forge ahead by exploiting the latest technologies such as genome editing and nanotechnology, but deployment rests with agronomists who must successfully integrate these technologies into wheat systems in a manner that is sustainable and results in full expression of the trait. Wheat systems will only become more complex on farms as integration is needed of emerging precision tools such as autonomous tractors, UAVs and robots equipped with technologies to intelligently map and navigate fields in order to manage the crop on a by-plant basis and to discriminate and remove herbicide-resistant weeds.

Research is also moving beyond nutrient management to study the plant–soil biome. While significant knowledge has been generated around soil microbial biomass or arbuscular mycorrhizal fungi interactions with wheat as influenced by preceding crop phases, soil scientist have yet to fully unlock the potential of biologicals. Innovations in this area could eventually lead to seed inoculants that could possibly improve the dismal NUE rates of 50% that currently plague wheat production. There is also the possibility that advances could lead to protection against soil-borne diseases described earlier or even abiotic stress. Advances in these areas will spawn the need for further research into rotational phases and break crops in wheat-based systems.

All of these areas of research have a common element – the generation of 'big data'. An ICM system for wheat in the future must be capable of managing the reams of data that each of these technologies generate, which will require nimble decision support systems for rapid synthesis and utilization of the data. Wheat producers are already generating significant amounts of data but is it really being processed and used efficiently or to its potential? For example, yield monitor data is generated but it can overwhelm the producer, and there will only be more data to digest in the future. Excellent work is currently being conducted to analyse spatial data and inference points within the field to increase the precision of making sound managements decisions per unit area of large-scale fields (Bakare, 2015).

Synergies between agronomy and the seed developer, trait/technology company and other innovators will need to be forged at the onset of discovery. This not only ensures delivery of desired yield gains but could result in multiplicative effects of simply additive effects (Robertson et al., 2016). Moreover, the society as a whole needs to be engaged so that the end-users of the wheat production chain trust that the flour, pasta, bread, pastry, etc., on their plate have been produced in a safe, sustainable manner with minimal to no consequences to the environment. After all, farmers, researchers and more and more the public, in general, realize that we are only temporary custodians of the land and the place where we reside – it is imperative we leave it in as good or in a better state than when we inherited it.

5 Where to look for further information

Addressing the challenges that lie ahead for wheat agronomy research requires greater collaboration on both a regional and an international scale. This will require enhanced partnerships between public, private and producer entities. This was recognized at the 2011 meeting of the G20 Ministries of Agriculture, which established the Wheat Initiative. Core themes outlined in the Wheat Initiative's Strategic Research Agenda are (1) increased yield potential, (2) protect yield potential, (3) protect the environment and increase the sustainability of wheat production systems and (4) ensure the supply of high-quality, safe wheat. Agronomy and ICM systems research will fall largely under Priority 3, but will also be cross-cutting. Expert Working Groups (EWGs) have been established for all core themes and sub-themes. An agronomy EWG is being assembled through the leadership of Dr John Kirkegaard (CSIRO), Dr Jerry Hatfield (USDA-ARS), Dr Sanford Eigenbrode (Univ of Idaho), Dr Bill Pan (Washington State University) and Dr Brian Beres (AAFC). This EWG have held meetings since November 2016 to initiate the process of establishing research inventories by country/region and developing research priorities by country that will shape future research and enhance collaboration globally to address common objectives around wheat agronomy research. Updates on this process and general information for the Wheat Initiative can be found at http://www.wheatinitiative.org/. A 'Wheat Initiative Agronomists Community' has also been established within the Communities framework of the American Society of Agronomy (https://www.agronomy.org/membership/communities/wheat-initiative-agronomists-community).

6 References

Andrew, I. K. S., J. Storkey and D. L. Sparkes. 2015. A review of the potential for competitive cereal cultivars as a tool in integrated weed management. *Weed Research* 55:239–48.

Angus, J. F., J. A. Kirkegaard, J. R. Hunt, M. H. Ryan, L. Ohlander and M. B. Peoples. 2015. Break crops and rotations for wheat. *Crop and Pasture Science* 66:523–52.

Anonymous. 2007. Crop land in cultivated summer fallow [Online]. Available by United States Department of Agriculture http://www.agcensus.usda.gov/Publications/2007/Full_Report/index.asp (verified 20 December).

Anonymous. 2009. Estimated summerfallow areas, annual [Online]. Available by Government of Canada http://www.statcan.gc.ca/cgi-bin/imdb/p2SV.pl?Function=getSurvey&SDDS=3401&lang=en&db=imdb&adm=8&dis=2 (verified 15 November).

Anonymous. 2016a. Estimated summerfallow areas, annual [Online]. Available by Government of Canada http://www.statcan.gc.ca/cgi-bin/imdb/p2SV.pl?Function=getSurvey&SDDS=3401&lang=en&db=imdb&adm=8&dis=2 (verified 15 December).

Anonymous. 2016b. FOASTAT Production Crops - estimates of production of crops worldwide [Online]. Available by United Nations http://faostat3.fao.org/home/E (posted 28 September 2016; verified 28 September 2016).

Anonymous. 2016c. Field crop reporting series – estimates of production of principal field crops [Online]. Available by Government of Canada http://www.canadagrainscouncil.ca/subscriber/CGCSection/www_show_handbook (verified 28 September 2016).

Asseng, S., F. Ewert, P. Martre, R. P. Rotter, D. B. Lobell, D. Cammarano, B. A. Kimball, M. J. Ottman, G. W. Wall, J. W. White, M. P. Reynolds, P. D. Alderman, P. V. V. Prasad, P. K. Aggarwal, J. Anothai, B. Basso, C. Biernath, A. J. Challinor, G. De Sanctis, J. Doltra, E. Fereres, M. Garcia-Vila, S. Gayler, G. Hoogenboom, L. A. Hunt, R. C. Izaurralde, M. Jabloun, C. D. Jones, K. C. Kersebaum, A. K. Koehler, C. Muller, S. Naresh Kumar, C. Nendel, G. O/'Leary, J. E. Olesen, T. Palosuo, E. Priesack, E. Eyshi Rezaei, A. C. Ruane, M. A. Semenov, I. Shcherbak, C. Stockle, P. Stratonovitch, T. Streck, I. Supit, F. Tao, P. J. Thorburn, K. Waha, E. Wang, D. Wallach, J. Wolf, Z. Zhao and Y. Zhu. 2015. Rising temperatures reduce global wheat production. *Nature Climate Change* 5:143–7.

Bakare, M. A. 2015. *Geostatistical Analysis of Yield Monitor Data for Precision Agriculture*, University of Alberta, Edmonton, AB.

Bell, L. W., A. D. Moore and J. A. Kirkegaard. 2014. Evolution in crop–livestock integration systems that improve farm productivity and environmental performance in Australia. *European Journal of Agronomy* 57:10–20.

Beres, B. L. 2011. Integrating the building blocks of agronomy and biocontrol into an integrated pest management strategy for wheat stem sawfly. Dissertation, University of Alberta, Edmonton, Alberta.

Beres, B. L., H. A. Cárcamo and J. R. Byers. 2007. Effect of wheat stem sawfly damage on yield and quality of selected Canadian spring wheat. *Journal of Economic Entomology* 100:79–87.

Beres, B. L., H. A. Cárcamo, R.-C. Yang and D. M. Spaner. 2011a. Integrating spring wheat sowing density with variety selection to manage wheat stem sawfly. *Agronomy Journal* 103:1755–64.

Beres, B. L., K. N. Harker, G. W. Clayton, R. E. Blackshaw and R. J. Graf. 2010a. Weed competitive ability of spring and winter cereals in the Northern Great Plains. *Weed Technology* 24:108–16.

Beres, B. L., L. M. Dosdall, D. K. Weaver, D. M. Spaner and H. A. Cárcamo. 2011b. The biology and integrated management of wheat stem sawfly, *Cephus cinctus* (Hymenoptera: Cephidae), and the need for continuing research. *Canadian Entomologist* 143:105–25.

Beres, B. L., G. W. Clayton, K. N. Harker, C. F. Stevenson, R. E. Blackshaw and R. J. Graf. 2010b. A sustainable management package to improve winter wheat production and competition with weeds. *Agronomy Journal* 102:649–57.

Beres, B. L., H. A. Cárcamo, L. M. Dosdall, M. L. Evenden, R. C. Yang and D. M. Spaner. 2011c. Do interactions between residue management and direct seeding affect wheat stem sawfly and grain yield? *Agronomy Journal* 103:1635–44.

Beres, B. L., K. N. Harker, G. W. Clayton, E. Bremer, T. T. O'Donovan, R. E. Blackshaw and A. M. Smith. 2010c. Influence of N fertilization method on weed growth, grain yield and grain protein concentration in no-till winter wheat. *Canadian Journal of Plant Science* 90:637–44.

Beres, B. L., R. H. McKenzie, H. A. Cárcamo, L. M. Dosdall, M. L. Evenden, R. C. Yang and D. M. Spaner. 2012. Influence of seeding rate, nitrogen management and micronutrient blend applications on pith expression in solid-stemmed spring wheat. *Crop Science* 52:1316–29.

Beres, B. L., H. A. Cárcamo, D. K. Weaver, L. M. Dosdall, M. L. Evenden, B. D. Hill, R. C. Yang, R. H. McKenzie and D. M. Spaner. 2011d. Integrating the building blocks of agronomy and biocontrol into an IPM strategy for wheat stem sawfly. *Prairie Soils & Crops: Scientific Perspectives for Innovative Management* 4:54–65.

Beres, B. L., T. K. Turkington, H. R. Kutcher, B. Irvine, E. N. Johnson, J. T. O'Donovan, K. N. Harker, C. B. Holzapfel, R. Mohr, G. Peng and D. M. Spaner. 2016. Winter wheat cropping system response to seed treatments, seed size, and sowing density. *Agronomy Journal* 108:1101–11.

Beres, B. L., B. D. Hill, H. A. Cárcamo, J. J. Knodel, D. K. Weaver, R. D. Cuthbert, 2017. An artificial neural network model to predict wheat stem sawfly cutting in solid-stemmed wheat cultivars. *Canadian Journal of Plant Science* 97(2). In press. doi:10.1139/cjps-2016-0364.

Blackshaw, R. E. 1994. Differential competitive ability of winter wheat cultivars against downy brome. *Agronomy Journal* 86:649–54.

Bollinger, A., J. Magid, J. T. Carneiro Amado, F. Skora Neto, M. F. S. Ribeiro, A. Calegari, R. Ralisch and A. de Neergaard. 2005. Taking stock of the Brazilian 'Zero-Till Revolution': A review of landmark research and farmer's practice. *Advances in Agronomy* 91:47–110.

Bradley, B. D., M. Christodoulou, C. Caspari and P. Di Luca. 2002. Integrated crop management systems in the EU. In *Environment_Directorate-General_European_Commission* (ed.), European Commission, Brussels, Belgium, pp. 157.

Brandt, S. A. 2010. Systems approach to agronomic innovation. In C. W. Lindwall and B. Sonntag, (eds), *Landscapes Transformed: The History of Conservation Tillage and Direct Seeding.* Knowledge Impact in Society, Saskatoon, SK, pp. 131–9.

Buteler, M., and D. K. Weaver. 2012. Host selection by the wheat stem sawfly in winter wheat and the role of semiochemicals mediating oviposition preference. *Entomologia Experimentalis et Applicata* 143:138–47.

Buteler, M., R. K. Peterson, M. L. Hofland and D. K. Weaver. 2015. A multiple decrement life table reveals that host plant resistance and parasitism are major causes of mortality for the wheat stem sawfly. *Environmental Entomology* 44:1571–80.

Cárcamo, H., B. Beres, T. Larson, C. Klima and X.-H. Wu. 2016. Effect of Wheat Cultivars and Blends on the Oviposition and Larval Mortality of Cephus cinctus (Hymenoptera: Cephidae) and Parasitism by Bracon cephi (Hymenoptera: Braconidae). *Environmental entomology* 45:397–403.

Cárcamo, H., B. Beres, F. Clarke, R. Byers, H.-H. Mündel, K. May and R. DePauw. 2005. Influence of plant host quality on fitness and sex ratio of the wheat stem sawfly (Hymenoptera: Cephidae). *Environmental Entomology* 34:1579–92.

Criddle, N. 1922. The western wheat stem sawfly and its control. Dominion of Canada Department of Agriculture Pamphlet No. 6 New Series:1–8.

Darwinkle, A. 1978. Patterns of tillering and grain production of winter wheat at a wide range of plant densities. *Netherlands Journal of Agricultural Science* 26:383–98.

Derksen, D. A., R. L. Anderson, R. E. Blackshaw and B. Maxwell. 2002. Weed dynamics and management strategies for cropping systems in the northern Great Plains. *Agronomy Journal* 94:174–85.

Dove, H., and J. Kirkegaard. 2014. Using dual-purpose crops in sheep-grazing systems. *Journal of the Science of Food and Agriculture* 94:1276–83.

Duvick, D. N. 2005. Genetic progress in yield of United States maize (Zea mays L.). *Maydica* 50:193–202.

Engel, R., C. Jones and R. Wallander. 2011. Ammonia volatilization from urea and mitigation by NBPT following surface application to cold soils. *Soil Science Society of America Journal* 75:2348–57.

Entz, M. H., V. S. Baron, P. M. Carr, D. W. Meyer, S. R. Smith and W. P. McCaughey. 2002. Potential of forages to diversify cropping systems in the northern Great Plains. *Agronomy Journal* 94:240–50.

Fernandez, M. R., R. P. Zentner, P. Basnyat, D. Gehl, F. Selles and D. Huber. 2009. Glyphosate associations with cereal diseases caused by Fusarium spp. in the Canadian Prairies. *European Journal of Agronomy* 31:133–43.

Fowler, D. B. 1983. The effect of management practices on winter survival and yield of winter wheat produced in regions with harsh winter climates. In L. V. Gusta, et al., (eds), *New Frontiers in Winter Wheat Production*. Div. of Ext. and Comm. Rel., University of Saskatchewan, Saskatoon, SK, pp. 238–82.

Fulton, F., and B. Sonntag. 2010. History of early western Canadian agriculture – setting the stage for conservation. In C. W. Lindwall and B. Sonntag, (eds), *Landscapes Transformed: The History of Conservation Tillage and Direct Seeding*. Knowledge Impact in Society, Saskatoon, SK, pp. 1–14.

Fulton, M. 2010. Foreward. In C. W. Lindwall and B. Sonntag, (eds), *Landscapes Transformed: The History of Conservation Tillage and Direct Seeding*. Knowledge Impact in Society, Saskatoon, SK, p. 219.

Giller, K. E., J. A. Andersson, M. Corbeels, J. Kirkegaard, D. Mortensen, O. Erenstein and B. Vanlauwe. 2015. Beyond conservation agriculture. *Frontiers in Plant Science* 6.

Harker, K. N., R. E. Blackshaw, H. J. Beckie and J. T. O'Donovan. 2009. Weed invasions in western Canada cropping systems – management of invasive weeds. In I. Inderjit, ed., Vol. 5. Springer, The Netherlands, pp. 151–66.

Harker, K. N., G. W. Clayton, R. E. Blackshaw, J. T. O'Donovan and F. C. Stevenson. 2003. Seeding rate, herbicide timing and competitive hybrids contribute to integrated weed management in canola (Brassica napus). *Canadian Journal of Plant Science* 83:433–40.

Holen, D. L., P. L. Bruckner, J. M. Martin, G. R. Carlson, D. M. Wichman and J. E. Berg. 2001. Response of winter wheat to simulated stand reduction. *Agronomy Journal* 93:364–70.

Holmes, N. 1984. The effect of light on the resistance of hard red spring wheats to the wheat stem sawfly, Cephus cinctus (Hymenoptera: Cephidae). *The Canadian Entomologist* 116:677–84.

Holmes, N., W. Nelson, L. Peterson and C. Farstad. 1963. Causes of variations in effectiveness of Bracon cephi (Gahan)(Hymenoptera: Braconidae) as a parasite of the wheat stem sawfly. *The Canadian Entomologist* 95:113–26.

Hunt, J. R., A. D. Swan, N. A. Fettell, P. D. Breust, I. D. Menz, M. B. Peoples and J. A. Kirkegaard. 2016. Sheep grazing on crop residues do not reduce crop yields in no-till,controlled traffic farming systems in an equi-seasonal rainfallenvironment. *Field Crops Research* 196:22–32.

Janosky, J. S., D. L. Young and W. F. Schillinger. 2002. Economics of conservation tillage in a wheat fallow rotation. *Agronomy Journal* 94:527–31.

Janzen, H. H. 2001. Soil science on the Canadian prairies – Peering into the future from a century ago. *Canadian Journal of Soil Science* 81:489–503.

Kawate, M. K., S. G. Colwell, A. G. Ogg Jr. and J. M. Kraft. 1997. Effect of glyphosate-treated henbit (Lamium amplexicaule) and downy brome (Bromus tectorum) on *Fusarium solani* f. sp. *pisi* and *Pythium ultimum*. *Weed Science* 45:739–43.

Kirkegaard, J., and J. Hunt. 2010. Increasing production by matching farming system management and genotypes in water-limited environments. *Journal of Experimental Botany* 61:4129–43.

Kirkegaard, J. A., P. J. Hocking, J. F. Angus, G. N. Howe and P. A. Gardner. 1997. Comparison of canola, Indian mustard and Linola in two contrasting environments. II. Break-crop and nitrogen effects on subsequent wheat crops. *Field Crops Research* 52:179–91.

Kirkegaard, J. A., M. K. Conyers, J. R. Hunt, C. A. Kirkby, M. Watt and G. J. Rebetzke. 2014a. Sense and nonsense in conservation agriculture: Principles, pragmatism and productivity in Australian mixed farming systems. *Agriculture, Ecosystems & Environment* 187:133–45.

Kirkegaard, J. A., J. R. Hunt, T. M. McBeath, J. M. Lilley, A. Moore, K. Verburg, M. Robertson, Y. Oliver, P. R. Ward, S. Milroy and A. M. Whitbread. 2014b. Improving water productivity in the Australian Grains industry – a nationally coordinated approach. *Crop and Pasture Science* 65:583–601.

Lafond, G. P., and Y. Gan. 1999. Row spacing and seeding rate studies in no-till winter wheat for the Northern Great Plains. *Journal of Production Agriculture* 12:624–9.

Lafond, G. P., S. A. Brandt, G. W. Clayton, R. B. Irvine and W. E. May. 2011. Rainfed Farming Systems on the Canadian Prairies. In P. Tow, et al., eds. *Rainfed Farming Systems*. Springer, The Netherlands, Dordrecht, pp. 467–510.

Larney, F. J., C. W. Lindwall, R. C. Izaurralde and A. P. Moulin. 1994. Tillage systems for soil and water conservation on the Canadian Prairie. In *Conservation Tillage in Temperate Agroecosystems*. CRC Press, Boca Raton, FL, pp. 305–28.

Lawes, J. B., and J. H. Gilbert. 1894. Rotation of crops. *Journal of the Royal Agricultural Society of England* 55:585–646.

Lindwall, C. W., and D. T. Anderson. 1981. Agronomic evaluation of minimum tillage systems for summer fallow in southern Alberta. *Canadian Journal of Plant Science* 61:247–53.

Llewellyn, R. S., F. H. D'Emden and G. Kuehne. 2012. Extensive use of no-tillage in grain growing regions of Australia. *Field Crops Research* 132:204–12.

Mathews, O. R. 1945. Crop residue management in dry-land crop production. *Agronomy Journal* 37:297–306.

Meers, S. B. 2005. Impact of harvest operations on parasitism of the wheat stem sawfly, *Cephus cinctus* Norton (Hymenoptera: Cephidae). Master of Science, Montana State University, Bozemon, Montana.

Meers, S. B., and S. Barkley. 2015. 2016 wheat stem sawfly forecast [Online]. Available by Government of Alberta http://www1.agric.gov.ab.ca/$department/deptdocs.nsf/all/prm12975 (posted 24 November 2015; verified 5 October 2016).

Miller, P., R. Engel and J. Holmes. 2006. Cropping sequence effect of pea and pea management on spring wheat in the northern Great Plains. *Agronomy Journal* 98:1610–19.

Morrill, W. L., and G. D. Kushnak. 1996. Wheat stem sawfly (Hymenoptera: Cephidae) adaptation to winter wheat. *Environmental Entomology* 25:1128–32.

Morrill, W. L., and G. D. Kushnak. 1999. Planting date influence on the wheat stem sawfly (Hymenoptera: Cephidae) in spring wheat. *Journal of Agricultural Urban Entomological* 16:123–8.

Morrill, W. L., J. W. Gabor, E. A. Hockett and G. D. Kushnak. 1992. Wheat stem sawfly (Hymenoptera: Cephidae) resistance in winter wheat. *Journal of Economic Entomology* 85:2008–11.

Nilsen, K. T., J. M. Clarke, B. L. Beres and C. J. Pozniak. 2016. Sowing density and cultivar effects on pith expression in solid-temmed durum wheat. *Agronomy Journal* 108:219–28.

O'Donovan, J. T., K. N. Harker, G. W. Clayton and L. M. Hall. 2000. Wild oat (*Avena fatua*) interference in barley (*Hordeum vulgare*) is influenced by barley variety and seeding rate. *Weed Technology* 14:624–9.

Parry, M. L., C. Rosenzweig, A. Iglesias, M. Livermore and G. Fischer. 2004. Effects of climate change on global food production under SRES emissions and socio-economic scenarios. *Global Environmental Change* 14:53–67.

Peterson, G. A., A. J. Schlegel, D. L. Tanaka and O. R. Jones. 1996. Precipitation use efficiency as affected by cropping and tillage systems. *Journal of Production Agriculture* 9:180–6.

Rahe, J. E., C. A. Lévesque and G. S. Johal. 1990. Synergistic role of soil fungi in the herbicidal efficacy of glyphosate. In R. E. Hoagland, (ed.), *Biological Weed Control Mircobes and Microbial Produts as Herbicides*. Symposium, 9–14 April 1989. American Chemical Society, Washington, DC, pp. 260–75.

Rand, T. A., D. K. Waters, S. L. Blodgett, J. J. Knodel and M. O. Harris. 2014. Increased area of a highly suitable host crop increases herbivore pressure in intensified agricultural landscapes. *Agriculture, Ecosystems & Environment* 186:135–43.

Rebetzke, G. J., S. E. Bruce and J. A. Kirkegaard. 2005. Longer coleoptiles improve emergence through crop residues to increase seedling number and biomass in wheat (*Triticum aestivum* L.). *Plant and Soil* 272:87–100.

Robertson, M., J. Kirkegaard, G. Rebetzke, R. Llewellyn and T. Wark. 2016. Prospects for yield improvement in the Australian wheat industry: a perspective. *Food and Energy Security* 5:107–22.

Rochette, P., J. D. MacDonald, D. A. Angers, M. H. Chantigny, M. Gasser and N. Bertrand. 2009. Banding of urea increased ammonia volatilization in a dry acidic soil. *Journal of Environmental Qualilty* 38:1383–90.

Roget, D. K., N. R. Venn and A. D. Rovira. 1987. Reduction of *Rhizoctonia* root rot of direct-drilled wheat by short-term chemical fallow. *Australian Journal of Experimental Agriculture* 27:425–30.

Schellhorn, N. A., S. Macfadyen, F. J. Bianchi, D. G. Williams and M. P. Zalucki. 2008. Managing ecosystem services in broadacre landscapes: what are the appropriate spatial scales? *Animal Production Science* 48:1549–59.

Smith, E. G., H. H. Janzen and F. J. Larney. 2015a. Long-term cropping system impact on quality and productivity of a Dark Brown Chernozem in southern Alberta. *Canadian Journal of Soil Science* 95:177–86.

Smith, E. G., H. H. Janzen and R. Kroebel. 2015b. Yield and profitability of fallow and fertilizer inputs in long-term wheat rotation plots at Lethbridge, Alberta. *Canadian Journal of Plant Science* 95:579–87.

Smith, E. G., D. LeRoy, P. Maccallum and H. Janzen. 2016. Will summer fallow re-emerge in the Dark Brown soil zone of the Canadian Prairie as a response to net return risk? *Canadian Journal of Plant Science*.

Szczepaniec, A., K. D. Glover and W. Berzonsky. 2015. Impact of solid and hollow varieties of winter and spring wheat on severity of wheat stem sawfly (Hymenoptera: Cephidae) infestations and yield and quality of grain. *Journal of Economic Entomology* 108:2316–23.

Thomas, J. B., G. B. Schaalje and M. N. Grant. 1994. Height, competition and yield potential in winter wheat. *Euphytica* 74:9–17.

Veres, A., S. Petit, C. Conord and C. Lavigne. 2013. Does landscape composition affect pest abundance and their control by natural enemies? A review. *Agriculture, Ecosystems and Environment* 166:110–17.

Wallace, L. E., and F. H. McNeal. 1966. *Stem Sawflies of Economic Importance in Grain Crops in the United States Agricultural Research Service*, U. S. Dept. of Agriculture, Washington.

Watson, P. R., D. A. Derksen and R. C. VanAcker. 2006. The ability of 29 barley cultivars to compete and withstand competition. *Weed Science* 54:783–92.

Whaley, J. M., D. L. Sparkes, M. J. Foulkes, J. H. Spink, T. Semere and R. K. Scott. 2000. The physiological response of winter wheat to reductions in plant density. *Annals of Applied Biology* 137:165–77.

Zentner, R. P., C. A. Campbell, F. Selles, P. G. Jefferson, R. Lemke, B. G. McConkey, M. R. Fernandez, C. Hamel, Y. Gan and A. G. Thomas. 2006. Effect of fallow frequency, flexible rotations, legume green manure, and wheat class on the economics of wheat production in the Brown soil zone. *Canadian Journal of Plant Science* 86:413–23.

Zentner, R. P., D. D. Wall, C. N. Nagy, E. G. Smith, D. L. Young, P. R. Miller, C. A. Campbell, B. G. McConkey, S. A. Brandt, G. P. Lafond, A. M. Johnston and D. A. Derksen. 2002. Economics of crop diversification and soil tillage opportunities in the Canadian prairies. *Agronomy Journal* 94:216–30.

Chapter 8

Organic production of wheat and spelt

T. F. Döring, Humboldt-Universität zu Berlin, Germany

1 Introduction
2 Wheat in organic rotations
3 Organic wheat in species mixtures
4 Diversification of organic wheat at the genetic level
5 Organic wheat and farm biodiversity
6 Future trends and conclusion
7 Where to look for further information
8 Acknowledgements
9 References

1 Introduction

In late March 2011, field technicians in Bernburg, Germany, sowed an organic spring wheat variety experiment on their fertile loess soil, with the wheat following a pre-crop of faba beans (*Vicia faba*) (LLFG, 2011). On the same day, nearly 7000 km to the west, organic crop specialist Chris Reberg-Horton took pictures of winter wheat plots from an organic variety trial near Kinston, North Carolina, documenting the differences of crop development among the various wheat varieties. It was also the day when a conference on climate change and agriculture ended in Budapest, Hungary, where plant breeder Peter Mikó had presented work on the assessment of gene bank accessions of the tetraploid wheat species *Triticum timopheevii* under organic conditions (Mikó et al., 2011). Still on the same date, organic farmer John Pawsey posted a blog about how he manages his winter wheat crops on his farm in Suffolk, East Anglia, United Kingdom, with a camera-guided mechanical weeding system and a harrow comb (Pawsey, 2011).

These flashlights from around the globe illustrate the tremendous geographic range of organic wheat systems. Wheat is grown organically from Sweden to Spain, from Argentina to Australia and from Canada to Kazakhstan. Globally, organic wheat is produced on an area of approximately 1.2 Mio ha, according to figures from 2013; compared to 2004, this was an increase by almost 0.8 Mio ha; that is, within one decade, the organic wheat area had approximately tripled (FiBL, 2015). Countries with the largest organic wheat area (>100 000 ha) in the five-year period of 2009–13 were China, the United States, Turkey and

http://dx.doi.org/10.19103/AS.2016.0004.31

Italy. Further countries with substantial organic wheat area (>50 000 ha) in this period were, in descending order, Canada, Kazakhstan, Ukraine and Germany.

The importance of wheat for organic agriculture is also illustrated by its high proportion among organic cereals. In North America, for example, where the most important organic arable crops are cereals, wheat is the main organic cereal grown, with more than a third of the cereal area, followed by oats and maize. Globally, wheat made up 42% and 36% of the organic cereal area in 2012 and 2013, respectively (Willer and Lernoud, 2015). Such high importance of wheat for organic farming is not a recent phenomenon. Already in 1990, wheat was assessed to be 'probably the most important cereal cash crop in organic systems' (Lampkin, 1994, p. 135).

One reason for the importance of wheat in organic agriculture is its favourable economic performance. In comparison with other organically produced cereal crops such as barley, oats, rye and triticale, organic winter wheat frequently achieves the highest gross margins (Lampkin et al., 2011). This results from a combination of high yield potential of wheat and relatively high prices, in particular for bread wheat. Furthermore, the global expanse of organic wheat farming and, concurrently, its large significance for the organic sector, might be attributed to the versatility, diversity and plasticity of wheat as a plant species.

At the same time, the expansive distribution of organic wheat production is also associated with a large diversity of cropping systems, reflecting, on the one hand, the multitude of environmental conditions under which organic wheat is grown, and, on the other hand, the diversity of standards, regulations and policies of organic farming in different parts of the world. This diversity makes the subject of organic wheat cropping not only interesting but also complex and unwieldy. Inevitably, this chapter can therefore only give limited perspectives on selected organic wheat-cropping systems. Furthermore, because of the more extensive literature on organic wheat from Europe and North America than from other parts of the world, especially from Asia, this chapter will be geographically biased.

Is the diversity of wheat-cropping systems among organic farms held together by any commonalities? One common theme may be seen in the diversity *within* organic farms. Diversity on the farm is an important issue in principles, standards and practices. For example, according to UK organic farming standards, a leading principle of organic production is 'to foster biodiversity' by 'maintaining and increasing ecological diversity within and around cropped land' (Soil Association, UK, 2016, pp. 7 and 26). Diversity within organic farms is also central in the internationally agreed principles of organic farming. According to the principle of ecology, 'Organic agriculture should attain ecological balance through the design of farming systems, establishment of habitats and maintenance of genetic and agricultural diversity' (IFOAM, 2005) .

The structuring element of this chapter on organic wheat and spelt is therefore (on-farm) diversification. The chapter deals with four aspects of diversification, with particular emphasis on diversifying crops. The first of the following sections revolves around crop rotations as an important tool of diversification. Organic farming relies on crop rotations for plant nutrition, pest and disease control, and for the regulation of weeds. Crop rotations are therefore often seen as the backbone of organic cropping systems, and an understanding of organic wheat growing is impossible without detailed knowledge of associated rotations. This perspective on cycling wheat with other crop species in organic farming (section 2) is then complemented by looking at diversification through mixing wheat with other crop species in intercropping systems (section 3). In Section 4, the diversification of organic wheat production associated with seed systems and plant breeding is reviewed. Finally, the links between organic wheat production and farmland biodiversity are briefly discussed.

2 Wheat in organic rotations

2.1 Importance of crop rotations in organic agriculture

A crop rotation is the planned, quasi-cyclical sequence of different crop species together with their associated agronomic activities on a given piece of land, usually a field or group of fields. Crop rotations can be seen as the backbone of organic agriculture. The importance of crop rotations and their distinctness for organic systems was already emphasised by the early founding figures of organic farming (see review by Stinner, 2007). More recent studies have reiterated the central role of crop rotations for organic agriculture (Lampkin, 1990; Stockdale et al., 2001; Watson et al., 2002; Leoni et al., 2015; van Bruggen et al., 2016). Usually, different fields on a farm are in different stages of the crop rotation, so that the proportion of particular crop species in the sequence, that is, over time, is reflected in the proportion of the farm area given to that crop at any one time.

When planning a crop rotation, four key questions need to be considered: (1) Which crop species should be grown? (2) Which proportions of the land should these species take? (3) In which order are the different crops to be grown? (4) What are the associated management practices? How these questions are answered on a particular farm follows a number of general rules and principles, which can be grouped around the topics of economy, plant protection, nutrient management and soil quality. This section discusses these aspects briefly in the context of organic wheat. However, it must be noted that rotations in practice are rarely fixed. Which crop species is grown on a particular field from year to year depends increasingly on the markets. Thus, deviations from the rotation plan are common, not only in conventional agriculture, but partly also within organic farming (Kolbe, 2016).

2.2 Concentration of wheat in the rotation

Wheat, especially when grown for bread production, is characterised by a relatively high market price, combined with high yield potential (Lampkin et al., 2011). However, wheat is also a demanding crop species in that it requires high soil fertility and is comparatively sensitive to many pests and diseases. In comparison to many other crop species, wheat yields respond relatively strongly to increased inputs, for example, to increased levels of nitrogen supply (Heyn, 2005; Káš et al., 2010), or to intensive use of plant protection products. For conventional farmers – and for those selling agricultural inputs – wheat is therefore a highly attractive crop species.

Consequently, the concentration of wheat in conventional agriculture is often high, especially in areas with high soil quality that are particularly suitable for arable cropping. For example, in Northern Germany, where grain yields of conventional winter wheat are regularly above 11 t ha^{-1} in official variety trials (Stephan, 2015), the crop land is strongly dominated by wheat, which takes up a proportion among cereals of around 70% (SAHSH, 2010). In these areas, conventional winter wheat is often grown after wheat, and sometimes, when sowing conditions for alternative crops such as oilseed rape and winter barley are suboptimal, wheat is grown in the same field three to five years in succession (Obenauf, 2012). Also for Germany as a whole, wheat is the most dominant cereal species, taking up nearly half of the conventionally managed cereal area (FAOSTAT, 2014). In comparison, in organic systems in Germany, wheat is grown on only 25.7% of the organic cereal area (FiBL, 2015).

Even though this figure indicates high importance of wheat for organic growers (also see above), the comparison with the conventional sector suggests that wheat is relatively under-represented in organic systems. Under-representation of wheat among the cereals is not only evident from Germany but also from other countries, such as France, Spain, Switzerland and the United Kingdom (FiBL, 2015). Thus, in organic systems the place of wheat in the crop sequence is partly taken by other cereal species. For example, in comparison to wheat, relatively higher proportions of land are managed organically for rye and oats (Table 1). This is in accordance with these two cereal species being less responsive to input-driven intensification than wheat. Similarly, millet, a less (nutrient-) demanding crop species than wheat, has a substantially higher organic proportion of production than wheat (Smith et al., 2004).

2.3 Break years for reducing pests, diseases and weeds in wheat

One of the main reasons why crop rotations are so important for organic farming is the ability to regulate plant diseases and pests through crop rotation design (Leoni et al., 2015; van Bruggen et al., 2016). The basic idea is to pause cropping a particular species on a particular field for a number of years in order to withhold the host plant from specialised (mostly soil-borne) pests and pathogens. Thus, in order to avoid a population build-up of these pest and pathogen species in the soil, a number of break years is recommended. These break years are also referred to as the minimum return period between the same crop species (Castellazzi et al., 2008). In addition to reducing pests and diseases, pausing a crop for a number of seasons can also help with weed control. In this way, weed species associated with the growing pattern (sowing times) and agronomic management of a certain crop species will not build up excessively over time.

Illustrative of the importance of break years is the range of plant protection problems that have been observed in recent years in conventional wheat–wheat sequences. These include higher infestation with annual grasses such as blackgrass (*Alopecurus myosuroides*) and brome (*Bromus* spp.), increased occurrence of fungal diseases such as take-all (*Gaeumannomyces graminis* var. *tritici*), eyespot (*Tapesia yallundae*), *Fusarium* head blight (*Fusarium* spp.) and DTR/HTR (*Pyrenophora tritici-repentis*) (Obenauf, 2012). Recent microbiological work from Poland confirms that fungal diseases of the stem base are more severe in winter wheat monocultures than in organic and conventional rotations (Lenc et al., 2015). Furthermore, shortening the return period of wheat can lead to a problematic build-up of soil-borne pests such as the saddle gall midge (*Haplodiplosis equestris*), Hessian fly (*Mayetiola destructor*) (Weisz et al., 2014) or the cereal cyst nematode (*Heterodera avenae*).

These problems are not unexpected. Already more than a century ago, it was known that wheat does not grow well after wheat, and therefore it should only be grown on the same field after a pause of 3 to 4 years (Schlipf, 1898, p. 147), or at least that its frequency should not exceed 4 times in 10 years (von Rümker, 1911). By the 1960s, it had been well established that soil-borne fungal diseases in wheat (take-all and eye spot) cause problems when cereal and wheat concentrations are increased in the rotation (Klapp,

Table 1 Absolute area of organic wheat and spelt as well as percentage of wheat area managed organically for the 22 countries with the largest area of organic wheat in the years between 2009 and 2013; the percentage of barley, oats, rye and total land area managed organically is given for comparison

	Area [1000 ha]*		Percentage of area managed organically (2013)				
Country	Organic wheat	Organic spelt	Wheat	Barley	Oats	Rye	All crops
Argentina	13.6		0.4	0.0	0.4	0.2	2.3
Austria	30.0	7.9	10.3	5.9	30.4	34.3	19.4
Canada	96.4	2.8	0.8	1.6	14.1	8.4	1.3
Czech Republic	8.9		1.1	0.9	10.8	7.8	11.2
Denmark	9.7		1.3	2.2	26.6	12.7	6.4
Finland	6.6		2.8	1.1	7.1	15.4	9.0
France	47.5	4.4	1.0	1.0	9.9	19.2	3.9
Germany	53.6	20.0	1.7	1.4	17.5	7.6	6.4
Greece	20.3		3.4	4.9	9.1	10.5	4.6
Hungary	13.4		1.4	0.5	2.3	2.4	3.1
Italy	111.9		5.2	11.9	17.2	6.3	10.3
Kazakhstan	81.9	0.8	0.8	0.3	1.1	0.0	0.1
Lithuania	12.8		2.2	1.9	24.2	28.5	5.7
Poland	9.7		0.4	0.3	3.5	3.6	4.3
Romania	46.2		2.8	1.8	0.8	3.8	2.1
Slovakia	6.7		1.8	1.3	14.6	14.5	8.3
Spain	41.6		2.2	2.0	9.3	4.7	6.4
Sweden	27.2		7.4	5.7	16.4	9.6	16.3
Turkey	128.9		1.7	0.9	1.4	0.9	1.9
Ukraine	56.6	1.8	1.0	0.8	1.0	2.4	1.0
UK	17.6		0.6	1.6	9.4	69.1	3.3
USA	151.0	4.8	0.7	2.0	5.9	8.7	0.6

* Wheat: average of five years between 2009 and 2013 except for Spain where data for only four years was available; for China (162931 ha organic wheat in 2013) and Mongolia (6461 ha in 2013), data from only one year was available; in Latvia, 6737 ha of organic wheat were grown in 2009–12; but no data was available for 2013. For spelt, data for all five years (2009–13) was only available from Germany and Austria; averages of organic spelt area for the other five spelt-growing countries listed here are based on two years only.

Source: FiBL, 2015.

1967). In accordance with these early recommendations, Lampkin (1994, p. 135) states for organic rotations that wheat 'should not occur more frequently than every three or four years' because of cereal cyst nematodes and soil-borne fungal diseases such as take-all and eyespot. Similarly, Freyer (2003) recommends a break duration of 3 years for wheat when no chemical plant protection is used. In addition, most other cereals are seen as poor preceding crops for wheat partly because they are hosts of these pathogens. Where the cereal cyst nematode (*Heterodera avenae*) occurs, oats (as its main host species) is not recommended as a pre-crop for wheat.

Because no mineral nitrogen fertiliser is used in organic farming systems, winter wheat development in spring depends on the mineralisation of nitrogen from organic matter in the soil. Therefore, when temperature or soil moisture conditions are not conducive for mineralisation, wheat frequently shows relatively late spring growth, which can cause problems with weed infestation. Moreover, poor establishment of wheat can lead to severe weed infestation in organic systems. As mentioned above, rotational design is an important tool for weed management in organic wheat cropping. For example, a common strategy of organic farming systems is to alternate spring- and autumn-sown crops in the rotation. Thus the typical weed species associated with winter cereals (e.g. *Apera spica-venti*, *Centaurea cyanus* and *Papaver rhoeas* in Europe) can be kept under relatively successful control.

In addition, the typical grass-clover phase in organic rotations, mainly used for building soil fertility (see below), has a strong reducing effect on the occurrence of annual weed species (Döring et al., 2012). However, perennial weeds such as thistles (e.g. *Cirsium arvense*) remain a severe problem in organic rotations, both in the ley and in the arable phase (Niggli, 2007). On the other hand, weed communities in organic wheat also contribute to farm biodiversity. Furthermore, while trade-offs between weed biodiversity and wheat yields may be common, the relationship between wheat quality and weeds may be more complex. For example, in a comprehensive study on organic wheat quality, a positive correlation was found between grain protein content and both crop nitrogen status and weed density (Casagrande et al., 2009).

In summary, the concentration of wheat in organic cropping systems is partly constrained by rotational rules that aim to keep pests, diseases and weeds at bay. From this point of view, not more than about one-third of the rotation should be dedicated to wheat crops. However, as shown in the following section, the frequency of wheat in organic rotations is also constrained by nutrients.

2.4 Crop rotations for supplying nutrients to organic wheat

Wheat yields in organic systems have been shown to be nitrogen-limited (Berry et al., 2002). Furthermore, bread-making quality of wheat, when measured by grain protein content, is strongly dependent on nitrogen supply. In a study from France, it was shown that organic winter wheat production is characterised by comparatively low grain protein content (Casagrande et al., 2009), indicating that nitrogen supply may be limiting wheat quality. This view was also held in Germany (Neumann et al., 2003). Because wheat is a demanding crop species in terms of nitrogen supply, it is usually placed after leguminous crops in organic rotations. As pre-crops, the legumes are able to fix atmospheric nitrogen with symbiotic bacteria; the legumes are then incorporated into the soil, where mineralisation by soil microbes makes this nitrogen available for the following crop (Watson et al., 2002). This principle of building fertility through legumes was already well described more than two centuries ago (Thaer, 1809). Accordingly, long before the differentiation of conventional and organic agriculture, clover (*Trifolium* spp.), lucerne (*Medicago sativa*) or other perennial forage legumes were seen as suitable pre-crops for wheat (von Rümker, 1911). Grain legumes such as faba beans (*Vicia faba*) and peas (*Pisum sativum*) were recommended and practiced pre-crops before wheat (zu Putlitz and Meyer, 1914). Notably, on better soils, where the long-strawed wheat varieties of the time would tend to lodge, winter oilseed rape (*Brassica napus* ssp. *oleifera*) was recommended instead.

In current organic agriculture, pre-crop suitability for winter wheat is differentiated by the targeted end use of the wheat crop (Kolbe, 2016). When wheat is used for brewing beer, the grain is required to have low protein contents. Therefore, wheat for brewing should not be grown after grass-clover or grain legumes, but after other non-cereals such as potatoes, sunflower or beet. For bread wheat, two or more years of lucerne, clover and grass-legume mixtures are considered as good pre-crops, but caution must be taken under drought conditions, since forage legumes may reduce water availability for the following crop. In addition, shorter periods of forage legumes, for example 1–1.5 years of lucerne, clovers, serradella (*Ornithopus* spp.) or grass-legume mixes are considered as good pre-crops before bread wheat and are frequently used in this function by organic farmers.

Importantly, such organic legume-based leys fulfil many functions (Storkey et al., 2015) apart from providing nitrogen for the following wheat, including provision of fodder for ruminants, protection from soil erosion, supply of resources for the soil fauna, contribution to increasing soil organic matter (SOM) levels (Murata and Goh, 1997), reduction of arable weeds (Döring et al., 2012) and the support of pollinators. Although organic wheat requires pre-crops with high potential of supplying nitrogen, such as a grass-clover ley, the rotational sequence grass-clover – wheat can therefore also be viewed from the pre-crop's perspective. If the farm business requires the production of forage, but permanent grassland is not an option, wheat can be seen as one of the most efficient crops that are able to convert the high soil fertility after the grass-clover phase into a marketable product. Because rye, oats, spelt or millet respond much less to high nitrogen levels than wheat, they are not suitable to take the place after a fertility-building crop such as grass-clover. As these leys may be needed anyway for forage production, with positive effects on soil health, soil protection and biodiversity, their 'side effect' of high soil fertility needs to be used most efficiently, and wheat is the most efficient cereal species to do so.

However, even wheat may not always be able to make efficient use of the nitrogen from organic leys. As already noted by Lampkin (1994, p. 135), there is 'some doubt that the winter wheat crop can take up sufficient nitrogen in the autumn to prevent serious leaching losses especially on light sandy soils'. After ploughing up fertility-building grass-clover or lucerne ley, the young wheat plants often do not grow fast enough to prevent the valuable nitrogen from being translocated to levels where it is not available to the plant roots anymore. High contents of mineral nitrogen in the soil are typical after incorporation of the grass-clover ley, because aeration of the soil during the mechanical disturbance leads to an N mineralisation flush. For German organic farms, Kolbe (2016) found an average between about 120 and 140 kg N ha^{-1} in the autumn after incorporation of the ley, with some values well above 200 kg N ha^{-1}.

Grain legumes such as peas, faba beans, lentils (*Lens culinaris*) or lupins (*Lupinus* spp.), as well as non-legumes such as potatoes (*Solanum tuberosum*), are often considered as alternative pre-crops for wheat. In stockless systems, forage legumes such as clover may not be used directly anyway, so that grain legumes as cash crops are a natural alternative for providing N to the following wheat. However, the grain legume species may also require a relatively long rotational break (Döring, 2015). For pea, recent estimates of required break years under organic management range above six years. Shortening this period is likely to lead to the accumulation of soil-borne fungal diseases attacking the young pea plants. The recommended number of break years is slightly lower for faba beans, and even lower for soya beans (Döring, 2015). However, replacing forage legumes with grain legumes as pre-crops for wheat also has the disadvantage of worsening the SOM balance of the rotation. Grain legumes, though considered to make a moderately positive contribution to SOM, do not reach the high levels of a 2- or 3 year grass-legume ley. The contribution of wheat itself to the build-up of SOM is relatively small or slightly negative, depending on the fate of the wheat straw. In a long-term comparison between two organic rotations in Germany, it was recently reported that SOM levels significantly decreased over time when the legume-based ley was replaced by grain legumes (Schulz et al., 2014).

Although ample use of cover crops in the rotation, for example, after wheat and before the next spring crop, may partly counterbalance the loss of SOM, the sustainable provision of nitrogen – for wheat and other crops – in organic rotations remains a serious challenge. Rotational planning is further complicated by the fact that forage legumes such as red clover may also share pathogens with grain legumes. The minimum rotational pause required for successful legume cropping therefore poses an indirect but serious constraint for the concentration of the nitrogen-hungry wheat in organic cropping systems.

2.5 Typical organic crop rotations containing wheat

Several different organic crop rotations containing wheat are compiled in Table 2. Unfortunately, it is difficult to say to which degree these rotations, collected from published research papers and textbooks, are representative of actual practices of organic farming. However, because one of the general and often-stated aims of organic agriculture research is to be practice-oriented, it is unlikely that the reported rotations deviate considerably from organic farmers' practices. In addition, the rotations tabulated here show strong similarities when compared with a further set of wheat-containing organic rotations that were compiled from various European countries (David et al., 2012).

Given the large geographic range of the included studies, it is not surprising that there is a multitude of ways of placing wheat in organic crop rotations. This diversity of agronomic approaches refers to (a) the proportion of wheat crops in the rotation, (b) the length of the rotation and (c) the number of break years between wheat crops. In the tabulated organic crop rotations, soft wheat (*Triticum aestivum*), durum wheat (*T. durum*) and spelt (*T. spelta*) are subsumed under the heading of *Triticum* crops; because of their genetic relatedness and their similarities in terms of agronomic management, they are treated as equal for determining proportions and break years of 'wheat' crops in these rotations.

The rotation length varies from only 2 years to a maximum of 9 years, with a median of 6 years. Among the 25 rotations from Europe, reported rotation lengths are never below 4 years. The extreme case of the 2-year rice–wheat organic rotation from India follows the conventional regional practice, where wheat is alternated with various summer crops (e.g. rice, cotton, soya bean, maize, pulses and sugarcane), depending on climate and soil conditions, although it is unknown to which degree organic farmers in this region do indeed grow wheat every second year. Such simple rotations remain unconvincing from an organic farming perspective unless autochthonous N supply is ensured through the integration of biological nitrogen fixation, for example, by including leguminous green manures such as 'dhaincha' (*Sesbania aculeata*) (Ram et al., 2014). While short organic crop rotations have also been reported from elsewhere (e.g. soya bean – maize – wheat from Northern Italy, W. Vogt-Kaute, pers. comm. 2015), it is clear that there are multiple benefits of longer and more diverse rotations – not only for the health and quality of the wheat crop (see above). Indeed, a recent global meta-analysis of the financial performance of organic versus conventional systems found that organic-to-conventional financial performance was higher when crops in the organic system were grown in longer or more diverse rotations (Crowder and Reganold, 2015).

In the collated organic crop rotations, the highest proportion of wheat is 50% (i.e. in the 2-year wheat–rice rotation discussed above), that is, substantially higher than the usually recommended maximal proportion of one-third. Although such a high share of wheat is uncommon in organic rotations, it is by no means unique. In an organic field trial from Switzerland (not tabulated), one of the two six-year rotations representing stockless organic systems had winter wheat every second year, alternating it with grain legumes and a 1-year grass-clover ley (Taramarcaz and Clerc, 2013).

The lowest proportion of wheat in the organic rotations in Table 2 is 12.5%, that is, once every eight years, and the median is 29%, that is, just below a third of the rotation. Closely linked with these figures are break years between wheat crops (including wheat, durum and spelt), that is, the minimum number of years between wheat crops in the same rotation. This period ranges from 1 year to 7 years in the presented rotations, with a median of 2 years, that is, wheat being grown every third year. However, in some cases not tabulated here, break years in organic farming may be even lower, that is, with wheat directly following wheat. For Australia, it was reported that an organic farm 'had a typical rotation of 6 years of pasture followed by 2 years of cropping, usually wheat–oats or wheat–wheat' (Ryan et al., 2004, p. 208). As in other cases, short or absent break periods are compensated by longer periods within the same rotation.

There are many different ways to classify crop rotations; for the purpose of this chapter, rotations are classified according to the duration of the fertility-building phase of the rotation, the presence or absence of grain legumes and the position of wheat in the rotation. As already discussed above, organic wheat crops mostly directly follow a legume-based ley (groups A and B1) or a grain legume (B3, C1 and D1). Alternatively, wheat is the second main crop (B2, C2 and D2), often following potatoes or maize, although the sequence maize–wheat is often criticised for its potential to increase the risk of *Fusarium* head blight in wheat.

The tabulated organic crop rotations also reveal the large diversity of crop species that follow wheat in the rotation. Sometimes it is recommended that after wheat no grain legumes or leys should be grown (Kolbe, 2016). This recommendation makes sense in cases when wheat itself follows grain legumes or leys, because then the break period for the legumes themselves would be too short. As particularly suited crops after (winter) wheat, spring-sown row crops are often used, for example potatoes, maize or fodder beet. These have the advantage of cultural weed control that comes with alternating spring- and autumn-sown crops. When spring-sown crops follow wheat, cover crops need to be integrated into the rotation to protect the soil over the autumn and winter period. Mostly, such cover crops are sown into the wheat stubbles directly after the wheat harvest, but sometimes they are undersown into the growing wheat crop in spring. Further crops that frequently follow wheat in organic rotations are non-wheat cereals, that is, species that are less demanding than wheat such as oats and rye.

Crop diversification is often mentioned as a tool for improving the yield performance of organic systems (Østergård et al., 2009; Ponisio et al., 2015). One approach of diversification is to use longer and more diverse crop rotations. As argued above, wheat has its specific demands in terms of crop rotations, meaning that good organic practice, by following minimum return years of wheat, is already bound to use an appropriate length of the rotation (i.e. long enough). At the same time, however, there are also limitations to

Table 2 Organic crop rotations with various wheat crops; the rotation is set to begin with the strongest legume element of fertility building before the first wheat. Green manures before spring crops and short-term cover crops indicated with short term cover crops indicated with*; all *Triticum* crops in bold

Group (see Table 3)	Crops preceding first wheat					First wheat	Crops following first wheat						Wheat concentration	Length of rotation (yrs)	Min. break years (*Trit.*)	Source	Country
A	F3	F3	F3	F3	F3	**WW**	O						0.14	7	6	15	AU
B1			GC	GC	GC	**W**	R	GL1*	**W**	B*			0.25	8	2	9	UK
			GC	GC	F2	**WW**	BF	SB*					0.17	6	5	4	DK
				GC	GC	**WW***	SO	GL1*	WT*	SB			0.14	7	6	5	AT/DE
				GC	GC	**WW***	WR	**SW**	F1	**Spe**	SB		0.38	8	1	5	AT/DE
				GC	GC	**WW***	SO	**SW***	GL*	**Spe***	WR*		0.38	8	1	5	AT/DE
				Lu	Lu	**W**	R*	Pot	**W***				0.33	6	2	12	DE
				GC	GC	**W**	Pot	**W**	B*	GC	**W***	SM	0.33	9	1	9	UK
				GC	GC	**W**	Pot	**W**	R/O				0.33	6	1	9	UK
				Lu	Lu	**W**	Pot*	O	**W***				0.33	6	2	12	DE
				Lu	Lu	**W**	Lin						0.25	4	3	18	CA
B2		GC	GC	GC	Pot	**WW**	BR	**WW**					0.29	7	1	10	CH
			Lu	Lu	Pot	**WW**	WR	GL1	**WW**	SO			0.25	8	2	5	AT/DE
			GC	GC	Pot	**WW**	BR	**WW**	SB				0.29	7	1	10	CH
			GC	GC	Pot	**WW***	SO*	F1	**Spe**	SB			0.25	8	2	5	AT/DE
			GC	GC	SM	**WW**	GCX	T					0.20	5	4	16	DE
			GC	GC	Pot	**Spelt**	GL1	WR	**SW**				0.29	7	2	3	DE
B3					GL2	**W**	RC	RC	M				0.20	5	4	11	US
					GL1	**WW**	SQ	GC	GC	Pot	Veg	WR	0.13	8	7	5	AT/DE
C1					SC	**W**	GL1	Lin/ OSR					0.25	4	3	17	CA
					GC	**SW**	GL3	SB*					0.25	4	3	13	DK
					Fa	**WW***	GL1	WR*	SB				0.20	5	4	19	DE
					RC	**WW**	WR*	SO*	**Spe**	WR*	SB*		0.29	7	2	16	DE
					LuC	**WW***	SB*	**Spe***	WR*				0.40	5	1	16	DE
C2				GC	Veg	**SW**	K	WR*					0.20	5	4	7	DE

		RC	M	**W**					0.33	3	2	6	US
D1			GL1	**WW***	SCe*	Pot	**WW**	WR*	0.33	6	2	9	UK
			GL1	**WW***	SCe*	Pot	GCX	O*	0.33	6	5	9	UK
			GL1	**W**	*M	**Dur**	*SF		0.40	5	1	2	IT
			GL3	**W**	Pot	M	M	Veg	0.17	6	5	1	BO
			GL1	**W**	Lin	O			0.25	4	3	18	CA
			GL2	**W**	M				0.33	3	2	11	US
D2		GL2	M	**W**					0.33	3	2	6	US
		GL3	Ri	**W**					0.33	3	2	8	IN
	GL1*	Veg	Pot	**SW**	Veg				0.20	5	4	7	DE
D3				**W***	R*				0.50	2	1	14	IN

Abbreviations of crop species

Forage legumes/grass-clover leys	GL: Grain legumes	Non-cereals, maize and grain legume–cereal mix	Small grain cereals
F1: Various or unspecified forage crops; F2: Barley/pea/ryegrass mixture F3: Legume/grass pasture GC: Grass/clover ley (e.g. white clover *Trifolium repens* and ryegrass *Lolium* sp.) Lu: Lucerne (alfalfa) (*Medicago sativa*) RC: Red clover (*Trifolium pratense*) SC: Sweet clover (*Melilotus sp.*) LuC: Lucerne and/or white clover Fa: Green manure fallow with *T. pratense*, lucerne or other annual cover crops HV: Hairy vetch (*Vicia villosa*)	GL1: Faba beans (*Vicia faba*) or peas (*Pisum sativum*) GL2: Soya bean (*Glycine max*) GL3: Other grain legumes, for example, lupins (*Lupinus spp.*), mung beans (*Vigna radiata*); or unspecified	BF: Fodder beet (*Beta vulgaris* ssp. *vulgaris* var. *alba*) BR: Beetroot (*Beta vulgaris* ssp. *vulgaris* var. *rubra*) GCX: Grain legume/cereal mix Lin: Flax/linola (*Linum usitatissimum*) M: Maize (corn) (*Zea mays*) OSR: Oilseed rape (canola) (*Brassica napus*) Pot: Potatoes (*Solanum tuberosum*) SF: Sunflower (*Helianthus annuus*) SM: Silage maize (*Zea mays*) SQ: Seed squash (*Cucurbita pepo*) Veg: Field vegetables, for example, carrots (*Daucus carota*), leeks (*Allium ampeloprasum*), onions (*Allium cepa*), cabbages (*Brassica oleracea*)	Ce: Cereals; Dur: Durum wheat (*Triticum durum*) Ri: Rice (*Oryza sativa*) SB: Spring barley (*Hordeum vulgare*) SCe: Spring cereals SO: Spring oats (*Avena sativa*) Spe: Spelt wheat (*Triticum spelta*) SW: Spring wheat (*Triticum aestivum*) W: Wheat (*Triticum aestivum*) WR: Winter rye (*Secale cereale*) WW: Winter wheat (*Triticum aestivum*) WT: Winter triticale (×*Triticosecale*)

Sources: 1: (Augstburger, 1988); 2: (Bàrberi and Mazzoncini, 2006); 3: (Diepenbrock et al., 2012); 4: (Eriksen and Askegaard, 2000); 5: (Freyer, 2003, pp. 106–15); 6: (Heiniger et al., 2013); 7: (Hof-Kautz, 2011); 8: (Kumar, 2014); 9: (Lampkin, 1990, p. 151); 10: (Mäder et al., 2007); 11: (Mirsky et al., 2012); 12: (Neuerburg and Padel, 1992); 13: (Olesen et al., 2002); 14: (Ram et al., 2014); 15: (Ryan et al., 2004); 16: (Schmidt, 2010); 17: (Smith et al., 2004); 18: (Turmel et al., 2009); 19: (Vakali et al., 2011).

AT: Austria; AU: Australia; BO: Bolivia; CA: Canada; CH: Switzerland; DE: Germany; DK: Denmark; IN: India; IT: Italy; UK: United Kingdom; US: United States of America

Table 3 Classification of crop rotations in Table 2

Group	Duration of legume-grass ley or legume ley phase	Grain legumes	Wheat position
A	Very long, >4 years, making up >50% of rotation	Not present	In first year after legume-based ley
B1	Moderate proportion of legume-grass ley phase, 2–3 years, making up <50% of rotation	May be present	In first year after legume-based ley; a further wheat crop may follow later
B2			In second year after ley, following a non-cereal
B3		Present	In first year after grain legumes; ley is not followed by wheat
C1	Short legume-grass ley in rotation (1–1.5 years)	Present	In first year after legume-based ley
C2			In second year after legume-based ley
D1	No legume-based ley in rotation	Present	In first year after grain legumes
D2			In second year after grain legume
D3		Not present	In rotation with non-legume

diversification in crop rotations. If wheat occurs more than once in the rotation, replacing it partly with less demanding crops in order to diversify the rotation may be economically disadvantageous. From the perspective of *wheat* production, lengthening the rotation will simply dilute the proportion of wheat in the rotation, thereby decreasing the amount of wheat production from the rotation. Also, the strategy of replacing (a second) wheat by a different crop in the rotation may not always make efficient use of the soil fertility built during the ley phase, for the reasons mentioned above.

To summarise, organic wheat cultivation always has to be seen in conjunction with the entire crop rotation with its site-specific constraints and opportunities. In organic systems wheat is as an attractive but demanding crop. In most current organic systems, wheat cultivation is strongly dependent on the cultivation of legumes in the rotation, which entails constraints on rotational design.

2.6 Specific conditions of organic wheat cultivation

In comparison to conventional wheat, organic wheat cultivation is characterised by a number of technical and agronomic specificities. These include tillage, variety choice, sowing time, seed density, use of growth regulators, fertilisation, and the control of pests, diseases and weeds. Variety choice is discussed in Section 4.1.

In terms of tillage, organic wheat cultivation is presently characterised by a less frequent use of reduced tillage, that is, more prevalent usage of the plough. This is mainly because the conventional practice of combining reduced tillage or no-till systems with herbicides is not an option for organic farmers. Nevertheless, many organic wheat farmers do use reduced tillage successfully; a recent meta-analysis shows that organic yields in reduced tillage systems are not always substantially lower than in the ploughed organic control, although combinable crops did suffer the greatest yield reductions by reducing tillage, namely by 8% (Cooper et al., 2016).

As for sowing time, organic farmers frequently sow their winter wheat less early than is practiced in conventional wheat cultivation. Several risks are associated with early sowing time, including higher infestation with fungal and viral diseases, as well as higher weed pressure. Therefore, organic farmers in large parts of central and western Europe typically sow in early October, while conventional winter wheat sowing in the same area is often done in mid-September. Although later sowing may reduce plant health risks, it is also associated with potential yield losses due to shorter vegetation time. A study from Bulgaria showed that early sowing can reduce the yield gap between conventional and organic wheat (Milev et al., 2014). Seed density is often slightly higher in organic than in conventional agriculture, in order to achieve stronger competition against weeds in the early phases of crop development, and to compensate later sowing.

Growth regulators, that is, plant hormones widely applied in conventional wheat cultivation to shorten the stem, and to manipulate the relative strength of yield components, are not used in organic farming. However, the risk of lodging is still low in organic wheat, because of the use of modern short-straw varieties and the relatively lower nitrogen supply to the organic wheat plants. In terms of fertilisation, wheat receives nitrogen mostly via the rotation (see Section 2.4). In addition, organic manures are applied to wheat in the form of farmyard manure or preferably, liquid manures, such as slurry. Although nitrogen may be the most yield-limiting factor in many environments, phosphorus availability has been shown to limit organic wheat yields as well, especially under Australian growing conditions (Emmens, 2003).

Apart from cultural weed control as discussed above, organic practice also allows for measures of direct weed control in wheat. This is done mechanically, with harrow comb and tine weeder when the crop is still young. The superficial disturbance of the soil has also effects on nitrogen mineralisation and therefore improves nitrogen supply for the growing wheat plant. Plant protection against pests and diseases is not performed through direct control in organic wheat cultivation, but relies on indirect measures such as rotation (see Section 2.3), intercropping (Section 3.1), genetic disease resistance (Section 4.1) or the use of variety mixtures (Section 4.2).

3 Organic wheat in species mixtures

One of the most intensively studied tools of diversifying cropping systems is the use of crop species mixtures. Multiple agronomic and ecological benefits have been reported to be associated with the various ways of mixing crop species in the field (Malézieux et al., 2009). In this section, two practices of mixing wheat with other crop species are discussed that appear to be relatively common among organic farmers. These practices are (a) mixing wheat with other combinable crops and (b) undersowing small-grained legumes into wheat.

3.1 Intercropping wheat with other combinable crops

Intercropping can be defined as the practice of growing 'two or more crops [...] in association with one another' (Vandermeer, 1989, p. 1). It appears to be more widespread among organic farmers than in conventional agriculture, although actual adoption rates in the organic farming community are largely unknown. In any case, because of the reducing effects of intercropping on pests, diseases and weeds, intercropping has repeatedly been described as being particularly suited for organic and low-input systems (Reynolds et al., 1994). However, wheat appears to be less frequently used in crop species mixtures than other cereals, because the mixed product can best be used as an on-farm produced feed, without the need of post-harvest separation of the grains. Organic wheat, however, tends to be grown preferentially for bread making, rather than for feed.

Intercropping wheat with other species is not a new practice. Simultaneous cultivation of spring wheat and flax (*Linum usitatissimum*) was reported to have been practiced already in the nineteenth century (Arny et al., 1929). In addition, wheat–oat mixtures were widespread in the northwest United States at the beginning of the twentieth century (Bailey, 1914). Wheat–rye mixtures, known as 'meslin', 'maslin', 'mistling' or 'mun-corn', were grown in England in the early eighteenth century (Mortimer, 1708), and were common practice at the end of the nineteenth century in Germany (Wollny, 1885). Several decades later, a German handbook of practical agriculture still reported that among species mixtures used by farmers, one combination is early wheat with late-maturing rye (Zimmermann, 1958).

Compared with mono-cropping, several advantages have been stated for intercropping. Reduction of the risk of low yields, complementarity of resource use, higher resource use efficiency by the mixture, reduction of minimal return periods for each partner, reduced damage by plant pests and diseases and physical support of one partner by the other were all known to agriculturalists in the nineteenth century (Wollny, 1885). Further benefits revealed by later research include phosphorus mobilisation for wheat through the roots of white lupins (*Lupinus albus*) (Horst and Waschkies, 1987) and higher grain quality of the cereal partner (Jensen et al., 2006).

More than 120 years after Wollny's enthusiastic promotion of intercrops, organic farmers across five European countries named improved yield stability, better weed suppression and higher fodder qualities as the most prominent reasons for their use of intercropping (Jensen et al., 2006). Recent research, however, calls into question whether cereal–legume intercrops do in fact exhibit greater yield stability than sole crops (Hauggaard-Nielsen et al., 2008).

Improved weed control through intercropping has been widely reported (Hauggaard-Nielsen et al., 2008; Malézieux et al., 2009) and is particularly plausible when the intercrop is characterised by a higher total crop density than used in the respective monocultures. However, in wheat intercropping systems, wheat is usually the species that benefits less than the other partner from these weed reducing effects of intercropping. For example, in the spring wheat–flax system mentioned above (Arny et al., 1929), it was found that although the weight of weeds was mostly lower in the spring wheat–flax intercrop than in the flax monoculture, the intercrop was more weed infested than the sole spring wheat, due to the poor competitive ability of the flax. A more recent re-enactment of these early trials under certified organic conditions in Germany confirmed these findings (Paulsen et al., 2006). Similarly, in a study on cereal–pea intercropping in organic farming in Lithuania, Šarūnaitė et al. (2012) found that at two sites weed density was higher in the pea sole crop than in the pea–spring wheat intercrop, but weed density was even smaller in the wheat monoculture. However, weed biomass was lower in the pea–spring wheat intercrop than the average weed biomass of the two monocultures. Nevertheless, from the *wheat* cultivation perspective, intercropping with other combinable crops is unlikely to serve as an option for improving weed control, unless more competitive partners are chosen.

One of the main stated benefits of intercropping is the improved agronomic productivity. A common way of measuring the effect of intercropping is the use of the land equivalent ratio (LER), that is, the sum of the monoculture areas needed to achieve the same yields as obtained from the intercrop. LER above 1 signifies an advantage of intercropping over the monocultures. Extensive research on intercropping shows that LER values, although frequently above 1, are highly variable. In a study on intercropping wheat and faba beans under organic conditions, it was shown that LERs were strongly dependent on the relative plant densities of the two crops (Bulson et al., 1997). Here, the best combination (i.e. resulting in the highest LER of 1.29) was found at 75% density of the wheat monoculture plus 75% density of the bean monoculture, that is, a combined plant density of 150%. However, other combinations showed disadvantages for mixing the two species, for example, at relative densities of 25% for wheat, and 100% for bean, the LER was only 0.97.

Not surprisingly, the LER is also dependent on the identity of the intercropping partners. In Lithuania, organic field trials were conducted over three years to evaluate different grain legumes as partners of spring wheat (Šarūnaitė et al., 2010). Only in one case, namely the combination of wheat and vetch, measured LERs were greater than 1 in all three trial years (with an average of 1.28). In contrast, wheat–pea and wheat–lupin mixtures showed LER below 1 in two out of three trial years. This variability is in line with research on organic wheat intercrops in Canada, where several crop species were tested as partners of wheat (Pridham and Entz, 2008). No consistent improvement of yields was found in the intercrops, and results were often strongly dependent on the environment.

Thus, the success of wheat intercrops remains patchy. Even if total yields of an intercrop may be more stable than monoculture yields, yield fluctuations of *individual* yields (i.e. of components within the intercrop) tend to be higher than those of the respective monocultures. Correspondingly, ecological research shows that higher diversity is often associated with higher stability at community level but with lower stability at population level (Tilman, 1996).

Furthermore, it is sometimes argued that diversification (e.g. through intercropping) may help to reduce the yield gap between organic and conventional agriculture (Ponisio et al., 2015). However, in a recent comprehensive study in France, it was found that in fertilised treatments the relative yield difference between organic and conventional production was 41% for wheat monocrops (with 3.5 vs 5.9 t ha^{-1} for organic and conventional systems, respectively), but 42% for pea–wheat intercrops (with 2.9 vs. 5.0 t ha^{-1}). In unfertilised treatments, the yield difference between organic and conventional production was 3% for sole wheat, 5% for sole pea, but 9% for wheat–pea intercrops (Pelzer et al., 2012).

In conclusion, it appears that intercrops with organic wheat are neither fully understood nor sufficiently reliable in practice. Therefore, to make use of the potential benefits of intercrops in organic wheat production, more *systematic* research is needed that can help remove the obstacles to successful wheat intercropping, such as problems of unequal maturation of the intercrop components (Jensen et al., 2006), as well as low predictability of best crop partners and optimal plant densities.

3.2 Intercropping wheat with forage crops: living mulches

Wheat is not only intercropped with other combinable crops such as grain legumes and cereals, but also with a tremendous range of other species. These plants include so-called forage crops since they can be grazed or cut for feeding livestock, for example, various clover species. In some systems, these subsidiary crops are also referred to as 'cover crops', following their primary function to cover the soil and protect it against erosion. In most cases, leguminous species are used as subsidiary crops.

Because these companions of wheat are not produced for grain, the problem of unequal timing of maturity (that can occur when intercropping wheat with other combinable crops) is irrelevant. In comparison with the systems discussed in the previous section, intercropping with forage or cover crops is therefore characterised by larger flexibility. Advantages of intercropping wheat with various legumes include increased productivity, nitrogen output, ground cover and light capture (Reynolds et al., 1994). Intercropping wheat with forage legumes was also found to improve N availability for the wheat crop and to increase grain protein content in bread wheat (David et al., 2013).

A particularly well-studied system is the so-called wide-row system of wheat production (Neumann et al., 2006). Instead of the usual practice of 10–15 cm row width between the wheat rows, the wide-row system features row widths of 30 cm or even 50 cm, and clover (e.g. white clover, *Trifolium repens*) is grown in between these rows (Neumann et al., 2003; Neumann, 2005). The clover can then be mulched with specialised machinery. However, this can lead to reduced wheat yields (caused by wider rows).

A further well-established system of organic wheat cultivation is spring-time undersowing of clover into an established winter wheat; in many cases, this undersown short-term clover crop is followed by a spring cereal (Table 2). Alternatively, wheat can be directly sown into an established leguminous cover crop. In organic field trials in Switzerland, this latter system was shown to reduce weed infestation in comparison to a system without cover crops, but competition between the legumes and the wheat reduced wheat yields substantially (Hiltbrunner, 2006).

4 Diversification of organic wheat at the genetic level

4.1 Organic wheat breeding and variety choice

The choice of wheat cultivar is an important way to manage nutrient dynamics, to control weeds (e.g. Lampkin, 1990) and diseases, and to obtain higher grain quality in organic wheat (David et al., 2013). Given the contrasting management regimes of conventional and organic agriculture, especially with regard to plant protection and plant nutrition, it is obvious that organic farming requires crop genotypes with other traits than conventional agriculture. In particular, from the perspective of organic crop management, desirable traits include early vigour for good establishment, high competitive ability against weeds, good nutrient acquisition, high nutrient efficiency, in particular with respect to nitrogen, resistance against prevalent pests and diseases, and high end-use quality (Arterburn et al., 2012).

It is therefore sensible to ask whether organic farmers need different wheat varieties from what is being grown by their conventional colleagues. This question has two main aspects: First, should wheat varieties be tested on organically managed land for making recommendations about their use in organic agriculture, or is it sufficient to follow variety recommendations from conventional wheat trials? Second, does the wheat breeding process need to take place under organic conditions or can selection of genotypes for variety development be performed on conventionally managed land?

With respect to the first question, findings are partly inconclusive. No significant genotype x farming system interactions were observed in wheat in a study in Switzerland (Hildermann et al., 2009). In soft wheat varieties in Italy, correlations between the organic and conventional systems were high and it was concluded that genotype x location interactions were much more important than genotype x system interactions (Annicchiarico et al., 2010), which implies that information gained under conventional management can be transferred to organic conditions without substantial loss of reliability. Also, Wortman et al. (2013) who evaluated varieties of maize, soya bean and wheat under organic and conventional management did not find consistent interactions between genotype and farming system. An important aspect in this context is the fact that conventional wheat trials are conducted in any case, and the question is not only whether additional information from organic trials is necessary, but also whether conventional trial data can be ignored by organic farmers. With regard to the latter question, Przystalski et al. (2008, p. 417) concluded that 'ignoring information from non-organic trials when making decisions regarding performance under organic conditions is a sub-optimal strategy'. Thus, combining data from both organic and conventional trials seems to be useful. In any case, however, it is now well established that so-called modern wheat varieties perform better than older ones under organic conditions.

With regard to the question of organic wheat breeding, a study on spring wheat concluded that 'breeding spring wheat specific to organic agriculture should be conducted on organically managed land' (Reid et al., 2009). On the other hand, researchers investigating wheat cultivars in the long-term DOK trial in Switzerland found that organically bred cultivars did not outperform the conventionally bred cultivars in the organic systems, in terms of grain yield and nutrient use efficiency (Hildermann et al., 2010). However, for a conclusive answer to the question of organic wheat breeding, more time should be allowed for the intensive organic breeding activities to bear fruit; new organically bred varieties released over the next few years will need to be tested.

4.2 Variety mixtures and populations

A large number of studies on wheat has shown that using increased genetic diversity in the field by mixing cultivars reduces plant diseases (e.g. Brophy and Mundt, 1991; Finckh and Mundt, 1992, 1996; Finckh et al., 2000; Finckh and Wolfe, 2006; Gigot et al., 2013). Disease-reducing effects of cultivar mixtures are particularly strong when disease pressure is high (Gigot et al., 2013) and tend to be weaker, or disappear completely when plants are treated with fungicides (Finckh and Mundt, 1996). This latter observation is one reason why the strategy of mixing cultivars may be particularly suitable for low-input and organic agriculture (Finckh et al., 2000). Notably, however, in organic systems foliar diseases caused by fungal pathogens may be limited anyway. For example, no detectable negative effect of foliar diseases on wheat yield were observed in a study on organic wheat in England (Döring et al., 2010a). Nevertheless, variety mixtures will ensure against disease outbreaks in general by reducing the risk of genetic resistance breakdown (Finckh et al., 2000).

In addition, while the effect of mixing varieties on average yield levels may be low (Finckh et al., 2003; Döring et al., 2015) or non-significant (Flamm, 2009; Fischl et al., 2013), variety mixtures often show improved yield stability. For instance, greater yield stability was observed in wheat variety mixtures in comparison to the corresponding monocultures in spring wheat trials in Canada (Kaut et al., 2009). Similarly, winter wheat mixtures showed greater yield stability than monocultures of pure-line varieties in England (Döring et al., 2015), Austria (Flamm, 2009) and Germany (Finckh et al., 2003).

Finally, an expected benefit of using variety mixtures is an improvement of end-use quality, in particular the increase of protein content. Because of the plausibly better nitrogen use efficiency in the mixture, it is hoped that the mixture may achieve protein contents above the (weighted) average of the component varieties. While some trials confirm this expectation (Sarandon and Sarandon, 1995), others have not found any significant effects of mixing on protein content (Finckh et al., 2003; Flamm, 2009; Fischl et al., 2013).

An alternative way of increasing intraspecific diversity in wheat cultivation is the use of wheat composite cross populations (CCPs). In this approach called evolutionary plant breeding (Döring et al., 2011), parent varieties are crossed with each other and the entire offspring is sown in the field; from this crop, a proportion of the harvested grain is taken as seed for the next generation. This cycle, continuing for several generations is expected to lead to crop improvement by natural selection, in that those genotypes adapted to growing conditions increase in frequency within the population (Suneson, 1956). Following on from previous positive experience with this approach, for example, in terms of improved yield (Allard, 1988), renewed interest in cereal CCPs has led to a resurgence in research over the past few years (Murphy et al., 2005; Goldringer et al., 2006).

Trials performed in England show the advantage of CCPs over monocultures in terms of yield stability; however, CCPs were not consistently more stable than the genetically less diverse variety mixtures (Döring et al., 2015). Moreover, especially under low-yield conditions, relative yield advantage of populations over monocultures have been demonstrated (Döring et al., 2010b). Recent results with high relevance for organic wheat cultivation show that wheat CCPs grown under organic management for several generations increased in early vigour over time, measured as seedling root length and seedling root and shoot weight. In the CCPs under conventional management, no such differences were detected (Bertholdsson et al., 2016). However, despite the proven ability of CCPs to change over time, a pre-condition for their successful cultivation is the general suitability of the founding germplasm for the general growing conditions. For example, organic field trials in Italy showed that the UK-derived wheat populations were unsuitable under Mediterranean growing conditions; that is, the identity of the non-adapted parents was not compensated by the increased genetic diversity (Costanzo and Bàrberi, 2013; Costanzo and Bàrberi, 2014). Finally, the continuous use of saved seed in evolutionary plant breeding can be thwarted by seed-borne diseases unless special care is taken to ensure seed hygiene and seed health (Döring et al., 2011).

4.3 Wheat relatives: the example of spelt

So far, this chapter has concentrated on hexaploid common wheat (*Triticum aestivum*). One way of crop diversification not covered in the previous sections is the use of the various close relatives of common wheat. These relatives include the other species in the genus *Triticum*, namely the other hexaploid species spelt (*Triticum spelta*), Macha wheat (*T. macha*), club wheat (*T. compactum*) and Indian dwarf wheat (*T. sphaerococcum*); the tetraploid species durum wheat (*T. durum*; see Chapter Durum wheat: properties, cultivation and uses), emmer wheat (*T. dicoccum*), Polish wheat (*T. polonicum*) and rivet (*T. turgidum*); and the diploid species einkorn (*T. monococcum*). This section describes the properties and uses of as spelt in an organic farming context.

Spelt was an important crop in Neolithic time. Although it was continuously present in southwestern part of central Europe for millenia, it lost much of its importance in comparison to wheat over time. Although at the end of the nineteenth century spelt was mentioned as the most important bread cereal in southwest Germany (Schlipf, 1898, p. 142), its area waned during the intensification of agriculture (Klapp, 1967). While in the land of Württemberg after the First World War the areas grown with spelt and wheat were about equal in size, wheat became increasingly popular (Schindler, 1923), and already by 1935 the ratio had dropped to 1:2 for spelt:wheat (Brouwer, 1972). Later, during the early 1980s spelt was grown on only a few hundred hectares in the southern parts of Germany, although at the time relatively large areas remained in Switzerland and in Belgium, with about 6000 ha in each country (Körber-Grohne, 1995).

However, for several reasons cultivation of spelt experienced a renaissance during the last few decades. A reason for this rediscovery may be the appreciation of spelt for its nutritional quality and taste (Lacko-Bartošová and Rédlová, 2007; Kohajdová and Karovičová, 2008); spelt has a higher protein content than wheat (Bonafaccia et al., 2000) and a different gluten quality and higher gluten content (Kohajdová and Karovičová, 2008). Also responsible for the renaissance of spelt is the fact that spelt is particularly suited to organic agriculture. Thus, the growing organic markets around the world have increased demand for spelt products.

Currently, that is for the years 2009–13, the global organic spelt area is around 40 000 ha (Willer and Lernoud, 2015), with about half of this located in Germany (between 17 500 and 22 000 ha in years between 2009 and 2013), and larger areas (>1000 ha) also in Austria, France, the United States and Canada (Table 1). Within Germany, spelt is mainly grown in the southern part. From 2003 to 2013, the area of spelt grown in Baden-Württemberg, southwest Germany, increased from below 8000 ha to more than 13 000 ha (Reinsch, 2014). However, spelt is also grown further north (Krug, 2015).

In contrast to its close relative *T. aestivum*, spelt is not free-threshing. The hulls are firmly attached to the grains and are not separated at harvesting with the combine. De-hulling the grains has to be performed post-harvest with specialised machinery. With a plant length of around 110 cm and, in older varieties such as Oberkulmer Rotkorn, of up to ca. 140 cm (Reinsch, 2014), spelt has a considerably longer stem than common wheat. Therefore, spelt can be more prone to lodging than the dominating short-strawed wheat varieties, although modern spelt varieties are shorter and have good lodging resistance (Reinsch, 2014). Spelt is self-fertile, with an outcrossing rate between 0 and 5% (Körber-Grohne, 1995).

Nowadays, spelt is reportedly mostly used as an autumn-sown crop, because of its good winter hardiness and its tolerance to late frosts, although it seems that in earlier times spring-sown spelt was commonly grown in Germany (Zimmermann, 1958). In comparison to wheat, spelt has lower requirements in terms of soil fertility. In addition, spelt yields respond less to increased levels of nitrogen fertiliser than wheat. Therefore, it has become less popular among conventional farmers than the more demanding, but also more responsive wheat. Despite having lower nitrogen requirements than wheat, spelt achieves higher protein content; in terms of P and K requirements, spelt is similar to common wheat. Because of its cold tolerance, spelt is especially suitable for higher altitudes. Although it can be grown on shallower soils, spelt has higher water requirements than wheat and is therefore not suited for light sandy soils. Like winter wheat, autumn-sown spelt has a good tolerance for late sowing; for Germany, possible sowing times are reported to be from September to November (Reinsch, 2014).

In past literature, spelt was considered to be less susceptible to smuts and to bird damage than wheat and better suited to rougher climate than wheat (Schlipf, 1898, p. 142; Brouwer, 1972). More recent advisory literature emphasises similar susceptibility of spelt to mildew and brown rust as in wheat. A particular difficulty in growing organic spelt is the infection with soil-borne bunt (*Tilletia caries*) (Reinsch, 2014). Most spelt varieties are said to have good competitiveness against weeds. Mechanical weeding such as harrowing or hoeing is tolerated by spelt and pre-emergence harrowing is possible. The most important weeding pass is at the 3–4 leaf stage.

In terms of position in the rotation and suitable pre-crops, spelt is considered equivalent to wheat (Schindler, 1923; Krug, 2015), with a recommended break period of 2 years after wheat (Reinsch, 2014). In past literature, spelt had been recommended to follow after clover, lucerne, tobacco and oilseed rape with less suitable pre-crops being potatoes, cabbage, maize, rye, hemp and flax. Notably, also self-succession of spelt was deemed possible (Schlipf, 1898, p. 142; Schindler, 1923). Recent agriculture advice is not decided on the suitability of grain legumes and grass-clover as pre-crops. Although some recommend these crops to be followed by spelt, especially with modern spelt varieties (Reinsch, 2014), others see these pre-crops as too risky because of lodging (Kolbe, 2016). Alternative pre-crops before spelt include sunflower, potatoes and oilseed rape.

Spelt is known to have a high tillering ability (Reinsch, 2014). However, in comparison to other winter cereals, the ear density of spelt is relatively low (380–440 ears m^{-2} in spelt; 440–550 ears m^{-2} in winter wheat; 490–560 ears m^{-2} in winter durum; data from AGES (2014)). At the same time spelt does not compensate this low ear density by a high number of ears per grain – on the contrary, typical values range between 17.2 and 21.4 grains per ear, in comparison to winter wheat with 29.1–45.5 grains per ear. Morphologically, this is easily seen in that the ear of spelt is less dense than that of wheat. The resulting grain density of spelt is therefore low (6800–7900 m^{-2}), but its thousand grain weight (TGW) is above that of winter wheat: typical values of TGW are 45.7–52.1 g in spelt and 36.8–50.6 g in winter wheat (data from AGES, 2014).

In terms of grain yield under organic farming conditions, spelt performs slightly less well than common wheat. In organic winter spelt trials, reported average grain yields ranged from 3.56 t ha^{-1} to 3.93 t ha^{-1} in north Germany in 2015 (Krug, 2015) and from 5.85 t ha^{-1} to 7.21 t ha^{-1} (including hulls) in Slovakia in 2004–2005 (Lacko-Bartošová and Rédlová, 2007). Average yields in organic trials were at 4.5 t ha^{-1} (including hulls) in southwest Germany (Reinsch, 2014). Recent variety trials in Bavaria tested nine spelt varieties under organic cropping management. Here, spelt yields (hulled grains) varied strongly between 2.2 t ha^{-1} and 6.0 t ha^{-1}, depending on site conditions (Urbatzka and Cais, 2013). Yield variation was attributed by these authors to differences in nitrogen supply. These results are in contrast to the reputation of spelt to achieve high yield stability (Reinsch, 2014) and to be less responsive to nitrogen than wheat. In the state of New York, county average of spelt yields over four years (2006–10) ranged between 2.59 t ha^{-1} and 3.22 t ha^{-1}. In high-fertility plots, spelt achieved grain yields between 1.53 t ha^{-1} and 3.53 t ha^{-1} (Caldwell et al., 2014).

The nutritional quality and end uses of spelt have been reviewed by Kohajdová and Karovičová (2008). Spelt is partly used for animal feeding flour, but mainly for human consumption in the form of bread, breakfast cereals, pasta, crackers and biscuits, as well as a number of regional specialties. Most notable among these specialities is the 'Grünkern' of southern Germany. For the Grünkern, the grain is harvested with thrashers while it is still green; it is then dried in ovens. It is highly appreciated for its rich nutty and smoky taste.

Traditionally, Grünkern has been used not only in soups and sauces but also in puddings and pastries.

5 Organic wheat and farm biodiversity

In both public and scientific debates, biodiversity plays an important role for the assessment of the pros and cons of organic farming, in particular against the backdrop of an ongoing dramatic decline of farmland biodiversity. As part of the land-sharing versus land-sparing debate, the discussions about organic farming and its role in protection of farmland biodiversity have become more heated over the past decade. While evidence for the positive effects of organic farming for farmland biodiversity has become more differentiated, several recent reviews have also highlighted the yield gap between organic and conventional production (e.g. Ponisio et al., 2015). For this chapter, three aspects of the effects of organic wheat production on farm biodiversity (including plants, arthropods and birds) are briefly reviewed: (1) the comparison between organic and conventional wheat production, (2) the comparison between organic wheat and other organic crops and (3) effects of agronomic wheat management within organic systems.

5.1 Organic vs. conventional wheat

Many of the studies comparing the effects of organic and conventional production on biodiversity have been conducted in wheat fields, probably because this crop is common and economically important in both systems. With regard to higher plants, including arable weeds, the evidence is relatively clear-cut. In wheat production, abundance, richness and diversity of wild plants benefit from organic management. For example, Roschewitz et al. (2005) investigated 12 pairs of organic and conventional winter wheat fields in Germany and found higher weed cover in the organic system, with a median of 10.6% cover of broadleaf weeds compared to 1.3% in the conventional wheat fields; also higher numbers of weed seeds were found in the seedbanks of the organic fields. Further, species richness (α-diversity) was higher in the organic wheat fields, which also contained more red-list species. Broadleaved plants benefitted more from organic farming than grasses. A further study showed that among the investigated plants, insect-pollinated species benefitted more from organic farming than non-insect-pollinated plant species (Gabriel and Tscharntke 2007). Further evidence confirms the benefits of organic farming for plant biodiversity for southern Europe, where again higher plant species richness was found in organic wheat fields than in conventional fields (Caballero-López et al., 2010).

For arthropods, the effect of farming type in wheat is less clear than for plants. While some arthropod taxa were more abundant and more species rich in organic wheat, other taxa responded in the opposite way or were unaffected by farming type. Alvarez et al. (2001) studied epigaeic Collembola (springtails) in England in five organic and four conventional wheat fields and found inconsistent effects of the farming system on community composition of these arthropods. Collembolan species diversity and richness were also not significantly different between the farming systems. In another study conducted in England, significantly lower densities of some dipterans, aphids, aphid-specific predators and some beetles were found in organic than in conventional wheat fields, whereas the opposite effect was seen in weevils, spiders, springtails, plant hoppers and sawfly larvae

(Moreby et al. 1994). Similarly, mixed results were found in a study on diversity in organic and conventional wheat fields in three regions of Germany (Clough et al., 2007). For bees, α-diversity, between-site β-diversity and between-region β-diversity were higher in organic than in conventional fields, but α-diversity was not different between the two farming systems for epigaeic arthropods. However, an earlier review had shown that epigaeic beetles profit from organic production, as higher species richness of carabid beetles was found in organic compared to in conventional wheat fields (Döring and Kromp, 2003). The carabid species most benefitting from organic farming in wheat are those that are typical for arable areas, and open fields; this is probably because organic wheat crops are often less dense than conventional ones.

With birds, studies on the effect of farming type rarely attribute the findings to specific crops such as wheat. In general, however, higher bird abundances and bird species richness were found on organic than on conventional farms (Freemark and Kirk, 2001; Batáry et al., 2010).

5.2 Organic wheat vs. other organic crops

The choice of crop species strongly affects wild farmland biodiversity. The composition of the weed community changes with the crops in the rotation. Thus, longer, that is, more species-rich, crop rotations may also increase farm biodiversity through the associated flora and fauna. In this context, it is important to note that direct comparisons of crop species with regard to their weed flora are only really meaningful if both the vegetation and the seedbank are considered. Many arable weed species present in organic wheat fields are often 'lost' from the visible vegetation during subsequent row crops (e.g. potatoes) or during the ley phase in the rotation, but will re-appear from the seedbank in the following wheat or other cereal crop.

Birds may also be affected by the choice of crop. An example comes from a study on paired conventional and organic winter wheat fields and meadows in ten landscapes in Germany. Although more bird species were found in organic than in conventional fields *regardless* of land-use type (wheat fields vs. meadows), this pattern was different for the skylark (Batáry et al., 2010), a typical arable bird species with severely declining population in Germany. For this species, higher abundance was observed in the organic wheat than in the organic meadow.

5.3 Effects of wheat management within organic systems

Several factors in organic wheat cultivation can influence biodiversity. First, field selection will affect field size and geometry. Because edge effects are frequently observed in many taxa, that is, with higher abundance, richness and diversity at the field edge than in the centre, field size may indirectly influence biodiversity. Similar considerations apply to the location of a field within the farm, as biodiversity is positively influenced by neighbouring hedges and other landscape features. Apart from these indirect effects of field choice, biodiversity in organic wheat fields has also been shown to be affected by wheat genotype, tillage, crop sequence and rotation, seed density and sowing time. An interesting case of how diversity at the crop genetic level may affect farmland biodiversity was presented in a recently published two-year-long study from northern France (Chateil et al., 2013). It was demonstrated that crop genetic diversity (in the form of a complex mixture of varieties and landraces of wheat) had a positive impact on biodiversity of Collembola, spiders and predatory carabids in comparison to a pure-line wheat monoculture. As plant richness was unaffected by crop diversity treatment, potential reasons for the effects on arthropods may be that the mixture provided more diverse food resources or was more complex in terms of crop architecture.

Although these findings are encouraging as they show possible synergies in managing wheat crops, other studies highlight the need to consider trade-offs as well. Recent research has shown that agronomic management practices that optimise wheat yields in organic farms may come at a cost of decreased biodiversity. Gabriel et al. (2013, p. 359) concluded that 'wildlife benefits accrue in low-yielding conventional farms as much as they do in organic ones, and conversely, they disappear in the most intensive organic farms whose yields rival those of conventional practices'.

6 Future trends and conclusion

No crop species can be said to play a 'central' role in organic farming, but wheat is – even at a global scale – among the more important ones. This chapter has highlighted the diversity of cropping systems around organic wheat cultivation. Although this overview is incomplete and biased, a number of current challenges and opportunities emerge. Among the general challenges of organic wheat production is certainly the task to achieve sustainable nutrient management with particular respect to nitrogen. The constraints of nitrogen supply in organic rotations, for example, through the required break years for legumes, set relatively narrow boundaries for designing balanced and long-term sustainable production systems. The task of minimising nitrogen losses and optimising supply to the growing wheat plant is still unsolved. Furthermore, diversification in the form of intercropping and variety mixtures bears potential, but research in this area seems to make little progress. The complexity of species and variety mixtures requires more comprehensive testing regimes, and this is particularly true for a demanding partner species such as wheat. One way of generating a reliable stream of relevant data would be to integrate crop mixtures into official variety testing programmes.

As this chapter has shown, wheat is produced organically on a vastly distributed scale. Because of its near-global distribution under organic management, wheat provides an excellent opportunity for networking, that is, sharing data and practical experience, as well as exchange of germplasm across countries and continents. As part of a strengthened network around organic wheat cultivation, it is likely that a central role will be played by long-term experiments, because these will help to gauge the long-term sustainability and future of organic wheat cultivation.

7 Where to look for further information

As this chapter shows, the topic of organic wheat and spelt cultivation spans several scientific disciplines and can be approached from many different angles. Furthermore, wheat and spelt production in organic systems cannot be separated from other crops because of the importance of crop rotation. Finally, cropping systems around the cultivation of organic wheat and spelt are often highly specific to their locality. For these reasons, it is not surprising that available information is not organised in a centralised way around organic wheat. More generalised sources for organic agriculture research include the organic e-prints database (http://orgprints.org) and four scientific journals: *Biological Agriculture and Horticulture* (ISSN 2165-0616), *Journal of Organic Systems* (ISSN 1177-4258), *Organic Agriculture* (ISSN: 1879-4238) and *Organic Farming* (ISSN: 2297-6485). To obtain localised and practice-oriented information, organic farmers and advisors may be contacted through their regional and national organic farmers' organisations.

8 Acknowledgements

I thank Maria Finckh, Bhullar Gurbir, Samuel Knapp, Paul Mäder, Monika Messmer, Daniel Neuhoff and Martin Wolfe for sharing their insights into the cultivation of organic wheat.

9 References

AGES (2014). *Österreichische Sortenliste 2014 Republik Österreich*, Wien: Österreichische Agentur für Gesundheit und Ernährungssicherheit GmbH.

Allard, R. W. (1988). Genetic changes associated with the evolution of adaptedness in cultivated plants and their wild progenitors. *Journal of Heredity*, 79, 225–38.

Alvarez, T., Frampton, G. K. and Goulson, D. (2001). Epigeic Collembola in winter wheat under organic, integrated and conventional farm management regimes. *Agriculture, Ecosystems & Environment*, 83, 95–110.

Annicchiarico, P., Chiapparino, E. and Perenzin, M. (2010). Response of common wheat varieties to organic and conventional production systems across Italian locations, and implications for selection. *Field Crops Research*, 116, 230–8.

Arny, A. C., Stoa, T. E., McKee, C. and Dillmann, A. C. (1929). *Flax Cropping in Mixture with Wheat, Oats, and Barley*. Washington, DC: US Department of Agriculture.

Arterburn, M., Murphy, K. and Jones, S. S. (2012). Wheat: Breeding for organic farming systems. In E. T. Lammerts van Bueren and J. R. Myers (Eds), *Organic Crop Breeding*, pp. 163–74. Hoboken, NY, USA: Wiley-Blackwell.

Augstburger, F. (1988). Erfahrungen mit dem Transfer ökologischer Landbautechniken an Kleinbauern bolivianischer Hochtäler. In P. Rottach (Ed.), *Ökologischer Landbau in den Tropen – Ecofarming in Theorie und Praxis*, , 3 ed., pp. 183–90. Karlsruhe, Germany: Verlag C. F. Müller.

Bailey, C. (1914). The composition and quality of wheat grown in mixtures with oats. *Agronomy Journal*, 6, 215–17.

Bàrberi, P. and Mazzoncini, M. (2006). The MASCOT (Mediterranean Arable Systems COmparison Trial) long-term experiment (Pisa, Italy). In J. Raupp, C. Pekrun, M. Oltmanns and U. Köpke (Eds), *Long-term Field Experiments In Organic Farming*, pp. 1–14. Berlin: ISOFAR Scientific Series.

Batáry, P., Matthiesen, T. and Tscharntke, T. (2010). Landscape-moderated importance of hedges in conserving farmland bird diversity of organic vs. conventional croplands and grasslands. *Biological Conservation*, 143, 2020–7.

Berry, P. M., Sylvester-Bradley, R., Philipps, L., Hatch, D. J., Cuttle, S. P., Rayns, F. W. and Gosling, P. (2002). Is the productivity of organic farms restricted by the supply of available nitrogen? *Soil Use and Management*, 18, 248–55.

Bertholdsson, N., Weedon, O., Brumlop, S. and Finckh, M. (2016). Evolutionary changes of weed competitive traits in winter wheat composite cross populations in organic and conventional farming systems. *European Journal of Agronomy*, 79, 23–30.

Bonafaccia, G., Galli, V., Francisci, R., Mair, V., Skrabanja, V. and Kreft, I. (2000). Characteristics of spelt wheat products and nutritional value of spelt wheat-based bread. *Food Chemistry*, 68, 437–41.
Brophy, L. S. and Mundt, C. C. (1991). Influence of plant spatial patterns on disease dynamics, plant competition and grain-yield in genetically diverse wheat populations. *Agriculture Ecosystems & Environment*, 35, 1–12.
Brouwer, W. (1972). *Handbuch des Speziellen Pflanzenbaus – Band I*. Hamburg, Berlin: Paul Parey.
Bulson, H. A. J., Snaydon, R. W. and Stopes, C. E. (1997). Effects of plant density on intercropped wheat and field beans in an organic farming system. *Journal of Agricultural Science*, 128, 59–71.
Caballero-López, B., Blanco-Moreno, J., Pérez, N., Pujade-Villar, J., Ventura, D., Oliva, F. and Sans, F. (2010). A functional approach to assessing plant–arthropod interaction in winter wheat. *Agriculture, Ecosystems & Environment*, 137, 288–93.
Caldwell, B., Mohler, C. L., Ketterings, Q. M. and DiTommaso, A. (2014). Yields and profitability during and after transition in organic grain cropping systems. *Agronomy Journal*, 106, 871–80.
Casagrande, M., David, C., Valantin-Morison, M., Makowski, D. and Jeuffroy, M.-H. (2009). Factors limiting the grain protein content of organic winter wheat in south-eastern France: a mixed-model approach. *Agronomy for Sustainable Development*, 29, 565–74.
Castellazzi, M., Wood, G., Burgess, P. J., Morris, J., Conrad, K. and Perry, J. (2008). A systematic representation of crop rotations. *Agricultural Systems*, 97, 26–33.
Chateil, C., Goldringer, I., Tarallo, L., Kerbiriou, C., Le Viol, I., Ponge, J.-F., Salmon, S., Gachet, S. and Porcher, E. (2013). Crop genetic diversity benefits farmland biodiversity in cultivated fields. *Agriculture Ecosystems & Environment*, 171, 25–32.
Clough, Y., Holzschuh, A., Gabriel, D., Purtauf, T., Kleijn, D., Kruess, A., Steffan-Dewenter, I. and Tscharntke, T. (2007). Alpha and beta diversity of arthropods and plants in organically and conventionally managed wheat fields. *Journal of Applied Ecology*, 44, 804–12.
Cooper, J., Baranski, M., Stewart, G., Nobel-de Lange, M., Bàrberi, P., Fließbach, A., Peigné, J., Berner, A., Brock, C. and Casagrande, M. (2016). Shallow non-inversion tillage in organic farming maintains crop yields and increases soil C stocks: a meta-analysis. *Agronomy for Sustainable Development*, 36, 1–20.
Costanzo, A. and Bàrberi, P. (2013). Diversity to improve wheat cropping systems: Initial evidence on the potential of using evolutionary populations. In T. F. Döring, S. A. Howlett, L. R. Winkler and M. S. Wolfe (Eds), *International Symposium on Evolutionary Breeding in Cereals*, p. 6. Birmingham, UK: Organic Research Centre.
Costanzo, A. and Bàrberi, P. (2014). Functional agrobiodiversity and agroecosystem services in sustainable wheat production: A review. *Agronomy for Sustainable Development*, 34, 327–48.
Crowder, D. W. and Reganold, J. P. (2015). Financial competitiveness of organic agriculture on a global scale. *Proceedings of the National Academy of Sciences*, 112, 7611–16.
David, C., Abecassis, J., Carcea, M., Celette, F., Corre-Hellou, G., Friedel, J., Hiltbrunner, J., Mäder, P., Messmer, M. and Narducci, V. (2013). Améliorer la qualité technologique, nutritionnelle et sanitaire du blé biologique. Principaux leviers agronomiques et technologiques. *Innovations Agronomiques*, 32.
David, C., Abecassis, J., Carcea, M., Celette, F., Friedel, J. K., Hellou, G., Hiltbrunner, J., Messmer, M., Narducci, V., Peigné, J., Samson, M. F., Schweinzer, A., Thomsen, I. K. and Thommen, A. (2012). Organic bread wheat production and market in Europe. *Sustainable Agriculture Reviews*, 11, 43–62.
Diepenbrock, W., Ellmer, F. and Léon, J. (2012). *Ackerbau, Pflanzenbau und Pflanzenzüchtung*. Stuttgart: Verlag Eugen Ullmer.
Döring, T. F. (2015). Grain Legume Cropping Systems in Temperate Climates. In A. De Ron (Ed.), *Grain Legumes – Handbook of Plant Breeding*, pp. 401–34. New York: Heidelberg: Springer.
Döring, T. F., Annicchiarico, P., Clarke, S., Haigh, Z., Jones, H. E., Pearce, H., Snape, J., Zhan, J. and Wolfe, M. S. (2015). Comparative analysis of performance and stability among composite cross populations, variety mixtures and pure lines of winter wheat in organic and conventional cropping systems. *Field Crops Research*, 183, 235–45.

Döring, T. F., Jones, H., Clarke, S., Baddeley, J., Haigh, Z., Pearce, H., Crowley, O. and Wolfe, M. (2010a). Foliar diseases and their yield effects on organic wheat. *The Organic Research Centre Bulletin*, 99, 4.

Döring, T. F., Wolfe, M., Jones, H., Pearce, H. and Zhan, J. (2010b). Breeding for resilience in wheat – Nature's choice. In *Breeding for Resilience: A Strategy for Organic and Low-Input Farming sytems? Eucarpia 2nd Conference of the Organic and Low-Input Agriculture Section*, 1–3 December 2010, pp. 45–8. Goldringer I Paris: France.

Döring, T. F., Kovacs, G., Wolfe, M. S. and Murphy, K. (2011). Evolutionary plant breeding in cereals – into a new era. *Sustainability*, 3, 1944–71.

Döring, T. F. and Kromp, B. (2003). Which carabid species benefit from organic agriculture? – a review of comparative studies in winter cereals from Germany and Switzerland. *Agriculture, Ecosystems & Environment*, 98, 153–61.

Döring, T. F., Storkey, J., Baddeley, J. A., Crowley, O., Howlett, S., McCalman, H., Pearce, H., Roderick, S. and Jones, H. E. (2012). Legume based plant mixtures for delivery of multiple ecosystem services: Weed diversity and weed control. In *SAC-SEPA Biennial conference: Valuing Ecosystems: Policy, Economic and Management Interactions*, pp. 163–8. Edinburgh.

Emmens, J. (2003). Considerations for conversion to organic production for wheat-based farming systems. Food and Agriculture Organization Working Papers.

Eriksen, J. and Askegaard, M. (2000). Sulphate leaching in an organic crop rotation on sandy soil in Denmark. *Agriculture, Ecosystems & Environment*, 78, 107–14.

FAOSTAT (2014). Crop production data. http://faostat3.fao.org/.

FIBL (2015). The Organic-World.net website maintained by the Research Institute of Organic Agriculture (FiBL), Frick, Switzerland. Data available at http://www.organic-world.net/statistics/. Accessed 6 January 2017.

Finckh, M. R., Gacek, E. S., Goyeau, H., Lannou, C., Merz, U., Mundt, C. C., Munk, L., Nadziak, J., Newton, A. C., de Vallavieille-Pope, C. and Wolfe, M. S. (2000). Cereal variety and species mixtures in practice, with emphasis on disease resistance. *Agronomie*, 20, 813–37.

Finckh, M. R. and Mundt, C. C. (1992). Stripe rust, yield, and plant competition in wheat cultivar mixtures. *Ecology and Epidemiology*, 82, 905–13.

Finckh, M. R. and Mundt, C. C. (1996). Temporal dynamics of plant competition in genetically diverse wheat populations in the presence and absence of stripe rust. *Journal of Applied Ecology*, 33, 1041–52.

Finckh, M. R., Schulze-Schilddorf, G., Butz, A., Greiner, L., Knape, C., Lützkendorf, K. and Rabenstein, F. (2003). *Interaktionen zwischen Anbaubedingungen, Pilzbefall, Backqualität und Mykotoxinbelastung in der ökologischen Weizenproduktion*, Bonn, Germany: Bundesanstalt für Landwirtschaft und Ernährung (BLE), Bonn, Geschäftsstelle Bundesprogramm Ökologischer Landbau. .

Finckh, M. R. and Wolfe, M. S. (2006). Diversification Strategies. In B. M. Cooke, G. Jones and B. Kaye (Eds), *The Epidemiology of Plant Disease*, 2nd ed. Netherlands: Springer.

Fischl, M., Traudtner, F., Surböck, A. and Kranzler, A. (2013). *Qualitätsweizenerzeugung im Biolandbau: Mehrjährige Ergebnisse zu Kulturführungsmaßnahmen und Sortenwahl aus Bionet-Praxisversuchen*. Wien: Ländliches Fortbildungsinstitut Österreich.

Flamm, C. (2009). Wirkung eines Mischanbaues von Weizensorten auf Anbaueigenschaften, Krankheiten, Ertrag und Qualität unter den Bedingungen des Biolandbaues. In *Werte – Wege – Wirkungen: Biolandbau im Spannungsfeld zwischen Ernährungssicherung, Markt und Klimawandel Beiträge zur 10. Wissenschaftstagung Ökologischer Landbau, ETH Zürich*, 11–13. Februar 2009, pp. 189–92. Zürich: Verlag Dr. Köster.

Freemark, K. E. and Kirk, D. A. (2001). Birds on organic and conventional farms in Ontario: partitioning effects of habitat and practices on species composition and abundance. *Biological Conservation*, 101, 337–50.

Freyer, B. (2003). *Fruchtfolgen Konventionell, Integriert, Biologisch*. Stuttgart, Germany: Ulmer.

Gabriel, D., Sait, S. M., Kunin, W. E. and Benton, T. G. (2013). Food production vs. biodiversity: comparing organic and conventional agriculture. *Journal of Applied Ecology*, 50, 355–64.

Gabriel, D. and Tscharntke, T. (2007). Insect pollinated plants benefit from organic farming. *Agriculture, Ecosystems & Environment*, 118, 43–8.

Gigot, C., Saint-Jean, S., Huber, L., Maumené, C., Leconte, M., Kerhornou, B. and Vallavieille-Pope, C. (2013). Protective effects of a wheat cultivar mixture against splash-dispersed septoria tritici blotch epidemics. *Plant Pathology*, 62, 1011–19.

Goldringer, I., Prounin, C., Rousset, M., Galic, N. and Bonnin, I. (2006). Rapid differentiation of experimental populations of wheat for heading time in response to local climatic conditions. *Annals of Botany*, 98, 805–17.

Hauggaard-Nielsen, H., Jørnsgaard, B., Kinane, J. and Jensen, E. S. (2008). Grain legume–cereal intercropping: the practical application of diversity, competition and facilitation in arable and organic cropping systems. *Renewable Agriculture and Food Systems*, 23, 3–12.

Heiniger, R., Reberg-Horton, C. and Hamilton, M. (2013). Chapter 2. Organic Crop Production Systems. In *North Carolina Organic Grain Production Guide*, North Carolina Cooperative Extension Service of North Carolina State University (Ed). Raleigh, NC: USA.

Heyn, J. (2005). Die Anwendung der Ertragskurvenberechnung nach von Boguslawski und Schneider bei der Auswertung aktueller N-Steigerungsversuchsserien. *Archives of Agronomy and Soil Science*, 51, 487–99.

Hildermann, I., Messmer, M., Dubois, D., Boller, T., Wiemken, A. and Mäder, P. (2010). Nutrient use efficiency and arbuscular mycorrhizal root colonisation of winter wheat cultivars in different farming systems of the DOK long-term trial. *Journal of the Science of Food and Agriculture*, 90, 2027–38.

Hildermann, I., Thommen, A., Dubois, D., Boller, T., Wiemken, A. and Mäder, P. (2009). Yield and baking quality of winter wheat cultivars in different farming systems of the DOK long-term trial. *Journal of the Science of Food and Agriculture*, 89, 2477–91.

Hiltbrunner, J., Liedgens, M., Bloch, L., Stamp, P. and Streit, B. (2006). Legume cover crops as living mulches for winter wheat: Components of biomass and the control of weeds. *European Journal of Agronomy* (In press).

Hof-Kautz, C. (2011). Fruchtfolgeversuch unter den Bedingungen des Ökologischen Landbaus 2011 In *Versuch zum Ökologischen Landbau Nordrhein-Westfalen*. Köln-Auweiler: Germany.

Horst, W. J. and Waschkies, C. (1987). Phosphatversorgung von Sommerweizen (*Triticum aestivum* L.) in Mischkultur mit Weißer Lupine (*Lupinus albus* L.). *Zeitschrift für Pflanzenernährung und Bodenkunde*, 150, 1–8.

IFOAM (2005). The Principles of Organic Agriculture. Accessed 22 April 2013.

Jensen E. S., Ambus, P., Bellostas, N., Boisen, S., Brisson, N., Corre-Hellou, G., Crozat, Y., Dahlmann, C., Dibet, A. and von Fragstein, P. (2006). Intercropping of cereals and grain legumes for increased production, weed control, im-proved product quality and prevention of N–losses in European organic farming systems. In *Joint Organic Congress*. Odense, Denmark. http://orgprints.org/7659/. Accessed 18 November 2016.

Káš, M., Haberle, J. and Matějková, S. (2010). Crop productivity under increasing nitrogen rates and different organic fertilization systems in a long-term IOSDV experiment in the Czech Republic†. *Archives of Agronomy and Soil Science*, 56, 451–61.

Kaut, A., Mason, H., Navabi, A., O'donovan J. and Spaner, D. (2009). Performance and stability of performance of spring wheat variety mixtures in organic and conventional management systems in western Canada. *The Journal of Agricultural Science*, 147, 141–53.

Klapp, E. (1967). *Lehrbuch des Acker- und Pflanzenbaus*. Berlin: Paul Parey.

Kohajdová, Z. and Karovičová, J. (2008). Nutritional value and baking applications of spelt wheat. *Acta Scientiarum. Polonorum, Technologia Alimentaria*, 7, 5–14.

Kolbe, H. (2016). Fruchtfolge aus Sicht der Humusversorgung sowie des ökologischen Landbaus. In *Gute fachliche Praxis – Bodenfruchtbarkeit*. AID infodienst Ernährung: Verbraucherschutz, Bonn, pp. 85–9.

Körber-Grohne, U. (1995). *Nutzpflanzen in Deutschland von der Vorgeschichte bis heute*. Hamburg: Nikol Verlagsgesellschaft.

Krug, G.-U. (2015). Ökodinkel – Befriedigende Ergebnisse 2015. *Bauernblatt*, 10. Oktober 2015, 35–6.

Kumar, D. (2014). Influence of nutrient sources and inclusion of mungbean on productivity, soil fertility and profitability of organic rice-wheat cropping system. In G. Rahmann and U. Aksoy Istanbul (Eds), *Proceedings of the 4th ISOFAR Scientific Conference 'Building Organic Bridges', at the Organic World Congress 2014*, pp. 655–8. 13–15 October 2014, Turkey.

Lacko-Bartošová, M. and Rédlová, M. (2007). The significance of spelt wheat cultivated in ecological farming in the Slovak Republic. In *Proceeding of the Conference on Organic Farming*, pp. 79–81.

Lampkin, N. (1994). *Organic Farming*, Ipswich: Farming Press.

Lampkin, N., Measures, M. and Padel, S. (Eds) (2011). *2011/2012 Organic Farm Management Handbook*. Hamstead Marshall, UK: Organic Research Centre, Elm Farm.

Lenc, L., Kwaśna, H., Sadowski, C. and Grabowski, A. (2015). Microbiota in Wheat Roots, Rhizosphere and Soil in Crops Grown in Organic and Other Production Systems. *Journal of Phytopathology*, 163, 245–63.

Leoni, C., Rossing, W. and Van Bruggen, A. H. C. (2015). Crop rotation. In M. R. Finckh, A. H. C. Van Bruggen and L. Tamm (Eds), *Plant Diseases and Their Management in Organic Agriculture*, pp. 127–40. St. Paul, Minnesota: APS Press.

LLFG (2011). Versuchsfeldführer 2011 Standort Bernburg. *Fachinformationen 211* [Online]. Available: http://www.llg.sachsen-anhalt.de/fileadmin/Bibliothek/Politik_und_Verwaltung/MLU/LLFG/Dokumente/abt2_acker_pflanze/sortenpruefung/vff_feldtage/vff_bernburg.pdf.

Mäder, P., Hahn, D., Dubois, D., Gunst, L., Alföldi, T., Bergmann, H., Oehme, M., Amado, R., Schneider, H., Graf, U., Velimirov, A., Fließbach, A. and Niggli, U. (2007). Wheat quality in organic and conventional farming: results of a 21 year field experiment. *Journal of the Science of Food and Agriculture*, 87, 1826–35.

Malézieux, E., Crozat, Y., Dupraz, C., Laurans, M., Makowski, D., Ozier-Lafontaine, H., Rapidel, B., De Tourdonnet, S. and Valantin-Morison, M. (2009). Mixing plant species in cropping systems: concepts, tools and models: a review. In *Sustainable Agriculture*, pp. 329–53. Springer.

Mikó, P., Megyeri, M. and Kovács, G. (2011). Characterization of *Triticum timopheevii* gene bank accessions to gain useful materials for organic wheat breeding. In O. Veisz (Ed.), *Climate Change: Challenges and Opportunities in Agriculture AGRISAFE Final Conference*, pp. 90–3. Budapest: Agricultural Research Institute of the Hungarian Academy of Sciences.

Milev, G., Nankov, N., Iliev, I., Ivanova, A. and Nankova, M. (2014). Growing Wheat (*Trititcum aestivum* L.) by the Methods of Organic Agriculture Under the Conditions of Dobrudzha Region, Bulgaria. *Türk Tarım ve Doğa Bilimleri*, 6, 849–57.

Mirsky, S. B., Ryan, M. R., Curran, W. S., Teasdale, J. R., Maul, J., Spargo, J. T., Moyer, J., Grantham, A. M., Weber, D. and Way, T. R. (2012). Conservation tillage issues: Cover crop-based organic rotational no-till grain production in the mid-Atlantic region, USA. *Renewable Agriculture and Food Systems*, 27, 31–40.

Moreby, S., Aebischer, N., Southway, S. and Sotherton, N. (1994). A comparison of the flora and arthropod fauna of organically and conventionally grown winter wheat in southern England. *Annals of Applied Biology*, 125, 13–27.

Mortimer, J. (1708). *The Whole Art of Husbandry or, the Way of Managing and Improving of Land*. Printed by J. H. for H. Mortlock, 2nd ed., London.

Murata, T. and Goh, K. (1997). Effects of cropping systems on soil organic matter in a pair of conventional and biodynamic mixed cropping farms in Canterbury, New Zealand. *Biology and Fertility of Soils*, 25, 372–81.

Murphy, K., Lammer, D., Lyon, S., Carter, B. and Jones, S. (2005). Breeding for organic and low-input farming systems: An evolutionary-participatory breeding method for inbred cereal grains. *Renewable Agriculture and Food Systems*, 20, 48–55.

Neuerburg, W. and Padel, S. (1992). *Organisch-biologischer Landbau in der Praxis*. München: Verlagsunion Agrar.

Neumann, H., Loges, R. and Taube, F. (2003). Bicropping im ökologischen Winterweizenanbau-eine Alternative zum Anbausystem der Weiten Reihe *Beiträge zur 7. Wissenschaftstagung zum Ökologischen Landbau Ökologischer Landbau der Zukunft*, pp. 81–4.

Neumann, H., Loges, R. and Taube, F. (2006). The wide row system - Innovation for organic winter wheat cultivation? *Berichte Uber Landwirtschaft*, 84, 404–24.

Neumann, H. J. (2005). *Optimierungsstrategien für den Getreideanbau im ökologischen Landbau: System 'weite Reihe' und Direktsaat in ausdauernden Weißklee ('Bi-cropping')*. Christian-Albrechts Universität Kiel.

Niggli, U. (2007). The evolution of organic practice. In W. Lockeretz (Ed.), *Organic Farming: an International History*, pp. 73–92. CAB International.

Obenauf, U. (2012). Weniger Selbstfolgen zur nächsten Ernte. *Bauernblatt*, 29 September 2012, 37–43.

Olesen, J. E., Rasmussen, I. A., Askegaard, M. and Kristensen, K. (2002). Whole-rotation dry matter and nitrogen grain yields from the first course of an organic farming crop rotation experiment. *The Journal of Agricultural Science*, 139, 361–70.

Østergård, H., Finckh, M. R., Fontaine, L., Goldringer, I., Hoad, S. P., Kristensen, K., Lammerts van Bueren, E. T., Mascher, F., Munk, L. and Wolfe, M. S. (2009). Time for a shift in crop production: embracing complexity through diversity at all levels. *Journal of the Science of Food and Agriculture*, 89, 1439–45.

Paulsen, H. M., Schochow, M., Ulber, B., Kühne, S. and Rahmann, G. (2006). Mixed cropping systems for biological control of weeds and pests in organic oilseed crops. *Aspects of Applied Biology*, 79, 215–20.

Pawsey, J. (2011). News Flash: Hoe Action on Suffolk Farm. http://johnpawsey.blogspot.de/ [Online]. Accessed 19 January 2016.

Pelzer, E., Bazot, M., Makowski, D., Corre-Hellou, G., Naudin, C., Al Rifaï, M., Baranger, E., Bedoussac, L., Biarnès, V. and Boucheny, P. (2012). Pea–wheat intercrops in low-input conditions combine high economic performances and low environmental impacts. *European Journal of Agronomy*, 40, 39–53.

Ponisio, L. C., M'Gonigle, L. K., Mace, K. C., Palomino, J., de Valpine, P. and Kremen, C. (2015). Diversification practices reduce organic to conventional yield gap. *Proceedings of the Royal Society of London B: Biological Sciences*, 282, 20141396.

Pridham, J. C. and Entz, M. H. (2008). Intercropping spring wheat with cereal grains, legumes, and oilseeds fails to improve productivity under organic management. *Agronomy Journal*, 100, 1436–42.

Przystalski, M., Osman, A., Thiemt, E. M., Rolland, B., Ericson, L., Østergård, H., Levy, L., Wolfe, M., Büchse, A., Piepho, H.-P. and Krajewski, P. (2008). Comparing the performance of cereal varieties in organic and non-organic cropping systems in different European countries. *Euphytica*, 163, 417–33.

Ram, M., Davari, M. R. and Sharma, S. N. (2014). Direct, residual and cumulative effects of organic manures and biofertilizers on yields, NPK uptake, grain quality and economics of wheat (*Triticum aestivum* L.) under organic farming of rice-wheat cropping system. *Journal of Organic Systems*, 9, 16–30.

Reid, T. A., Yang, R. C., Salmon, D. F. and Spaner, D. (2009). Should spring wheat breeding for organically managed systems be conducted on organically managed land? *Euphytica*, 169, 239–52.

Reinsch, M. (2014). Dinkel. *Hinweise zum Pflanzenbau – Ökologischer Landbau* [Online].

Reynolds, M., Sayre, K. and Vivar, H. (1994). Intercropping wheat and barley with N-fixing legume species: a method for improving ground cover, N-use efficiency and productivity in low input systems. *Journal of Agricultural Science-Cambridge*, 123, 175.

Roschewitz, I., Gabriel, D., Tscharntke, T. and Thies, C. (2005). The effects of landscape complexity on arable weed species diversity in organic and conventional farming. *Journal of Applied Ecology*, 42, 873–82.

Ryan, M. H., Derrick, J. and Dann, P. (2004). Grain mineral concentrations and yield of wheat grown under organic and conventional management. *Journal of the Science of Food and Agriculture*, 84, 207–16.

SAHSH (2010). Landwirtschaftszählung 2010 – Landwirtschaftliche Betriebe mit Ackerland und Anbau von Getreide zur Körnergewinnung 2010 nach Getreidearten. Accessed 30 January 2016.

Sarandon, S. J. and Sarandon, R. (1995). Mixture of cultivars:pilot field trial of an ecological alternative to improve production or quality of wheat (*Triticum aestivum*). *Journal of Applied Ecology*, 32, 288–94.

Šarūnaitė, L., Arlauskienė, A., Deveikytė, I., Maikštėnienė, S. and Kadžiulienė, Ž. (2012). Intercropping of pea and Spring cereals for Weed control in an organic farming system. *Weed Control InTech, Croatia*, 3–30.

Šarūnaitė, L., Deveikytė, I. and Kadžiulienė, Ž. (2010). Intercropping spring wheat with grain legume for increased production in an organic crop rotation. *Žemdirbystė Agriculture*, 97, 51–8.

Schindler, F. (1923). *Handbuch des Getreidebaus auf wissenschaftlicher und praktischer Grundlage*. Berlin: Verlagsbuchhandlung Paul Parey.

Schlipf, J. A. (1898). *Handbuch der Landwirtschaft, 13 Auflage*. Berlin: Paul Parey.

Schmidt, H. (2010). *Öko-Ackerbau ohne tiefes Pflügen*. Berlin: Verlag Dr. Köster.

Schulz, F., Brock, C., Schmidt, H., Franz, K.-P. and Leithold, G. (2014). Development of soil organic matter stocks under different farm types and tillage systems in the Organic Arable Farming Experiment Gladbacherhof. *Archives of Agronomy and Soil Science*, 60, 313–26.

Smith, E. G., Jill Clapperton, M. and Blackshaw, R. E. (2004). Profitability and risk of organic production systems in the northern Great Plains. *Renewable Agriculture and Food Systems*, 19, 152–8.

Soil Association United Kingdom (2016). *Organic Standards Farming and Growing, Revision 17*, 4 August 2016, p. 248. https://www.soilassociation.org/media/1220/sa-farming-and-growing-standards.pdf. Accessed 18 November 2016.

Stephan, H. (2015). Ergebnisse der Landessortenversuche Winterweizen 2015 - Überraschend gute Erträge. *Bauernblatt*, 12 September 2015, 30–4.

Stinner, D. H. (2007). The science of organic farming. In W. Lockeretz (Ed.), *Organic Farming: an International History*, pp. 40–72. CAB International.

Stockdale E. A., Lampkin, N. H., Hovi, M., Keatinge, R., Lennartsson, E. K. M., Macdonald, D. W., Padel, S., Tattersall, F. H., Wolfe, M. S. and Watson, C. A. (2001). Agronomic and environmental implications of organic farming systems. *Advances in Agronomy*, 70, 261–327.

Storkey, J., Döring, T. F., Baddeley, J., Collins, R., Roderick, S., Jones, H. and Watson, C. A. (2015). Engineering a plant community to deliver multiple ecosystem services. *Ecological Applications*, 25, 1034–43.

Suneson, C. A. (1956). An evolutionary plant breeding method. *Agronomy Journal*, 48, 188–91.

Taramarcaz, J. and Clerc, M. (2013). Viehloser Bio-Ackerbau: Resultate aus zwölf Beobachtungsjahren des Versuchsbetriebs Mapraz. *Agrarforschung Schweiz*, 4, 124–31.

Thaer, A. D. (1809). *Grundsätze der rationellen Landwirthschaft - Erster Band. Begründung der Lehre und des Gewerbes*. Berlin: Realschulbuchhandlung. Faksimile-Ausgabe.

Tilman, D. (1996). Biodiversity: Population versus ecosystem stability. *Ecology*, 77, 350–63.

Turmel, M.-S., Entz, M. H., Bamford, K. and Thiessen Martens, J. R. (2009). The influence of crop rotation on the mineral nutrient content of organic vs. conventionally produced wheat grain: Preliminary results from a long-term field study. *Canadian Journal of Plant Science*, 89, 915–19.

Urbatzka, P. and Cais, K. (2013). Der Ökodinkel ist den Stickstoff wert. *Bayerisches Landwirtschaftliches Wochenblatt*, 38, 26–7.

Vakali, C., Zaller, J. G. and Köpke, U. (2011). Reduced tillage effects on soil properties and growth of cereals and associated weeds under organic farming. *Soil and Tillage Research*, 111, 133–41.

van Bruggen, A. H., Gamliel, A. and Finckh, M. R. (2016). Plant disease management in organic farming systems. *Pest Management Science*, 72, 30–44.

Vandermeer, J. (1989). *The Ecology of Intercropping*. Cambridge, UK: Cambridge University Press.

von Rümker, K. (1911). *Über Fruchtfolge*. Berlin: Paul Parey.

Watson, C. A., Atkinson, D., Gosling, P., Jackson, L. R. and Rayns, F. W. (2002). Managing soil fertility in organic farming systems. *Soil Use and Management*, 18, 239–47.

Weisz, R., Cowger, C. and Reisig, D. (2014). Chapter 4: Crop Production Management – Organic Wheat and Small Grains. In *North Carolina Organic Grain Production Guide*. North Carolina State University.

Willer, H. and Lernoud, J. (2015). *The World of Organic Agriculture. Statistics and Emerging Trends 2015 – FiBL IFOAM Report*. Frick, Switzerland and Bonn, Germany.

Wollny, E. (1885). *Saat und Pflege der landwirthschaftlichen Kulturpflanzen – Handbuch für die Praxis*. Berlin: Verlag Paul Parey.

Wortman, S. E., Francis, C. A., Galusha, T. D., Hoagland, C., Van Wart, J., Baenziger, P. S., Hoegemeyer, T. and Johnson, M. (2013). Evaluating cultivars for organic farming: maize, soybean, and wheat genotype by system interactions in Eastern Nebraska. *Agroecology and Sustainable Food Systems*, 37, 915–32.

Zimmermann, M. (1958). *Schlipf Praktisches Handbuch der Landwirtschaft, 32. Auflage*. Hamburg, Berlin: Paul Parey.

zu Putlitz, K. and Meyer, L. (1914). *Landlexikon – Ein Nachschlagewerk des allgemeinesn Wissens unter besonderer Berücksichtugung der Landwirtschaft, Forstwirtschaft, Gärtnerei, der ländlichen Industrien und der ländlichen Justiz- und Verwaltungspraxis*. Stuttgart: Deutsche Verlags-Anstalt.

Chapter 9

Durum wheat: production, challenges and opportunities

J. M. Clarke, K. Nilsen, D. Khitiri, X. Lin and C. J. Pozniak, University of Saskatchewan, Canada; and K. Ammar, International Maize and Wheat Improvement Center (CIMMYT), Mexico

1 Introduction
2 Durum wheat yield and agronomic traits
3 Abiotic stresses on durum wheat
4 Major diseases of durum wheat
5 Insect pests of durum wheat
6 End-use quality of durum wheat
7 Durum wheat breeding technologies
8 Future trends and conclusion
9 Where to look for further information
10 References

1 Introduction

Durum wheat (*Triticum turgidum* var. *durum* Desf.) is principally used for manufacture of pasta, couscous and, to lesser extent, in the production of bulgur and bread. The durum milling and pasta industries, the latter utilizing some 75% of the worldwide production, desire a grain with high yellow pigment concentration, good milling properties such as high semolina yield and low ash, and appropriate gluten strength. These requirements have shaped durum breeding programmes for the past several decades, particularly in nations that grow durum primarily as an export commodity, such as Canada, Mexico and Australia.

Half of the durum is produced in the Mediterranean Basin and 23% in North America (Canada, the United States and Mexico). Based on ten years' average (International Grains Council, 2013), the durum produced globally represents 5.8% of total wheat production and is sown on 8% of the worldwide total wheat area. While this indicates that durum wheat is globally minor, in some countries it is considered as a primary cereal and the main source of calories and protein (North Africa and Middle East) in human diets. In Italy, the largest durum consumer, durum represents about 50% of total wheat production, while in Canada, the largest durum producer and exporter, durum makes up nearly 20% of total wheat production.

http://dx.doi.org/10.19103/AS.2016.0004.32

Durum wheat is an allotetraploid (2*n* = 28: AABB) harbouring an A genome from *Triticum urartu* Tumanian and a B genome from *Aegilops speltoides* Tausch (Jauhar and Peterson, 2013). The A and B genomes of modern durum and those of hexaploid wheat (2*n* = 42: AABBDD) have evolved from those of a common tetraploid ancestor and traits can be readily transferred between the two species by conventional breeding approaches. Durum genetic improvement efforts began in the early twentieth century in Italy, France and the United States, with initial efforts focused on selection from landraces introduced from North Africa (Clarke et al., 2012). By the 1940s, cultivars produced by hybridization and subsequent selection appeared in Europe and North America as well as in North Africa. In industrialized countries, breeding and release of durum cultivars was instrumental in improving the end-use suitability of the crop by keeping pace with changes in consumer preferences, food safety regulations and processing technology. Improvements in agronomic traits and disease and insect resistance have increased productivity, enhanced sustainability and competitiveness of the crop, therefore its economic attractiveness for producers. Global wheat yields, including durum, increased nearly threefold from 1960 to 2015, increasing production by a similar proportion, while area sown remained relatively static (Ainslie, 2010). This trend is observed for durum wheat as supported by national and regional statistics in Mexico (SIAP) or Tunisia (Ammar et al., 2011). It is generally assumed that half of the yield improvement is due to genetic enhancement and half is due to improved agronomic management (Clarke et al., 2010b). Improved agronomic management includes appropriate nutrient application, better seeding equipment and practices, effective control of competition from weeds and, in the case of Canada, conservation tillage systems to increase soil water retention capacity and therefore moisture availability over time. Genetic gains in farmers' yields were achieved via better disease and insect resistance, optimization of adaptation to the environment and fitting plant architecture to intensive management, particularly with the adoption of semi-dwarf stature. Durum breeding continues to add new issues to address, such as consumer concerns about food safety and nutritional benefits.

In this chapter we attempt to address production issues considered to be globally important, describe some approaches that are used to mitigate production constraints and the progress realized through these approaches in breeding in recent years. Finally, we will touch briefly on emerging breeding technologies that will impact future breeding.

2 Durum wheat yield and agronomic traits

Maximizing grain yield potential in durum wheat requires optimizing the utilization of water, nutrients and light, while avoiding the negative effects of biotic and abiotic stresses (Royo et al., 2016). Increased grain yield and improvement of other agronomic traits such as straw strength, semi-dwarf height and adaptation to local environments are important objectives for breeders. In Canada, we estimate that the rate of genetic improvement in grain yield since the release of the first cultivar, 'Stewart 63', in 1963 has been 0.7% per year (Clarke et al., 2010b). Similar rates of gain were reported in Spain (Garcia del Moral et al., 2005) and Italy (Pecetti and Annicchiarico, 1998; De Vita et al., 2007). These genetic gains in yield realized in rainfed, often drought-prone conditions, are well below those achieved under irrigation and high-input conditions. In fact, in Mexico where the great majority of the durum wheat area is fully irrigated with highly favourable conditions, genetic gains in national yields have been in the order of 2.8% annually, based on official statistics from 2003 to 2013.

Research to dissect the genetic basis of grain yield and identify closely linked DNA markers has been ongoing in a number of related species including barley (Cattivelli et al., 2002), common wheat (Gupta et al., 1999; Gonzalez-Hernandez et al., 2004; Kirigwi et al., 2007), rice (Zeng et al., 2006), maize (Tuberosa et al., 2002) and in durum wheat itself (Maccaferri et al., 2008). However, development of markers for genetically complex traits such as yield is difficult due to the large number of genetic factors, each contributing in a very small manner to the phenotype. This, combined with complex genotype–environmental interactions, has resulted in the identification of quantitative trait loci (QTL) with only minor effects, often not expressed in all environments. Mathews et al. (2008), investigating genotype by QTL interactions for yield over six drought-stressed environments reported only one QTL related directly to yield that was consistent across all environments. Other QTLs were significant in only a subset of environments, or co-located with other traits such as plant height and flowering time. In durum wheat, two QTLs for grain yield with consistent effects across a range of environments were reported (Maccaferri et al., 2008). The first QTL on 2B, with additive effects of 0.38 to 1.7 q ha^{-1}, is novel in that it has not been previously described in wheat. The second on chromosome 3BS, explaining up to 21.5% of the phenotypic variation, appears to be collinear with other yield QTLs identified in common wheat (Maccaferri et al., 2008). To date, the usefulness of QTL identified as associated with yield is yet to be demonstrated for use in breeding programmes and therefore QTL mapping has yet to deliver impact in terms of genetic gains for yield.

Shorter conventional height or semi-dwarf cultivars were developed in the late 1960s in response to intensification in farming practices, particularly direct combine harvesting, and production of durum under irrigation or high rainfall regions where lodging resistance is essential. Currently, most durum cultivars in production globally carry the *RhtB1b* allele at the *Rht-B1* locus conferring semi-dwarf stature as well as increasing the harvest index and responsiveness to fertilization under intensive conditions. Notable exceptions to the worldwide generalized adoption of semi-dwarf durum types are seen in high-latitude dryland environments, where some degree of photoperiod sensitivity is also beneficial, such as the central plains region of North America, parts of Kazakhstan and Russia. Most cultivars from the North Dakota and Canadian breeding programmes are of intermediate height (Clarke et al., 2010b; Elias and Manthey, 2005). Recent Canadian cultivars are considerably shorter and stronger-strawed than the older types such as Kyle (Townley-Smith et al., 1987), the predominant cultivar until 2002. Current major Canadian cultivars, such as Strongfield (Clarke et al., 2005) and CDC Verona (Pozniak et al., 2009) are of intermediate height, with only one semi-dwarf, namely AC Navigator (Clarke et al., 2000) which is in limited production. However, *Rht-B1* has been shown to have undesirable pleiotropic effects such as root biomass (Subira et al., 2016), test weight in some environments (McClung et al., 1986) and short coleoptile length, the latter undesirable for establishment in dry seedbeds. In Australia, breeding populations are being converted to the gibberellic acid-sensitive height-controlling genes, such as *Rht18*. *Rht18* allows for semi-dwarf height but expresses coleoptiles of normal length (Condon et al., 2004), theoretically circumventing the problems related to short coleoptile in rainfed environments with problems in soil moisture at sowing.

Phenological development, and in particular, timing and relative length of vegetative and reproductive growth stages are also important factors for maximizing durum wheat yield in a given environment. The reproductive phase is particularly critical in determining yield potential as it determines the duration of spike formation and allocation of resources to grain production (Royo et al., 2016). Modulating flowering time is an important breeding target for breeders and is an important stress-avoidance mechanism to limit exposure of the plant to abiotic stresses during flowering. Indeed, avoidance of high temperatures during flowering is critical, as even brief temperatures >30°C can significantly reduce fertility and yield (Wheeler et al., 2000). Flowering time in wheat is genetically complex, and is a function of the allelic state at three groups of loci: photoperiod sensitivity (*Ppd*), vernalization response (*Vrn*) and Earliness per se (*Eps*) (Sukumaran et al., 2016). Vernalization requirement is largely controlled by the *Vrn-1* loci, whose combined expression determines the length of the cold period necessary for a wheat plant to undergo flowering and complete its life cycle under optimal conditions. True 'winter' durum types, requiring obligatory and relatively long cold periods during their vegetative stage for proper flowering and grain production, are grown only in parts of Northern and Eastern Europe and parts of Turkey. Most durum cultivars grown worldwide are 'spring' types, with practically no significant vernalization requirements, having the ability to flower and complete their cycle with hardly any colder period. Intermediate or 'facultative' types exist and are grown in some areas in Europe and central Asia. In the central great plains of North America, there is little variation for vernalization response (Clarke et al., 1998a), which is not surprising given that most commercial durum wheat is planted to avoid periods of cold temperatures. In contrast, photoperiod sensitivity is a trait that drastically distinguishes different gene pools within spring durum wheat. Variation at the *Ppd-1* loci has been reported in durum wheat (Royo et al., 2016; Clarke et al., 1998a). For photoperiod-insensitive types, flowering will occur regardless of day length as the result of non-functional *Ppd-1* alleles at both homologous loci. In low-latitude environments, particularly in Mediterranean conditions, photoperiod-insensitive genotypes have been selected by breeders to escape the combined burden of terminal drought and heat stress that frequently occurs in those areas (Maccaferri et al., 2008). However, with increasing latitude, varying and generally increasing levels of photoperiod sensitivity become beneficial for durum wheat to optimally fit to its growing environment. At high latitudes, above 50°, photoperiod sensitivity becomes a definitive advantage as indicated by the predominance of photoperiod-sensitive types in the Canadian/North Dakota or Kazakhstan/Russia germplasm. In the Northern Great Plains of North America, most cultivars are photoperiod-sensitive, a strategy to delay flowering until mid-season rains.

3 Abiotic stresses on durum wheat

Drought is common in many durum-growing areas, including the Mediterranean region, Australia, Canada and the Western US, especially as the crop nears maturity. Breeders, therefore, target development of high-yielding cultivars that are also able to maintain acceptable performance when water is limited, avoiding very 'drought-tolerant' types that may not have superior yield potential when rainfall becomes favourable (non-responsive types). This is generally, and most effectively, done through multi-year, multi-location testing to assess performance stability over a wide range of rainfall regimes, including the response in dry years/environments (Hurd, 1974). In dry irrigated environments, drought-tolerant but responsive genotypes can be selected through the evaluation of their performance under both full and reduced irrigation (artificially generate hydric stress by withholding irrigation water). Early maturity can partially avoid terminal drought stress, but this earliness-driven drought escape capacity may come at the expense of grain yield. It is therefore important to select for both traits simultaneously and ideally try to identify genotypes with earliness or grain filling rate that vary on the basis of growing conditions. There is also hope that QTL associated with water-use efficiency or other drought-tolerance-related traits can be validated in different genetic backgrounds and then, when suitable markers are available, used in marker-assisted selection (MAS). Australian research shows that differences in carbon isotope discrimination can reveal genetic variability for stomatal conductance and hence water-use efficiency (Condon et al., 2004). Durum genotypes also display heritable genetic variability for osmotic adjustment, the control of osmotic pressure and therefore cellular turgor pressure against increasing water deficits (Morgan et al., 1986). Both of these traits have been used for selection in practical breeding programmes, in contrast to most putative drought-response traits suggested for indirect selection for grain yield.

Physiological approaches to improving yield on the basis of selection of proxy traits, traits with higher heritability and associated with lower G × E interactions than yield itself have the potential to complement traditional breeding approaches and possibly accelerate improvement in drought-prone environments (Richard et al., 2015). For example, leaf photosynthetic rate, leaf conductance and canopy temperature were all associated with yield progress in a set of eight spring bread wheat lines selected for improved yield potential (Fischer et al., 1998). Numerous searches for QTL for putative drought-related traits have been the focus of physiological genetic studies. Isidro et al. (2012a), for example, investigated QTL associated with leaf angle of durum and identified QTL associated with brassinosteroid concentrations in durum wheat leaves (Isidro et al., 2012b). This phytohormone is thought to be related to biomass production and stress tolerance. Other research (Peleg et al., 2009a; Merchuk-Ovnat et al., 2016) suggests that wild relatives such as emmer wheat, *Triticum turgidum* ssp. *dicoccoides*, have potential to contribute drought-adaptive QTLs to durum wheat. Graziana et al. (2014) found major QTLs on 2B and 3B related to peduncle length, a proxy for water availability, and 1000-kernel weight and grain yield in diverse environments. Validation of these reported QTLs could lead to their use in practical breeding programmes.

Root physiology and morphology are known to play an important role in drought adaptation in a number of species, including wheat (Richard et al., 2015; Manschadi et al., 2006). In particular, the ability of a plant to have its roots penetrate deeper in the soil profile where water is more readily available is a crucial drought response (Sanguineti et al., 2007). Kimurto et al. (2005) reported that water stress applied to wheat at seedling stage induced deeper roots and a larger root surface area. Also, the angle and number of seminal roots (i.e. those that develop directly from the embryo) has been suggested as targets for selection to improve drought tolerance in wheat breeding programmes (Manschadi et al., 2010). However, investigation of root morphology under field conditions is technically difficult and resource demanding (Cane et al., 2014), but this is expected to be less of a challenge as large-scale, high-throughput visualization technologies are developed (Richard et al., 2015; Atkinson et al., 2015).

From a genetic perspective, several studies identified QTLs for root traits in hexaploid wheat (Atkinson et al., 2015; Bai et al., 2013; Christopher et al., 2013; Petrarulo et al., 2015) and durum (Sanguineti et al., 2007; Cane et al., 2014). In nearly all of these studies, several dozen QTLs were identified, indicating complex genetic inheritance. In durum wheat, at least 30 QTLs for root growth angle and 20 QTL clusters for root length were identified using linkage and association mapping (Cane et al., 2014; Maccaferri et al., 2016). In a separate study, eight QTLs were identified using a bi-parental mapping population (Petrarulo et al., 2015). In hexaploid wheat, Atkinson et al. (2015) reported 29 QTLs for seedling root traits in a bi-parental mapping population, 11 of which localized to a common interval on chromosome 6D. Because this QTL resides on 6D, its utility in durum wheat breeding will be limited using conventional breeding approaches. However, there is emerging evidence that some of these QTLs show homology with hexaploid wheat (Petrarulo et al., 2015; Maccaferri et al., 2016), and these may be suggestive of a common genetic architecture that could be the target of MAS. Maccaferri et al. (2016) suggested those root-related QTLs most associated with grain yield could be the target of marker-assisted breeding, and reported three QTL clusters that may be of value for breeders. However, detailed evaluation of these QTLs in near-isogenic lines under stressed and non-stressed environments is still required to validate their use in wheat breeding programmes.

Boron (Bo) is an essential micronutrient for plants, and its deficiency can lead to induced male sterility in wheat (Rerkasem and Jamjod, 1997). Also, excessive soil boron concentration is a significant production problem in many parts of the world, particularly in regions with low rainfall and alkaline or volcanic soils (Rashid and Ryan, 2004; Schnurbusch et al., 2008). In wheat, boron toxicity decreases root growth, above-ground biomass and yield (Pallotta et al., 2014). Genetic mechanisms for boron tolerance have been identified on chromosomes 4A (*Bo4*) and 7B (*Bo1*) (Pallotta et al., 2014; Jefferies et al., 2000). Both genes have been cloned and shown to be a boron transporter-like gene (Pallotta et al., 2014). Current data show that tetraploid wheat lines are the likely sources of *Bo1* and *Bo4* in Bo-tolerant hexaploid bread wheat, stressing the importance of durum wheat as source of useful alleles for hexaploid wheat improvement. In both durum and common wheat, MAS for boron tolerance is effective and DNA markers are available to breeding programmes (Schnurbusch et al., 2008; Pallotta et al., 2014).

Soil salinity causes significant reductions in plant productivity and economic losses associated with reduced grain yield and quality of agricultural crops (Pitman and Lauchli, 2002). More than 6% of global agricultural land is affected by salinity or sodicity (Munns, 2005). Irrigated land is particularly prone to salinization, with about half the existing irrigation systems of the world now affected by salinization or waterlogging, due to low-quality irrigation water or to excessive leaching and subsequent rising water tables (Szabolcs, 1994). Durum is more sensitive to salinity than hexaploid wheat (Rawson et al., 1988), prompting research efforts to provide a solution. There is significant genetic variation for salt tolerance in durum based on Na^+ exclusion. The Na^+ exclusion trait eliminates sodium toxicity symptoms and premature senescence of leaves, allowing the flag leaf to remain active until maturity (Husain et al., 2003). The Na^+ exclusion trait is controlled by two interacting dominant genes (Munns et al., 2003), designated *Nax1* and *Nax2* (Lindsay et al., 2004; Huang et al., 2006). However, it remains to be seen if selection for the right alleles at these loci is effective in increasing actual durum yield under salinity conditions and is neutral to the expression of yield potential when salinity is not an issue.

4 Major diseases of durum wheat

4.1 The wheat rusts

The wheat rusts include leaf rust, yellow or stripe rust and stem rust, and all have historically been important biotic constraints in wheat-growing areas worldwide (Eversmeyer and Kramer, 2000). All three pathogens are capable of causing substantial economic losses, but their incidence varies with climate conditions, earliness and severity of the attack; the degree of resistance of the predominant cultivars; and the diverse ecological requirements of the pathogens (Kolmer, 1996; Vida et al., 2009; Huerta-Espino et al., 2011). Stem rust, caused by *Puccinia graminis* f. sp. *tritici*, is considered to be the most damaging of the wheat rusts because it attacks all the above-ground parts of the plant (Knott, 1989) and directly interferes with nutrients translocation in the stem. Disease spread is favoured by humid conditions and warmer temperatures of 15 to 35°C and may result in considerable amounts of kernel shrivelling and yield loss (Roelfs et al., 1992). Historically, stem rust was observed on durum wheat, especially when and where tall and late landraces were extensively grown (Mediterranean region primarily), providing green tissue for the pathogen to colonize late in the season when high temperatures prevailed. However, most modern durum germplasm, including the North American, CIMMYT/ICARDA and European germplasm groups, is not affected by stem rust in most growing areas worldwide, when sown at optimal planting times, with the notable exception of Ethiopia and Turkey. Even the appearance of the Ug99 lineage of races in Eastern Africa did not affect to a great extent durum wheat, as most germplasm groups have high frequencies of resistant genotypes when screened under artificial epidemics of TTKSK (one of the predominant *Ug99* races) in Kenya (Ammar, unpublished; Pozniak et al., 2008). However, when durum wheat is exposed to the Ethiopian races of stem rust, very little resistance is observed in any germplasm group. Olivera et al. (2012) reported that only 5.2% of a widely variable collection of 996 tetraploid wheats, including durum wild relatives, was found to be resistant to the Ethiopian races TRTTF and JRCQC, which have been shown to have combined virulence on *Sr13* and *Sr9e*, the former of which is still effective against the *Ug99* lineage. Virulence on *Sr13* of stem rust races affecting durum wheat have also been reported in Turkey (Mert et al., 2012) and very recently in Sicily (Hodson et al., 2017). On the basis of these results and others, international stem rust resistance breeding efforts are focusing primarily on breeding germplasm resistant to the Ethiopian races, as these possess the widest virulence spectrum.

Stripe or yellow rust, caused by the fungal pathogen *P. striiformis* Westend. f. sp. *tritici* Eriks, is reportedly adapted to cooler climates (2–15°C) and can be as destructive as stem rust under epidemic conditions. Yield losses up to 50% may occur due to shrivelled grain and damaged tillers (Roelfs et al., 1992). Randhawa et al. (2012) report that the incidence and severity of stripe rust increased dramatically in North America during 1999 because of the occurrence of new *P. striiformis* isolates showing significant adaptation to warmer temperatures and enhanced aggressiveness. Whereas durum wheat has been generally and globally highly resistant to stripe rust, it is risky to assume that the crop is reliably and durably protected from this pathogen. A relative increase in incidence of stripe rust was observed during 2015 in the CIMMYT breeding germplasm in Mexico (some 10% of advanced lines showing some susceptibility as opposed to less than 2% previously) and in North Africa, Spain and Italy (Ammar, pers. comm.). While there is no evidence that these observations are related to the appearance of durum-specific virulence (as in the case of leaf rust, see below), it indicates that, under highly favourable environmental conditions, the newly predominant races affecting bread wheat are starting to become more of a problem on durum wheat. Whatever the cause, it is important that the durum wheat community worldwide starts addressing the potential stripe rust problem in a preventative manner. Most of the numerous studies on the genetics of stripe rust resistance focused on hexaploid with very few addressing durum wheat (Bansal et al., 2014). Among 67 officially named Yr genes, only three, *Yr53*, *Yr64* and *Yr65*, were detected in durum wheat (Xu et al., 2013; Cheng et al., 2014). Bansal et al. (2014) detected QTL for stripe rust on chromosomes 2A, 3B and 5B of a bi-parental cross. Xue Lin et al. (unpublished) identified genomic regions for stripe rust resistance in durum wheat on 5A and 7B by association mapping and linkage mapping, with the 7B region explaining the greatest proportion of the variation. Interestingly, the location of the 7B resistance is coincident with Qlr.ubo-7B.2 and the location of a meta-QTL (Soriano and Royo, 2015), which suggests that this region is a genetic hotspot for rust resistance in durum wheat.

Leaf rust, caused by *Puccinia triticina* Eriks (formerly *P. recondita* f. sp. *tritici*), is the most regularly occurring of the three rusts found on wheat (Kolmer, 1996; Bolton et al., 2008). This disease poses a major threat to wheat production over large geographical areas. Common and durum wheat can suffer significant yield losses due to leaf rust, ranging from 1 to 40% (Ordonez and Kolmer, 2007). Total crop losses because of leaf rust are rare but yield reductions up to 62% are reported under conditions favourable for disease build-up (Sayre et al., 1998). The primary damage results from premature defoliation of the durum wheat plant, which leads to decreased numbers of kernels per head and to shrivelling of kernels. McCallum et al. (2007) reported that Canadian durum wheat cultivars, produced in dry areas, have traditionally been resistant to the local races of leaf rust; however, bread wheat cultivars have generally been susceptible with yield losses up to 20%.

The most cost-effective and environmentally safe means of combating these diseases is through the use of resistance in commercial cultivars (Vida et al., 2009; Ellis et al., 2007; Rosewarne et al., 2008). Generally, host resistance to the wheat rusts can be classified into two broad categories on the basis of the mode of inheritance and/or how resistance is expressed. Seedling resistance genes are effective for the whole lifecycle of the plant. Adult plant resistance (APR) is commonly detected at the post-seedling stage and often described as field resistance. APR genes have been shown to be an important part of durable leaf rust resistance (Todorovska et al., 2009). Of the few APR genes that have been studied in wheat, the largest proportion appear to be race-non-specific, but a limited number are clearly race-specific. Those of the race-non-specific class with partial resistance are associated with a slow rusting phenotype, showing longer latent periods, fewer uredinia and smaller uredinia sizes within the first two weeks post inoculation, when compared to susceptible plants (Rosewarne et al., 2008). The selection of wheat genotypes with a combination of non-race-specific genes defining durable resistance over years, as well as race-specific genes at the seedling stage, is a task of high importance for wheat breeding programmes.

To date, 72 designated genes for resistance to leaf rust (Lr) have been catalogued (McIntosh et al., 2013) and characterized for differential reaction to single leaf rust isolates. However, relatively little is known on the occurrence and nature of leaf rust resistance in durum wheat, with information being only recently generated in response to the dramatic increase in incidence on durum worldwide and the breakdown of resistance in several countries. For example, leaf rust epidemics during 2001 to 2003 severely affected an irrigated durum wheat production area of 250 000 hectares in north-western Mexico, causing estimated losses of at least US$32 million (Singh et al., 2004b). Among the known designated leaf rust resistance genes, only *Lr14a* and *Lr23* are reported to have originated from tetraploid wheat (Herrera-Foessel et al., 2005). Other Lr genes have been postulated to be present in durum and emmer wheat such as *Lr1, Lr25, Lr3, Lr10, Lr13, Lr16, Lr17a* and *Lr33* (Singh et al., 2004a; Shynbolar and Aralbek, 2010). Genes *Lr3* and *Lr14a* are effective against *P. triticina* race BBG/BN found in Mexico as well as against races with similar avirulence/virulence patterns prevailing in other durum wheat-producing countries. Virulence to *Lr10* and *Lr23* is frequent among *P. triticina* races collected from durum wheat (Herrera-Foessel et al., 2008a). A new leaf rust resistance gene, designated *Lr61*, was identified and mapped on the short arm of chromosome 6B of the Chilean cultivar Guayacan INIA, originating from the CIMMYT programme (Herrera-Foessel et al., 2008b). Maccaferri et al. (2010) used an association mapping approach on a collection of 164 elite durum wheat accessions to reveal that the major gene/gene cluster present in the *Lr14*/QLr.ubo-7B.2 region is the most important source of leaf rust resistance currently exploited by durum wheat breeders. A recent analysis identified 35 meta-QTL for leaf rust resistance on at least 17 wheat chromosomes (Soriano and Royo, 2015).

Early studies suggest that leaf rust resistance in Canadian durum wheat cultivars relies largely on single, race-specific gene resistance. This has been effective because leaf rust is rarely a concern in the semi-arid durum-producing regions of western Canada. However, the emergence of BBG/BN and BBG/BP (a BBG/BN mutant with added virulence on *Lr27+31*) in Mexico and the definitive trend towards a more dynamic and mutating *P. triticina* populations in the southern part of North America are of concern because nearly all commercially grown and newly released durum wheat cultivars from Canada and the United States are highly susceptible to these races. This stresses the importance of a proactive preventative breeding to introgress effective and preferably novel *Lr* genes into Canadian and North American germplasm to protect the North American crop from possible migration northwards of the Mexican races or the appearance of local ones with virulence on durum wheat. Marker-assisted breeding has provided a valuable approach for durum wheat breeders to monitor the incorporation and/or stacking of known and molecularly marked *Lr* genes into elite germplasm and tools to identify markers linked to putatively novel *Lr* genes. While most marker identification approaches are costly and time-consuming, identification and validation of single-nucleotide polymorphisms (SNPs) has rapidly improved in recent years mostly due to reduction of sequencing costs and increased throughput. Next-generation sequencing and SNP genotyping technologies are increasingly enhancing the viability and usefulness of MAS and marker-based breeding approaches as additional tools for breeders to use, particularly in breeding for rust resistance, with leaf resistance being a major target, thanks to the continuous identification and marking of new resistance genes from durum or tetraploid sources.

4.2 Leaf spotting complex

The leaf spot diseases, tan spot (*Pyrenophora tritici-repentis* (Died.) Drechs.), *Stagonospora nodorum* blotch (caused by *Phaeosphaeria nodorum* [anamorph: *Stagonospora nodorum* (Berk.) Castellani and Germano]) and *Septoria tritici* blotch (caused by *Mycosphaerella graminicola* [anamorph: *Septoria tritici* Rob. ex Desm.]), are prevalent in many durum production areas. Tan spot tends to be the most widespread and may cause yield reductions as high as 50% (Rees et al., 1982) in hexaploid wheat. Using near-isogenic pairs of durum lines, Fernandez et al. (2010) demonstrated that a 16% reduction in tan spot symptoms increased durum grain yield by 17%. Yield reduction is caused by factors such as reduced kernel weight (Fernandez et al., 2002). Increased incidence of tan spot is attributed to minimum tillage and retention of stubble (Strelkov and Lamari, 2003), making breeding for resistance a priority in many breeding programmes. Potential sources of resistance to tan spot and other leaf spot diseases exist in the major durum germplasm groups worldwide as well as in related tetraploid species and therefore usable genetic variability for resistance is not limiting (Singh et al., 2006a–c; Chu et al., 2008).

However, there are at least eight races of tan spot (Strelkov and Lamari, 2003), complicating breeding efforts and requiring local confirmation of effectiveness of sources identified from different environments. Therefore, DNA markers may play an important role in breeding for resistance. Relatively few QTL have been reported in durum compared to hexaploid wheat. Sensitivity to tan spot ToxA produced by race 2 isolates was shown to be controlled by a single dominant gene *Tsn1* on chromosome 5B (Faris et al., 1996). Singh et al. (2006a) reported a QTL for resistance to tan spot race 3 on the long arm of chromosome 3B of durum and QTL for races 3 and 5 on 3B (Singh et al., 2008). Chu et al. (2010) reported QTL for resistance to races 1 and 2 on chromosomes 3AS, 3BL, 5AL and 7BL. Our own unpublished results of association mapping in a durum diversity panel and a set of elite breeding lines detected marker–trait associations on all chromosomes, most of which corresponded to published QTL regions for hexaploid wheat.

Little is known about the genetics of resistance to *S. nodorum* in durum wheat. However, usable levels of resistance should be available in modern germplasm as indicated by results from screening of elite CIMMYT material (Singh and Ammar, unpublished).

Septoria tritici is a major disease of bread wheat globally but, to date, it is only locally important on durum wheat. While it has always been a very significant problem in Tunisia and Eastern Algeria, possibly the main yield-limiting biotic constraint in favourable years (Berraies et al., 2014b), only recently has it been reported as significant on durum wheat in other parts of North Africa and Southern Europe (Ammar, pers. comm.). Extensive testing of wide collection of modern and old durum germplasm from different geographical areas has yielded very few sources of resistance to the Tunisian strains of *S. tritici*. In fact, less than 4% of the modern CIMMYT or ICARDA germplasm tested showed acceptable levels of resistance under those conditions (Gharbi and Ammar, unpublished). In terms of the genetic basis of the inheritance of the very few durum sources characterized to date, results from research conducted in Tunisia indicate that it could be either qualitative or semi-quantitative. Ferjaoui et al. (2011) reported a monogenic inheritance of the complete resistance observed in the Tunisian landrace Agili. Berrais et al. (2014a) identified two major QTLs from the resistant cultivar Selim, one on 3BS and the other on 6BS, together explaining some 60% of the genetic variability for reaction to *Septoria*.

4.3 *Fusarium* head blight

Fusarium head blight (FHB), caused by *Fusarium* spp., is a significant problem in Europe and North America. FHB is a concern due to mycotoxins such as the trichothecene deoxynivalenol (DON) which is a significant food safety hazard. Many countries have recently tightened regulations for DON content in grain and food products. For example, the European Union regulation 856/2005 sets the maximum level of DON allowed in durum at 1.750 µg kg^{-1}. *Fusarium* damage also has negative effects on pasta quality, particularly colour (Dexter et al., 1997). Seventeen *Fusarium* species have shown association with the occurrence of FHB and several species co-exist (Parry et al., 1995), but *Fusarium graminearum* Schwabe (teleomorph *Gibberella zeae* Schwein. Petch) tends to be the most prevalent, followed by *F. culmorum*, and these two are the focus of most research. Twenty years ago FHB resistance was not a breeding objective in North America (Joppa and Williams, 1988), indicating how quickly the disease became commercially important.

Resistance to FHB in wheat is polygenic and strongly influenced by environment, both of which make evaluation and selection for resistance rather difficult. Relative to hexaploid wheat, durum is generally more susceptible to FHB under field conditions (Fernandez et al., 2002), and for the most part, breeding efforts to improve resistance have been slow (Prat et al., 2014), primarily due to the following two reasons. Firstly, FHB reaction is quantitatively expressed and genotype-by-environment interactions are significant for FHB-related traits, resulting in intermediate heritability across environments and low correlation among environments (Buerstmayr et al., 2009; Ruan et al., 2012). Secondly, no highly effective resistance to FHB has been identified in durum (Ban et al., 2005; Elias and Manthey, 2005) and none of those with moderate levels of resistance provides adequate protection under epidemic conditions (Clarke et al., 2010b).

There have been numerous attempts to introgress FHB resistance from various sources of hexaploid wheat to durum because most of the resistance QTL reported in hexaploid wheat are located on the A and B genomes (Buerstmayr et al., 2009). However it has been largely unsuccessful because of resistance suppression and/or lack of proper expression of the bread wheat genes in the durum background, and the transfer process is also complicated by the difference in ploidy levels, linkage drag between the two species and/or chromosome asynapsis (Prat et al., 2014; Cai et al., 2005). A possible explanation for the reduced expression of FHB resistance genes from bread wheat in durum could be due to the absence of the D-genome, believed to influence expression of FHB resistance (Fakhfakh et al., 2011). Nevertheless, Prat et al. (2016) recently confirmed the transfer of *Fhb1* from bread wheat to durum. Similarly, Giancaspro et al. (2016) reported reduced expression of FHB incidence and severity in recombinant inbred lines (RILs) developed from the cross of a resistant hexaploid and a susceptible durum. Most of the loci associated with the reduced symptoms corresponded to loci reported in the literature. The recent positional cloning of *Fhb1* will facilitate the design of perfect markers that will be useful in future breeding efforts (Rawat et al., 2016) and should be the target of studies to resolve its potential suppression in tetraploid wheat backgrounds.

Several studies have identified host resistance to FHB in tetraploid wheat that could be combined in an additive manner to improve resistance (Buerstmayr et al., 2012). To date, at least 11 QTLs with small to moderate effects were reported in tetraploid wheat (Prat et al., 2014 for review), many of which appear to be collinear with the QTL identified in hexaploid wheat. Stack et al. (2002) noted that substitution of chromosome 3A from *T. dicoccoides* into the durum wheat cultivar Langdon produced a moderate level of resistance. Somers et al. (2006) identified an FHB resistance QTL on 6B deriving from *T. carthlicum* accession 'Blackbird' and another on 2B derived from the durum parent. Ruan et al. (2012) identified two QTLs, one from *T. diccoccum*, that when combined reduce FHB infection by up to 80% under endemic conditions. While deployment of the secondary gene pool of durum wheat is expected to provide useful alleles for breeding, their use is likely to be slowed by the need to get rid of linkage drag with unfavourable agronomic characteristics (Prat et al., 2014).

Recent QTL mapping studies suggest that some durum wheat cultivars possess favourable alleles that could be useful for FHB resistance breeding. Ghavami et al. (2011)

reported a resistance source within Tunisian cultivated durum, with a major QTL on 5B, and possible suppressors of expression of resistance on 2A. Clarke et al. (2010b) reported intermediate resistance in a Canadian durum line 'DT696'. Stacking these alleles through a combination of phenotypic and MAS appears to be a promising approach to obtain transgressive segregants for FHB resistance in durum wheat breeding programmes. Alternatively, recurrent selection strategies could be employed to rapidly accumulate medium- to small-effect resistance alleles in breeding populations (Prat et al., 2014).

4.4 Other diseases

Viruses are a production constraint in some regions. The soil-borne wheat mosaic virus causes severe yield losses in central and northern Italy (Vallega and Rubies Autonell, 1985). Effective genetic resistance is available mostly conditioned by a major QTL on 2BS (Maccaferri et al., 2011). Barley yellow dwarf virus, transmitted by aphids, and wheat streak mosaic virus which is spread by curl mites, can occur throughout most major durum production regions, but the extent of damage tends to be highly variable.

5 Insect pests of durum wheat

The orange wheat blossom midge (OWBM), *Sitodiplosis mosellana* (Gehin), is a widespread problem in northern temperate wheat production worldwide, and is becoming increasingly problematic for durum produced in those regions. Midge larvae feed on developing wheat kernels, causing shrivelling, cracking and deformation (Lamb et al., 2000). In addition, wheat damaged by OWBM is generally characterized by kernels with dark flour colour, high flour ash, weak sticky dough properties and reduction of flour yield (Dexter et al., 1997). This results in economic losses because of reduced yield, crop quality and grade (Lamb et al., 2000). The surface discolouration of kernels is of particular concern for durum wheat because it reduces the aesthetic appearance in commercial pasta products (Dexter and Edwards, 1998). Damaged kernels produce semolina with high speck counts and reduced gluten strength in the case of nearly mature kernels damaged by feeding.

Genetic resistance to OWBM was identified for both antixenosis and antibiosis, which inhibit oviposition and larval growth, respectively (Thomas et al., 2005). Oviposition deterrence to female wheat midge was confirmed in cultivars of durum (Lamb et al., 2001) and hexaploid wheat (Lamb et al., 2002), but the genetic control and physiological basis of this deterrence is largely unknown. In contrast, antibiosis resistance to the OWBM has been well studied over the past decade, but to date, only a single gene (*Sm1*) has been identified conferring antibiosis-based resistance. *Sm1* has been mapped to chromosome 2BS and was identified in the hexaploid winter wheat cultivars Caldwell, Howell, Augusta and Seneca (Thomas et al., 2005). At the physiological level, *Sm1* appears to confer resistance by elevating levels of phenolic acids in the seed coat of infected seeds induced by feeding larvae (Ding et al., 2000).

To our knowledge, no antibiosis resistance has been identified in surveys of diverse durum germplasm (Lamb et al., 2001). Therefore, *Sm1* was transferred from the hexaploid wheat cultivar Caldwell to durum wheat using a combination of phenotypic selection and MAS, and durum cultivars with midge resistance have now been commercialized in Canada (Pozniak and Clarke, 2015; Singh et al., 2015). To manage virulence, these cultivars are marketed as varietal blends consisting of 90% of the midge resistance type and 10% of a susceptible type to reduce the chance of mutation in the insect to overcome the single resistance gene *Sm1* (Vera et al., 2013). Recently, a robust DNA marker for *Sm1* was reported in hexaploid wheat (Kassa et al., 2016), and we have determined that it is effective in selecting for *Sm1* in durum (Pozniak, unpublished results).

The wheat stem sawfly (*Cephus cinctus*) is a destructive insect pest of durum wheat in the Northern Great Plains of North America and serious infestations have been reported in Syria, Turkey, Jordan and North Africa (Damania et al., 1997). In Canada, severe infestations of sawfly have been reported in southern Alberta, Saskatchewan and eastern Manitoba since the early 1920s (Criddle, 1923). In the United States, areas most prone to sawfly damage include north and eastern Montana, North Dakota, northern South Dakota and western Minnesota (Beres et al., 2011). Growing solid-stemmed wheat cultivars that develop pith in the culm lumen has been the most effective strategy to minimize yield losses in North America for the greater part of the last century. Pith increases the occurrence of egg mortality through mechanical crushing (Holmes and Peterson, 1961) and acts as a barrier restricting larval movement inside the stem to within one or two internodes from the point of egg deposition (Holmes and Peterson, 1962). The survivorship of sawfly is negatively influenced by stem solidness and those insects that do survive have lower female mass and fecundity compared to those recovered from hollow-stemmed cultivars (Carcamo et al., 2005).

There are multiple sources of stem solidness in wheat and it remains unclear whether the underlying genetics are similar among sources. Most hexaploid wheat cultivars in North America derive their stem solidness from the Portuguese landrace S-615. The underlying genetics of stem solidness in the S-615 source are complex, and may include the action of a major gene, coupled with four or more additional recessive genes. In durum wheat, the most well-known source is derived from the South African cultivar Golden Ball. The German cultivar Biodur is a second source, which was bred into CDC Fortitude (Pozniak et al., 2015) and AAC Raymore (Singh et al., 2014) for production in high sawfly-pressure regions. Damania et al. (1997) reported sources of stem solidness in a collection of durum wheat landraces from Turkey, but it is not clear if these sources have been utilized in breeding programmes.

The genetics of solid stem expression is well characterized in both hexaploid and durum wheat. In hexaploid wheat, a major QTL that explains at least 76% of the variation for stem solidness was reported on chromosome 3BL (Cook et al., 2004; Talbert et al., 2014). A second QTL conferring stem solidness, Qss.msub-3DL, was localized to chromosome 3DL (Lanning et al., 2006). Genome-wide association mapping identified novel QTL for stem solidness on chromosomes 2A, 3A, 5B and 5D (Varella et al., 2015). Durum wheat has superior stem solidness to many hexaploid wheat cultivars (Nilsen et al., 2016), and phenotypic expression is controlled by a single dominant gene (Clarke et al., 2002a). To date, the only mapping work in durum wheat identified a single locus (*SSt1*) responsible for conferring stem solidness in the doubled-haploid population Kofa/W9262-260D3, and RIL populations Golden Ball/DT379//STD65 and G9580B-FE1C/AC Navigator (Houshmand et al., 2007). The authors suggest that W9262-260D3 (Kyle*2/Biodur) and Golden Ball carry the same source for stem solidness, although they did note differences in polymorphism for certain markers flanking the *SSt1* locus (Cook et al., 2004; Houshmand et al., 2007).

The Hessian fly (*Mayetiola destructor* Say) occurs throughout many durum-growing areas in North America, North Africa and Southwestern Europe (Nsarellah et al., 2005). The insect causes substantial economic losses in North Africa (Nsarellah et al., 2005), but is generally not a problem in North American durum production regions. There are numerous biotypes of the insect and single resistance genes are frequently overcome by mutation. Genes for resistance can be found in durum, hexaploid wheat and wild relatives, and combinations of these are deployed in areas prone to Hessian fly damage (Nsarellah et al., 2005). Several DNA markers were reported that will facilitate resistance breeding efforts, especially in areas where the insect is not reliably present (Dweikat et al., 2002, Liu et al., 2005).

Sunn pest (*Eurygaster* spp.) is widespread in wheat in the Mediterranean region and reduces grain yield as well as dough strength through secretion of proteolytic enzymes into kernels. There is evidence of variable resistance to the pest in durum (Arzani et al., 2001; El Bouhssini et al., 2009) and potential for selection in breeding programmes.

6 End-use quality of durum wheat

Durum wheat quality can be broadly categorized into physical and chemical quality attributes, and rheological and processing characteristics. Physical grain quality traits include test/hectolitre weight, kernel weight and proportion of vitreous kernels, and freedom from black point (Clarke et al., 1998b). Chemical grain quality traits include the content of yellow pigments, protein concentration and gluten quality. The latter two traits greatly influence the rheological properties of dough.

The physical quality of grain is an important grading factor because it can influence the semolina yield or milling potential of durum grain. Test weight is a measure of grain density and soundness or proper grain fill, and is widely utilized as a grading factor because it is easy to measure and has some relationship to semolina yield; however, there is no consensus on the use of test weight as a predictor of milling yield in wheat (Dexter and Edwards, 1998). Kernel weight is often associated with test weight, particularly in water-stressed or diseased environments where kernel shrivelling occurs. Larger kernels show negative association with protein concentration and gluten content and thus can have a pleiotropic effect on the rheological properties of semolina-produced dough. Both kernel weight and test weight are complexly inherited (Clarke et al., 2009a), but appear to be at least moderately heritable. Lebsock and Amaya (1969) found average heritabilities of 0.67 for test weight and 0.72 for kernel weight in four durum crosses. Marker–trait associations were reported on most chromosomes in a durum diversity panel (Maccaferri et al., 2011). In bi-parental populations, Elouafi and Nachit (2004) found QTL on 6B and 7A for test weight and on 3B, 4B and 6B for kernel weight, while Zhang et al. (2008) reported QTL on 1B and 2B for test weight and on 1A and 5A for kernel weight.

The yellow colour of semolina and pasta is an important end-use quality trait in international markets (Troccoli et al., 2000). The bright yellow colour of durum pasta is a function of the concentration of carotenoid pigments, mainly lutein in free ester form, present in the grain (Hentschel et al., 2002). In durum grain, the rate and duration of lutein accumulation is variable among genotypes expressing high, intermediate and low yellow pigment. Recent evidence suggests that the ratio of *trans*-zeaxanthin to *trans*-lutein is inversely correlated with total pigment and is consistent with the fact that the β-branch of lycopene cyclization leading to α-carotene and thus lutein synthesis is limiting in low-pigment genotypes (Ramachandran et al., 2010).

The genetics of yellow pigment are well understood in durum wheat and the trait is highly heritable (Elouafi et al., 2001; Clarke et al., 2006). QTL for yellow pigment were identified on most chromosomes, but a QTL with a large effect was identified in several mapping populations on the distal region of the long arm of chromosome 7B (Zhang et al., 2008; Elouafi et al., 2001; Pozniak et al., 2007). A homeologous QTL with a similar effect was also reported on 7AL (Singh et al., 2009). Additional QTL were identified on chromosomes 5A (Hessler et al., 2002), 3B (Blanco et al., 2011), 2A, 4B and 6B (Pozniak et al., 2007). The QTL on the group 7 chromosomes is most strongly associated with allelic variation in genes coding for phytoene synthase (Zhang et al., 2008; Pozniak et al., 2007; Singh et al., 2009), the first major enzyme in carotenoid biosynthesis, and this was confirmed using association mapping (Reimer et al., 2008). However, the final colour of processed pasta depends not only on total pigment content of the grain, but also on activity of the enzyme lipoxygenase during processing (Borrelli et al., 1999). More recently, Carrera et al. (2007) identified a QTL on chromosome 4B that is linked to a deletion at the *Lpx-B1* locus, which was associated with a 4.5-fold reduction in lipoxygenase activity and improved pasta colour, particularly under low-temperature drying of pasta. High-temperature drying (>90°C) of pasta is typically associated with improved pasta colour, likely because these temperatures are sufficient to neutralize the effects of lipoxygenase activity during pasta manufacture. In our research, we have shown that the partial deletion at Lpx-B1 improves pasta colour by as much as 20% when the pasta is dried at lower temperatures (Fu et al., 2011). The deletion can be selected with a DNA marker or by a simple test measuring the colour change of meal samples over time (Fu et al., 2011).

Grain protein concentration (GPC) and gluten quality are recognized as the most important determinants of the cooking quality of pasta products (D'Egidio et al., 1990) because these two factors influence the rheological properties of semolina dough. Both traits have been studied intensively (Pagnotta et al., 2005) and generally high GPC is associated with reduced pasta stickiness and increased tolerance to overcooking (Marchylo et al., 1998). An increase in GPC is usually associated with an increase in the amount of gluten proteins (Dexter and Dronzek, 1975). However, an increase in GPC does not guarantee good gluten strength (Ciaffi et al., 1991). High GPC is most important when high-temperature drying technology is used to improve the cooking quality, colour and nutritional quality of pasta, because high-temperature drying of pasta is known to modify the physicochemical characteristics of the gluten proteins (Destefanis and Sgrulletta, 1990). At low-temperature processing, both GPC and gluten quality have similar importance in determining pasta cooking quality (Dexter and Matsuo, 1980; Novaro et al., 1993). In breeding programmes, protein concentration can be readily measured on whole or ground grain using near-infrared spectroscopy (NIRS), but genetic gain from selection tends to be low because of the complexity of inheritance (Clarke et al., 2009b) and the large effect of environment on expression of protein accumulation and the need for waiting until the later stage of breeding for selection to be effective. Molecular approaches to improvement of GPC have been proposed, and QTL for elevated grain protein have been reported on at least 14 chromosomes in hexaploid and durum wheat (Suprayogi et al., 2009). In durum wheat, Blanco et al. (1996) reported QTL for GPC on 4BS, 5AL, 6AS, 6BS and 7BS. Additional QTL were reported on chromosome 7A (Suprayogi et al., 2009). In particular, introgression of the functional allele of *Gpc-B1* has been suggested as one strategy to increase GPC, given that this gene has been cloned and functional markers are available (Uauy et al., 2006). Indeed, introgression of *Gpc-B1* into adapted breeding lines has resulted in significant increases in GPC, wet gluten content and spaghetti firmness (Brevis et al., 2010). Perfect

SNP markers exist to reliably assist fast and effective introgression of this locus into elite germplasm. However, practical results obtained from breeding programmes in Canada (Pozniak et al. unpublished) and Mexico (Ammar, unpublished) indicate that the presence of *Gpc-B1* in elite durum wheat backgrounds is often associated with significant reductions in grain yield, grain weight, test weight and semolina yield and significant increases in ash concentration. This is consistent with other reports in the literature (Brevis et al., 2010).

In contrast to protein content, gluten strength is relatively easy to select for in durum breeding programmes. The SDS sedimentation test is widely used to predict gluten strength in early generation selection and requires as little as 1 g of whole meal (Dick and Quick, 1983). The gluten index (Cubadda et al., 1992) requires a larger sample and has a slightly larger labour requirement than the SDS sedimentation test. However, Clarke et al. (2010a) found that the gluten index was not affected by GPC and therefore provides a better measure of gluten strength across diverse environments compared to SDS sedimentation volume. Gluten strength is generally less affected by the environment than many of the other quality attributes, making it an objective of choice in breeding programmes. Because of this, most germplasm groups from programmes which have given any emphasis to this trait (North and South America, Europe, Australia, CIMMYT) have levels of gluten strength varying from acceptable to outstanding. This achievement is also due to the fact that gluten strength in durum is heavily influenced by a single locus, *Glu-B3*, on the short arm of chromosome 1B and coding for the low-molecular-weight glutenin (LMWG) subunits of the gluten polymeric protein, and that systematic selection against the allele of *Glu-B3* conferring weak gluten (LMWG-1 form) has been implemented by most programmes. Gluten strength is also affected by allelic variation at other loci, coding for other LMWG subunits (*Glu-A3*) as well as for high-molecular-weight subunits (*Glu-A1* and *Glu-B1*), but these have relatively limited effects in durum wheat. Gluten strength in bread wheat, which has two additional glutenin loci on the D-genome, is affected by allelic variation at these loci in a more balanced manner.

High yield of semolina with low ash concentration and good colour is important to the economic returns of the durum wheat milling industry. Matsuo et al. (1982) reported that year and location are the primary factors associated with variance in semolina yield, with genotype accounting for less than 15% of variance. In contrast, Chaurand et al. (1999) found that genetic differences in semolina yield exceeded those for environments, and suggested that breeding for semolina yield and other milling properties was worthwhile. However, semolina yield is difficult to select for in breeding programmes because of the high labour and large sample requirement for milling. Consequently, various methods were proposed to predict semolina yield, including micro-milling procedures (Vasiljevic et al., 1977) and other non-destructive tests such as NIRS (Ripetti-Ballester et al., 2000) and image analysis. McCaig et al. (2005) found that NIRS explained on average less than 60% of the variation for semolina yield. Similarly, Sissons et al. (2006) found NIRS to be a poor predictor of semolina yield of a large set of durum breeding lines. Sissons et al. (2000) evaluated the single kernel characterization system for prediction of semolina yield and suggested that it predicted a relatively small portion of the variation ($r = 0.25$). Novaro et al. (2001) assessed image analysis to predict semolina yield of durum and found that kernel shape and volume parameters in conjunction with test weight and kernel weight could explain about 30% of the variation in semolina yield.

Another approach to selection for semolina yield and other milling properties would be discovery of DNA markers that could be applied to breeding material. Several such studies in hexaploid wheat have identified QTL on the A and B genomes of hexaploid wheat. In various studies consensus has developed to indicate the importance of chromosomes 1B, 2A, 2B, 3B, 5A, 5B, 6A, 7A and 7B (Breseghello and Sorrells, 2006; Cabrera et al., 2015; Carter et al., 2012; Fox et al., 2013; Ishikawa et al., 2014; Kuchel et al., 2006; Kunert et al., 2007; Lehmeniek et al., 2006; McCartney et al., 2006; Parker et al., 1999; Raman et al., 2009; Smith et al., 2001; Tsilo et al., 2011) to flour yield of hexaploid wheat. However, there are no comparable reports in the literature of QTL for durum semolina yield, although our own unpublished research indicated that many of the regions identified in hexaploid wheat are also related to milling parameters in durum.

While not a quality trait as such, concentration of the non-essential heavy metal cadmium in foods including cereal grains is a long-standing food safety concern (Wiebe et al., 2010). Discussions under the FAO Codex Alimentarius seek to establish international standards, and, in 2003, the European Union set a limit of 0.2 $\mu g\ g^{-1}$ of cadmium in grains. Elevated natural levels of cadmium in soils of the North American durum production region posed a threat to durum exports. Efforts to provide a genetic solution to the problem were started in Canada in the early 1990s. Genetic variation for grain cadmium concentration was found in durum, with low concentration controlled by a single dominant gene (Clarke et al., 1997), *Cdu-B1*, located on the long arm of chromosome 5B (Wiebe et al., 2010; Knox et al., 2009). Incorporation of *Cdu-B1* into cultivars reduced average grain cadmium to concentrations well below international limits, averaging 50% lower than conventional durum cultivars. The allele for low concentration is specific for cadmium and has no effect on concentrations of other ions, and does not appear to be detrimental to any major economic traits (Clarke et al., 2002b). A DNA marker (Wiebe et al., 2010; Penner et al., 1995) is used to select for low grain cadmium concentration in Canadian and US durum breeding programmes. The low cadmium cultivar Strongfield (Clarke et al., 2005) entered commercial production in Canada in 2006 and covered more than 50% of the seeded area by 2008. All new durum cultivars registered for production in Canada and the Southwest of the United States must now be low cadmium types.

There are nutritional reasons to breed for improved micronutrient levels in crops (Welch and Graham 2004). Metals such as zinc and iron are important in human and animal nutrition, with good potential for manipulation through breeding (Cakmak et al., 2010) to enhance concentrations when deficient in local diets. Wild emmer wheat exhibits high levels of many micronutrients and could be used as a genetic resource for cultivated wheats (Cakmak et al. 2004). Peleg et al. (2009b) identified QTL associated with enhanced levels of minerals including Zn and Fe originating from wild emmer wheat to facilitate such efforts.

7 Durum wheat breeding technologies

7.1 Doubled haploids

Conventional breeding methods in northern hemisphere breeding programmes produce new cultivars in 10–16 years, depending on whether contra-season nurseries are utilized. Doubled haploids created from haploid wheat plants developed by anther culture or fertilization with maize pollen could reduce the development time further to eight years and are useful for genetic research. Widespread production of doubled-haploid hexaploid wheat by the wheat X maize system was reported some years ago (Howes et al., 1998;

Snape, 1998). However, efficiency of production of doubled haploids in durum to date is very low in comparison with that of bread wheat. Likewise, the generation of doubled haploids using microspore culture systems has been reported (Cistue et al., 2009), but these techniques are not widely deployed in durum wheat because of unpredictable performance and very limited research development efforts specific to this crop.

Maize-pollen-derived haploids were generally more successful than anther culture in durum wheat (Cherkaoui et al., 2000), although Cistue et al. (2009) reported some success with anther culture. The rate of doubled-haploid plant production in the durum X maize system was improved (Knox et al., 2009), bringing the system to practical usage. The hormone treatment was altered to use dicamba rather than 2,4-D, which gave a twofold increase in the number of doubled-haploid plants per emasculated spike. However, this rate is still about half that achieved for hexaploid wheat with the same system. There is also some effect of the durum genotype on doubled-haploid success rate (David et al., 1999; Knox et al., 2000).

Whereas cost still prohibits routine use of doubled haploids in some durum breeding programmes, they are acceptable for the production of genetic study populations or for generating progenies from crosses involving complex traits or traits that are difficult to measure. The combination of doubled haploids and MAS can be a very useful tool for efficient incorporation of complex traits (Howes et al., 1998). An anther culture-derived durum cultivar was commercialized in Spain (Cistue et al., 2009) and a maize-pollen-derived cultivar was commercialized in Canada (Singh et al., 2012).

7.2 Marker-assisted selection

The main advantage of marker-assisted breeding is to obtain genotypic data to predict phenotypic performance without the need for extensive phenotypic evaluation. Although molecular markers have been available for more than 25 years, they are only now being deployed widely in MAS programmes, largely to support introgression, stacking and maintenance of single-gene traits in wheat breeding (DePauw et al., 2011; Randhawa et al., 2013). For example, the Canadian durum wheat breeding programme routinely uses MAS for leaf rust resistance, resistance to insect pests, improved pasta colour and reduced grain cadmium concentrations (DePauw et al., 2011; Randhawa et al., 2013). More recently, the fine mapping of genes controlling quantitative traits and the application of high-throughput marker technologies (Jordan et al., 2015) have greatly expanded the applicability of MAS in plant breeding because markers tightly linked to the phenotype are becoming increasingly available (Randhawa et al., 2013). In most programmes, MAS is used to improve selection efficacy, and to assist in assessment of purity, parental selection, marker-assisted backcrossing and marker-assisted gene pyramiding of disease resistance (Collard and Mackill, 2008).

SNPs are the marker of choice to support marker-assisted breeding programmes (Rafalski, 2002). Recent advances in next-generation sequencing have significantly facilitated the discovery of single SNPs by whole-genome or reduced-representation sequencing in wheat (Jordan et al., 2015; Winfield et al., 2012). However, sequencing in crops with highly repetitive, complex genomes like that of durum wheat can still be challenging and costly. Genotyping-by-sequencing can be used for *de novo* genotyping and SNP discovery, and has been used effectively for the polyploid wheat genome (Poland et al., 2012). The recent development of high-density SNP arrays, including the 90K iSelect SNP assay (Wang et al., 2014), has resulted in the construction of a high-density SNP maps of durum wheat (Avni et al., 2014; Maccaferri et al., 2015) which is expected to accelerate marker and gene discovery in durum wheat.

Given the unprecedented ability to rapidly genotype large mapping populations with high marker density (Tuberosa and Pozniak, 2014), the number of studies reporting QTL in wheat are increasing manifold (Salvi and Tuberosa, 2015). Despite this, few have yielded results that are actually applied in practical durum wheat breeding programmes. Indeed, most quantitative traits of agronomic importance are controlled by many genes with strong environmental influence, and the identified QTLs explain only a small portion of the phenotypic variation present in breeding programmes. Secondly, most studies do not have sufficient phenotypic precision to accurately localize large-effect QTL (Collard and Mackill, 2008) or to detect small-effect QTL or QTL × QTL interactions (Salvi and Tuberosa, 2015). The role of epistasis in QTL variation is still poorly understood (Morgante and Salamini, 2003) and likely explains ascertainment bias where the effect of a QTL may be reduced in some genetic backgrounds (Holland, 2010). In addition, effective population sizes in breeding programmes increase by 50% with each additional single-marker selection, and selection for multiple QTLs can quickly result in unmanageable population sizes or a rapid loss of genetic variability if small populations are used. Managing large population sizes, however, will become routine as cost-effective, high-throughput DNA extraction systems are adopted (Zheng et al., 2015).

Genomic selection (GS) is a new approach to MAS designed to exploit high-throughput, inexpensive DNA marker information to accurately predict the genetic value of breeding candidates for complex traits. It is suggested as a novel approach to increase selection gain in crop breeding (Meuwissen et al., 2001; Heffner et al., 2009; Jannink et al., 2010). The GS method uses the joint effects of all marker loci to calculate a genomic estimated breeding value (GEBV) and enables capture of total genetic variation without phenotyping (Meuwissen et al., 2001). It is also used to improve accuracy of selection for phenotyped individuals even if entry-mean heritability is low (Endelman et al., 2014; Heslot et al., 2015). It is expected to be more effective than conventional MAS due to the inclusion of small-effect QTL that may otherwise be undetectable. The prediction of GEBV for selection candidates without phenotyping is proposed to save cost (Heffner et al., 2009) and speed up selection with the use of off-season nurseries, increasing genetic gain per unit time (Heffner et al., 2010). Complex, low heritability or negatively correlated traits can have greater genetic gains with GS over phenotypic methods if multiple selection cycles per year are used (Heffner et al., 2009).

GS shows promise to improve genetic gain per unit time and to reduce cost in crop improvement programmes (Heffner et al., 2010, 2011b) and is generating intense scientific research activity. Genomic prediction studies were conducted in many plant species, including wheat (Poland et al., 2012; Heffner et al., 2011a,b; Heslot et al., 2013; Charmet et al., 2014) and other cereals. GS has been applied to predict genotypic values for resistance to FHB and stem rust in wheat (Rutkoski et al., 2011). Meidaner et al. (2013) demonstrated the superiority of GS over association mapping for elucidating the genetics of resistance to *S. tritici* blotch. To our knowledge, however, there are no published results specifically dealing with GS in durum wheat.

Implementation of GS proceeds in two steps: 1. estimation of the effects of markers in a reference or training population and 2. prediction of GEBVs for individuals in the breeding population. The training population consists of well-adapted breeding lines and cultivars that are phenotyped in the target population of environments, whereas the breeding population consists of individuals that are genotyped but not phenotyped. To predict

GEBVs for individuals with genotypes but no phenotypes, the effect of markers they carry can be summed across the genome. GS can proceed using single markers, haplotypes of markers or an identical by descent approach. Most empirical studies used a prediction model that attempts to capture the total additive genetic variance to estimate breeding value of individuals on the basis of the sum of all marker effects. A number of GS models have been proposed for predicting phenotypes using large numbers of markers (Goddard and Hayes, 2007; Daetwyler et al., 2010; Perez et al., 2010; Thavamanikumar et al., 2015). These models mainly differ in the assumptions of marker effects contributing to total variance. The prediction model is then used to impose multiple generations of selection. The selected individuals can be recycled in the crossing programme and/or evaluated in advanced replicated field trials. Eventually, the best lines are added to the training population and the prediction model is updated (Sorrells, 2015). The accuracy of GEBVs is affected by training population size, the level and distribution of linkage disequilibrium between markers and QTLs, trait heritability, marker density, effective population size of breeding population (genetic diversity of breeding population), genetic relationship between training population and selection candidates, and statistical model.

Implementation of GS in durum wheat breeding will be contingent on the cost of genotyping the training and breeding populations, and must be competitive to the cost of phenotyping and the associated financial return on increased genetic gain via GS (Bassi et al., 2016). Because marker technology is continuously reducing the cost per data point and increasing the number of available markers, genotyping is currently less costly than phenotyping in an applied plant breeding programme. Therefore, it is believed that alternating progeny field testing with selection on the basis of only markers should reduce the expense and years involved in field testing, thereby greatly cutting the time needed to complete plant breeding cycles and bring new cultivars to market. These considerations, however, need to be empirically confirmed through a relevant and systematic comparison of selection effectiveness, actual genetic gains and actual costs of the GS-based and conventional breeding strategies.

8 Future trends and conclusion

The sequences of many crop genomes have been reported, and these provide a useful starting point to explore genome organization and evolution, and provide insight into genetic variation which can be manipulated by breeders. Indeed, next-generation sequencing technologies make it now possible to resequence entire plant genomes economically (Varshney et al., 2014; Bolger et al., 2014). However, a contiguous reference genome sequence anchored to genetic maps is still lacking for polyploid wheat (Eversole et al., 2014), largely because of the difficulties related to assembling a large (16 Gb) and highly repetitive (80%) genome. However, assemblies of the gene space are available (International Wheat Genome Sequencing, 2014) and a full-genome sequence of hexaploid wheat is expected in 2017 (IWGSC, pers. comm.). More recently, the first genome assembly of wild emmer (*Triticum turgidum* ssp. *dicoccoides*), the tetraploid ancestor species of durum wheat, was generated by the International Wild Emmer Wheat Genome Sequencing Consortium. The 10.1-Gigabase *de novo* assembly was generated using whole-genome shotgun sequencing coupled with a novel assembly algorithm. This assembly provides detailed information on gene content, genome architecture and global gene expression patterns (A. Distelfeld, pers. comm.). Using a similar approach to that for wild emmer wheat, the Durum Wheat Genome Sequencing Consortium has recently completed a high-quality reference sequence of the durum wheat cultivar Svevo (L. Catavelli, pers. comm.). Svevo is a semi-dwarf, spring habit adapted to some Mediterranean conditions and this cultivar has been used extensively for genetic analysis in durum wheat (Maccaferri et al., 2008; Avni et al., 2014; Maccaferri et al., 2015). The availability of these reference sequences will further marker applications, allow allele mining, QTL cloning, and support marker discovery for GS or sequence-based association mapping approaches (Bolger et al., 2014; Morgante and Salamini, 2003). Indeed, reduced-representation sequencing using exome capture of TILLING populations, such as those developed in the durum wheat cultivar Kronos (Uauy et al., 2009), is expected to provide a useful resource not only for functional genomic studies in durum wheat, but also for functional allele mining. A logical follow-up to the Svevo sequencing project is to generate whole-genome assemblies for multiple durum cultivars and selecting those that represent the genetic and functional diversity available (Barabaschi et al., 2016) in durum wheat worldwide. This will allow full characterization of the pan genome of durum. In *Arabidopsis*, sequencing of over 1100 accessions supported sequence-based association mapping approaches (Alonso-Blanco et al., 2016). A similar approach is being used in rice, where an initiative dedicated to the resequencing of 3000 rice accessions has been initiated to fully characterize the pan genome (Li et al., 2014).

Genome editing is a promising new breeding technology that will allow precise generation of new allelic variants for use in breeding programmes. Several systems of genome editing have been developed, including engineered homing endonucleases or meganucleases, zinc finger nucleases, transcription activator-like effector nucleases and clustered regularly interspersed short palindromic repeats (CRISPR)/Cas9 (see Voytas and Gao (2014) for review). All are based on cellular DNA repair pathways that direct repair of double-strand breaks at a locus where a DNA sequence modification is desired (Voytas and Gao, 2014). With CRISPR/Cas9, targeting for double-strand breaks is achieved through a guide RNA that pairs with a specific target sequence with high fidelity. Subsequent genome modification is via non-homologous end-joining or through homologous recombination.

The potential to develop alternate alleles for breeding using CRISPR/Cas9 has only been recently demonstrated for disease resistance in wheat (Wang et al., 2014) and increased grain size in both common and durum wheat (Zhang et al., 2016). However, there are still a number of factors that challenge its full implementation in wheat breeding programmes. Firstly, delivery of CRISPR/Cas9 expression complexes into host tissues can be problematic because durum wheat is generally recalcitrant to genetic transformation systems (Zhang et al., 2016). Secondly, genome editing relies on accurate sequence annotation for the precise determination of the site targeted for mutation. This can be particularly challenging if the target gene is part of a multigene family or if homeologus copies are present (Curtin et al., 2013), such as the case in the allotetraploid durum. Fortunately, the availability of annotated genome sequences for wild emmer and durum wheat (see above) should allow precise guiding of RNAs that can target a specific copy responsible for a phenotype. However, significant effort to link gene annotation to phenotypic function is still a massive task required before genome editing will become routine (Barabaschi et al., 2016). Indeed, the availability of TILLING mutants in durum wheat (see above) will aid this work. Thirdly, the final product of gene editing must be accepted by consumers and it is still not clear if the products of genome editing will be regulated as similar to transgenic plants (Voytas and Gao, 2014). This is particularly true if CRISPR/Cas9 DNA integrates into plant chromosomes, which can trigger legislative/political concerns (Jones, 2015). However, Araki and Ishii (2015) suggest that genome-editing events not based on the integration of DNA (vs. single-point mutations) should not be regulated as GMOs.

Hybrid wheat breeding is suggested as a viable approach to improving durum wheat yields (Gowda et al., 2010). To be effective, hybrid breeding requires an effective pollination control system and sufficient heterosis to outperform conventionally bred inbred cultivars (Duvick and Cassman, 1999). Durum wheat is an inbreeding species, but genetic systems of pollination control are available via alien cytoplasmic systems (Simons et al., 2003). However, because of the compact nature of the durum wheat inflorescence, coupled with a pseudo-cleistogamic nature of flowering and low pollen production (Pfeiffer et al., 2008), large-scale production of durum wheat hybrids is still challenging (Gowda et al., 2010). Early studies on heterosis in durum wheat have reported heterosis for grain yield up to 20% higher than the mid-parental value. Gowda et al. (2010) reported that on average, durum wheat hybrids yielded 10% higher than the mid-parent performance, and is similar to recent reports for common wheat and triticale (Dreisigacker et al., 2005; Fischer et al., 2010). While these results are promising, full implementation of hybrid breeding in durum will likely require methods to predict heterotic performance and develop contrasting heterotic groups, strategies for large-scale assessment of general and specific combining ability, and strategies that deal with seed production problems. In the case of hybrid performance, GS has been suggested as one method to not only predict hybrid performance, but as a means to predict contrasting heterotic groups for inbred line development (Zhao et al., 2015).

9 Where to look for further information

For further information on durum wheat, please see *Durum Wheat Breeding: Current Approaches and Future Strategies*, Volumes 1 and 2 (Conxita Royo, Miloudi M. Nachit, Natale Di Fonzo, José Luis Araus, Wolfgang H. Pfeiffer, Gustavo A. Slafer, Editors). This comprehensive resource summarizes the genetics and molecular biology, physiology and breeding strategies of durum wheat. Another useful resource is *Durum Wheat – Chemistry and Technology* – Second Edition (Michael J. Sissons, Joël Abecassis, Brian Marchylo, and Marina Carcea, Editors). This book covers all aspects of durum wheat production, agronomy and the chemical composition of the durum wheat grain. It is also an excellent resource for information on industrial approaches to processing durum wheat.

10 References

Ainslie, C. N. 2010. The Western Grass-stem Sawfly. Washington, DC., U.S. Dept. of Agriculture.

Alonso-Blanco, C., Andrade, J., Becker, C., Bemm, F., Bergelson, J., Borgwardt, K. M., et al. 2016. 1,135 Genomes reveal the global pattern of polymorphism in T thaliana. *Cell*, 166, 481–91.

Ammar, K., Gharbi, M. S. and Deghaies, M. 2011. Wheat in Tunisia. In: Bonjean, A. P., Angus, W. J. and Ginkel, M. V. (eds.), *The World Wheat Book*. Paris, France: Lavoisier Publishing.

Araki, M. and Ishii, T. 2015. Towards social acceptance of plant breeding by genome editing. *Trends in Plant Science*, 20, 145–9.

Arzani, A., Mohammadi, A. and Hatami, B. 2001. Evaluation of the resistance of wheat (bread and durum) and Aegilops germplasm to sunn pest (Eurygaster integriceps Put.). *Wheat in a Global Environment*, 9, 415–19.

Atkinson, J. A., Wingen, L. U., Griffiths, M., Pound, M. P., Gaju, O., Foulkes, M. J., Le Gouis, J., Griffiths, S., Bennett, M. J., King, J. and Wells, D. M. 2015. Phenotyping pipeline reveals major seedling root growth QTL in hexaploid wheat. *Journal of Experimental Botany*, 66, 2283–92.

Avni, R., Nave, M., Eilam, T., Sela, H., Alekperov, C., Peleg, Z., Dvorak, J., Korol, A. and Distelfeld, A. 2014. Ultra-dense genetic map of durum wheat × wild emmer wheat developed using the 90K iSelect SNP genotyping assay. *Molecular Breeding*, 34, 1549–62.

Bai, C. H., Liang, Y. L. and Hawkesford, M. J. 2013. Identification of QTLs associated with seedling root traits and their correlation with plant height in wheat. *Journal of Experimental Botany*, 64, 1745–53.

Ban, T., Kishii, M., Ammar, K., Murakami, J., Lewis, J., William, M., Pena, R. J., Payne, T., Singh, R. and Trethowan, R. CIMMYT'S challenges for global communications and germplasm enhancement for FHB resistance in durum and bread wheat. National Fusarium Head Blight Forum, 11–13 December 2005 Milwaukee, WI, pp. 6–10.

Barabaschi, D., Tondelli, A., Desiderio, F., Volante, A., Vaccino, P., Vale, G. and Cattivelli, L. 2016. Next generation breeding. *Plant Science*, 242, 3–13.

Bassi, F. M., Bentley, A. R., Charmet, G., Ortiz, R. and Crossa, J. 2016. Breeding schemes for the implementation of genomic selection in wheat (Triticum spp.). *Plant Science*, 242, 23–36.

Beres, B. L., Dosdall, L. M., Weaver, D. K., Carcamo, H. A. and Spaner, D. M. 2011. Biology and integrated management of wheat stem sawfly and the need for continuing research. *Canadian Entomologist*, 143, 105–25.

Berraies, S., Ammar, K., Salah Gharbi, M., Yahyaoui, A. and Rezgui, S. 2014a. Quantitative inheritance of resistance to Septoria tritici blotch in durum wheat in Tunisia. *Chilean Journal of Agricultural Research*, 74, 35–40.

Berraies, S., Salah Gharbi, M., Rezgui, S. and Yahyaoui, A. 2014b. Estimating grain yield losses caused by septoria leaf blotch on durum wheat in Tunisia. *Chilean Journal of Agricultural Research*, 74, 432–7.

Blanco, A., Colasuonno, P., Gadaleta, A., Mangini, G., Schiavulli, A., Simeone, R., Digesu, A. M., De Vita, P., Mastrangelo, A. M. and Cattivelli, L. 2011. Quantitative trait loci for yellow pigment concentration and individual carotenoid compounds in durum wheat. *Journal of Cereal Science*, 54, 255–64.

Bolger, A. M., Lohse, M. and Usadel, B. 2014. Trimmomatic: a flexible trimmer for Illumina sequence data. *Bioinformatics*, 30, 2114–20.

Bolton, M. D., Kolmer, J. A. and Garvin, D. F. 2008. Wheat leaf rust caused by Puccinia triticina. *Molecular Plant Pathology*, 9, 563–75.

Borrelli, G. M., Troccoli, A., Di Fonzo, N. and Fares, C. 1999. Durum wheat lipoxygenase activity and other quality parameters that affect pasta color. *Cereal Chemistry*, 76, 335–40.

Breseghello, F. and Sorrells, M. E. 2006. Association mapping of kernel size and milling quality in wheat (Triticum aestivum L.) cultivars. *Genetics*, 172, 1165–77.

Brevis, J. C., Morris, C. F., Manthey, F. and Dubcovsky, J. 2010. Effect of the grain protein content locus Gpc-B1 on bread and pasta quality. *Journal of Cereal Science*, 51, 357–65.

Buerstmayr, H., Ban, T. and Anderson, J. A. 2009. QTL mapping and marker-assisted selection for Fusarium head blight resistance in wheat: a review. *Plant Breeding*, 128, 1–26.

Buerstmayr, M., Huber, K., Heckmann, J., Steiner, B., Nelson, J. C. and Buerstmayr, H. 2012. Mapping of QTL for Fusarium head blight resistance and morphological and developmental traits in three backcross populations derived from Triticum dicoccum × Triticum durum. *Theoretical and Applied Genetics*, 125, 1751–65.

Cabrera, A., Guttieri, M., Smith, N., Souza, E., Sturbaum, A., Hua, D., Griffey, C., Barnett, M., Murphy, P., Ohm, H., Uphaus, J., Sorrells, M., Heffner, E., Brown-Guedira, G., Van Sanford, D.

and Sneller, C. 2015. Identification of milling and baking quality QTL in multiple soft wheat mapping populations. *Theoretical and Applied Genetics*, 128, 2227–42.

Cai, X., Chen, P. D., Xu, S. S., Oliver, R. E. and Chen, X. 2005. Utilization of alien genes to enhance Fusarium head blight resistance in wheat – A review. *Euphytica*, 142, 309–18.

Cakmak, I., Torun, A., Millet, E., Feldman, M., Fahima, T., Korol, A.,Nevo, E., Braun, H. J., and Ozkan, H. 2004. Triticum dicoccoides: An important genetic resource for increasing zinc and iron concentration in modern cultivated wheat. *Soil Science and Plant Nutrition*, 50, 1047–54.

Cakmak, I. Pfeiffer, W. H. and McClafferty, B. 2010. Biofortification of durum wheat with zinc and iron. *Cereal Chemistry*, 87, 10–20.

Cane, M. A., Maccaferri, M., Nazemi, G., Salvi, S., Francia, R., Colalongo, C. and Tuberosa, R. 2014. Association mapping for root architectural traits in durum wheat seedlings as related to agronomic performance. *Molecular Breeding*, 34, 1629–45.

Carcamo, H. A., Beres, B. L., Clarke, F., Byers, R. J., Mundel, H. H., May, K. and DePauw, R. 2005. Influence of plant host quality on fitness and sex ratio of the wheat stem sawfly (Hymenoptera: Cephidae). *Environmental Entomology*, 34, 1579–92.

Carrera, A., Echenique, V., Zhang, W., Helguera, M., Manthey, F., Schrager, A., Picca, A., Cervigni, G. and Dubcovsky, J. 2007. A deletion at the Lpx-B1 locus is associated with low lipoxygenase activity and improved pasta color in durum wheat (Triticum turgidum ssp durum). *Journal of Cereal Science*, 45, 67–77.

Carter, A. H., Garland-Campbell, K., Morris, C. F. and Kidwell, K. K. 2012. Chromosomes 3B and 4D are associated with several milling and baking quality traits in a soft white spring wheat (Triticum aestivum L.) population. *Theoretical and Applied Genetics*, 124, 1079–96.

Cattivelli, L., Baldi, P., Crosatti, C., Di Fonzo, N., Faccioli, P., Grossi, M., Mastrangelo, A. M., Pecchioni, N. and Stanca, A. M. 2002. Chromosome regions and stress-related sequences involved in resistance to abiotic stress in Triticeae. *Plant Molecular Biology*, 48, 649–65.

Charmet, G., Storlie, E., Oury, F. X., Laurent, V., Beghin, D., Chevarin, L., Lapierre, A., Perretant, M. R., Rolland, B., Heumez, E., Duchalais, L., Goudemand, E., Bordes, J. and Robert, O. 2014. Genome-wide prediction of three important traits in bread wheat. *Molecular Breeding*, 34, 1843–52.

Chaurand, M., Lempereur, I., Roulland, T. M., Autran, J. C. and Abecassis, J. 1999. Genetic and agronomic effects on semolina milling value of durum wheat. *Crop Science*, 39, 790–5.

Cherkaoui, S., Lamsaouri, O., Chlyah, A. and Chlyah, H. 2000. Durum wheat × maize crosses for haploid wheat production: Influence of parental genotypes and various experimental factors. *Plant Breeding*, 119, 31–6.

Christopher, J., Christopher, M., Jennings, R., Jones, S., Fletcher, S., Borrell, A., Manschadi, A. M., Jordan, D., Mace, E. and Hammer, G. 2013. QTL for root angle and number in a population developed from bread wheats (Triticum aestivum) with contrasting adaptation to water-limited environments. *Theoretical and Applied Genetics*, 126, 1563–74.

Chu, C. G., Chao, S., Friesen, T. L., Faris, J. D., Zhong, S. and Xu, S. S. 2010. Identification of novel tan spot resistance QTLs using an SSR-based linkage map of tetraploid wheat. *Molecular Breeding*, 25, 327–38.

Chu, C. G., Friesen, T. L., Faris, J. D. and Xu, S. S. 2008. Evaluation of seedling resistance to tan spot and Stagonospora nodorum blotch in tetraploid wheat. *Crop Science*, 48, 1107–16.

Cistue, L., Romagosa, I., Batlle, F. and Echavarri, B. 2009. Improvements in the production of doubled haploids in durum wheat (Triticum turgidum L.) through isolated microspore culture. *Plant Cell Reports*, 28, 727–35.

Clarke, F. R., Clarke, J. M., Ames, N. A., Knox, R. E. and Ross, R. J. 2010a. Gluten index compared with SDS-sedimentation volume for early generation selection for gluten strength in durum wheat. *Canadian Journal of Plant Science*, 90, 1–11.

Clarke, F. R., Clarke, J. M. and Knox, R. E. 2002a. Inheritance of stem solidness in eight durum wheat crosses. *Canadian Journal of Plant Science*, 82, 661–4.

Clarke, F. R., Clarke, J. M., McCaig, T. N., Knox, R. E. and DePauw, R. M. 2006. Inheritance of yellow pigment concentration in seven durum wheat crosses. *Canadian Journal of Plant Science*, 86, 133–41.

Clarke, F. R., Clarke, J. M., Pozniak, C. J. and Knox, R. E. 2009a. Inheritance of test weight and kernel weight in eight durum wheat crosses. *Canadian Journal of Plant Science*, 89, 1047–57.

Clarke, F. R., Clarke, J. M., Pozniak, C. J., Knox, R. E. and McCaig, T. N. 2009b. Protein concentration inheritance and selection in durum wheat. *Canadian Journal of Plant Science*, 89, 601–12.

Clarke, J., De Ambrogio, E., Hare, R. and Roument, P. 2012. Genetics and breeding of durum wheat. In: Sissons, M., Abecassis, J., Marchylo, B. and Carea, M. (eds.), *Durum Wheat: Chemistry and Technology*. 2nd Ed. ed. St. Paul, MN, USA: AACC International.

Clarke, J. M., Clarke, F. R. and Pozniak, C. J. 2010b. Forty-six years of genetic improvement in Canadian durum wheat cultivars. *Canadian Journal of Plant Science*, 90, 791–801.

Clarke, J. M., DePauw, R. M. and Thiessen, L. L. 1998a. Registration of wheat genetic stocks near-isogenic for photoperiod sensitivity. *Crop Science*, 38, 882.

Clarke, J. M., Leisle, D. and Kopytko, G. L. 1997. Inheritance of cadmium concentration in five durum wheat crosses. *Crop Science*, 37, 1722–6.

Clarke, J. M., Marchylo, B. A., Kovacs, M. I. P., Noll, J. S., McCaig, T. N. and Howes, N. K. 1998b. Breeding durum wheat for pasta quality in Canada (Reprinted from Wheat: Prospects for global improvement, 1998). *Euphytica*, 100, 163–70.

Clarke, J. M., McCaig, T. N., DePauw, R. M., Knox, R. E., Clarke, F. R., Fernandez, M. R. and Ames, N. P. 2005. Strongfield durum wheat. *Canadian Journal of Plant Science*, 85, 651–4.

Clarke, J. M., McLeod, J. G., DePauw, R. M., Marchylo, B. A., McCaig, T. N., Knox, R. E., Fernandez, M. R. and Ames, N. 2000. AC Navigator durum wheat. *Canadian Journal of Plant Science*, 80, 343–5.

Clarke, J. M., Norvell, W. A., Clarke, F. R. and Buckley, W. T. 2002b. Concentration of cadmium and other elements in the grain of near-isogenic durum lines. *Canadian Journal of Plant Science*, 82, 27–33.

Collard, B. C. Y. and Mackill, D. J. 2008. Marker-assisted selection: an approach for precision plant breeding in the twenty-first century. *Philosophical Transactions of the Royal Society B-Biological Sciences*, 363, 557–72.

Condon, A. G., Hare, R. and Rebetzke, G. J. A breeding program to improve the water use efficiency of durum wheat production. 11th Wheat Breeders Assembly, 21–24 September 2004 Canberra, Australia, pp. 108–11.

Cook, J. P., Wichman, D. M., Martin, J. M., Bruckner, P. L. and Talbert, L. E. 2004. Identification of microsatellite markers associated with a stem solidness locus in wheat. *Crop Science*, 44, 1397–402.

Criddle, N. 1923. The Life Habits of Cephus Cinctus Norton in Manitoba. *The Canadian Entomologist*, 55, 1–4.

Cubadda, R., Carcea, M. and Pasqui, L. A. 1992. Suitability of the gluten index method for assessing gluten strength in durum-wheat and semolina. *Cereal Foods World*, 37, 866–9.

Curtin, S. J., Anderson, J. E., Starker, C. G., Baltes, N. J., Mani, D., Voytas, D. F. and Stupar, R. M. 2013. Targeted mutagenesis for functional analysis of gene duplication in legumes. *Legume Genomics: Methods and Protocols*, 1069, 25–42.

D'Egidio, M., Mariani, B., Nardi, S., Novaro, P. and Cubadda, R. 1990. Chemical and technolgical variables and their relationships: a predictive value equation for pasta cooking quality. *Cereal Chemistry*, 67, 275–81.

Daetwyler, H. D., Pong-Wong, R., Villanueva, B. and Woolliams, J. A. 2010. The impact of genetic architecture on genome-wide evaluation methods. *Genetics*, 185, 1021–31.

Damania, A. B., Pecetti, L., Qualset, C. O. and Humeid, B. O. 1997. Diversity and geographic distribution of stem solidness and environmental stress tolerance in a collection of durum wheat landraces from Turkey. *Genetic Resources and Crop Evolution*, 44, 101–8.

David, J. L., Dusautoir, J. C., Raynaud, C. and Roumet, P. 1999. Heritable variation in the ability to produce haploid embryos via pollination with maize and embryo rescue in durum wheat. *Genome*, 42, 338–42.

De Vita, P., Nicosia, O. L., Nigro, F., Platani, C., Riefolo, C., Di Fonzo, N. and Cattivelli, L. 2007. Breeding progress in morpho-physiological, agronomical and qualitative traits of durum wheat cultivars released in Italy during the 20th century. *European Journal of Agronomy*, 26, 39–53.

DePauw, R. M., Knox, R. E., Humphreys, D. G., Thomas, J. B., Fox, S. L., Brown, P. D., Singh, A. K., Pozniak, C., Randhawa, H. S., Fowler, D. B., Graf, R. J. and Hucl, P. 2011. New breeding tools impact Canadian commercial farmer fields. *Czech Journal of Genetics and Plant Breeding*, 47, S28–S34.

Destefanis, E. and Sgrulletta, D. 1990. Effects of high-temperature drying on technological properties of pasta. *Journal of Cereal Science*, 12, 97–104.

Dexter, J. E. and Dronzek, B. L. 1975. Amino-acid composition of maturing endosperm from Hexaploid triticale and its spring rye and durum-wheat parents. *Canadian Journal of Plant Science*, 55, 537–46.

Dexter, J. E. and Edwards, M. N. 1998. The implications of frequently encountered grading factors on the processing quality of hexaploid wheat. Association of Operative Millers – Bulletin (June). Winnipeg, Manitoba: Canadian Grain Commission, Grain Research Laboratory.

Dexter, J. E., Marchylo, B. A., Clear, R. M. and Clarke, J. M. 1997. Effect of Fusarium head blight on semolina milling and pasta-making quality of durum wheat. *Cereal Chemistry*, 74, 519–25.

Dexter, J. E. and Matsuo, R. R. 1980. Relationship between durum-wheat protein-properties and pasta dough rheology and spaghetti cooking quality. *Journal of Agricultural and Food Chemistry*, 28, 899–902.

Dick, J. W. and Quick, J. S. 1983. A modified screening-test for rapid estimation of gluten strength in early-generation durum-wheat breeding lines. *Cereal Chemistry*, 60, 315–18.

Ding, H., Lamb, R. J. and Ames, N. 2000. Inducible production of phenolic acids in wheat and antibiotic resistance to Sitodiplosis mosellana. *Journal of Chemical Ecology*, 26, 969–85.

Dreisigacker, S., Melchinger, A. E., Zhang, P., Ammar, K., Flachenecker, C., Hoisington, D. and Warburton, M. L. 2005. Hybrid performance and heterosis in spring bread wheat, and their relations to SSR-based genetic distances and coefficients of parentage. *Euphytica*, 144, 51–9.

Duvick, D. N. and Cassman, K. G. 1999. Post-green revolution trends in yield potential of temperate maize in the north-central United States. *Crop Science*, 39, 1622–30.

Dweikat, I., Zhang, W. and Ohm, H. 2002. Development of STS markers linked to Hessian fly resistance gene H6 in wheat. *Theoretical and Applied Genetics*, 105, 766–70.

El Bouhssini, M., Street, K., Joubi, A., Ibrahim, Z. and Rihawi, F. 2009. Sources of wheat resistance to Sunn pest, Eurygaster integriceps Puton, in Syria. *Genetic Resources and Crop Evolution*, 56, 1065–9.

Elias, E. M. and Manthey, F. 2005. Durum wheat breeding at North Dakota State University. In: Royo, C., Nachit, M., di Fonzo, N., Araus, J. L., Pfeiffer, W. and Slafer, G. A. (eds.), *Durum Wheat Breeding: Current Approaches and Strategies*. New York: The Haworth Press.

Ellis, J. G., Mago, R., Kota, R., Dodds, P. N., McFadden, H., Lawrence, G., Spielmeyer, W. and Lagudah, E. 2007. Wheat rust resistance research at CSIRO. *Australian Journal of Agricultural Research*, 58, 507–11.

Elouafi, I. and Nachit, M. M. 2004. A genetic linkage map of the Durum × Triticum dicoccoides backcross population based on SSRs and AFLP markers, and QTL analysis for milling traits. *Theoretical and Applied Genetics*, 108, 401–13.

Elouafi, I., Nachit, M. M. and Martin, L. M. 2001. Identification of a microsatellite on chromosome 7B showing a strong linkage with yellow pigment in durum wheat (Triticum turgidum L. var. durum). *Hereditas*, 135, 255–61.

Endelman, J. B., Atlin, G. N., Beyene, Y., Semagn, K., Zhang, X. C., Sorrells, M. E. and Jannink, J. L. 2014. Optimal design of preliminary yield trials with genome-wide markers. *Crop Science*, 54, 48–59.

Eversmeyer, M. G. and Kramer, C. L. 2000. Epidemiology of wheat leaf and stem rust in the central Great Plains of the USA. *Annual Review of Phytopathology*, 38, 491–513.

Eversole, K., Feuillet, C., Mayer, K. F. X. and Rogers, J. 2014. Slicing the wheat genome. *Science*, 345, 285.

Fakhfakh, M. M., Yahyaoui, A., Rezgui, S., Elias, E. M. and Daaloul, A. 2011. Inheritances of Fusarium head blight resistance in a cross involving local and exotic durum wheat cultivars. *Crop Science*, 51, 2517–24.

Faris, J. D., Anderson, J. A., Francl, L. J. and Jordahl, J. G. 1996. Chromosomal location of a gene conditioning insensitivity in wheat to a necrosis-inducing culture filtrate from Pyrenophora tritici-repentis. *Phytopathology*, 86, 459–63.

Ferjaoui, S., Sbei, A., Aouadi, N. and Hamza, S. 2011. Monogenic inheritance of resistance to Septoria tritici blotch in durum wheat 'Agili'. *International Journal of Plant Breeding*, 5, 17–20.

Fernandez, M. R., Clarke, F. R., Knox, R. E., Clarke, J. M. and Singh, A. K. 2010. Quantification of effects of leaf spotting diseases on grain yield and market quality of durum wheat using near-isogenic lines. *Canadian Journal of Plant Pathology-Revue Canadienne De Phytopathologie*, 32, 177–87.

Fernandez, M. R., Clarke, J. M. and DePauw, R. M. 2002. The effect of plant height on tan spot on durum wheat in southern Saskatchewan. *Crop Science*, 42, 159–64.

Fischer, R. A., Rees, D., Sayre, K. D., Lu, Z. M., Condon, A. G. and Saavedra, A. L. 1998. Wheat yield progress associated with higher stomatal conductance and photosynthetic rate, and cooler canopies. *Crop Science*, 38, 1467–75.

Fischer, S., Maurer, H. P., Wurschum, T., Mohring, J., Piepho, H. P., Schon, C. C., Thiemt, E. M., Dhillon, B. S., Weissmann, E. A., Melchinger, A. E. and Reif, J. C. 2010. Development of Heterotic Groups in Triticale. *Crop Science*, 50, 584–90.

Fox, G. P., Martin, A., Kelly, A. M., Sutherland, M. W., Martin, D., Banks, P. M. and Sheppard, J. 2013. QTLs for water absorption and flour yield identified in the doubled haploid wheat population Lang/QT8766. *Euphytica*, 192, 453–62.

Fu, B. X., Schlichting, L., Pozniak, C. J. and Singh, A. K. 2011. A fast, simple, and reliable method to predict pasta yellowness. *Cereal Chemistry*, 88, 264–70.

Garcia del Moral, L., Royo, C. and Slafer, G. A. 2005. Genetic improvement effects on durum wheat yield physiology. In *Durum Wheat Breeding: Current Approaches and Strategies*. The Haworth Press, New York.

Ghavami, F., Elias, E. M., Mamidi, S., Ansari, O., Sargolzaei, M., Adhikari, T., Mergoum, M. and Kianian, S. F. 2011. Mixed model association mapping for Fusarium head blight resistance in Tunisian-Derived durum wheat populations. *G3-Genes Genomes Genetics*, 1, 209–18.

Giancaspro, A., Giove, S. L., Zito, D., Blanco, A. and Gadaleta, A. 2016. Mapping QTLs for fusarium head blight resistance in an interspecific wheat population. *Frontiers in Plant Science*, 7, 1381.

Goddard, M. E. and Hayes, B. J. 2007. Genomic selection. *Journal of Animal Breeding and Genetics*, 124, 323–30.

Gonzalez-Hernandez, J. L., Elias, E. M. and Kianian, S. F. 2004. Mapping genes for grain protein concentration and grain yield on chromosome 5B of Triticum turgidum (L.) var. dicoccoides. *Euphytica*, 139, 217–25.

Gowda, M., Kling, C., Wurschum, T., Liu, W., Maurer, H. P., Hahn, V. and Reif, J. C. 2010. Hybrid breeding in durum wheat: Heterosis and combining ability. *Crop Science*, 50, 2224–30.

Graziani M, Maccaferri M, Royo C, Salvatorelli F, Tuberosa R. 2014. QTL dissection of yield components and morpho-physiological traits in a durum wheat elite population tested in contrasting thermo-pluviometric conditions. *Crop and Pasture Science*, 65, 80–95.

Gupta, P. K., Varshney, R. K., Sharma, P. C. and Ramesh, B. 1999. Molecular markers and their applications in wheat breeding. *Plant Breeding*, 118, 369–90.

Heffner, E. L., Jannink, J. L., Iwata, H., Souza, E. and Sorrells, M. E. 2011a. Genomic selection accuracy for grain quality traits in Biparental wheat populations. *Crop Science*, 51, 2597–606.

Heffner, E. L., Jannink, J. L. and Sorrells, M. E. 2011b. Genomic selection accuracy using multifamily prediction models in a wheat breeding program. *Plant Genome*, 4, 65–75.

Heffner, E. L., Lorenz, A. J., Jannink, J. L. and Sorrells, M. E. 2010. Plant breeding with genomic selection: Gain per unit time and cost. *Crop Science*, 50, 1681–90.

Heffner, E. L., Sorrells, M. E. and Jannink, J. L. 2009. Genomic se ection for crop improvement. *Crop Science*, 49, 1–12.

Hentschel, V., Kranl, K., Hollmann, J., Lindhauer, M. G., Bohm, V. and Bitsch, R. 2002. Spectrophotometric determination of yellow pigment content and evaluation of carotenoids by high-performance liquid chromatography in durum wheat grain. *Journal of Agricultural and Food Chemistry*, 50, 6663–8.

Herrera-Foessel, S. A., Djurle, A., Yuen, J., Singh, R. P., William, H. M., Garcia, V. and Huerta-Espino, J. 2008a. Identification and molecular characterization of leaf rust resistance gene Lr14a in durum wheat. *Plant Disease*, 92, 469–73.

Herrera-Foessel, S. A., Singh, R. P., Huerta-Espino, J., William, H. M., Djurle, A. and Yuen, J. 2008b. Molecular mapping of a leaf rust resistance gene on the short arm of chromosome 6B of durum wheat. *Plant Disease*, 92, 1650–4.

Herrera-Foessel, S. A., Singh, R. P., Huerta-Espino, J., Yuen, J. and Djurle, A. 2005. New genes for leaf rust resistance in CIMMYT durum wheats. *Plant Disease*, 89, 809–14.

Heslot, N., Jannink, J. L. and Sorrells, M. E. 2015. Perspectives for genomic selection applications and research in plants. *Crop Science*, 55, 1–12.

Heslot, N., Rutkoski, J., Poland, J., Jannink, J. L. and Sorrells, M. E. 2013. Impact of marker ascertainment bias on genomic selection accuracy and estimates of genetic diversity. *PLoS ONE*, 8.

Hessler, T. G., Thomson, M. J., Benscher, D., Nachit, M. M. and Sorrells, M. E. 2002. Association of a lipoxygenase locus, Lpx-B1, with variation in lipoxygenase activity in durum wheat seeds. *Crop Science*, 42, 1695–700.

Holland, J. B. 2010. *Epistasis and Plant Breeding*. Plant Breeding Reviews. John Wiley and Sons, Inc.

Holmes, N. D. and Peterson, L. K. 1961. Resistance of spring wheats to the wheat stem sawfly, Cephus cinctus Nort. (Hymenoptera: Cephidae): I. Resistance to the egg. *The Canadian Entomologist*, 93, 250–60.

Holmes, N. D. and Peterson, L. K. 1962. Resistance of spring wheats to the wheat stem sawfly, Cephus ductus Nort. (Hymenoptera: Cephidae) II. Resistance to the Larva. *The Canadian Entomologist*, 94, 348–65.

Houshmand, S., Knox, R. E., Clarke, F. R. and Clarke, J. M. 2007. Microsatellite markers flanking a stem solidness gene on chromosome 3BL in durum wheat. *Molecular Breeding*, 20, 261–70.

Howes, N. K., Woods, S. M. and Townley-Smith, T. F. 1998. Simulations and practical problems of applying multiple marker assisted selection and doubled haploids to wheat breeding programs. *Euphytica*, 100, 225–30.

Huang, S. B., Spielmeyer, W., Lagudah, E. S., James, R. A., Platten, J. D., Dennis, E. S. and Munns, R. 2006. A sodium transporter (HKT7) is a candidate for Nax1, a gene for salt tolerance in durum wheat. *Plant Physiology*, 142, 1718–27.

Huerta-Espino, J., Singh, R. P., German, S., McCallum, B. D., Park, R. F., Chen, W. Q., Bhardwaj, S. C. and Goyeau, H. 2011. Global status of wheat leaf rust caused by Puccinia triticina. *Euphytica*, 179, 143–60.

Hurd, E. A. 1974. Phenotype and drought tolerance in wheat. *Agricultural Meteorology*, 14, 39–55.

Husain, S., Munns, R. and Condon, A. G. 2003. Effect of sodium exclusion trait on chlorophyll retention and growth of durum wheat in saline soil. *Australian Journal of Agricultural Research*, 54, 589–97.

International Wheat Genome Sequencing, C. 2014. A chromosome-based draft sequence of the hexaploid bread wheat (Triticum aestivum) genome. *Science* (New York, N.Y.), 345, 1251788.

Ishikawa, G., Nakamura, K., Ito, H., Saito, M., Sato, M., Jinno, H., Yoshimura, Y., Nishimura, T., Maejima, H., Uehara, Y., Kobayashi, F. and Nakamura, T. 2014. Association mapping and validation of QTLs for flour yield in the soft winter wheat variety Kitahonami. *PLoS ONE*, 9.

Isidro, J., Knox, R., Clarke, F., Singh, A., DePauw, R., Clarke, J. and Somers, D. 2012a. Quantitative genetic analysis and mapping of leaf angle in durum wheat. *Planta*, 236, 1713–23.

Isidro, J., Knox, R., Singh, A., Clarke, F., Krishna, P., DePauw, R., Clarke, J. and Somers, D. 2012b. Brassinosteroid leaf unrolling QTL mapping in durum wheat. *Planta*, 236, 273–81.

Jannink, J. L., Lorenz, A. J. and Iwata, H. 2010. Genomic selection in plant breeding: from theory to practice. *Briefings in Functional Genomics*, 9, 166–77.

Jauhar, P. P. and Peterson, T. S. 2013. Synthesis and characterization of advanced durum wheat hybrids and addition lines with Thinopyrum Chromosomes. *Journal of Heredity*, 104, 428–36.

Jefferies, S. P., Pallotta, M. A., Paull, J. G., Karakousis, A., Kretschmer, J. M., Manning, S., Islam, A. K. M. R., Langridge, P. and Chalmers, K. J. 2000. Mapping and validation of chromosome regions conferring boron toxicity tolerance in wheat (Triticum aestivum). *Theoretical and Applied Genetics*, 101, 767–77.

Joppa, L. and Williams, N. 1988. Genetics and breeding of durum in the United States. In: Fabriani, G. and Lintas, C. (eds.), *Durum Wheat Chemistry and Technology*. St. Paul, MN: Amer. Assoc. Cereal Chemists.

Jordan, K. W., Wang, S. C., Lun, Y. N., Gardiner, L. J., MacLachlan, R., Hucl, P., Wiebe, K., Wong, D., Forrest, K. L., Sharpe, A. G., Sidebottom, C. H. D., Hall, N., Toomajian, C., Close, T., Dubcovsky, J., Akhunova, A., Talbert, L., Bansal, U. K., Bariana, H. S., Hayden, M. J., Pozniak, C., Jeddeloh, J. A., Hall, A. and Akhunov, E. 2015. A haplotype map of allohexaploid wheat reveals distinct patterns of selection on homoeologous genomes. *Genome Biology*, 16.

Kassa, M. T., Haas, S., Schliephake, E., Lewis, C., You, F. M., Pozniak, C. J., Kramer, I., Perovic, D., Sharpe, A. G., Fobert, P. R., Koch, M., Wise, I. L., Fenwick, P., Berry, S., Simmonds, J., Hourcade, D., Senellart, P., Duchalais, L., Robert, O., Forster, J., Thomas, J. B., Friedt, W., Ordon, F., Uauy, C. and McCartney, C. A. 2016. A saturated SNP linkage map for the orange wheat blossom midge resistance gene Sm1. *Theoretical and Applied Genetics*, 129, 1507–17.

Kimurto, P. K., Kinyua, M. G., Ogola, J. B. O., Macharia, J. M. and Njau, P. N. 2005. Morphological traits associated with draught tolerance in bread wheat under tropical conditions. *Asian Journal of Plant Sciences*, 4, 284–92.

Kirigwi, F. M., Van Ginkel, M., Brown-Guedira, G., Gill, B. S., Paulsen, G. M. and Fritz, A. K. 2007. Markers associated with a QTL for grain yield in wheat under drought. *Molecular Breeding*, 20, 401–13.

Knott, D. R. 1989. *The Wheat Rusts: Breeding for Resistance*, Springer-Verlag, Berlin, Heidelberg.

Knox, R. E., Clarke, J. M. and DePauw, R. M. 2000. Dicamba and growth condition effects on doubled haploid production in durum wheat crossed with maize. *Plant Breeding*, 119, 289–98.

Knox, R. E., Pozniak, C. J., Clarke, F. R., Clarke, J. M., Houshmand, S. and Singh, A. K. 2009. Chromosomal location of the cadmium uptake gene (Cdu1) in durum wheat. *Genome*, 52, 741–7.

Kolmer, J. A. 1996. Genetics of resistance to wheat leaf rust. *Annual Review of Phytopathology*, 34, 435–55.

Kuchel, H., Langridge, P., Mosionek, L., Williams, K. and Jeffries, S. P. 2006. The genetic control of milling yield, dough rheology and baking quality of wheat. *Theoretical and Applied Genetics*, 112, 1487–95.

Kunert, A., Naz, A. A., Dedeck, O., Pillen, K. and Leon, J. 2007. AB-QTL analysis in winter wheat: I. Synthetic hexaploid wheat (T-turgidum ssp dicoccoides × T tauschii) as a source of favourable alleles for milling and baking quality traits. *Theoretical and Applied Genetics*, 115, 683–95.

Lamb, R. J., McKenzie, R. I. H., Wise, I. L., Barker, P. S., Smith, M. A. H. and Olfert, O. O. 2000. Resistance to Sitodiplosis mosellana (Diptera: Cecidomyiidae) in spring wheat (Gramineae). *Canadian Entomologist*, 132, 591–605.

Lamb, R. J., Smith, M. A. H., Wise, I. L., Clarke, P. and Clarke, J. 2001. Oviposition deterrence to Sitodiplosis mosellana (Diptera: Cecidomyiidae): a source of resistance for durum wheat (Gramineae). *Canadian Entomologist*, 133, 579–91.

Lamb, R. J., Wise, I. L., Smith, M. A. H., McKenzie, R. I. H., Thomas, J. and Olfert, O. O. 2002. Oviposition deterrence against Sitodiplosis mosellana (Diptera: Cecidomyiidae) in spring wheat (Gramineae). *Canadian Entomologist*, 134, 85–96.

Lanning, S. P., Fox, P., Elser, J., Martin, J. M., Blake, N. K. and Talbert, L. E. 2006. Microsatellite markers associated with a secondary stem solidness locus in wheat. *Crop Science*, 46, 1701–3.

Lebsock, K. L. and Amaya, A. 1969. Variation and covariation of agronomic traits in durum wheat. *Crop Science*, 9, 372–5.

Lehmensiek, A., Eckermann, P. J., Verbyla, A. P., Appels, R., Sutherland, M. W., Martin, D. and Daggard, G. E. 2006. Flour yield QTLs in three Australian doubled haploid wheat populations. *Australian Journal of Agricultural Research*, 57, 1115–22.

Li, J. Y., Wang, J. and Zeigler, R. S. 2014. The 3,000 rice genomes project: new opportunities and challenges for future rice research. *Gigascience*, 3.

Lindsay, M. P., Lagudah, E. S., Hare, R. A. and Munns, R. 2004. A locus for sodium exclusion (Nax1), a trait for salt tolerance, mapped in durum wheat. *Functional Plant Biology*, 31, 1105–14.

Liu, X. M., Fritz, A. K., Reese, J. C., Wilde, G. E., Gill, B. S. anc Chen, M. S. 2005. H9, H10, and H11 compose a cluster of Hessian fly-resistance genes in the distal gene-rich region of wheat chromosome 1AS. *Theoretical and Applied Genetics*, 110, 1473–80.

Maccaferri, M., El-Feki, W., Nazemi, G., Salvi, S., Cane, M. A., Colalongo, M. C., Stefanelli, S. and Tuberosa, R. 2016. Prioritizing quantitative trait loci for root system architecture in tetraploid wheat. *Journal of Experimental Botany*, 67, 1161–78.

Maccaferri, M., Ricci, A., Salvi, S., Milner, S. G., Noli, E. et al. 2015. A high-density, SNP-based consensus map of tetraploid wheat as a bridge to integrate durum and bread wheat genomics and breeding. *Plant Biotechnology Journal*, 13, 648–63.

Maccaferri, M., Sanguineti, M. C., Corneti, S., Ortega, J. L. A., Ben Salem, M., Bort, J., DeAmbrogio, E., del Moral, L. F. G., Demontis, A., El-Ahmed, A., Maalouf, F., Machlab, H., Martos, V., Moragues, M., Motawaj, J., Nachit, M., Nserallah, N., Ouabbou, H., Royo, C., Slama, A. and Tuberosa, R. 2008. Quantitative trait loci for grain yield and adaptation of durum wheat (Triticum durum Desf.) across a wide range of water availability. *Genetics*, 178, 489–511.

Maccaferri, M., Sanguineti, M. C., Demontis, A., El-Ahmed, A., del Moral, L. G., Maalouf, F., Nachit, M., Nserallah, N., Ouabbou, H., Rhouma, S., Royo, C., Villegas, D. and Tuberosa, R. 2011. Association mapping in durum wheat grown across a broad range of water regimes. *Journal of Experimental Botany*, 62, 409–38.

Maccaferri, M, Ratti, C, Rubies-Autonell, C, Vallega, V, Demontis, A, Stefanelli, S, Tuberosa, R and Sanguineti, MC. 2011. Resistance to soil-borne cereal mosaic virusin durum wheat is controlled by a major QTL on chromosome arm 2BS and minor loci. *Theoretical and Applied Genetics*, 123, 527–44.

Maccaferri, M., Sanguineti, M. C., Mantovani, P., Demontis, A., Massi, A., Ammar, K., Kolmer, J. A., Czembor, J. H., Ezrati, S. and Tuberosa, R. 2010. Association mapping of leaf rust response in durum wheat. *Molecular Breeding*, 26, 189–228.

Manschadi, A. M., Christopher, J., Devoil, P. and Hammer, G. L. 2006. The role of root architectural traits in adaptation of wheat to water-limited environments. *Functional Plant Biology*, 33, 823–37.

Manschadi, A. M., Christopher, J. T., Hammer, G. L. and Devoil, P. 2010. Experimental and modelling studies of drought-adaptive root architectural traits in wheat (Triticum aestivum L.). *Plant Biosystems*, 144, 458–62.

Marchylo, B., Dexter, J. E., Clark, J. and Ames, N. Effects of protein content on CWAD quality. In: Fowler, B., Geddes, W. E., Johnston, A. M. and Preston, K. R. (eds.), *Wheat Protein Production and Marketing*. Wheat Protein Symposium, 9–10 March 1998 Saskatoon, Saskatchewan. University of Extension Press. University of Saskatchewan.

Matsuo, R. R., Dexter, J. E., Kosmolak, F. G. and Leisle, D. 1982. Statistical evaluation of tests for assessing spaghetti-making quality of durum-wheat. *Cereal Chemistry*, 59, 222–8.

McCaig, T. N., Clarke, J., Dexter, J. E. and DePauw, R. Potential of VIS/NIR Spectroscopy to predict milling yield and other quality traits in a durum wheat breeding program. Cereals – the future challenge, ICC Jubilee Conference, 3–6 July 2005 Vienna, Austria, p. 120.

McCartney, C. A., Somers, D. J., Lukow, O., Ames, N., Noll, J., Cloutier, S., Humphreys, D. G. and McCallum, B. D. 2006. QTL analysis of quality traits in the spring wheat cross RL4452 × 'AC Domain'. *Plant Breeding*, 125, 565–75.

McClung, A. M., Cantrell, R. G., Quick, J. S. and Gregory, R. S. 1986. Influence of the Rht1 Semidwarf gene on yield, yield components, and grain protein in durum-wheat. *Crop Science*, 26, 1095–9.

McIntosh, R. A., Dubcovsky, J., Rogers, J., Morris, C. F., Appels, R. and Xia, X. C. Catalogue of gene symbols for wheat: 2013–14 Supplement. 12th International Wheat Genetics Symposium, 2013 Yokohama, Japan.

Merchuk-Ovnat, L., Barak, V., Fahima, T., Ordon, F., Lidzbarsky, G. A., Krugman, T. and Saranga, Y. 2016. Ancestral QTL alleles from wild emmer wheat improve drought resistance and productivity in modern wheat cultivars. *Frontiers in Plant Science*, 7.

Mert, Z., Karakaya, A., Dusunceli, F., Akan, K. and Cetin, L. 2012. Determination of Puccinia graminis f. sp. tritici races of wheat in Turkey. *Turkish Journal of Agriculture and Forestry*, 36, 107–20.

Meuwissen, T. H. E., Hayes, B. J. and Goddard, M. E. 2001. Prediction of total genetic value using genome-wide dense marker maps. *Genetics*, 157, 1819–29.

Miedaner, T., Zhao, Y. S., Gowda, M., Longin, C. F. H., Korzun, V., Ebmeyer, E., Kazman, E. and Reif, J. C. 2013. Genetic architecture of resistance to Septoria tritici blotch in European wheat. *Bmc Genomics*, 14.

Morgan, J. M., Hare, R. A. and Fletcher, R. J. 1986. Genetic-variation in osmoregulation in bread and durum wheats and its relationship to grain-yield in a range of field environments. *Australian Journal of Agricultural Research*, 37, 449–57.

Morgante, M. and Salamini, F. 2003. From plant genomics to breeding practice. *Current Opinion in Biotechnology*, 14, 214–19.

Munns, R. 2005. Genes and salt tolerance: bringing them together. *New Phytologist*, 167, 645–63.

Munns, R., Rebetzke, G. J., Husain, S., James, R. A. and Hare, R. A. 2003. Genetic control of sodium exclusion in durum wheat. *Australian Journal of Agricultural Research*, 54, 627–35.

Nilsen, K. T., Clarke, J. M., Beres, B. L. and Pozniak, C. J. 2016. Sowing density and cultivar effects on pith expression in solid-stemmed durum wheat. *Agronomy Journal*, 108, 219–28.

Novaro, P., Colucci, F., Venora, G. and D'Egidio, M. G. 2001. Image analysis of whole grains: A noninvasive method to predict semolina yield in durum wheat. *Cereal Chemistry*, 78, 217–21.

Novaro, P., Degidio, M. G., Mariani, B. M. and Nardi, S. 1993. Combined effect of protein-content and high-temperature drying systems on pasta cooking quality. *Cereal Chemistry*, 70, 716–19.

Nsarellah, N., Lhaloui, S., El Bouhssini, M., Amri, A. and Nachit, M. 2005. Basis for insect pest resistance in durum wheat. In: Royo, C., Nachit, M., di Fonzo, N., Araus, J. L., Pfeiffer, W. and Slafer, G. A. (eds.), *Durum Wheat Breeding: Current Approaches and Strategies*. New York: The Haworth Press.

Ordonez, M. E. and Kolmer, J. A. 2007. Virulence phenotypes of a worldwide collection of Puccinia triticina from durum wheat. *Phytopathology*, 97, 344–51.

Pagnotta, M. A., Blanco, A., Gadaleta, A. and Fares, C. 2005. Functional determinants of grain quality. In: Royo, C., Nachit, M., Di Fonzo, N., Araus, J. L., Pfeiffer, W. and Slafer, G. A. (eds.), *Durum Wheat Breeding: Current Approaches and Future trategies*. CRC Press, Boca Raton.

Pallotta, M., Schnurbusch, T., Hayes, J., Hay, A., Baumann, U., Paull, J., Langridge, P. and Sutton, T. 2014. Molecular basis of adaptation to high soil boron in wheat landraces and elite cultivars. *Nature*, 514, 88-+.

Parikh K. S., Mehrota, A. K., Russo M. J., Lang RM, Anderson A et al. Percutaneous transcatheter aortic valve closure successfully treats left ventricular assist device associated aortic insufficiency and improves cardiac hemodynamics. JACC Cardiovasc. Interv. 2013;6:84-9.

Parker, G. D., Chalmers, K. J., Rathjen, A. J. and Langridge, P. 1999. Mapping loci associated with milling yield in wheat (Triticum aestivum L.). *Molecular Breeding*, 5, 561–8.

Parry, D. W., Jenkinson, P. and Mcleod, L. 1995. Fusarium Ear Blight (Scab) in Small-Grain Cereals – a review. *Plant Pathology*, 44, 207–38.

Pecetti, L. and Annicchiarico, P. 1998. Agronomic value and plant type of Italian durum wheat cultivars from different eras of breeding. *Euphytica*, 99, 9–15.

Peleg, Z., Fahima, T., Krugman, T., Abbo, S., Yakir, D., Korol, A. B. and Saranga, Y. 2009a. Genomic dissection of drought resistance in durum wheat × wild emmer wheat recombinant inbreed line population. *Plant Cell and Environment*, 32, 758–79.

Peleg, Z., Cakmak I Ozturk L Yazici A Jun Y Budak H Fahima, T.D., Korol, A. B. and Saranga, Y. 2009b. Quantitative trait loci conferring grain mineral nutrient concentrations in durum wheat × wild emmer wheat RIL population. *Theoretical and Applied Genetics* 119, 353–69.

Penner, G. A., Clarke, J., Bezte, L. J. and Leisle, D. 1995. Identif cation of Rapd markers linked to a gene governing cadmium uptake in durum-wheat. *Genome*, 38, 543–7.

Perez, P., de los Campos, G., Crossa, J. and Gianola, D. 2010. Genomic-enabled prediction based on molecular markers and pedigree using the Bayesian linear regression package in R. *Plant Genome*, 3, 106–16.

Petrarulo, M., Marone, D., Ferragonio, P., Cattivelli, L., Rubiales, D., De Vita, P. and Mastrangelo, A. M. 2015. Genetic analysis of root morphological traits in wheat. *Molecular Genetics and Genomics*, 290, 785–806.

Pfeiffer, W., Trethowan, R., Ammar, K. and Sayre, K. D. 2008. Increasing yield potential and stability in durum wheat. In: Royo, C. and di Fonzo, N. (eds.), *Durum Wheat Breeding: Current Approaches and Future Strategies*. CRC Press, Boca Raton.

Pitman, M. G. and Lauchli, A. 2002. Global impacts of salinity and agricultural ecosystems. In: Lauchli, A. and Luttge, U. (eds.), *Salinity: Evnironment-Plants-Molecules*. Springer, The Netherlands.

Poland, J., Endelman, J., Dawson, J., Rutkoski, J., Wu, S. Y., Manes, Y., Dreisigacker, S., Crossa, J., Sanchez-Villeda, H., Sorrells, M. and Jannink, J. L. 2012. Genomic selection in wheat breeding using genotyping-by-sequencing. *Plant Genome*, 5, 103–13.

Pozniak, C., Reimer, S., Fetch, T., Clark, J., Clarke, F., Somers, D., Knox, R. and Singh, A. Association mapping of Ug99 resistance in diverse durum wheat population. In: Rudi, A., Russell, E., Peter, L., Michael, M., Lynn, M. and Peter, S. (eds.), Proceedings of the 11th International Wheat Genetics Symposium, 24–29 August 2008 Brisbane, Australia, pp. 809–11.

Pozniak, C. J. and Clarke, J. M. 2015. CDC Carbide durum wheat. *Canadian Journal of Plant Science*, 95, 1007–12.

Pozniak, C. J., Fox, S. L. and Knott, D. R. 2009. CDC Verona durum wheat. *Canadian Journal of Plant Science*, 89, 321–4.

Pozniak, C. J., Knox, R. E., Clarke, F. R. and Clarke, J. M. 2007. Identification of QTL and association of a phytoene synthase gene with endosperm colour in durum wheat. *Theoretical and Applied Genetics*, 114, 525–37.

Pozniak, C. J., Nilsen, K., Clarke, J. M. and Beres, B. L. 2015. CDC Fortitude durum wheat. *Canadian Journal of Plant Science*.

Prat, N., Buerstmayr, M., Steiner, B., Robert, O. and Buerstmayr, H. 2014. Current knowledge on resistance to Fusarium head blight in tetraploid wheat. *Molecular Breeding*, 34, 1689–99.

Prat, N., Guilbert, C., Prah, U., Wachter, E., Steiner, B., Langin, T., Robert, O. and Buerstmayr, H. 2016. QTL mapping of Fusarium head blight resistance in three related durum wheat populations. *Theoretical and Applied Genetics*.

Rafalski, A. 2002. Applications of single nucleotide polymorphisms in crop genetics. *Current Opinion in Plant Biology*, 5, 94–100.

Ramachandran, A., Pozniak, C. J., Clarke, J. M. and Singh, A. K. 2010. Carotenoid accumulation during grain development in durum wheat. *Journal of Cereal Science*, 52, 30–8.

Raman, R., Allen, H., Diffey, S., Raman, H., Martin, P. and McKelvie, K. 2009. Localisation of quantitative trait loci for quality attributes in a doubled haploid population of wheat (Triticum aestivum L.). *Genome*, 52, 701–15.

Randhawa, H., Puchalski, B. J., Frick, M., Goyal, A., Despins, T., Graf, R. J., Laroche, A. and Gaudet, D. A. 2012. Stripe rust resistance among western Canadian spring wheat and triticale varieties. *Canadian Journal of Plant Science*, 92, 713–22.

Randhawa, H. S., Asif, M., Pozniak, C., Clarke, J. M., Graf, R. J., Fox, S. L., Humphreys, D. G., Knox, R. E., DePauw, R. M., Singh, A. K., Cuthbert, R. D., Hucl, P. and Spaner, D. 2013. Application of molecular markers to wheat breeding in Canada. *Plant Breeding*, 132, 458–71.

Rashid, A. and Ryan, J. 2004. Micronutrient constraints to crop production in soils with Mediterranean-type characteristics: A review. *Journal of Plant Nutrition*, 27, 959–75.

Rawat, N., Pumphrey, M. O., Liu, S., Zhang, X., Tiwari1, V. K., Ando, K., Trick, H. N., Bockus, W. W., Akhunov, E., Anderson, J. A. and Gill, B. S. 2016. Wheat Fhb1 encodes a chimeric lectin with agglutinin domains and a pore-forming toxin-like domain conferring resistance to Fusarium head blight. *Nature Genetics*. DOI:10.1038/ng.3706

Rawson, H. M., Richards, R. A. and Munns, R. 1988. An Examination of selection criteria for salt tolerance in wheat, barley and triticale genotypes. *Australian Journal of Agricultural Research*, 39, 759–72.

Rees, R. G., Platz, G. J. and Mayer, R. J. 1982. Yield losses in wheat from yellow spot – comparison of estimates derived from single tillers and plots. *Australian Journal of Agricultural Research*, 33, 899–908.

Reimer, S., Pozniak, C. J., Clarke, F. R., Clarke, J. M., Somers, D. J., Knox, R. E. and Singh, A. K. 2008. Association mapping of yellow pigment in an elite collection of durum wheat cultivars and breeding lines. *Genome*, 51, 1016–25.

Rerkasem, B. and Jamjod, S. 1997. Genotypic variation in plant response to low boron and implications for plant breeding. *Plant and Soil*, 193, 169–80.

Richard, C. A., Hickey, L. T., Fletcher, S., Jennings, R., Chenu, K. and Christopher, J. 2015. High-throughput phenotyping of seminal root traits in wheat. *Australian Journal of Agricultural Research*, 33, 899–908.

Ripetti-Ballester, V., Chaurand, M. and Roumet, P. 2000. Prediction du rendement en semoule par spectroscopie proche infrarouge sur grains entiers. In: Royo, C., Nachit, M., di Fonzo, N. and Araus, J. L. (eds.), *Durum Wheat Improvement in the Mediterranean Region: New Challenges*. Zaragoza: International Centre for Advanced Mediterranean Agronomic Studies (CIHEAM).

Roelfs, A. P., Singh, R. and Saari, E. E. 1992. *Rust Dieseases of Wheat: Concepts and Methods of Disease Management*, Mexico, CIMMYT.

Rosewarne, G. M., Singh, R. P., Huerta-Espino, J. and Rebetzke, G. J. 2008. Quantitative trait loci for slow-rusting resistance in wheat to leaf rust and stripe rust identified with multi-environment analysis. *Theoretical and Applied Genetics*, 116, 1027–34.

Royo, C., Dreisigacker, S., Alfaro, C., Ammar, K. and Villegas, D. 2016. Effect of Ppd-1 genes on durum wheat flowering time and grain filling duration in a wide range of latitudes. *Journal of Agricultural Science*, 154, 612–31.

Ruan, Y. F., Comeau, A., Langevin, F., Hucl, P., Clarke, J. M., Brule-Babel, A. and Pozniak, C. J. 2012. Identification of novel QTL for resistance to Fusarium head blight in a tetraploid wheat population. *Genome*, 55, 853–64.

Rutkoski, J. E., Heffner, E. L. and Sorrells, M. E. 2011. Genomic selection for durable stem rust resistance in wheat. *Euphytica*, 179, 161–73.

Salvi, S. and Tuberosa, R. 2015. The crop QTLome comes of age. *Current Opinion in Biotechnology*, 32, 179–85.

Sanguineti, M. C., Li, S., Maccaferri, M., Corneti, S., Rotondo, F., Chiari, T. and Tuberosa, R. 2007. Genetic dissection of seminal root architecture in elite durum wheat germplasm. *Annals of Applied Biology*, 151, 291–305.

Sayre, K. D., Singh, R. P., Huerta-Espino, J. and Rajaram, S. 1998. Genetic progress in reducing losses to leaf rust in CIMMYT-derived Mexican spring wheat cultivars. *Crop Science*, 38, 654–9.

Schnurbusch, T., Langridge, P. and Sutton, T. 2008. The Bo1-specific PCR marker AWW5L7 is predictive of boron tolerance status in a range of exotic durum and bread wheats. *Genome*, 51, 963–71.

Shynbolar, R. and Aralbek, R. 2010. Evaluation of leaf rust resistance genes in durum wheat varieties in Kazakhstan. *The Asian and Australiasian Journal of Plant Science and Biotechnology*, 4, 77–80.

SIAP. Servicio de Informacion Agroalimentaria y Pesquera [Online]. Available: http://www.gob.mx/siap [Accessed].

Simons, K. J., Gehlhar, S. B., Maan, S. S. and Kianian, S. F. 2003. Detailed mapping of the species cytoplasm-specific (scs) gene in durum wheat. *Genetics*, 165, 2129–36.

Singh, A., Reimer, S., Pozniak, C. J., Clarke, F. R., Clarke, J. M., Knox, R. E. and Singh, A. K. 2009. Allelic variation at Psy1–A1 and association with yellow pigment in durum wheat grain. *Theoretical and Applied Genetics*, 118, 1539–48.

Singh, A. K., Clarke, J. M., Knox, R. E., DePauw, R. M., McCaig, T. N., Cuthbert, R. D., Clarke, F. R. and Fernandez, M. R. 2014. AAC Raymore durum wheat. *Canadian Journal of Plant Science*, 94, 1289–96.

Singh, A. K., Clarke, J. M., Knox, R. E., DePauw, R. M., McCaig, T. N., Fernandez, M. R. and Clarke, F. R. 2012. Transcend durum wheat. *Canadian Journal of Plant Science*, 92, 809–13.

Singh, A. K., Clarke, J. M., Knox, R. E., DePauw, R. M., Wise, I., Thomas, J., McCaig, T. N., Cuthbert, R. D., Clarke, F. R. and Fernandez, M. R. 2015. AAC Marchwell durum wheat. *Canadian Journal of Plant Science*, 95, 189–95.

Singh, P. K., Gonzalez-Hernandez, J. L., Mergoum, M., Ali, S., Adhikari, T. B., Kianian, S. F., Elias, E. M. and Hughes, G. R. 2006a. Identification and molecular mapping of a gene conferring resistance to Pyrenophora tritici-repentis race 3 in tetraploid wheat. *Phytopathology*, 96, 885–9.

Singh, P. K., Mergoum, M., Ali, S., Adhikari, T. B., Elias, E. M., Anderson, J. A., Glover, K. D. and Berzonsky, W. A. 2006b. Evaluation of elite wheat germ plasm for resistance to tan spot. *Plant Disease*, 90, 1320–5.

Singh, P. K., Mergoum, M., Ali, S., Adhikari, T. B., Elias, E. M. and Hughes, G. R. 2006c. Identification of new sources of resistance to tan spot, Stagonospora nodorum blotch, and Septoria tritici blotch of wheat. *Crop Science*, 46, 2047–53.

Singh, P. K., Mergoum, M., Gonzalez-Hernandez, J. L., Ali, S., Adhikari, T. B., Kianian, S. F., Elias, E. M. and Hughes, G. R. 2008. Genetics and molecular mapping of resistance to necrosis inducing race 5 of Pyrenophora tritici-repentis in tetraploid wheat. *Molecular Breeding*, 21, 293–304.

Singh, R. P., Datta, D., Priyamvada, Singh, S. and Tiwari, D. 2004a. Marker-assisted selection for leaf rust resistance genes Lr19 and Lr24 in wheat (Triticum aestivum L.). *Journal of Applied Genetics*, 45, 399–403.

Singh, R. P., Huerta-Espino, J., Pfeiffer, W. and Figueroa-Lopez, P. 2004b. Occurrence and impact of a new leaf rust race on durum wheat in northwestern Mexico from 2001 to 2003. *Plant Disease*, 88, 703–8.

Sissons, M., Osborne, B. and Sissons, S. 2006. Application of near infrared reflectance spectroscopy to a durum wheat breeding programme. *Journal of near Infrared Spectroscopy*, 14, 17–25.

Sissons, M. J., Osborne, B. G., Hare, R. A., Sissons, S. A. and Jackson, R. 2000. Application of the single-kernel characterization system to durum wheat testing and quality prediction. *Cereal Chemistry*, 77, 4–10.

Smith, A. B., Cullis, B. R., Appels, R., Campbell, A. W., Cornish, G. B., Martin, D. and Allen, H. M. 2001. The statistical analysis of quality traits in plant improvement programs with application to the mapping of milling yield in wheat. *Australian Journal of Agricultural Research*, 52, 1207–19.

Snape, J. W. 1998. Golden calves or white elephants? Biotechnologies for wheat improvement. *Euphytica*, 100, 207–17.

Somers, D. J., Fedak, G., Clarke, J. and Cao, W. G. 2006. Mapping of FHB resistance QTLs in tetraploid wheat. *Genome*, 49, 1586–93.

Soriano, J. M. and Royo, C. 2015. Dissecting the genetic architecture of leaf rust resistance in wheat by QTL meta-analysis. *Phytopathology*, 105, 1585–93.

Sorrells, M. 2015. Genomic selection in plants: Empirical results and implications for wheat breeding. In: Ogihara, Y. (ed.), *Advances in Wheat Genetics: From Genome to Field*. Springer, Berlin.

Stack, R. W., Elias, E. M., Fetch, J. M., Miller, J. D. and Joppa, L. R. 2002. Fusarium head blight reaction of langdon durum – Triticum dicoccoides chromosome substitution lines. *Crop Science*, 42, 637–42.

Strelkov, S. E. and Lamari, L. 2003. Host-parasite interactions in tan spot [Pyrenophora tritici-repentis] of wheat. *Canadian Journal of Plant Pathology-Revue Canadienne De Phytopathologie*, 25, 339–49.

Subira, J., Ammar, K., Alvaro, F., del Moral, L. F. G., Dreisigacker, S. and Royo, C. 2016. Changes in durum wheat root and aerial biomass caused by the introduction of the Rht-B1b dwarfing allele and their effects on yield formation. *Plant and Soil*, 403, 291–304.

Sukumaran, S., Lopes, M. S., Dreisigacker, S., Dixon, L. E., Zikhali, M., Griffiths, S., Zheng, B., Chapman, S. and Reynolds, M. P. 2016. Identification of earliness per se flowering time locus in spring wheat through a genome-wide association study. *Crop Science*, 56, 2962–672.

Suprayogi, Y., Pozniak, C. J., Clarke, F. R., Clarke, J. M., Knox, R. E. and Singh, A. K. 2009. Identification and validation of quantitative trait loci for grain protein concentration in adapted Canadian durum wheat populations. *Theoretical and Applied Genetics*, 119, 437–48.

Szabolcs, I. 1994. Soils and salinisation. In: Pessarakali, M. (ed.), *Handbook of Plant and Crop Stress*. Marcel Dekker, New York.

Talbert, L. E., Sherman, J. D., Hofland, M. L., Lanning, S. P., Blake, N. K., Grabbe, R., Lamb, P. F., Martin, J. M. and Weaver, D. K. 2014. Resistance to Cephus cinctus Norton, the wheat stem sawfly, in a recombinant inbred line population of wheat derived from two resistance sources. *Plant Breeding*, 133, 427–32.

Thavamanikumar, S., Dolferus, R. and Thumma, B. R. 2015. Comparison of genomic selection models to predict flowering time and spike grain number in two Hexaploid wheat doubled haploid populations. *G3-Genes Genomes Genetics*, 5, 1991–8.

Thomas, J., Fineberg, N., Penner, G., McCartney, C., Aung, T., Wise, I. and McCallum, B. 2005. Chromosome location and markers of Sm1: a gene of wheat that conditions antibiotic resistance to orange wheat blossom midge. *Molecular Breeding*, 15, 183–92.

Todorovska, E., Christov, N., Slavov, S., Christova, P. and Vassilev, D. 2009. Biotic stress resistance in wheat – breeding and genomic selection implications. *Biotechnology and Biotechnological Equipment*, 23, 1417–26.

Townleysmith, T. F., Depauw, R. M., Lendrum, C. W. B., Mccrystal, G. E. and Patterson, L. A. 1987. Kyle durum-wheat. *Canadian Journal of Plant Science*, 67, 225–7.

Troccoli, A., Borrelli, G. M., De Vita, P., Fares, C. and Di Fonzo, N. 2000. Durum wheat quality: A multidisciplinary concept. *Journal of Cereal Science*, 32, 99–113.

Tsilo, T. J., Hareland, G. A., Chao, S. and Anderson, J. A. 2011. Genetic mapping and QTL analysis of flour color and milling yield related traits using recombinant inbred lines in hard red spring wheat. *Crop Science*, 51, 237–46.

Tuberosa, R. and Pozniak, C. J. 2014. Durum wheat genomics comes of age Introduction to the special issue on durum wheat genomics. *Molecular Breeding*, 34, 1527–30.

Tuberosa, R., Salvi, S., Sanguineti, M. C., Landi, P., MacCaferri, M. and Conti, S. 2002. Mapping QTLs regulating morpho-physiological traits and yield: Case studies, shortcomings and perspectives in drought-stressed maize. *Annals of Botany*, 89, 941–63.

Uauy, C., Distelfeld, A., Fahima, T., Blechl, A. and Dubcovsky, J. 2006. A NAC gene regulating senescence improves grain protein, zinc, and iron content in wheat. *Science*, 314, 1298–301.

Uauy, C., Paraiso, F., Colasuonno, P., Tran, R. K., Tsai, H., Berardi, S., Comai, L. and Dubcovsky, J. 2009. A modified TILLING approach to detect induced mutations in tetraploid and hexaploid wheat. *Bmc Plant Biology*, 9.

Vallega, V., Rubies-Autonell, C., 1985. Reactions of Italian Triticum durum cultivars to Soilborne wheat mosaic. *Plant Disease*, 69, 64–6.

Varella, A. C., Weaver, D. K., Sherman, J. D., Blake, N. K., Heo, H. Y., Kalous, J. R., Chao, S., Hofland, M. L., Martin, J. M., Kephart, K. D. and Talbert, L. E. 2015. Association analysis of stem solidness and wheat stem sawfly resistance in a panel of North American spring wheat Germplasm. *Crop Science*, 55, 2046–55.

Varshney, R. K., Terauchi, R. and McCouch, S. R. 2014. Harvesting the promising fruits of enomics: Applying genome sequencing technologies to crop breeding. *Plos Biology*, 12.

Vasiljevic, S., Banasik, O. J. and Shuey, W. C. 1977. Micro unit for producing durum semolina. *Cereal Chemistry*, 54, 397–404.

Vera, C. L., Fox, S. L., DePauw, R. M., Smith, M. A. H., Wise, I. L., Clarke, F. R., Procunier, J. D. and Lukow, O. M. 2013. Relative performance of resistant wheat varietal blends and susceptible wheat cultivars exposed to wheat midge, Sitodiplosis mosellana (Gehin). *Canadian Journal of Plant Science*, 93, 59–66.

Vida, G., Gal, M., Uhrin, A., Veisz, O., Syed, N. H., Flavell, A. J., Wang, Z. L. and Bedo, Z. 2009. Molecular markers for the identification of resistance genes and marker-assisted selection in breeding wheat for leaf rust resistance. *Euphytica*, 170, 67–76.

Voytas, D. F. and Gao, C. X. 2014. Precision genome engineering and agriculture: opportunities and regulatory challenges. *Plos Biology*, 12.

Wang, S. C., Wong, D. B., Forrest, K., Allen, A., Chao, S. M., Huang, B. E., Maccaferri, M., Salvi, S., Milner, S. G., Cattivelli, L., Mastrangelo, A. M., Whan, A., Stephen, S., Barker, G., Wieseke, R., Plieske, J., Lillemo, M., Mather, D., Appels, R., Dolferus, R., Brown-Guedira, G., Korol, A., Akhunova, A. R., Feuillet, C., Salse, J., Morgante, M., Pozniak, C., Luo, M. C., Dvorak, J., Morell, M., Dubcovsky, J., Ganal, M., Tuberosa, R., Lawley, C., Mikoulitch, I., Cavanagh, C., Edwards, K. J., Hayden, M., Akhunov, E. and Sequencing, I. W. G. 2014. Characterization of polyploid wheat genomic diversity using a high-density 90 000 sing e nucleotide polymorphism array. *Plant Biotechnology Journal*, 12, 787–96.

Welch, R. M. and Graham, R. D. 2004. Breeding for micronutrients in staple food crops from a human nutrition perspective. *Journal of Experimental Botany*, 55, 353–64.

Wheeler, T. R., Craufurd, P. Q., Ellis, R. H., Porter, J. R. and Prasad, P. V. V. 2000. Temperature variability and the yield of annual crops. *Agriculture Ecosystems and Environment*, 82, 159–67.

Wiebe, K., Harris, N. S., Faris, J. D., Clarke, J. M., Knox, R. E., Taylor, G. J. and Pozniak, C. J. 2010. Targeted mapping of Cdu1, a major locus regulating grain cadmium concentration in durum wheat (Triticum turgidum L. var durum). *Theoretical and Applied Genetics*, 121, 1047–58.

Winfield, M. O., Wilkinson, P. A., Allen, A. M., Barker, G. L. A., Coghill, J. A., Burridge, A., Hall, A., Brenchley, R. C., D'Amore, R., Hall, N., Bevan, M. W., Richmond, T., Gerhardt, D. J., Jeddeloh, J. A. and Edwards, K. J. 2012. Targeted re-sequencing of the allohexaploid wheat exome. *Plant Biotechnology Journal*, 10, 733–42.

Zhang, W., Chao, S., Manthey, F., Chicaiza, O., Brevis, J. C., Echenique, V. and Dubcovsky, J. 2008. QTL analysis of pasta quality using a composite microsatellite and SNP map of durum wheat. *Theoretical and Applied Genetics*, 117, 1361–77.

Zhang, Y., Liang, Z., Zong, Y., Wang, Y. P., Liu, J. X., Chen, K. L., Qiu, J. L. and Gao, C. X. 2016. Efficient and transgene-free genome editing in wheat through transient expression of CRISPR/Cas9 DNA or RNA. *Nature Communications*, 7.

Zhao, Y. S., Mette, M. F. and Reif, J. C. 2015. Genomic selection in hybrid breeding. *Plant Breeding*, 134, 1–10.

Zheng, X. T., Hoegenauer, K. A., Maeda, A. B. V., Wang, F., Stelly, D. M., Nichols, R. L. and Jones, D. C. 2015. Non-destructive high-throughput DNA extraction and genotyping methods for cotton seeds and seedlings. *Biotechniques*, 58, 234–43.

Part 3

Improving wheat cultivation in the developing world

Chapter 10

Supporting smallholders in improving wheat cultivation

Tinashe Chiurugwi, Simon Kerr, Ian Midgley and Lesley Boyd, National Institute of Agricultural Botany (NIAB), UK; Johnson Kamwaga, Food Crops Research Centre – Njoro, Kenya; Peter Njau, Highlands Agri-consult Services Ltd, Kenya; Terry van Gevelt, University of Cambridge, UK; and Claudia Canales and Max Marcheselli, the Malaysian Commonwealth Studies Centre (MCSC) and the Cambridge Malaysian Education and Development Trust (CMEDT), UK

1 Introduction: smallholder wheat cultivation

2 Limitations to smallholder wheat cultivation

3 How research addresses limitations to smallholder wheat cultivation

4 Case study: Agri-transfer

5 Future trends

6 Where to look for further information

7 References

1 Introduction: smallholder wheat cultivation

In the context of this chapter, smallholder wheat farmers are defined as farmers with limited access to resources (e.g. land, inputs, labour, machinery and finance) and services (e.g. input and output markets, information and advice) necessary for optimal commercial wheat cultivation (Beyene, Franzel and Mwangi, 1990; Dixon, Tanyeri-Abur and Wattenbach, 2004; Dixon et al., 2009; Nazli et al., 2014). Smallholder wheat farmers are also generally more reliant on off-farm income for livelihood and keep a significant proportion of their produce for their own consumption.

Consumption of wheat products is increasing rapidly among both rural and urban communities in low-income and middle-income countries (World Bank, 2016), with increases of between 19 and 38% expected in China, India, Pakistan, West Asia, North Africa and sub-Saharan Africa by 2023 (relative to 2011–13 levels; OECD and FAO, 2014; Valin et al., 2014; Mason et al., 2015; Alexandratos and Bruinsma, 2012). The main drivers for this rapid increase have been cited as improved household incomes, rising total population and changes in diets (Mason, Jayne and Shiferaw, 2015). The rate of increase in wheat demand is much higher than that for maize and rice, the other staple cereals, and wheat is now an important cereal, supplying a fifth of dietary protein and calories globally (Shiferaw et al., 2013). Over the same period (2013–23), growth in wheat production

http://dx.doi.org/10.19103/AS.2016.0004.37

is expected to be about 12% in developing countries (OECD and FAO, 2014). This is mainly because wheat yields are generally lower (compared to the global average) in most developing countries (Fig. 1). In addition, developing world wheat acreage is expected to grow by less than 2% between 2013 and 2023 (Fischer, Byerlee and Edmeades, 2014; OECD and FAO, 2014; FAOSTAT, 2016). Rising wheat demand in these countries will therefore create large internal deficits.

Projections are that yield-limiting factors such as drought, flooding and disease outbreaks are predicted to increase in intensity and frequency in the coming years (Fig. 2; Fischer et al., 2014; CIMMYT, 2014). This will make it even more difficult to increase productivity to meet the growing demands from the food and feed chain.

On the other hand, global wheat prices have not been favourable for producers: following decades of decline, they rose to record levels in 2008 and have remained turbulent with a downward trend since then (Fig. 3; Dixon et al., 2009; FAOSTAT, 2016). This is likely to keep growth in wheat production depressed in the foreseeable future (Shiferaw et al., 2013).

In 2014, just over half of the 800 million tonnes of global wheat production came from about 150 million hectares cultivated in 75 low-income and middle-income countries, mainly in South, South East and West Asia, North Africa, Middle East, sub-Saharan Africa

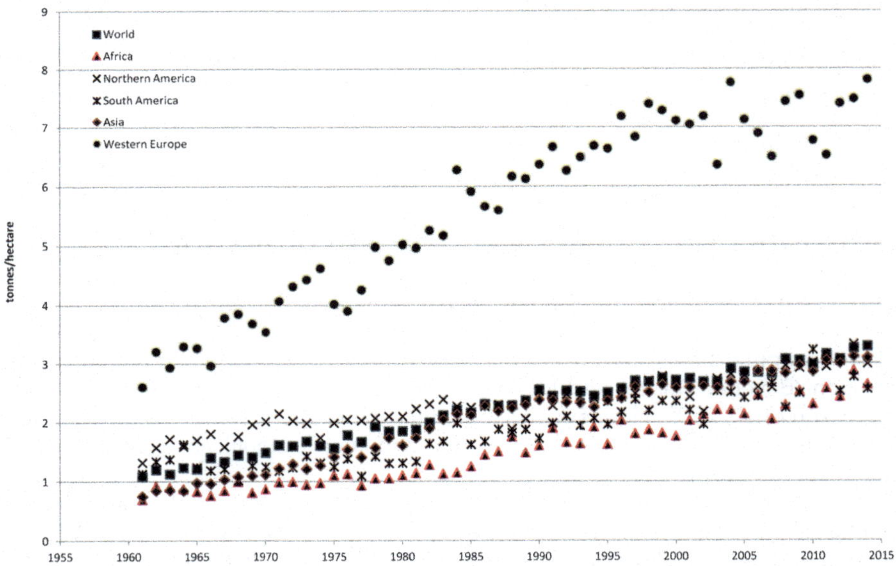

Figure 1 Global yield trends for wheat between 1961 and 2014. For the past 10–15 years, on-farm yields have stagnated or declined in most wheat-producing areas; the main reasons for this are often cited as the shift of established wheat farmers to more profitable crops (e.g. maize and oil crops) and increased impact of biotic/abiotic factors (Dixon et al., 2009; Shiferaw et al., 2013; Fischer, Byerlee and Edmeades, 2014; FAOSTAT, 2016). The difference between Western Europe and everywhere else is likely due to a combination of factors, including the use of better varieties, production infrastructure and agronomy and longer growing seasons.

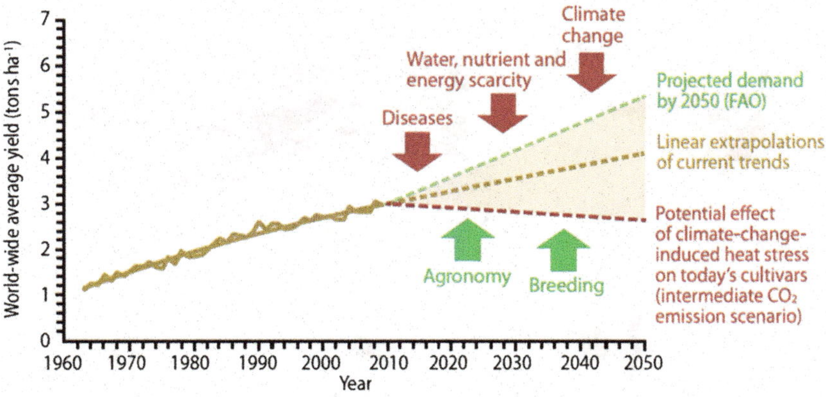

Figure 2 Projected wheat demand and yield trends under varying scenarios (CIMMYT, 2014). There is an urgent need to improve wheat varieties and agronomic practices to increase wheat production and meet growing demand. Used with permission from CIMMYT.

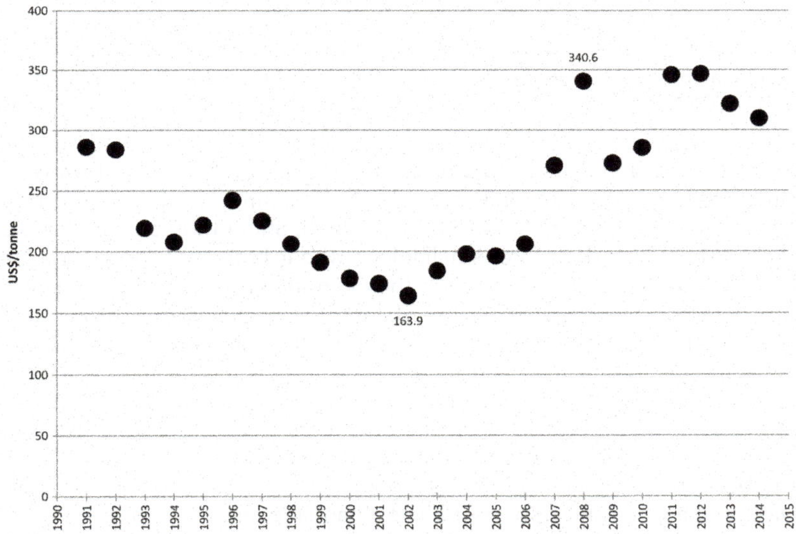

Figure 3 Global wheat prices between 1991 and 2014 (FAOSTAT, 2016). There was a general decline in wheat prices up to the early 2000s followed by a spike in the late 2000s (mainly due to falling stock levels, following poor weather and a significant shift of acreage to other crops), and another downward trend since 2012.

and Latin America (Fig. 4; FAOSTAT, 2016; World Bank, 2016; CIMMYT, 2016; Hodson and White, 2007; Rajaram et al., 1995).

Wheat production in these regions is mainly carried out by smallholder farmers, except in sub-Saharan Africa (excluding Ethiopia), where production by a few large-scale

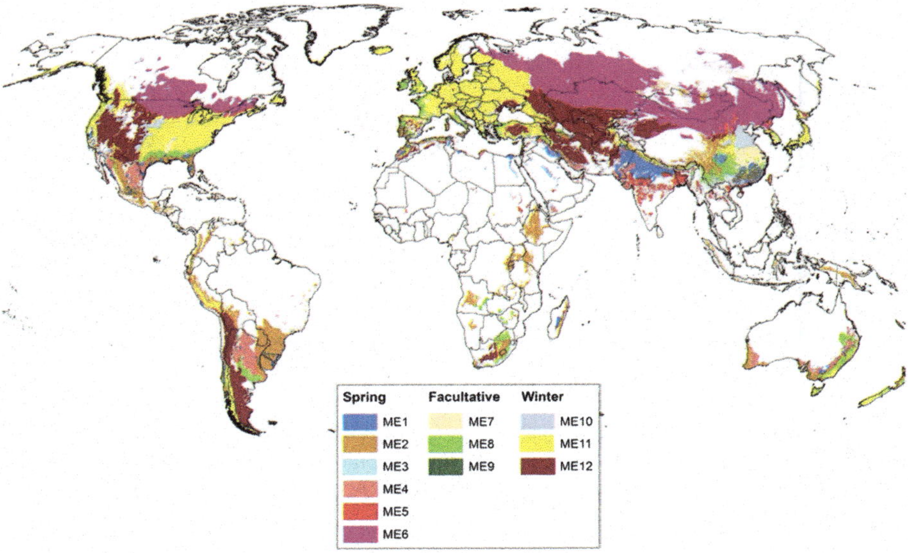

Figure 4 Global distribution of wheat production, classed into 12 agro-ecologies, referred to as the 'wheat mega-environments' (Rajaram, van Ginkel and Fischer, 1995; CIMMYT, 2016b). Used with permission from CIMMYT.

commercial farmers dominates (Mason, Jayne and Shiferaw, 2015). However, imports are still required to meet wheat demand in these countries, with wheat being their largest (43%) food import (Fig. 5; Mason et al., 2015; Shiferaw et al., 2013; Dixon et al., 2009; FAOSTAT, 2016). Therefore, smallholder wheat farmers have an opportunity to improve their incomes and livelihoods if they can increase their productivity to meet in-country deficits. Unfortunately, a lot of factors constrain their productivity, chief among them being (Beyene, Franzel and Mwangi, 1990; Shiferaw et al., 2013; CIMMYT, 2014):

- Biophysical factors: adverse climatic conditions, declining soil fertility, pests and diseases, poor access to seeds of improved varieties
- Socio-economic factors: poor access to land, farm machinery, labour, information/advice, and finance and poor post-harvest handling and marketing.

Research and development (R&D) into these biophysical and socio-economic constraints is needed to support smallholder wheat cultivation. The challenge is to increase yield gains from the current <1% to about 1.6% per year (through improved varieties and agronomy) and to narrow yield gaps (differences between observed and potential yields in a given region), which currently stand at 26–69% (Fischer, Byerlee and Edmeades, 2014). Additionally, any increases in production will need to come from improving productivity per unit of existing farmland, as increasing the area under wheat cultivation is likely to have adverse effects on the environment and other non-wheat land use purposes (Fischer, Byerlee and Edmeades, 2014).

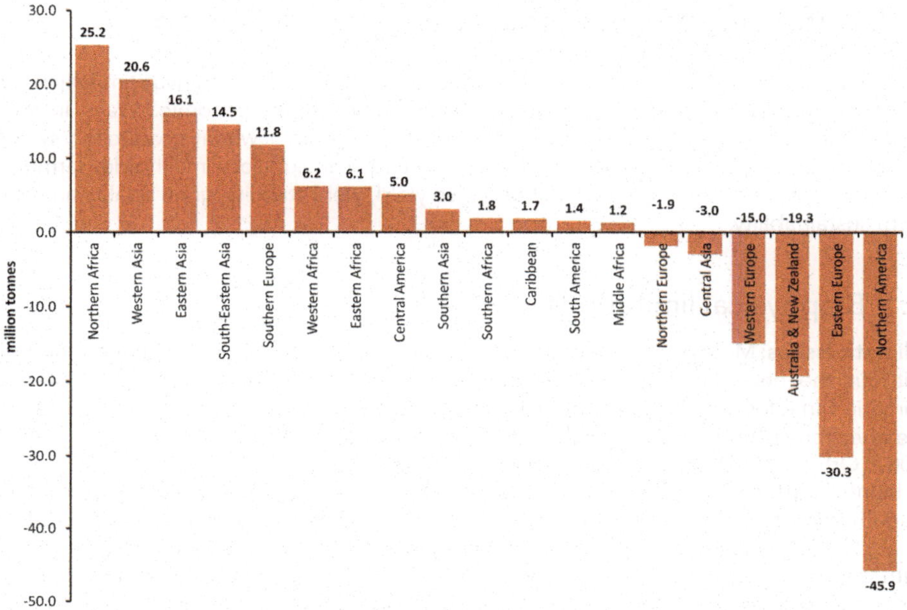

Figure 5 Average annual net imports of wheat for 2011, 2012 and 2013 (FAOSTAT, 2016). Most low-income to medium-income countries are net importers of wheat. The deficit exposes these economies to the vagaries of global markets, while at the same time presenting an opportunity for smallholder farmers.

With few exceptions, R&D that supports smallholder wheat production has had limited impact on farmer productivity, mainly because the dissemination of R&D outputs depends on poorly resourced public programmes and institutions; that is, there is very little private sector involvement in wheat R&D and dissemination thereof in developing countries (Reynolds and Borlaug, 2006). As a result, most smallholder farmers missed out on the Green Revolution (the large increase in global crop productivity achieved in the second half of the twentieth century through the use of advanced technologies and practices, such as agrochemicals and improved varieties).

Smallholder wheat cultivation also faces specific gender and social inclusion challenges that constrain productivity progress in many developing countries. Generally, women are the main actors in wheat production and processing, yet they have limited input in planning, decision-making and management of wheat-based farming enterprises (CGIAR Research Program on Wheat, 2015b). Differences in gender for each target smallholder farming community therefore have to be considered when assessing and addressing the constraints female and male wheat farmers face (Tiruneh et al., 1999; Doss, 2001; Nelson, 2013).

This chapter will summarise some of the main issues/challenges that constrain smallholder wheat cultivation and discuss ways in which R&D could address them. We will highlight some ongoing work that is supporting smallholder wheat farmers in Kenya in a case study and conclude with future trends in this area and sources of further information.

2 Limitations to smallholder wheat cultivation

This section looks at the main factors limiting smallholder wheat cultivation, which have been categorised into two groups: biophysical and socio-economic. Ways in which research is addressing these issues/challenges in smallholder wheat production systems are also discussed. We outline how research findings can support smallholder farmers to create wheat production systems that optimise yield, yield stability, grain quality and long-term sustainability.

2.1 Biophysical limitations

Climatic issues: Most smallholder farmers cultivate wheat in suboptimal climatic conditions, without the necessary support mechanisms for system resilience. Environmental and climatic variables are often extreme and unpredictable, and farmers do not have access to weather forecasting, decision-making tools or agricultural insurance (Hodson and White, 2007; Ortiz et al., 2008; Lobell, Schlenker and Costa-Roberts, 2011).

Rainfall patterns have become increasingly erratic in many wheat-growing regions, with negative impacts on wheat productivity (Beyene, Franzel and Mwangi, 1990; Shiferaw et al., 2013; Yami et al., 2013). Yield can be significantly reduced when droughts coincide with important developmental stages, such as flowering and grain filling. Erratic rainfall also exacerbates the effects of pest and diseases, while water logging can make it impossible to carry out essential farm operations such as land preparation, planting, weeding or harvest.

In places where smallholder farmers have access to irrigation, the water sources are increasingly overdrawn, and depletion of irrigation water is likely to cause a shift to rain-fed cultivation as scarce water is used on higher-value and less water-efficient crops (Sayre and Govaerts, 2009; Shiferaw et al., 2013; Aryal et al., 2015). Additionally, climate change is projected to increase incidence of drought and heat stress, particularly among smallholder wheat farming communities in South and West Asia, and North Africa (Kosina et al., 2007; Ortiz et al., 2008). With increasing water scarcity, it is also expected that soil salinity will increase in intensity in irrigated wheat farms located in drier regions (Shiferaw et al., 2013).

Soil fertility: Low soil fertility is a major production constraint in many smallholder wheat systems. Most smallholder wheat farmers have cultivated the same piece of land continuously for long periods without using the right type or amount of fertiliser, fallowing the land or rotating wheat with nitrogen-fixing crops (Beyene, Franzel and Mwangi, 1990; Sayre, 2002). These practices, coupled with burning of crop residues or removing them for feed or fuel, have led to significant nutrient mining, low soil organic matter content and degradation of soil physical properties (Beyene, Franzel and Mwangi, 1990; Sayre, 2002). Moreover, most farmers do not test their soils and some have continuously applied fertilisers following blanket recommendations gazetted by government advisers (Schnier et al., n.d.). In some cases, this has led to deterioration in the soils; for example, applying diammonium phosphate in Kenya has increased soil acidity in some locations (Andae, 2014).

Depletion of phosphorous and potassium reserves, and the rising costs of energy and fossil fuels are expected in the foreseeable future, further reducing fertiliser affordability for smallholder farmers (Shiferaw et al., 2013). Micronutrient deficiencies or toxicities are also a major constraint to smallholder productivity in several countries, for example, copper in parts of Kenya, manganese in parts of Tanzania, boron in Zambia and zinc in Zimbabwe (Tanner and Mwangi, 1992).

Pests and diseases: Global annual wheat yield losses due to pests and diseases are estimated at 160 million tonnes or about USD 30 billion (2014 figures), ranging between 10 and 30% (higher in smallholder systems where control is absent or ineffective) (Oreke, 2006, 2007; FAOSTAT, 2016). Important diseases that affect wheat production are stem, stripe and leaf rust, viral diseases, Fusarium head blight, *Septoria tritici* blotch, spot blotch, tan spot, wheat blast, smuts, bunts and various root diseases. Important pests include nematodes, Sunn pest, Hessian fly, aphids, cutworms, rodents, birds, snails and wheat stem sawfly (Oerke, 2007; Shiferaw et al., 2013). Rusts caused by *Puccinia triticina* (leaf rust), *P. striiformis* (stripe rust) and *P. graminis* (stem rust) are the most important wheat diseases worldwide.

Although varieties that are resistant or tolerant to some of these diseases exist, most smallholder farmers still use old and susceptible varieties, or adopt the resistant varieties late, when the resistance has broken down. In addition, smallholder farmers have little or no information, and limited access to control practices and chemical treatments that would otherwise limit damage on susceptible varieties. As a result, the emergence and spread of new diseases (e.g. wheat blast in Brazil and South America) and new strains of diseases (e.g. the Ug99 race of stem rust, to which 80% of wheat varieties grown worldwide were susceptible in 2007) is a constant threat to smallholder producers (Singh et al., 2008, 2011; Tufan et al., 2009). Pest and disease damage in smallholder farming situations is relatively worse because their crops experience significantly more physiological stress due to factors such as late sowing, inadequate nutrition and moisture stress (Duveiller, Singh and Nicol, 2007).

Weeds are the most important yield limiting constraint and the most costly biotic factor to control in wheat production (Anderson and Impiglia, 2002; Oerke, 2007). Common weed control methods in smallholder wheat systems involve a combination of herbicide applications and removal of weeds by hand. Most smallholder wheat farmers struggle with weed control due to shortage of labour, especially when recommended weeding time coincides with other priority activities. They also lack access to herbicides and are not always aware of their proper use (Beyene, Franzel and Mwangi, 1990; Oerke, 2007).

Varieties and seeds: While improved varieties that would suit the requirements of almost all major wheat-producing regions are available, there are problems with seed supply to smallholder farmers. Most wheat smallholders use seeds of old varieties from the informal sector, that is, uncertified seed that is saved on farm, exchanged among farmers or bought from local markets (Alemu and Bishaw, 2015). Among those that use certified seed of improved varieties, the majority obtain it through the various input support programmes operated by relief NGOs and national governments (Matuschke, Mishra and Qaim, 2007; Nazli et al., 2014; Alemu and Bishaw, 2015). These programmes have the unfortunate effect of distorting and further crippling the commercial seed industry in most countries, and the formal seed sector can only provide for a fraction of the national seed requirements.

In general, smallholder wheat farmers are often the last to adopt improved varieties in developing countries. The main reasons for the limited adoption rates include inadequate breeder and/or foundation seed production, lack of suitable certified seed growers, insufficient promotion of improved varieties and unaffordable (for most smallholder farmers) prices of certified seeds (Tiruneh et al., 1999; Mwanga et al., 2001; Matuschke, Mishra and Qaim, 2007; Nazli et al., 2014).

2.2 Socio-economic limitations

Land: Globally, small farms are getting smaller and larger farms are getting larger due to socio-economic changes such as population increase, more farmers wanting access to land and increased global flow of capital (van Vliet et al., 2015). Additionally, most smallholder wheat production is carried out in the land-scarce regions of Asia, Near East and North Africa. As a result, smallholder farmers have increasingly poorer access to land and a significant proportion of smallholder wheat producers cultivate wheat on leased land (Nazli et al., 2014). This can increase the farmers' costs, as some landowners demand rent several seasons in advance, and makes it difficult for them to put in place long-term plans, owing to the uncertainty associated with the land hire agreements (Mahayagu et al., 2007; Birungi and Hassan, 2010).

Farm machinery and labour: Farm machinery is generally unavailable or unaffordable to smallholder wheat farmers. Where smallholder farmers can access machinery, it is costly and oversubscribed, with most farming operations being delayed as farmers wait for their turn, usually after more-resourced farmers have finished their operations. Most farm machinery is also of inappropriate scale for small farms, or configured in ways that disadvantage smallholders (Tiruneh et al., 2001). This can range from poorly calibrated planters and sprayers, right through to harvesters fitted with compartments that hide grain and steal from unwitting farmers. Therefore, smallholders tend to use less advanced methods (e.g. draught animal-powered implements, knapsack sprayers and sickles) and plant seed or apply fertiliser by broadcasting (Beyene, Franzel and Mwangi, 1990; Nazli et al., 2014). Shortage and/or the high cost of labour in most smallholder farming communities often means that agronomic processes are not carried out effectively or efficiently, leading to reduction in yields, quality and profitability.

Post-harvest handling, processing and marketing: The majority of smallholder farmers' produce is traded soon after harvest through brokers or government institutions because they lack appropriate storage facilities (Mahayagu et al., 2007; Nazli et al., 2014). In general, smallholder farmers do not produce large enough volumes to control how they market their harvest. Therefore, the farmers do not have the opportunity to grade their produce to fetch better prices, cannot hold on to their produce to sell when the market is not glutted and can be at the mercy of brokers (Mahagayu et al., 2007). All these factors limit smallholder farmers' income. This in turn reduces future investment into the farms, particularly in bought-in inputs, further constraining the farmers' productivity (Birungi and Hassan, 2010). When farmers store their harvest for household consumption, the storage conditions are not always ideal, and a significant proportion of stored grain is lost post-harvest.

Information and advice: Smallholder farmers mainly rely on agricultural information and advice available through government agencies, development organisations or neighbours. This information is not always available or reliable because of limited resources, and poorly equipped and uninformed advisers (Mahayagu et al., 2007; Gildemacher and Mur, 2013; Gildemacher and Wongtschowski, 2015). As a result, smallholder wheat farmers often lag behind their more-resourced counterparts in terms of adopting new technologies and improved agronomic practices. This further limits their productivity and chances of reducing the environmental impact of their cultivation, through modern conservation agricultural practices.

Finance: Smallholder farmers generally lack access to credit and this stops some farmers from getting involved in wheat production altogether or from effectively running or improving existing operations (Matuschke, Mishra and Qaim, 2007; Birungi and Hassan, 2010). Many governments in developing countries offer subsidised finance to farmers and farmer groups; however, these schemes suffer from poor repayment rates and when the loans are provided in the form of inputs, the inputs often fail to meet the farmers' requirements (in terms of input type and quantity), reach the farmers later than required or end up in the hands of more-resourced farmers (Nazli et al., 2014).

2.3 Women in smallholder wheat farming

While it is generally accepted that female smallholder wheat farmers would be more negatively affected by the constraints listed above compared to male farmers, few studies have systematically assessed the implications of these gender differences (World Bank, 2008; Meinzen-dick et al., 2010; CGIAR Research Program on Wheat, 2015b). However, the reports that exist on women's role in smallholder agriculture in general, and a few wheat-specific studies, have proved useful to wheat researchers. They have allowed wheat researchers to start considering gender inequality among smallholder wheat farmers, so that their research outcomes reverse inequality or at least do not make it worse (Doss, 2001; CGIAR Research Program on Wheat, 2015a,b).

Research has shown that there are differences in the roles men and women play in wheat production systems depending on the household, socio-economic status and the location (Doss, 2001). Therefore, it is difficult to generalise or transfer results from one location to another, or even one household to another in the same locality (Doss, 2001). Additionally, given the complex nature of gender relations, simply comparing male and female farmers does not fully dissect the underlying factors behind the relationships. Armed with this understanding, researchers can be more conscious about the way they conduct their work and formulate solutions to smallholder farmers' problems. That is, they stand a better chance of narrowing gender gaps or avoiding making them worse.

It has been reported that if women had the same access to resources and services as men this would close the productivity gap and could increase general agricultural productivity by 2.5–4% (Food and Agriculture Organization, 2011). Addressing these gaps also delivers other benefits to society, for example, enhanced economic efficiency in agri-food systems, reduced hunger and malnutrition and improved access to education and other services (Fafchamps, Kebede and Quisumbing, 2009; Food and Agriculture Organization, 2011). It should also be noted that most smallholder wheat farmers live in patriarchal societies whose strong cultural and social norms support gender disparity, adding a significant level of challenge (CGIAR Research Program on Wheat, 2015b; CIMMYT, 2016).

Wheat researchers and policymakers can help close the gender gap and address gender-related limitations to smallholder wheat cultivation, but more research remains to be done to enable this. For example, while wheat researchers have long been working on improving various traits in wheat, few studies have specifically looked at the gendered dimensions of these traits, that is, the differential implications of these traits for men and women (Tiruneh et al., 1999; Ortiz-Ferrara et al., 2001; Tsegaye and Berg, 2007; Nelson, 2013; CGIAR Research Program on Wheat, 2015b; Tesfaye, Mesay and Bedada, 2015). For example, studies have shown that while men prefer yield-enhancing traits to maximise returns from sales, women may be more interested in storability and quality. Additionally, men and women tend to use different information sources when looking to find out about new technologies and agricultural practices, while different (depending on gender) socio-economic factors influence their adoption decisions (Ortiz-Ferrara et al., 2001; Tiruneh et al., 2001; Tsegaye and Berg, 2007; Stevenson et al., 2013; Tesfaye, Mesay and Bedada, 2015).

3 How research addresses limitations to smallholder wheat cultivation

Wheat research targeted at smallholder farmers is mainly carried out by public research organisations in developing countries working in collaboration with public research organisations in developing and developed countries. Much of this work is coordinated by two Consortium of International Agricultural Research (CGIAR) centres: the International Maize and Wheat Improvement Center (CIMMYT) and International Center for Agricultural Research in the Dry Areas (ICARDA) (Reynolds and Borlaug, 2006; Matuschke, Mishra and Qaim, 2007; Renkow and Byerlee, 2010; Stevenson et al., 2013; CGIAR, 2015; CGIAR Research Program on Wheat, 2015a; CIMMYT, 2015). There is limited private sector involvement in smallholder-focused wheat R&D. CGIAR-led research contributed 96% of annual on-farm wheat productivity gains in developing countries since 1965 (Renkow and Byerlee, 2010), with an annual estimated value of USD 2.8–3.8 billion (2015 estimate; CGIAR, 2015; Lantican et al., 2005). Over 60% of modern wheat varieties grown in developing countries (covering 80–90% of the area sown) have CGIAR parentage (Renkow and Byerlee, 2010; CIMMYT, 2015).

3.1 Addressing biophysical constraints

Research that addresses biophysical constraints reaches farmers in the form of improved technologies, including improved wheat varieties and agronomic practices. The aims are to improve agronomic practices and the quality and application of bought-in inputs. Adoption of improved varieties and crop management practices has improved and stabilised yields, and it has reduced environmental impact of smallholder wheat production (Gollin, 2006; Reynolds and Borlaug, 2006; Renkow and Byerlee, 2010; Stevenson et al., 2013). This has been achieved through modifying the crop's physiology (e.g. high-yielding dwarf varieties) and tolerance to abiotic and biotic stress (e.g. disease-resistant or disease-tolerant varieties) (Reynolds and Borlaug, 2006; Renkow and Byerlee, 2010; Joshi et al., 2011; Munns et al., 2012; Nia et al., 2012; Stevenson et al., 2013).

Research impact on smallholder farmers' productivity is often limited by low adoption rates, as well as limited success rates of improved technologies and practices. Work has been done to explore ways of improving adoption rates and performance of adopted research outputs within the complex environments smallholder farmers operate. To improve adoption rates, researchers need to turn their findings into information resources that famers can act on, and address the relevant incentives or disincentives for adoption

(Matuschke, Mishra and Qaim, 2007; Giller et al., 2011). One study has shown that farm size or subsistence levels did not influence new technology adoption decisions by Indian wheat farmers, whereas access to information and credit did (Matuschke, Mishra and Qaim, 2007). Similar socio-economic studies also explored the influence of factors such as differences influenced by gender and landownership on the performance of improved technologies and agronomic practices adopted by smallholder farmers (Doss, 2001; Matuschke, Mishra and Qaim, 2007; Birungi and Hassan, 2010; Alemu and Bishaw, 2015; Tesfaye, Mesay and Bedada, 2015). For example, male-headed households (MHH) are twice more likely to adopt improved varieties than female-headed households (FHH), and while access to extension workers influences adoption rates among MHH, radio ownership is more important in FHH (CGIAR Research Program on Wheat, 2015b). Moreover, it has been reported that uncertainty over landownership discourages the adoption of conservation agricultural practices (Birungi and Hassan, 2010).

One solution for increasing adoption of improved varieties is participatory variety selection, that is, involving farmers in the breeding process. For example, this is done through mother and baby trials where researcher-managed (mother) trials are conducted alongside farmer-managed (baby) trials (Snapp, 1999; Ortiz-Ferrara et al., 2001; Joshi et al., 2011; Pandit et al., 2011). The baby trials can also incorporate farmer-managed seed multiplication. This participatory approach speeds up identification of farmer-preferred traits and varieties, and improves farmers' knowledge about the varieties and seed production, while informally disseminating the varieties (Joshi et al., 2011; Pandit et al., 2011). The same methods have also been used to develop and deploy other technologies and agronomic practices.

Researchers have also looked at the impact of smallholder farmers' access to land in order to identify ways of improving productivity, while slowing down land degradation and enhancing soil health (Birungi and Hassan, 2010). This work goes alongside studies that seek to optimise smallholder wheat production systems through sustainable and climate-smart management of land, water and biodiversity (Thomas et al., 2012; Mekasha et al., 2014; Kassie et al., 2015). Key to this is a systems-based approach that takes into account the fact that smallholder wheat farmers use their land for several other crop or non-crop enterprises.

Alongside promoting improved agronomy, researchers have also developed and deployed conservation agricultural practices to smallholder wheat farmers and assessed the impact of their adoption on resource use efficiency (nutrients, water, labour, energy etc.), yields and productivity (Jat et al., 2009, 2014; Erenstein et al., 2012; Aryal et al., 2015). For example, laser land levelling has been used to improve irrigation efficiency in India (Jat et al., 2009, 2014; Erenstein et al., 2012; Aryal et al., 2015). Impact studies have shown that while the primary interest in conservation agriculture was soil and moisture conservation and enhancing soil health, other benefits (such as reduced energy and labour costs, improved crop stands and yields, improved nutrient and water use efficiency) also accrue from its adoption. The challenge now lies in understanding which conservation agricultural practices are best suited for different production conditions and the best ways to promote their adoption among smallholder farmers.

CIMMYT has defined 12 mega-environments in which wheat is grown (Fig. 3; Rajaram et al., 1995; Braun et al., 1996). These mega-environments have guided the development of wheat varieties and agronomic practices suited to a number of countries. Working this way researchers have identified mega-environment-specific traits that should help speed up adaptation to climate change as the mega-environments in specific regions transform from one to another (Hodson and White, 2007; Ortiz et al., 2008; Shiferaw et al., 2013).

3.2 Addressing socio-economic factors

Work has been done to improve policymakers' understanding of developing world wheat value chains, looking at the limitations and opportunities within them. The interventions that address socio-economic factors are mainly in the form of policies, services and institutions. This research has looked at issues like land shortages and competing land uses, demography (including those pertaining to women and young farmers) and global wheat trade (CGIAR Research Program on Wheat, 2015a).

To address land shortages, researchers have cautioned that in the few land-surplus developing countries (e.g. parts of sub-Saharan Africa and Latin America), increasing the area under smallholder wheat production is unlikely to improve local production levels or improve smallholder farmers' income (Negassa et al., 2013; Shiferaw et al., 2013). This is mainly because the cost involved (investment in research, irrigation, infrastructure, support services etc.) in such expansion initiatives will be prohibitive. Moreover, the benefits will take long to accrue and their long-term sustainability is highly uncertain (Negassa et al., 2013; Mason, Jayne and Shiferaw, 2015). In the majority of countries, there is need for further work to find ways of increasing wheat acreage in existing farming systems, alongside other cash and food crops, and improving productivity on existing acreage by addressing prevailing socio-economic and biophysical limitations (Shiferaw et al., 2013).

In most developing countries, research has shown that poor markets for items such as machinery, labour, services and credit stifle chances of improving smallholder wheat farmers' productivity (Mason, Jayne and Shiferaw, 2015). There have been suggestions that this problem could be addressed through promoting contract farming with the millers or large-scale commercial farmers among smallholder farmers. This way, the millers and more-resourced commercial farmers could help provide leadership, services, knowhow and a bargaining voice to smallholder farmers. There are reports which show that such approaches have had mixed successes in Africa and Asia (Sharma, 2008; Sriboonchitta and Wiboonpoongse, 2008; Kazunga, 2015). It appears that strict government regulation of private sector participation (e.g. through registered associations of wheat producers, marketers and processors) could increase success rates (Sriboonchitta and Wiboonpoongse, 2008; Kazunga, 2015; Mason, Jayne and Shiferaw, 2015). Further work is needed to formulate the policies and regulations that would support such initiatives.

Research has shown that wheat grain prices have a direct impact on farmers' willingness to grow wheat and improve yields, both among smallholder and better-resourced farmers (Tanner and Mwangi, 1992; Shiferaw et al., 2013; Mason, Jayne and Shiferaw, 2015). Wheat farmers are likely to use less inputs (e.g. reduced pest or disease control or fertiliser), shift to producing other crops or stop farming altogether if the ratio between wheat grain price and production cost is not attractive. Wheat grain prices in most developing countries are often government controlled (e.g. through tariffs and taxes) and researchers have looked at how government policy could be used to drive up smallholder production. For example, when it comes to tariffs on imported wheat grain, governments often struggle to balance the interests of millers and urban dwellers, who usually lobby for reduced tariffs, and large-scale wheat producers, who would prefer higher rates (Mason, Jayne and Shiferaw,

2015). Smallholder producers often have limited say in these policy discussions and most government policies do little to improve their production. As a result, imports continue to dominate the developing world wheat markets. Research has suggested that instead of trying to control markets using tariffs/taxes, governments could be better off investing in rural infrastructure, input subsidies, irrigation, research, extension and market information systems and so on, to support smallholder wheat farmers (Fan, Gulati and Thorat, 2008). Such an approach would raise both agricultural productivity for wheat and other staples, and rural incomes in general (Fan, Gulati and Thorat, 2008; Mason, Jayne and Shiferaw, 2015).

Socio-economic researchers have looked into novel ways of disseminating information about improved technologies and agronomic practices to smallholder farmers. Most solutions coming out of this work advocate for adopting innovation system approaches. In this approach, researchers are no longer the main players, but participants in a system that involves policymakers, agri-business, advisory services, financial services, farmers and so on (Gildemacher and Wongtschowski, 2015). This will facilitate communication among the multidisciplinary stakeholders along the wheat value chain to capture various end-user requirements and identifying what product or service could address those needs (CGIAR Research Program on Wheat, 2015a). Such a change in approach would be a significant step forward from the traditional linear innovation chain (running from researcher to farmer through the advisory service provider), which has so far had mixed success in promoting early adoption of R&D outputs by smallholder farmers (Negassa et al., 2013; Shiferaw et al., 2013; Gildemacher and Wongtschowski, 2015). All participants in the systems would therefore become potential sources of and targets/entry points for innovation. Further work is required to identify how this systems-based approach would work to support smallholder wheat farmers. However, there is a general consensus that it should include greater public–private sector collaboration in promoting improved skills, ways of interaction (including use of information and communications technology (ICT)) and incentives to act for all participants (Gildemacher and Mur, 2013; Gildemacher and Wongtschowski, 2015).

In coming up with policies to support smallholder wheat cultivation, researchers will need to work with developing world governments as they balance support for smallholder wheat farming and other priorities. There is need to identify what alternative national development initiatives, including supporting smallholder production of other staple crops, will be foregone as developing countries support smallholder wheat production (Tanner and Mwangi, 1992; Fan, Gulati and Thorat, 2008; Shiferaw et al., 2013; Mason, Jayne and Shiferaw, 2015).

Researchers also need to objectively demonstrate the competitiveness of smallholder wheat production, for example, competitiveness of wheat production over production of other crops, local production over imports and smallholder wheat cultivation over large-scale commercial farming. In some cases, smallholder wheat production may be shown to be supportive of other crops or industries. For example, blending other staples like sweet potato and cassava with wheat could potentially increase local demand and production of these crops by smallholder farmers. This could have additional benefits beyond improved productivity and household incomes, for example, improved nutrition when micronutrient-rich staples are combined with wheat (Shittu et al., 2008; Bibiana, Grace and Julius, 2014; Jenkins, Byker Shanks and Houghtaling, 2015). Further work is needed to ascertain long-term impact and sustainability of these initiatives, which have already been piloted in some developing countries.

4 Case study: Agri-transfer

4.1 Background

New, improved and disease-resistant wheat varieties developed by the Kenya Agricultural & Livestock Research Organisation (KALRO) do not always find their way into the fields of smallholder farmers (Fig. 6). Part of the problem is a lack of awareness and understanding by smallholder farmers on how to grow a good wheat crop, poor access to affordable, genuine, certified seed and inadequate dissemination platforms to educate and engage smallholder farmers. Adoption of these new varieties requires the implementation of suitable seed multiplication systems (e.g. community-based, farmer-managed systems), education about the innovation and best practice to optimise wheat production and suitable methods of disseminating information.

Agri-transfer is a 2-year technology transfer project supported by the UK Biotechnology and Biological Sciences Research Council (BBSRC) Flexible Interchange Programme and two UK charities: the Malaysian Centre for Commonwealth Studies (MCSC) and the Cambridge Malaysian Education and Development Trust (CMEDT). It involves National Institute of Agricultural Botany (NIAB), KALRO, MCSC and CMEDT staff working with the extension officers in Narok and Nakuru counties. The overall aim of this project, which started in March 2015, is to engage with smallholder farmer communities in Kenya to

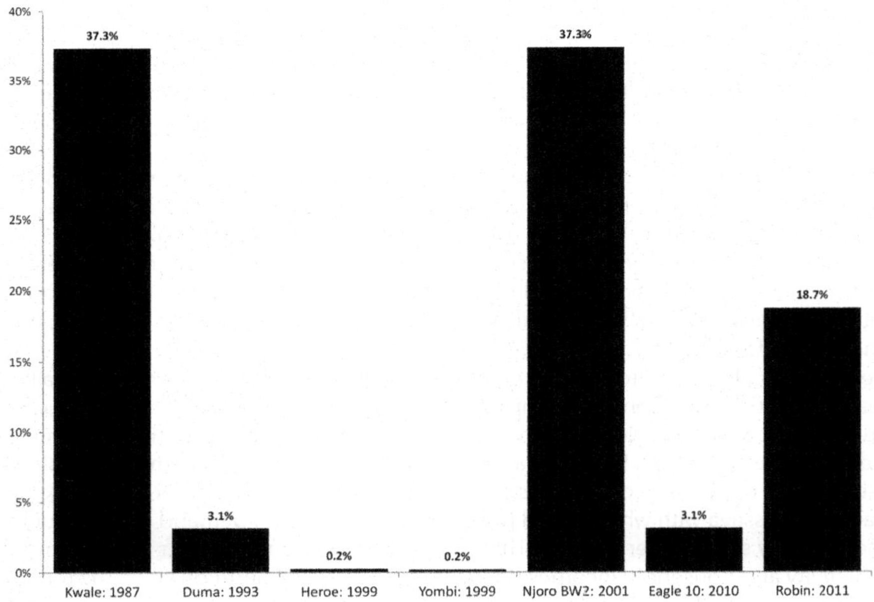

Figure 6 Estimates of relative variety acreage in Kenya in 2012, variety names are listed with the year of release. Varieties released since 2010 accounted for about 20% of the acreage, while varieties released more than 10 years earlier accounted for 75%. Varieties released in 2012 (Tae, Sunbird, Wren, Korongo, Kingbird, Hawk 12) were completely missing from farmers' fields (CIMMYT, 2016a).

improve wheat production, while finding best practice with regard to knowledge transfer to smallholder farmers. In the long term, it is hoped that this pilot project will generate useful information that can inform good practices in knowledge transfer and training to improve smallholder production of not only wheat, but also other appropriate smallholder farmer crops in communities across Kenya, as well as other developing countries.

4.2 Farmer-managed trials and demonstrations

Demonstration plots were grown on a five-acre farm in the Njoro Ward of Nakuru County, about 4 km from KALRO's centre in Njoro. The farm was jointly leased and managed by a smallholder farmers' group called Tuinuane (Fig. 7); they paid for the land and provided labour for manual operations on the farm, while the project provided all bought-in inputs and machinery. The demonstration plots were established during the main rainy season of 2015. The aim was to test and demonstrate agronomic practices (i.e. land preparation methods; pest, weed and disease control; fertiliser choice and application rates; impact of seed quality) and new wheat varieties.

The demonstration plots were set up to train the farmer group and to showcase to other farmers as part of an open field day. The response from the local farmers was one of considerable interest and enthusiasm, with a strong desire to learn. Unfortunately, the opportunity for a formal open day was lost because of late planting due to heavy rainfall at the start of the season and erratic crop development due to high temperatures and drought. However, despite these erratic weather conditions, the Tuinuane farmer group made a profit from the sale of the harvest. They sowed a wheat crop the following season,

Figure 7 Members of the Tuinuane farmer group with Agri-transfer team members in the wheat demonstration plots in Njoro Ward, September 2015.

on adjacent land to take advantage of the predicted enhanced rainfalls during the rainfall season of October–December 2015. The decision to sow was taken and implemented without assistance from KALRO, applying the knowledge and skills learnt within Agri-transfer. The group is also planning to register as a grower of certified wheat seed, contracted by the KALRO Seed Unit, during the season starting May–June 2016.

4.3 Wheat production handbook

All the major operations and practices carried out on the demonstration farm rented by the Tuinuane group were documented to create audio, video and photographic learning/training material. This material was developed into a digital wheat production guide (including information on off-farm activities such as agro-economic planning, marketing, processing and value addition) in the form of a mobile phone application. At the same time, the project team has developed a printed wheat production guide, an updated version of an old KALRO production guide last updated in 2003. The next phase of the project will compare the efficacy of the digital and the printed production guides as training platforms to identify the optimal approach for instructing smallholder wheat farmers in the practices of wheat production.

4.4 Farmer-managed certified seed production

Seed multiplication plots of the KALRO wheat variety Eagle 10 were sown in Mau Narok Ward of Nakuru County, about 30 km from KALRO's centre in Njoro. This was carried out in collaboration with the smallholder farming group called Pambazuka, teaching them how to produce certified wheat seed, while multiplying seed for their own use in the 2016 season.

Figure 8 Pambazuka farmer group with Agri-transfer team members in the seed multiplication plot in Mau Narok, January 2016.

5 Future trends

Although some of the issues affecting smallholder wheat production systems may fall solely within historically defined disciplines (i.e. agronomy, plant breeding, mechanisation, post-harvest handling, marketing etc.), most issues have multidisciplinary and interdisciplinary ramifications (Godfray et al., 2010a,b). Addressing these issues requires multi-partner, systems-level approaches to transform the rapid developments in wheat research into products and services for development. Smallholder wheat agri-food systems are highly interconnected with other socio-economic activities, locally and globally, such that no one aspect can be considered in isolation. Therefore, any future improvements in smallholder wheat cultivation should consider sustainable intensification of the systems. That is, there is need to sustainably increase yields and reduce yield gaps for wheat farming systems and improve post-harvest handling and marketing. This calls for improved efficiency in using agricultural inputs, minimising negative environmental impact and reducing post-harvest loss.

In future, smallholder wheat production systems will also need to withstand biophysical changes. Future research should produce climate-smart varieties that can, for example, withstand rising temperatures, use water more efficiently and withstand emerging pests and diseases while producing more per unit area. This calls for systems-based research methods that incorporate information from varied sources to clearly define research goals and pathways to impact, objectively quantify research impacts and come up with improved approaches fast.

Future wheat varieties have to support multiple products, and provide nutrition benefits and accelerated yield gains under smallholder farming conditions. Going forward, wheat breeding will need to modernise and employ innovative approaches, including marker-assisted breeding, genomic selection, high-throughput genotyping and phenotyping and use new methods of creating variation (e.g. transgenes and genome editing). These improved breeding methods offer a chance to speed up the breeding process and help better link product development and deployment with the dynamic conditions smallholder farmers operate in.

Smallholder farming will have to supply more sophisticated markets as consumer numbers and quality requirements increase, this will require better communication and integrated action in the wheat value chain, that is, from among technology developers, right through to farmers, processors and consumers. Thus, the future calls for changes in services, institutions and policies that better support smallholder farmers to ensure that new and appropriate technologies and agronomic practices are constantly developed and deployed to them. There is need for researchers to find out what influences smallholder farmer adoption of new technologies (particularly new varieties) and practices (including conservation agriculture) in order to achieve sustainable productivity improvements without worsening environmental impacts of wheat cultivation.

As the global consumption and demand for wheat increases, the world will need to call upon smallholder wheat farmers to contribute to this demand. At present, they are too poorly resourced and vulnerable to fluctuations in the agricultural environment, both biophysical (e.g. unpredictable climatic fluctuations, diseases and pests and water and land availability) and socio-economic (e.g. access to land, increases in costs of labour and agrochemicals and varying wheat prices) to make a significant contribution. The wheat R&D community – public and private – needs to take action to address the issues faced by smallholder farmers, ensuring knowledge transfer and technology uptake that will support smallholder farming communities on the road to commercial wheat production.

6 Where to look for further information

Good introductions

Reviews

Kosina, P., Reynolds, M., Dixon, J. and Joshi, A. (2007). 'Stakeholder perception of wheat production constraints, capacity building needs, and research partnerships in developing countries', *Euphytica*, 157(3), pp. 475–83.

Shiferaw, B., Smale, M., Braun, H. J., Duveiller, E. and Muricho, G. (2013). 'Crops that feed the world 10. Past successes and future challenges to the role played by wheat in global food security', *Food Security*, 5(3), pp. 291–317.

Books

Curtis, B., Rajaram, S. and Macpherson, H. eds. (2002). *Bread Wheat: Improvement and Production*. Rome: FAO. Available at: http://www.fao.org/docrep/006/y4011e/y4011e00.htm#Contents.

Dixon, J., Braun, H. J., Kosina, P. and Crouch, J. H. eds. (2009). *Wheat Facts and Futures 2009*, Mexico, DF: CIMMYT. Available at: http://repository.cimmyt.org/xmlui/handle/10883/1265.

Key societies, projects

- Genetic Resources Information System for Wheat and Triticale (GRIS): http://wheatpedigree.net/
- CGIAR Research Program on Wheat: http://wheat.org/

Wheat initiative

- The International Research Initiative for Wheat Improvement: http://www.wheatinitiative.org/
- The International Wheat Yield Partnership (IWYP): http://iwyp.org/

The Borlaug Global Rust Initiative (BGRI)

- The Durable Rust Resistance in Wheat Project: http://www.wheatrust.cornell.edu/
- Global Rust Reference Center: http://wheatrust.org/
- Marker Assisted Selection in Wheat: http://maswheat.ucdavis.edu/
- Rust Tracker: http://rusttracker.cimmyt.org/
- Wheat Atlas: http://wheatatlas.org/

Key conferences

- BGRI Workshop
- International Cereal and Bread Congress

A few research centres to watch

- The International Maize and Wheat Improvement Center (CIMMYT): http://www.cimmyt.org/en/
- The International Center for Agricultural Research in the Dry Areas (ICARDA): http://www.icarda.org/
- Chinese Academy of Agricultural Sciences (CAAS): http://www.caas.cn/en/
- Brazilian Agricultural Research Corporation (EMBRAPA): https://www.embrapa.br/en/trigo
- Ayub Agricultural Research Institute (AARI): http://www.aari.punjab.gov.pk/institutes-sections/wheat-research-institute
- KALRO Food Crops Research Institute-Njoro: http://www.kalro.org/food_crops_research_institute
- Ethiopian Institute of Agricultural Research (EIAR): http://www.eiar.gov.et/
- Indian Institute of Wheat and Barley Research: http://www.dwr.res.in/node/1
- National Institute of Agricultural Botany (NIAB): www.niab.com

7 References

Alemu, D. and Bishaw, Z. (2015). 'Commercial behaviours of smallholder farmers in wheat seed use and its implication for demand assessment in Ethiopia', *Development in Practice*, 25(6), pp. 798–814. doi: 10.1080/09614524.2015.1062469.

Alexandratos, N. and Bruinsma, J. (2012). *World Agriculture Towards 2030/2050: The 2012 Revision, ESA Working Paper No. 12-03*. Rome. doi: 10.1016/S0264-8377(03)00047-4.

Andae, G. (2014). 'State mulls controlling use of DAP fertiliser', *Business Daily Africa*, 19 February. Available at: http://www.businessdailyafrica.com/State-mulls-how-to-control-use-of-DAP-fertiliser/-/539546/2213608/-/bjkemg/-/index.html.

Anderson, W. K. and Impiglia, A. (2002). 'Management of dryland wheat', in B. C. Curtis, S. Rajaram and H. G. Macpherson (eds), *Bread Wheat: Improvement and Production*. FAO Production and Protection Series Rome: FAO. Available at: http://www.fao.org/docrep/006/y4011e/y4011e0s.htm#bm28.

Aryal, J. P., Mehrotra, M. B., Jat, M. L. and Sidhu, H. S. (2015). 'Impacts of laser land leveling in rice–wheat systems of the north–western indo-gangetic plains of India', *Food Security*, 7(3), pp. 725–38. Available at: http://link.springer.com/article/10.1007%2Fs12571-015-0460-y.

Beyene, H., Franzel, S. and Mwangi, W. (1990). 'Constraints to increasing wheat production in Ethiopia's small-holder sector', in D. G. Tanner, M. van Ginkel and W. Mwangi (eds), *Sixth Regional Wheat Workshop for Eastern, Central, and Southern Africa*. Mexico: CIMMYT, pp. 284–92. Available at: http://libcatalog.cimmyt.org/download/cis/24128.pdf.

Bibiana, I., Grace, N. and Julius, A. (2014). 'Quality evaluation of composite bread produced from wheat, maize and orange fleshed sweet potato flours', *American Journal of Food Science and Technology*. Science and Education Publishing, 2(4), pp. 109–15. doi: 10.12691/ajfst-2-4-1.

Birungi, P. and Hassan, R. (2010). 'Poverty, property rights and land management in Uganda', *African Journal of Agricultural and Resource Economics*. African Association of Agricultural Economists, 4(1), pp. 48–69.

Braun, H.-J., Rajaram, S. and Ginkel, M. (1996). 'CIMMYT's approach to breeding for wide adaptation', *Euphytica*, 92(1), pp. 175–83. doi: 10.1007/BF00022843.

CGIAR (2015). *Impacts of International Wheat Improvement Research in the World, 1994-2014, and the Need for Secure Funding*. Mexico, D.F.: CGIAR. Available at: http://repository.cimmyt.org/xmlui/handle/10883/4467.

CGIAR Research Program on Wheat (2015a). *CGIAR Research Program on Wheat, AFS-WHEAT: Pre-Proposal for Phase II (2017–2022)*. Mexico, D.F.: CGIAR. Available at: https://library.cgiar.org/bitstream/handle/10947/2915/Pre-Proposal_for_CRP_2nd_Call-CGIAR_Research_Program_on_Wheat.pdf.

CGIAR Research Program on Wheat (2015b). *CRP WHEAT Gender Strategy*. Mexico, D.F.: CGIAR Research Program on Wheat.

CIMMYT (2014). *Wheat Matters*. Mexico, D.F.: CGIAR.

CIMMYT (2015). *Impacts of International Wheat Improvement Research: 1994–2014*. Available at: http://www.cimmyt.org/en/what-we-do/wheat-research/item/impacts-of-international-wheat-improvement-research-1994-2014 (Accessed 22 February 2016).

CIMMYT (2016). *Wheat Atlas: Global Distribution of Wheat Megaenvironments*. Available at: http://wheatatlas.org/me-global-distribution (Accessed 22 January 2016).

Dixon, J., Braun, H. J., Kosina, P. and Crouch, J. H. (2009). 'Transitioning wheat research to meet the future demands of the developing world', in J. Dixon, H. J. Braun, P. Kosina and J. Crouch (eds), *Wheat Facts and Futures 2009*. Mexico, D.F.: CIMMYT. Available at: https://books.google.co.ke/books?id=LV83xluWFiYC.

Dixon, J., Tanyeri-Abur, A. and Wattenbach, H. (2004). 'Framework for analysing impacts of globalization on smallholders', in J. Dixon, K. Taniguchi, H. Wattenbach and A. Tanyeri-Arbur (eds), *Smallholders, Globalization and Policy Analysis*. Food and Agriculture. Rome: FAO. Available at: http://www.fao.org/docrep/007/y5784e/y5784e02.htm#bm02.

Doss, C. R. (2001). 'Designing agricultural technology for African Women farmers: lessons from 25 Years of Experience', *World Development*, 29(12), pp. 2075–92. doi: http://dx.doi.org/10.1016/S0305-750X(01)00088-2.

Duveiller, E., Singh, R. P. and Nicol, J. M. (2007). 'The challenges of maintaining wheat productivity: pests, diseases, and potential epidemics', *Euphytica*, 157(3), pp. 417–30. doi: 10.1007/s10681-007-9380-z.

Erenstein, O., Sayre, K., Wall, P., Hellin, J. and Dixon, J. (2012). 'Conservation agriculture in maize- and wheat-based systems in the (Sub)tropics: lessons from adaptation initiatives in South Asia, Mexico, and Southern Africa', *Journal of Sustainable Agriculture*. Taylor & Francis, 36(2), pp. 180–206. doi: 10.1080/10440046.2011.620230.

Fafchamps, M., Kebede, B. and Quisumbing, A. R. (2009). 'Intrahousehold welfare in rural ethiopia', *Oxford Bulletin of Economics and Statistics*, 71(4), pp. 567–99. doi: 10.1111/j.1468-0084.2009.00553.x.

Fan, S., Gulati, A. and Thorat, S. (2008). 'Investment, subsidies, and pro-poor growth in rural India', *Agricultural Economics*. Blackwell Publishing Inc., 39(2), pp. 163–70. doi: 10.1111/j.1574-0862.2008.00328.x.

FAOSTAT (2016). *FAOSTAT Database*. Available at: http://faostat3.fao.org/home/E (Accessed 23 January 2016).

Fischer, T., Byerlee, D. and Edmeades, G. (2014). 'Crop yields and global food security. Will yield increase continue to feed the world?', *ACIAR Monograph Series*, p. 660. doi: ISBN 978-1-925133-06-6.

Food and Agriculture Organization (2011). *The State of Food and Agriculture. Women in Agriculture: Closing the Gap for Development, Lancet*. doi: ISSN 0081-4539.

Gildemacher, P. and Mur, R. (2013). *Bringing New Ideas into Practice; Experiments with Agricultural Innovation. Learning from Research into Use in Africa, Overview*. Amsterdam: KIT publishers.

Gildemacher, P. and Wongtschowski, M. (2015). *Catalysing Innovation: From Theory to Action, KIT Working Papers*. Amsterdam: KIT Publishers (KIT Working Papers). Available at: http://213ou636sh0ptphd141fqei1.wpengine.netdna-cdn.com/sed/wp-content/uploads/sites/2/2015/06/WPS1_2015_online.pdf.

Giller, K. E., Tittonell, P., Rufino, M. C., van Wijk, M. T., Zingore, S., Mapfumo, P., Giller, K. E., Tittonell, P., Rufino, M. C., van Wijk, M. T., Zingore, S., Mapfumo, P., Adjei-Nsiah, S., Herrero, M., Chikowo, R., Corbeels, M., Rowe, E. C., Baijukya, F. Mwijage, A., Smith, J. Yeboah, E., van der

Burg, W. J., Sanogo, O. M., Misiko, M., de Ridder, N., Karanja, S., Kaizzi, C., K'ungu, J., Mwale, M., Nwaga, D., Pacini, C. and Vanlauwe, B. (2011). 'Communicating complexity: Integrated assessment of trade-offs concerning soil fertility management within African farming systems to support innovation and development', *Agricultural Systems*, 104(2), pp. 191–203. doi: http://dx.doi.org/10.1016/j.agsy.2010.07.002.

Godfray, H. C. J., Beddington, J. R., Crute, I. R., Haddad, L., Lawrence, D., Muir, J. F., Pretty, J., Robinson, S., Thomas, S. M. and Toulmin, C. (2010a). 'Food security: the challenge of feeding 9 billion people', *Science*, 327(5967), pp. 812–8. doi: 10.1126/science.1185383.

Godfray, H. C. J., Beddington, J. R., Crute, I. R., Haddad, L., Lawrence, D., Muir, J. F., Pretty, J., Robinson, S., Thomas, S. M. and Toulmin, C. (2010b). 'Theme Issue "Food security: feeding the world in 2050"', *Philosophical Transactions of the Royal Society B Biological Sciences*, 365, pp. 2765–3097.

Gollin, D. (2006). *Impacts of International Research on Intertemporal Yield Stability in Wheat and Maize: An Economic Assessment*. Mexico, D.F.: CIMMYT. Available at: http://repository.cimmyt.org/xmlui/handle/10883/1053.

Hodson, D. P. and White, J. W. (2007). 'Use of spatial analyses for global characterization of wheat-based production systems: Paper Presented at International Workshop on Increasing Wheat Yield Potential, CIMMYT, Obregon, Mexico, 20–24 March 2006', *The Journal of Agricultural Science*, 145(02), pp. 115–25.

Jat, M. L., Gathala, M. K., Ladha, J. K., Saharawat, Y. S., Jat, A. S., Kumar, V., Sharma, S. K., Kumar, V. and Gupta, R. (2009). 'Evaluation of precision land leveling and double zero-till systems in the rice–wheat rotation: Water use, productivity, profitability and soil physical properties', *Soil and Tillage Research*, 105(1), pp. 112–21. doi: http://dx.doi.org/10.1016/j.still.2009.06.003.

Jat, R. K., Sapkota, T. B., Singh, R. G., Jat, M. L., Kumar, M. and Gupta, R. K. (2014). 'Seven years of conservation agriculture in a rice–wheat rotation of Eastern Gangetic Plains of South Asia: Yield trends and economic profitability', *Field Crops Research*, 164, pp. 199–210. doi: http://dx.doi.org/10.1016/j.fcr.2014.04.015.

Jenkins, M., Byker Shanks, C. and Houghtaling, B. (2015). 'Orange-fleshed sweet potato: successes and remaining challenges of the introduction of a nutritionally superior staple crop in Mozambique', *Food and Nutrition Bulletin*. doi: 10.1177/0379572115597397.

Joshi, A. K., Azab, M., Mosaad, M., Moselhy, M., Osmanzai, M., Gelalcha, S., Joshi, A. K., Azab, M., Mosaad, M., Moselhy, M., Osmanzai, M., Gelalcha, S., Bedada, G., Bhatta, M. R., Hakim, A., Malaker, P. K., Haque, M. E., Tiwari, T. P., Majid, A., Jalal Kamali, M. R., Bishaw, Z., Singh, R. P., Payne, T. and Braun, H. J. (2011). 'Delivering rust resistant wheat to farmers: A step towards increased food security', *Euphytica*, 179(1), pp. 187–96. doi: 10.1007/s10681-010-0314-9.

Kassie, M., Teklewold, H., Jaleta, M., Marenya, P. and Erenstein, O. (2015). 'Understanding the adoption of a portfolio of sustainable intensification practices in eastern and southern Africa', *Land Use Policy*, 42, pp. 400–11. doi: 10.1016/j.landusepol.2014.08.016.

Kazunga, O. (2015). 'Contract wheat farming suspended', *Zimbabwe Chronicle*, 26 November. Available at: http://www.chronicle.co.zw/contract-wheat-farming-suspended/.

Kosina, P., Reynolds, M., Dixon, J. and Joshi, A. (2007). 'Stakeholder perception of wheat production constraints, capacity building needs, and research partnerships in developing countries', *Euphytica*, 157(3), pp. 475–83. doi: 10.1007/s10681-007-9529-9.

Lantican, M. A., Dubin, H. J. and Morris, M. L. (2005). *Impacts of International Wheat Breeding Research in the Developing World, 1988–2002*. Mexico, D.F.: CIMMYT.

Lobell, D. B., Schlenker, W. and Costa-Roberts, J. (2011). 'Climate trends and global crop production since 1980', *Science*, 333(6042), pp. 616–20. doi: 10.1126/science.1204531.

Mahayagu, M. C., Kamwaga, J., Ndiema, A. C., Kamundia, J. and Gamba, P. (2007). 'Wheat productivity, constraints associated in the eastern parts of Kenya Timau Division', in *African Crop Science Conference Proceedings*, pp. 1211–14.

Mason, N. M., Jayne, T. S. and Shiferaw, B. (2015). 'Africa's rising demand for wheat: trends, drivers, and policy implications', *Development Policy Review*, 33(5), pp. 581–613. doi: 10.1111/dpr.12129.

Matuschke, I., Mishra, R. R. and Qaim, M. (2007). 'Adoption and impact of hybrid wheat in India', *World Development*, 35(8), pp. 1422–35. doi: http://dx.doi.org/10.1016/j.worlddev.2007.04.005.

Meinzen-dick, R., Quisumbing, A., Behrman, J., Biermayr-Jenzano, P., Wilde, V., Noordeloos, M., Ragasa, C. and Beintema, N. (2010). 'Engendering agricultural research', *IFPRI Disscussion Paper 973* (May), p. 72. Available at: http://www.indiaenvironmentportal.org.in/files/ifpri-may2010.pdf.

Mekasha, A., Gerard, B., Tesfaye, K., Nigatu, L. and Duncan, A. J. (2014). 'Inter-connection between land use/land cover change and herders'/farmers' livestock feed resource management strategies: A case study from three Ethiopian eco-environments', *Agriculture, Ecosystems and Environment*, 188, pp. 150–62. doi: 10.1016/j.agee.2014.02.022.

Munns, R., James, R. A., Xu, B., Athman, A., Conn, S. J., Jordans, C., Byrt, C. S., Hare, R. A., Tyerman, S. D., Tester, M., Plett, D. and Gilliham, M. (2012). 'Wheat grain yield on saline soils is improved by an ancestral Na+ transporter gene', *Nat Biotech*. Nature Publishing Group, a division of Macmillan Publishers Limited. All Rights Reserved, 30(4), pp. 360–4. Available at: http://dx.doi.org/10.1038/nbt.2120.

Mwanga, J., Mussei, A., Mwangi, W. and Verkuijl, H. (2001). 'Adoption of improved wheat technologies by small scale farmers in Mbeya district of Southern Highlands, Tanzania', *Proceedings of the Tenth Regional Wheat Workshop for Eastern Central and Southern Africa University of Stellenbosch South Africa 1418 September 1998* (February), pp. 39–45. Available at: http://repository.cimmyt.org/xmlui/bitstream/handle/10883/1021/74246.pdf.

Nazli, H., Asare-Marfo, D., Smale, M., Malik, S. J. and Birol, E. (2014). *Smallholder Farming and Crop Variety Choice: Wheat Variety Choice in Pakistan*, 2. Washington, DC.

Negassa, A., Shiferaw, B., Koo, J., Sonder, K., Smale, M., Braun, H. J., Gbegbelegbe, S., Guo, Z., Hodson, D. P., Wood, S., Payne, T. S. and Bekele, G. A. (2013). *The Potential for Wheat Production in Africa: Analysis of Biophysical Suitability and Economic Profitability*. Mexico, D.F.: CIMMYT.

Nelson, K. (2013). *Analysis of Farmer Preferences for Wheat Variety Traits in Ethiopia: A Gender-Responsive Study*. Ithaca, NY: Cornell University.

Nia, S. H., Zarea, M. J., Rejali, F. and Varma, A. (2012). 'Yield and yield components of wheat as affected by salinity and inoculation with Azospirillum strains from saline or non-saline soil', *Journal of the Saudi Society of Agricultural Sciences*, 11(2), pp. 113–21. doi: http://dx.doi.org/10.1016/j.jssas.2012.02.001.

OECD and FAO (2014). 'OECD-FAO Agricultural Outlook'. Paris: OECD Publishing. doi: http://dx.doi.org/10.1787/agr_outlook-2014-en.

Oerke, E.-C. (2006). 'Crop losses to pests', *The Journal of Agricultural Science*, 144(01), pp. 31–43. doi: 10.1017/S0021859605005708.

Oerke, E.-C. (2007). 'Crop losses to animal pests, plant pathogens, and weeds', in *Encyclopedia of Pest Management, Volume II*. CRC Press, pp. 116–20. doi: doi:10.1201/9781420068467.ch29.

Ortiz, R., Sayre, K. D., Govaerts, B., Gupta, R., Subbarao, G. V., Ban, T., Hodson, D., Dixon, J. M., Iván Ortiz-Monasterio, J. and Reynolds, M. (2008). 'Climate change: Can wheat beat the heat?', *Agriculture, Ecosystems and Environment*, 126(1–2), pp. 46–58. doi: http://dx.doi.org/10.1016/j.agee.2008.01.019.

Ortiz-Ferrara, G., Bhatta, M. R., Pokharel, T., Mudwari, A., Thapa, D. B., Joshi, A. K., Chand, R., Muhammad, D., Duveiller, E. and Rajaram, S. (2001). 'Farmer Participatory Variety Selection in South Asia', in CIMMYT (ed.), *Research Highlights of the CIMMYT Wheat Program, 1999–2000*. Mexico, D.F.: CIMMYT.

Pandit, D. B., Mandal, M. S. N., Hakim, M. A., Barma, N. C. D., Tiwari, T. P. and Joshi, A. K. (2011). 'Farmers' preference and informal seed dissemination of first Ug99 tolerant wheat variety in Bangladesh', *Czech Journal of Genetics and Plant Breeding*, 47(SPEC. ISSUE 1).

Rajaram, S., van Ginkel, M. and Fischer, R. A. (1995). 'CIMMYT's wheat breeding mega-environments (ME)', in Z. S. Li and Z. Y. Xin (eds), *Proceedings of the Eighth International Wheat Genetics Symposium: Held at Beijing, China, 20–25 July 1993*. Beijing: China Agricultural Scientech Press. Available at: https://books.google.co.uk/books?id=8lseAAAACAAJ.

Renkow, M. and Byerlee, D. (2010). 'The impacts of CGIAR research: A review of recent evidence', *Food Policy*, 35, pp. 391–402. doi: 10.1016/j.foodpol.2010.04.006.

Reynolds, M. P. and Borlaug, N. E. (2006). 'Impacts of breeding on international collaborative wheat improvement', *The Journal of Agricultural Science*, 144, p. 3. doi: 10.1017/S0021859606005867.

Sayre, K. D. and Govaerts, B. (2009). 'Conservation agriculture for sustainable wheat production', in J. Dixon, H.-J. Braun, P. Kosina and J. Crouch (eds), *Wheat Facts and Futures 2009*. Mexico, D.F.: CIMMYT, pp. 62–9.

Sayre, K. D. (2002). 'Management of irrigated wheat', in B. Curtis, S. Rajaram and H. G. Macpherson (eds), *Bread Wheat: Improvement and Production*. FAO Plant. Rome: FAO. Available at: http://www.fao.org/docrep/006/y4011e/y4011e0s.htm#bm28.

Schnier, H. F., Recke, H., Muchena, F. N. and Muriuki, A. W. (n.d.). 'Towards a practical approach to fertilizer recommendations for food crop production in smallholder farms in Kenya', *Nutrient Cycling in Agroecosystems*, 47(3), pp. 213–26. doi: 10.1007/BF01986276.

Sharma, V. P. (2008). 'India's Agrarian crisis and corporate-led contract farming: socio-economic implications for smallholder producers', *International Food and Agribusiness Management Review*, 11(8), pp. 25–48.

Shiferaw, B., Smale, M., Braun, H. J., Duveiller, E., Reynolds, M. and Muricho, G. (2013). 'Crops that feed the world 10. Past successes and future challenges to the role played by wheat in global food security', *Food Security*, 5(3), pp. 291–317. doi: 10.1007/s12571-013-0263-y.

Shittu, T. A., Dixon, A., Awonorin, S. O., Sanni, L. O. and Maziya-Dixon, B. (2008). 'Bread from composite cassava–wheat flour. II: Effect of cassava genotype and nitrogen fertilizer on bread quality', *Food Research International*, 41(6), pp. 569–78. doi: http://dx.doi.org/10.1016/j.foodres.2008.03.008.

Singh, R. P., Hodson, D. P., Huerta-Espino, J., Jin, Y., Bhavani, S., Njau, P., Herrera-Foessel, S., Singh, P. K., Singh, S. and Govindan, V. (2011). 'The emergence of Ug99 races of the stem rust fungus is a threat to world wheat production', *Annual Review of Phytopathology*. United States, 49, pp. 465–81. doi: 10.1146/annurev-phyto-072910-095423.

Singh, R. P., Hodson, D. P., Huerta-Espino, J., Jin, Y., Njau, P., Wanyera, R., Herrera-Foessel, S. A. and Ward, R. W. (2008). 'Will stem rust destroy the world's wheat crop?', in B. T.-A. Agronomy (ed.), Academic Press, pp. 271–309. doi: http://dx.doi.org/10.1016/S0065-2113(08)00205-8.

Snapp, S. (1999). 'Mother and baby trials: a novel trial design being tried out in Malawi', *Target: Newsletter of the Southern Africa Soil Fertility Network*, 17, p. 8.

Sriboonchitta, S. and Wiboonpoongse, A. (2008). *Overview of Contract Farming in Thailand: Lessons Learned. ADBI Discussion Paper 112*. Tokyo: Asian Development Bank Institute. Available at: http://www.adb.org/publications/overview-contract-farming-thailand-lessons-learned.

Stevenson, J. R., Villoria, N., Byerlee, D., Kelley, T. and Maredia, M. (2013). 'Green Revolution research saved an estimated 18 to 27 million hectares from being brought into agricultural production', *Proceedings of the National Academy of Sciences*, 110 (21), pp. 8363–8. doi: 10.1073/pnas.1208065110.

Tanner, D. G. and Mwangi, W. (1992). 'Current issues in wheat research and production in Eastern, Central and Southern Africa: Constraints and achievements', in D. Tanner and W. Mwangi (eds), *Seventh Regional Wheat Workshop for Eastern, Central and Southern Africa*. CIMMYT, pp. 17–36.

Tesfaye, S., Mesay, Y. and Bedada, B. (2015). 'Gender differentials in scale and technical efficiencies among improved wheat variety growers in wheat belt areas of Ethiopia: A Data Envelopment Analysis (DEA)', *Journal of Economics and Sustainable Development*, 6(1), pp. 91–100.

Thomas, R. J., Akhtar-Schuster, M., Stringer, L. C., Marques, M. J., Escadafal, R., Abraham, E. and Enne, G. (2012). 'Fertile ground? Options for a science-policy platform for land', *Environmental Science and Policy*, 16, pp. 122–35. doi: 10.1016/j.envsci.2011.11.002.

Tiruneh, A., Mwangi, W., Verkuijl, H. and Tesfaye, T. (1999). 'Gender differentials in adoption of improved wheat varieties in Ada, Lume and Gimbichu woredas of the central highlands of Ethiopia', *Proceedings of the Tenth Regional Wheat Workshop for Eastern Central and Southern Africa University of Stellenbosch South Africa 14–18 September 1998*, pp. 31–8.

Tiruneh, A., Tesfaye, T., Mwangi, W. and Verkuijl, H. (2001). *Gender Differentials in Agricultural Production and Decision-Making among Smallholders in Ada, Lume, and Gimbichu Woredas of the Central Highlands of Ethiopia, Development*. Available at: http://apps.cimmyt.org/research/economics/map/impact_studies/impstudea_list/CentralHEth.pdf.

Tsegaye, B. and Berg, T. (2007). 'Utilization of durum wheat landraces in East Shewa, central Ethiopia: are home uses an incentive for on-farm conservation?', *Agriculture and Human Values*, 24(2), pp. 219–30. doi: 10.1007/s10460-006-9055-8.

Tufan, H. A., McGrann, G. R. D., Magusin, A., Morel, J.-B., Miche, L. and Boyd, L. A. (2009). 'Wheat blast: histopathology and transcriptome reprogramming in response to adapted and nonadapted Magnaporthe isolates', *The New Phytologist*. England, 184(2), pp. 473–84. doi: 10.1111/j.1469-8137.2009.02970.x.

Valin, H., Sands, R. D., van der Mensbrugghe, D., Nelson, G. C., Ahammad, H., Blanc, E., Bodirsky, B., Fujimori, S., Hasegawa, T., Havlik, P., Heyhoe, E., Kyle, P., Mason-D'Croz, D., Paltsev, S., Rolinski, S., Tabeau, A., van Meijl, H., von Lampe, M. and Willenbockel, D. (2014). 'The future of food demand: understanding differences in global economic models', *Agricultural Economics*, 45(1), pp. 51–67. doi: 10.1111/agec.12089.

van Vliet, J. A., Schut, A. G. T., Reidsma, P., Descheemaeker, K., Slingerland, M., van de Ven, G. W. J. and Giller, K. E. (2015). 'De-mystifying family farming: Features, diversity and trends across the globe', *Global Food Security*, 5, pp. 11–18. doi: http://dx.doi.org/10.1016/j.gfs.2015.03.001.

World Bank (2008). *Gender in Agriculture Sourcebook, Agriculture and Rural Development*. The World Bank. doi: 10.1596/978-0-8213-7587-7.

World Bank (2016). *Country and Lending Groups*. Available at: http://data.worldbank.org/about/country-and-lending-groups (Accessed 23 January 2016).

Yami, M., Solomon, T., Begna, B., Fufa, F., Alemu, T. and Alemu, D. (2013). 'Source of technical inefficiency of smallholder wheat farmers in selected watwerlogged areas of Ethiopia', *African Journal of Agricultural Research*, 8(29), pp. 3930–40.

Chapter 11

Improving wheat cultivation in Asia

Rajiv Kumar Sharma, Global Wheat Improvement Program – CIMMYT, India

1 Introduction
2 Improving and exploiting genetic diversity
3 Improving breeding techniques
4 Dealing with biotic stresses
5 Dealing with abiotic stresses
6 Improving aspects of quality
7 Improving cultivation practices
8 Case study: improving cultivation in India
9 Summary
10 Where to look for further information
11 References

1 Introduction

With annual production of about 315 million tonnes in Asia (FAO, 2016), wheat is second only to rice in importance. Since the start of the 'Green Revolution' around 1970, wheat production has seen an average annual growth rate of over 3% (Fig. 1), whilst wheat yield has grown at an annual average rate of 2.1% (FAO, 2016). However, yields vary significantly between developed and developing countries. Lopes et al. (2012) studied 26 wheat genotypes released between 1977 and 2008 and estimated annual increases in yields of between 0.5% and 0.9% in Asia and Africa. Other recent estimates of genetic gain are less than or around 1% per annum (Sharma et al., 2012; Manes et al., 2012). Wheat yield stagnation is currently being faced in several parts of the world (Knight et al., 2012). It has been estimated that wheat production in South Asia needs to grow at the rate of 2–2.5% annually until the middle of twenty-first century to feed its population (Chatrath et al., 2007). Singh et al. (2007) also reported that world wheat production needed to grow at 2% annually to meet future demands, half of which could come from breeding better varieties and half from improvements in cultivation techniques.

China exemplifies both the successes and the continuing challenges of improving wheat yields.

http://dx.doi.org/10.19103/AS.2016.0004.41

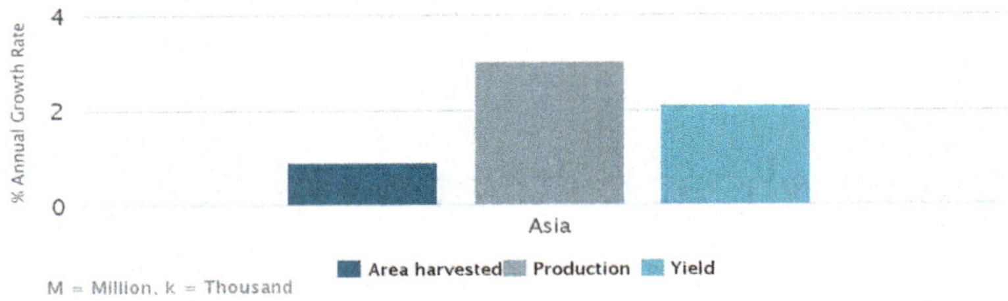

Figure 1 Annual growth rate of wheat area (0.9%), production (3.03%) and yield (2.1%) in Asia during the period 1970–2014.

Chinese wheat production experienced rapid growth after the beginning of hybridization work in 1914. One study found that grain yield of cultivars released between 1937 and 2004 in the Loess Plateau of China increased seven fold during this period (Chen and Hao, 2015). Qin Xiao Liang et al. (2015) investigated 1850 Chinese wheat cultivars from 1920 to 2014 and reported average grain yield gains ranging from 0.52% to 1.5% for different types and regions. The study emphasized concentrating on thousand kernel weight (TKW) and kernel number per spike for future increases in yield. Another Chinese study of historical yield gains revealed a similar figure of 1% average yield gain per year at the national level (Wu Wei, 2014) and also emphasized grain weight and spike weight to be the most promising traits for increasing grain yield. These increases also relied on cultivation practices such as the extensive use of fertilizers and irrigation. However, more intensive cultivation also led to serious problems such as decline in soil fertility, environmental pollution and concerns about remaining water resources (Wang et al., 2009). The review recommended conservation agriculture (CA) techniques to stabilize wheat production. Other studies have shown significant yield variations amongst farmers using new varieties (Li et al., 2014b).

This chapter reviews a number of key themes in improving wheat cultivation in Asia:

- Improving and exploiting genetic diversity
- Improving breeding techniques
- Dealing with biotic stresses
- Dealing with abiotic stresses
- Improving aspects of quality
- Improving cultivation practices

2 Improving and exploiting genetic diversity

Genetic improvement in wheat, like in any other crop, depends on utilizable variability available in the germplasm. Analysis of wheat origins indicates that only a limited number of individuals of the donor species contributed to the evolution of hexaploid wheat. As a result, allelic diversity of bread wheat is relatively narrow. This narrow diversity limits the improvements that can be achieved in common wheat.

Though wheat has already benefited from crossing with at least 52 species belonging to 13 genera (Wulff and Moscou, 2014), access to variability present in wheat's relatives depends on the genetic distance between wheat and these relatives and is still limited. It has been found that related species such as rye (*Secale cereale*) offer useful genes for the improvement of bread wheat (Bagherikia et al., 2014). Deepmala et al. (2015) studied variability among 1423 spring bread wheat accessions and concluded that synthetic hexaploids are genetically more diverse. The study identified new allelic variation for vernalization among landraces originating from Iraq, Iran, Afghanistan, Pakistan, Uzbekistan and Turkmenistan. A collection of 446 wheat landraces collected from Afghanistan was studied using 30 000 diversity array technology genetic markers of which 15 817 were found to be polymorphic. Results revealed high diversity among these landraces compared to those from other regions (Sohail et al., 2015).

To tackle limited utilizable variability in bread wheat germplasm, synthetic hexaploid wheats (SHW) have been created using wild relatives (van Ginkel and Ogbonnaya, 2007; Hao et al., 2013). Primary SHW have been used in China to develop new high-yielding wheat varieties (Li et al., 2014a). These new varieties had elite traits derived from tetraploid parents of SHW. The elite traits included disease resistance, more spikes per plant, more grains per spike and larger grains which led to higher grain yield potential. A three-year field experiment showed SHW-derived cultivars yielding on an average 1.5% higher than local elite nonsynthetic-derived cultivars (NSCs). The traits that mainly contributed to this higher yield were number of grains per unit area and a higher kernel weight (Tang et al., 2015).

The International Maize and Wheat Improvement Centre (CIMMYT) is the global custodian of wheat germplasm and has been the leading wheat research centre providing elite germplasm and other wheat production technologies to wheat-growing countries throughout the developing world. Its success has been documented by Sharma et al. (2012) who reported an annual yield gain of 27.8 kg/ha based on testing conducted over 919 environments in 69 countries during 1995–2009. Since wheat is so crucial to global food security, CIMMYT has recently forged a global alliance for improving food security and the livelihood of the resource-poor farmers in the developing world (CIMMYT, 2011). Investments into wheat research and development have been growing to forge better linkages between researchers and smallholders in particular (Braidotti, 2012).

CIMMYT has used synthetics and employed backcrossing to develop synthetic backcross-derived lines, which are agronomically similar to the improved parents but retain the introgressed novel variability (Dreisigacker et al., 2008). Mujeeb-Kazi et al. (2013) classified useful variability present among wheat relatives as primary, secondary and tertiary variability. They recommended strategies such as synthetic wheats, direct crosses with parental diploids and employing chromosomal exchanges called translocations to transfer useful novel variability into cultivated wheat. A number of SHW having high TKW were screened for other traits and several of them were resistant to diseases such as Karnal Bunt and stripe rust (Kazi et al., 2012).

A wild relative of wheat *Agropyron cristatum* has been exploited by Ye et al. (2015) to enrich fertile tiller number traits in bread wheat. They pinpointed location of fertile tiller number regulator sequence on arm 6PS and 6PL of *A. cristatum*. A segment of chromosome 6P from the same *Agropyron* species carrying enhanced TKW and longer spikes was introgressed into cultivated wheat by Zhang et al. (2015b). As has been noted, *S. cereale* or rye has also contributed to enrich bread wheat gene pool through translocations. 1A.1R and 1B.1R are now commonly found in wheat cultivars (Bagherikia et al., 2014). Another translocation involving substitution of 1B in Chinese Spring by the $1S^l$ from *Aegilops longissimia* ($2n = 2x = 14$, S^lS^l) resulted in higher amounts of A-type starch granules and greater diameter of both A and B granules (Cao et al., 2015). Within *Triticum* species, hulled tetraploids found in mountainous Central Zagros offer a potential source of useful variability (Sheibanirad et al., 2014). Cytogenetic studies confirmed their tetraploid nature with $2N = 28$ and these produced normal crosses with durum wheat.

Another *Triticum* species, viz. *Triticum dicoccoides*, harbours a range of useful traits (Peng et al., 2013) that need further exploration and utilization for future wheat breeding programmes. To further enrich the working genepool, Gupta et al. (2013) conducted a multicrop germplasm collection trip in the Far-Western Development Region of Nepal and collected 24 accessions of wheat among others. Rasheed et al. (2016) studied allelic diversity at 31 loci among 107 landraces (WLR) and 121 historical wheat cultivars (HWC). This study revealed higher frequency of favourable alleles influencing TKW in HWC and ample variability among WLR at most loci.

Diversifying *Rht* genes among wheat cultivars remains an objective for wheat breeders. Wang et al. (2015b) reported *Rht13* as an alternative to GA-responsive dwarfing genes in wheat. They compared the lone GA-responsive *Rht8* with these new genes and found that *Rht13* shortened the length of all internodes (except the basal node) as compared to *Rht8*, which mainly reduced peduncle length alone. They recommended utilizing *Rht8* and *Rht13* alone or in combination for different breeding objectives.

As an alternative strategy for generating variation, gamma rays have been used to create mutants of two wheat varieties, viz. Khirman and NIA-Saarang (Arain et al., 2015). The mutants were compared with mother varieties and treatments were found to be significantly different in spike length, grains per spike and grain yield per plant. Cheng et al. (2015) employed mutation breeding to create useful novel variability. They developed Fu4185, a high TKW mutant of the cultivar Shi4185. The mutant showed higher phenotypic stability across five different environments.

Zhao et al. (2015a) transferred stripe rust resistance and dough quality into bread wheat from the related species *Dasypyrum villosum*. Another translocation from this species with translocation line carrying longer spikes, more spikelets and more grains per spike was characterized by Zhang et al. (2015b). The translocation line with its morphological and co-dominant molecular markers could be useful as a novel germplasm for breeding high-yielding wheat varieties.

Two wheat–*Thinopyrum* substitution lines X479 and X482 were characterized by seed storage protein electrophoresis, genomic *in situ* hybridization, fluorescence *in situ* hybridization and PCR-based molecular markers (Li et al., 2015b). The substitution lines were fully fertile which indicated that rust resistance and novel agronomic traits from these substitution lines could be easily introgressed into bread wheat cultivars. From the same species, Garg et al. (2014) were able to develop a higher dough strength trait and bread-making quality.

3 Improving breeding techniques

Evans and Fischer (1999) defined potential yield as the one obtainable from the best adapted variety under best management conditions in the absence of all stresses. The Wheat Yield Consortium (Reynolds et al., 2012) has suggested that several physiological and biochemical traits such as radiation use efficiency, catalytic rate and regulation of rubisco, light and nitrogen distribution of canopies, spike photosynthesis and so on could be modified to maximize yields. Other potential traits include C4 photosynthesis, improved nutrient use efficiency, better adaptation to CA and traits improving productivity in drier, warmer environments.

There have been continued improvements in breeding techniques to develop these and other traits and maximize yields. One path-breaking development contributing to the spread of high-yielding spring wheat germplasm throughout the world was the introduction of dwarfing genes to increase the harvest index (HI), combined with improving selection for yield and disease resistance through shuttle breeding (Borlaug, 2007). However, conventional breeding can take six to eight years to develop a homozygous pure breeding line, whereas newer emerging threats warrant quicker solutions. One technique for accelerating breeding is chromosomal doubling of haploids to generate homozygous diploids in a single stroke. Among new emerging tools, dihaploid induction technology has wider acceptance, which promises to accelerate the breeding process (Tadesse et al., 2012). Niroula and Bimb (2009) have found the wheat x maize system of crossing better for its higher yield with less genotype dependency. Ever since heterosis breeding contributed quantum yield gains in crops such as maize, hybrid wheat has also been a target of several wheat breeders around the world (Sharma, 2013; Whitford et al., 2013).

Sun and He (2014) have highlighted potential constraints in further yield improvements, arguing that, through standard recombination and selection, current breeding approaches have led to the accumulation of unnecessary adaptive genes which act as a barrier to further improvement (encumbrance). Breeding has also placed undue emphasis on wide (geographically) adaptation, thereby compromising niche adaptation or the best fit of genotype (G) and environment (E) which is needed to get the best possible yield in local conditions.

One way of improving breeding techniques to address problems such as this is through greater understanding and exploitation of wheat genetics. The wheat genome is large, highly repetitive and complex because of its allohexaploid nature. The challenges of the complex, highly repetitive wheat genome are being overcome through recent advances in sequencing technology (Wang Meng et al., 2015). They suggest four strategies, viz. positional cloning, 'omics approaches, combining forward and reverse genetics, and epigenetics to exploit new tools to improve wheat. To tackle this challenge, the International Wheat Genome Sequencing (IWGS) Consortium decided to construct bacterial artificial chromosome (BAC)-based physical maps of individual chromosomes and then perform BAC by BAC sequencing. As an example, Kobayashi et al. (2015) recently reported construction of a physical map of chromosome 6B with the objective of revealing its structural features.

Mochida and Shinozaki (2013) have reviewed approaches to utilize information originating from genome sequencing of four major Triticeae genomes, viz. barley, wheat, *T. urartu* and *Aegilops tauschii*. They recommended innovative approaches and tools such as analysis of next-generation populations, evolutionary genomics and systems approaches with mathematical modelling to discover novel useful alleles. Muthamilarasan et al. (2014) recommended implementing genome sequence data in breeding and engineering transgenic wheat with enhanced biotic and abiotic stress tolerance.

Several researchers have developed and/or validated molecular markers for important traits. Examples include:

- Bagherikia et al. (2014) for 1A/1R translocation
- Li et al. (2015a) for grain yield and yield components
- Kim et al. (2014) for pre-harvest sprouting tolerance

- Cui Fa et al. (2016) for kernel size and quality
- Cheng et al. (2015) for kernel weight
- Zhang et al. (2015d) for kernel weight and spike length
- Liang et al. (2015) for high-molecular-weight glutenin subunits
- Jin et al. (2015) for insoluble glutenin
- Muhammad Faheem et al. (2015) for drought tolerance
- Zhao et al. (2014) for potassium deficiency
- Li et al. (2016) for *Fusarium* head blight (FHB); Wang et al. (2016), Yu et al. (2015) and Lu et al. (2014) for powdery mildew resistance.

As an illustration of the use of this technique, Ma Meng et al. (2016) unravelled the genetics and biochemical mechanism of seed weight taking a cue from *Arabidopsis*. They isolated *TaCYP78A5*, the orthologue of *Arabidopsis* CYP78A5/KLUH in wheat cultivar Shaan 512. They demonstrated that silencing *TaCYP78A5* reduced cell numbers in seed coat by 10%, whereas an overexpression increased the cell numbers in *Arabidopsis*. This direct evidence makes *TaCYP78A5* a potential target for yield improvement in wheat.

4 Dealing with biotic stresses

One of the main diseases faced by wheat is rust, which is classified into three types: leaf (also known as brown), stem (also known as black) and stripe (also known as yellow) rust. These types are present in varying degrees in different wheat-growing regions. Regular surveillance is conducted throughout countries such as India to detect and manage threats such as yellow rust (Suman Lata et al., 2015). A total of 2579 samples of this rust pathogen were collected during 2004–11 from nine affected states of India, revealing eight pathotypes in India (Prashar et al., 2015), although only three of them were widespread.

Lately, new virulent races of these diseases, for example, Ug99 of stem rust (Sharma et al., 2012, 2013) and the Warrior type of yellow rust (Hovmoller et al., 2016), have appeared. 'Rust never sleeps' was Dr Norman Borlaug's often-repeated admonition. The more wheat breeders develop resistant varieties, the more the fungus evolves into new and more virulent forms. As an example, a total of 2124 brown rust (*Puccinia triticina*) samples collected between 2008 and 2013 from across India, Bangladesh, Bhutan and Nepal comprised 37 pathotypes, including one new pathotype which had not been previously identified (Bhardwaj et al., 2016). This five-year study did not find any pathotype virulent on *Lr9, Lr24, Lr25, Lr32, Lr39* and *Lr45*, suggesting that these genes confer resistance to leaf rust in these countries and provide a potential avenue for new varieties. In addition to national efforts targeting rust resistance breeding, screening of 103 Iranian and Syrian wheat lines for stem rust resistance genes revealed *Sr2* to be the most common stem rust resistance gene (Patpour et al., 2014). Developments like this have allowed good progress in breeding for resistance to Ug99 stem rust race in Asia (Joshi et al., 2011; Sharma et al., 2013; Nagarajan et al., 2014).

Other diseases include powdery mildew, which is a major fungal disease of wheat in many parts of the world, including Asia. Spot blotch is a serious disease of wheat in warm and humid areas, particularly South Asia. Cereal cyst nematodes can also cause significant economic losses (Dababat et al., 2015). Wheat blast, first detected in mid-1980s, has been causing considerable losses in South America (Maciel, 2011). This disease has now been reported in Bangladesh (Callaway, 2016), threatening about 135 million tonnes of wheat grown in this part of the world.

There has been a number of initiatives to tackle these diseases. Yu et al. (2015) located an important powdery mildew resistance gene *Pm07J126* on the distal end of chromosome 6DS. Lu et al. (2014) isolated a powdery mildew immune wheat line carrying a rye translocation. Singh et al. (2015) reported two wheat lines, viz. CHUKU#1 and VAYI#1 as promising sources of spot blotch resistance for use in breeding. Mehrabi et al. (2014) screened 128 CIMMYT-bred SHW lines against 11 isolates of *Septoria tritici* blotch (STB). Nine genotypes were resistant to all the isolates. They concluded that some SHWs possessed broad spectrum of resistance gene(s) against various STB isolates. Amandeep et al. (2011) have identified key genes conferring resistance to Karnal Bunt. Najafi-Mirak (2012) has screened wheat accessions for resistance to sunn pest (*Eurygaster integriceps* Put.). Cultivar Falat was found to be the most resistant followed by durum line D-81-15 and *Triticale* line Triticale-1.

There has been an upsurge of FHB linked in part to the practice of leaving crop residues in the soil and minimal tillage in CA. Li et al. (2016) have identified sources of FHB resistance in Chinese and Japanese germplasm. CIMMYT already has an FHB screening nursery, which has identified several promising lines that could serve as donor in resistance breeding programmes (Osman et al., 2015). Sources of resistance have also been identified for eyespot disease, which is one of the major stem diseases in temperate regions (Wei et al., 2011).

5 Dealing with abiotic stresses

Climate change has the potential to seriously impact wheat production globally (Sharma et al., 2015). Using historic data Lobell et al. (2011) have estimated a 5.5% drop in potential wheat production linked to higher global temperatures. It has been estimated that by 2050 as much as 51% of the Indo-Gangetic Plains (IGP), which accounts for 15% of global production, might be reclassified as heat stressed (Ortiz, 2008). This would necessitate cultivar change and crop management adjustments to offset the adverse impact.

A Central Asian study on the impact of climate change on wheat productivity (Sommer et al., 2013) concluded that areas of rainfed spring wheat in the north and some irrigated winter wheat areas in the south of Central Asia were facing hotter temperatures during flowering and thus a potential reduction in yield. However, the study also suggested that the adverse effect of higher temperatures might be compensated to some extent by sowing date adjustment and changes in cultivar traits and production inputs. Chen et al. (2014) studied winter wheat over an entire growth period of winter wheat using a passive warming facility for three years. The 1.1°C increase in mean nighttime temperature in fact promoted wheat development. It reduced pre-anthesis duration by six days and increased post-anthesis period by five days. Changes in several other traits were recorded all finally culminating into 12.3% higher biomass and 12% higher yield. Similar yield-enhancing (16.3%) effect of increase in wheat canopy temperature (1.5°C) was reported by Tian et al. (2012). He et al. (2015) studied impact of climate warming on cultivar change, wheat phenology and crop management. They reported that changes in phenological stages and phases were significantly negatively correlated with a temperature increase during the period 1981–2009. They concluded that during this period, later sowing dates and the introduction of new cultivars with longer thermal time requirement had compensated for some of the increased temperature.

One of the potential effects of higher temperature is a higher risk of drought. Water deficit stress is one of the major problems affecting wheat production globally (Sayyah et al., 2015; Mahpara et al., 2014; Pask and Reynolds, 2013). Arid regions such as Central Asia are particularly vulnerable to global warming (Sharma et al., 2015; Aleksandrova et al., 2014). As an example, Tajikistan grows wheat on about 35 million hectares with an annual production of about 0.8 million tonnes. However, poor yield, low quality and abiotic stresses such as drought and salinity are threatening production (Muminjanov et al., 2014).

Research into dealing with drought included investigations into determining traits for yield under drought stress, which suggested that HI, number of kernels per spike and spike length contributed most to yield under drought (Sayyah et al., 2015). Differentially expressed proteins (DEP) in response to water deficit stress in embryo and endosperm during grain development revealed 155 and 130 DEPs in embryo and endosperm, respectively (Gu et al., 2015). Ghasemi and Farshadfar (2015) developed a drought tolerance score which they used to target genotypes in breeding more resilient varieties. Farshadfar et al. (2015) identified genotypes having higher values for traits such as chlorophyll content, proline content, cell membrane stability and relative water content under water deficit stress. Crosses displaying heterobeltiosis for important traits were also identified. A wild relative of wheat *Ae. biuncialis* has been shown to demonstrate higher drought tolerance and to be a suitable candidate for future breeding programmes (Zhao et al., 2013). Safarian and Abdolshahi (2014) recommended using pedigree and backcrossing methods to improve root traits for increased tolerance to drought. Pask and Reynolds (2013) found that genetic improvement of the yield of irrigated spring wheat was associated with additional water uptake from deeper soil profiles (60–120 cm), especially during grain filling. Most varieties possessing cold hardiness also showed good salinity tolerance and drought tolerance (You GuangXia et al., 2015).

Salinity is one of the most severe abiotic stresses and has a significant negative effect on plant biomass and crop yield. The complexity of plant response has made breeding more resilient varieties challenging (Shahbaz and Ashraf, 2013). Takahashi et al. (2015) succeeded in developing lines with adaptation to mild salinity conditions through selection of specific traits with various salinity-inducible genes. To demonstrate usefulness of a wheat NAC gene *TaNAC29* in stress tolerance, Xu et al. (2015) generated transgenic *Arabidopsis* overexpressing *TaNAC29*. When subjected to salt tolerance, this transgenic had enhanced tolerance to salt stress, reduced H_2O_2 accumulation, strengthened cell membrane stability and other tolerance-specific traits.

A related problem in saline soils is sodium content. Sodicity tolerance studies in three cereal species, viz. *Hordeum vulgare*, *Triticum aestivum* and *T. turgidum* ssp. durum with and without the Na^+ exclusion gene *Nax2* (Genc et al., 2016) revealed that under sodicity bread wheat and durum with *Nax2* gene had higher yield than barley and durum without the *Nax2* gene. However, under salinity all species yielded similarly, indicating that osmotic stress negated the benefits of Na^+ exclusion. They concluded that Na^+ exclusion can be an effective mechanism for sodicity tolerance, while osmoregulation and tissue tolerance to Na^+ and/or Cl^- need to be the main foci for further improvement of salinity tolerance research in cereals.

Further improvement in salinity tolerance requires variation in the germplasm. Almost 150 wheat genotypes were tested for seedling stage salinity tolerance under hydroponic conditions (Babar Hussain et al., 2015). The DH-3, 9436, DH-14, Chenab 2000 and six otherlines were ranked most tolerant and useful for further breeding work. Ibrahim et al. (2015) developed salt-tolerant genotypes through conventional breeding and then characterized them vis-à-vis local cultivar Tamooz-2. They concluded that tolerant lines had higher concentrations of K^+/Na^+ and Ca^{2+}/Na^+ in upper leaves as compared to local cultivar which had lower concentration in upper leaves. Another report from Pakistan (Ahmad Ali et al., 2014) reported synthetic-derived advance lines to be more tolerant to osmotic stress than rainfed bread wheat lines. Muhammad Ilyas et al. (2015) explored SHW for additional resources for boron tolerance. They reported existence of sequences in *Ae. tauschii* orthologous to available boron-tolerant genes in cereals. The SHW derived using *Ae. tauschii* can therefore be a potential source for additional boron tolerance genes.

Genetic transformation in wheat has been seen as having potential in improving tolerance to abiotic stress (Li et al., 2012). However, transgenic wheat still suffers from issues such as low yield and genotype dependency. Zhang et al. (2015c) reported constructing plant expression vector of glutathione peroxidase (*GPX*) to be utilized for *Agrobacterium*-mediated transformation of wheat for improving stress tolerance.

6 Improving aspects of quality

Increasing grain yield has traditionally been the major breeding objective. Improving both grain yield and protein content simultaneously is complicated by a strong negative correlation and genotype by environment (G X E) interaction, though some researchers have been successful in simultaneously improving both traits (Niu et al., 2010). There is increasing interest in identifying and improving quality traits. As an example, using DNA markers, 117 Pakistani spring wheat varieties were screened for specific quality determinants (Niaz et al., 2013) revealing 77 with 1B.1R translocation, 49 with grain hardness locus Pinb-D1b and 48 with Wx-B1b allele conferring superior noodle quality. They concluded that early generation marker-assisted selection can help efficiently utilize early generation variability of a breeding programme.

High-molecular-weight (HMW) glutenin subunits (GS) (HMWGS) determine dough elasticity and viscosity, the two critical determinants of bread-making quality. Peng et al. (2015) reported two novel y-type HMWGS, viz. 1Dy12.6 and 1Dy12.7, having potential to strengthen gluten–polymer interactions. They reported cloning of these two genes and claimed that these would be a valuable genetic resource for wheat quality improvement. Yasmeen et al. (2015) screened 242 wheat lines including landraces from Punjab and Baluchistan of Pakistan. The study revealed that commercial cultivars possessed favourable alleles such as 2*, 5+10 and 17+18, making them a good source for improving bread-making quality. Zhen et al. (2014) concluded that LMWGS allele Glu-A3a encodes a specific LMW-I type B-subunit that significantly affects bread-making quality. Yang et al. (2014) studied dough rheological and flour quality traits in 330 Chinese wheat varieties. For various dough rheological properties, they found higher coefficient of variation for development time (40.5%), stability time (58.1%) and farinograph quality number (42.4%) as compared to low variation for protein content (9.1%), wet gluten content (10.1%) and sedimentation value (15.3%). They concluded that there has been marked improvement in dough rheological properties in Chinese varieties released since 1986, while flour quality has remained steady. The 1RS chromosome of rye has contributed to wheat improvement in many countries; however, 1BL.1RS is also known for its adverse effect on quality (Li et al.,

2016). These quality defects are partly attributed to the ω-secalins, encoded by genes at Sec-1 locus. Li et al. (2016) reported an expression deletion mutation in the Sec-1 genes which markedly improved dough quality of the line, especially for gluten index and dough stickiness ($P < 0.05$).

Malnutrition is still one of the most serious challenges facing developing countries. Subhani et al. (2015) screened 58 old and new wheat cultivars of Pakistan for yield, protein and minerals, and found that mineral concentration of Zn^{2+}, Fe^{2+}, Cu^{2+} and NO_3^- and protein was significantly lower in newer varieties. Amiri et al. (2015) studied similar traits in 80 varieties in Iran released during the last 70 years and found that, whilst yield increased, protein, iron and zinc concentrations decreased. Range of micronutrients among 82 wheat accessions was investigated by Badakhshan et al. (2013). The study revealed that iron, zinc, protein and beta-carotene ranged from 41.36 to 67.67 mg/kg, from 36.37 to 73.8 mg/kg, from 7.5 to 15.6% and from 0.96 to 1.69 μg/g, respectively. They also found that landraces contained higher zinc concentration than modern varieties.

In response to these challenges, CIMMYT is developing and disseminating high Zn and Fe containing wheat varieties initially in IGP of South Asia, the region having high micronutrient malnutrition (Velu et al., 2011b). CIMMYT screened 600 core collection accessions for Fe and Zn concentration. The study revealed that grain Zn concentration ranged from 16.85 to 60.77 mg/kg, whereas Fe concentration varied from 26.26 to 68.78 mg/kg (Velu et al., 2011a). The advance lines carrying increased Zn and Fe were tested at nine locations in South Asia and Mexico. ANOVA revealed significant genotypic, environmental and genotype X environment (GXE) effects for Fe, Zn and grain yield (Velu et al., 2012). Pooled analysis over locations revealed an average superiority of 28 and 25% over checks for Zn and Fe, respectively. CIMMYT furthered its work by identifying two novel QTLs of large effect for grain zinc content (Hao et al., 2014), which together explained 10 to 15% of total phenotypic variation.

7 Improving cultivation practices

Since the Green Revolution, fertilizer consumption has been continuously increasing but productivity has stagnated. Site-specific nutrient management has been recommended to improve nutrient use efficiency (Jat et al., 2014a). CA also holds the promise to make crop raising more profitable and environmentally sustainable. Simulation studies on impact of water saving irrigation and CA practices for rice–wheat systems in the irrigated semi-arid dry lands of Central Asia (Devkota et al., 2015) revealed that for both the crops, water use efficiency could be improved by lowering evaporation losses. Space available to plants for growth also determines how much will it grow and produce. Nazim et al. (2014) assessed five wheat varieties under three different row spacing and broadcast sowing. They concluded 20 cm to be the optimum spacing that allows plants to produce maximum.

Inorganic nitrogen along with other elements is crucial for successful crop cultivation. Leaf proteome patterns revealed (Ruby Chandna and Altaf Ahmad, 2015) significant changes in protein pattern under both nitrogen-deficit and nitrogen-excess conditions. The study suggested that genes functioning in many physiological processes coordinated the response to availability of nitrogen and also for the improvement of N use efficiency (NUE) of crops. Genotypes with high nitrogen (N) uptake efficiency (UpE) and those with high N utilization efficiency (UtE) exhibited higher plant biomass, yield and N contents under N deficiency than those with medium and low NUE (Zhang et al., 2015a). High UpE genotypes were found to possess more of root biomass, root length, surface area and volume than other NUE types.

It is important that results of research reach farmers and that they adopt them. The ultimate success of research depends on decision of millions of farmers, which in turn is greatly influenced by the time lag between technology development and its large-scale availability to end users (Dixon et al., 2006; Reynolds and Borlaug, 2006b). S Erenstein (2010) adopted a triangulation approach to assess diffusion of zero-tillage technology in South Asia's irrigated plains, viz. India's Haryana state and Pakistan's Punjab province. This showed a much slower uptake and subsequent stagnation in the Pakistan's Punjab region.

8 Case study: improving cultivation in India

Formal wheat research in India started during the first decade of twentieth century with the Imperial Agricultural Research Institute (Sharma and Mishra, 1999). A major post-war initiative was the All India Coordinated Wheat Improvement Project which started in 1965 and coordinates research at over 150 research locations. To focus on research, India was divided into six wheat agro-climatic zones with varietal improvement targeted at these zones. The project has contributed to the release of over 400 varieties since 1965. Domestic efforts are ably supported by international centres such as CIMMYT in Mexico: 162 varieties out of 212 bred since 1965 had CIMMYT germplasm in their parentage (Sharma and Jagshoran, 2000).

The big turning point in India's wheat research came towards the end of 1960s when introductions from CIMMYT ushered in quantum jumps in wheat yield now known as the Green Revolution. India's wheat production increased from about 20 million tonnes in 1970 to over 94 million tonnes in 2014 (Fig. 2). The wheat yield increased from about one tonne per hectare to currently over three tonne per hectare. Singh et al. (2006) analysed morpho-physiological traits associated with growth and productivity in 15 promising wheat cultivars developed between 1960s and 1990. They found that maximum productivity in semi-dwarf long-duration varieties came from increased biological yield, leaf area index, grains per spike, days to flowering and grain growth rate. On the other hand, increased productivity of semi-dwarf short and medium duration was attributed to improved HI.

The breakthroughs that contributed to Green Revolution now seem to have been fully exploited. More recent developments have led to more modest progress with around 1.1% annual yield gains (Dixon et al., 2009). The challenge is now what is needed for the next leap forward (Rajbir et al., 2010). As has been seen, problems include new virulent strains of rust, increasing problems of drought and dwindling water sources as well as continuing wide variations in yield between regions (Joshi et al., 2007).

Indian researchers have been investigating various avenues to improve yields. In the area of genetic variability, for example, Rajender Singh et al. (2015) employed the target-induced local lesions in genome (TILLING) technique to generate usable novel variability in bread wheat, resulting in increased seeds per spike, number of tillers per plant and thousand grain weight (TGW). Sehgal et al. (2011) have also transferred useful genes from *Ae. tauschii* to bread wheat through direct hybridization.

There have also been advances in breeding such as the use of molecular markers. As an example, Vandana Jaiswal et al. (2015) identified four SNPs in the promoter region

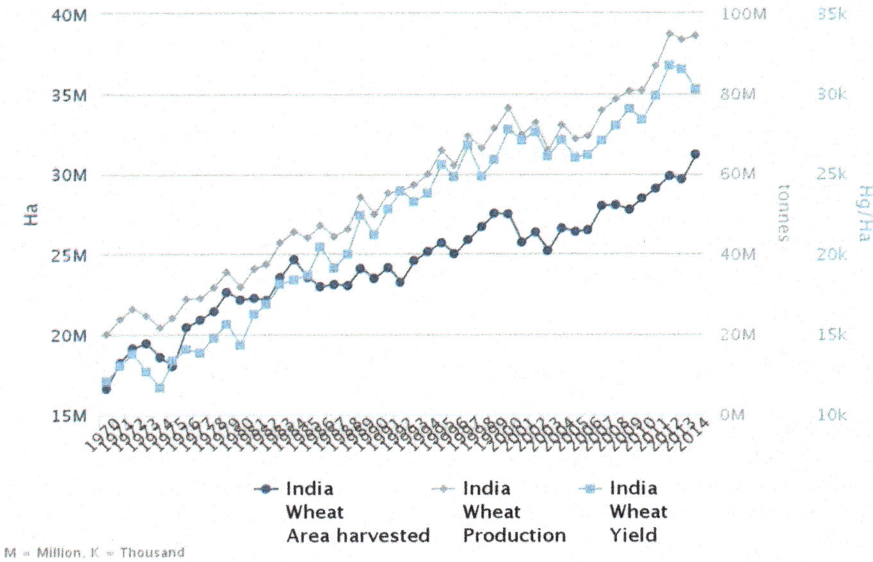

Figure 2 Wheat production, productivity and area in India from 1974 to 2014. Source: FAO, 2016.

of orthologue TaGW2-6A. They found one of the SNPs associated with higher TGW. Other research groups in India have been using molecular markers in crop improvement programme, for example:

- Vishwakarma et al. (2015) for grain quality traits
- Kulwal et al. (2010) for seed dormancy and pre-harvest sprouting (PHS)
- Mir et al. (2012) for grain weight
- Singh et al. (2013) for vernalization and photoperiod genes
- Chandna et al. (2010) for nitrogen use efficiency
- Pradeep Sharma et al. (2014) and Davinder Sharma et al. (2015) for heat tolerance
- Naik et al. (2015) for leaf rust resistance gene
- Gaganpreet Kaur et al. (2012) for grain protein gene *GpcB1*
- Umesh Goutam et al. (2015) for HMW alleles.

There have also been developments in transgenic techniques. As an example, Amit Kumar et al. (2015) studied the effect of plant growth regulators on *in vitro* callus induction and plant regeneration from mature wheat embryos. They found the effect of growth regulators to be genotype dependent and that genotype Raj3765 was best both in terms of callus induction and plant regeneration. Ali et al. (2015) studied variation in callus induction and regeneration efficiency among 30 Indian genotypes. They reported 80 to 100% callusing efficiency among genotypes studied. The bombardment caused transformation efficiency ranged from 0.29% to 46%.

Heat stress during the grain filling period is said to be one of the most important abiotic stresses in South Asia, reducing production by millions of tonnes. Simarjit Kaur et al. (2013) found that heat stress reduced yield by increasing chlorophyll loss (41.5%) and membrane injury (6.3%) and by reducing grain filling period (27.4%) and TGW (21.4%). One study estimated that mean wheat yields during 1980–2011 declined by 7% (204Kg/ha) with a 1°C rise in minimum mean temperature (Tn) (Rao et al., 2015). Shivani Nagar et al. (2015) assessed terminal heat stress tolerance of 40 Indian wheat genotypes and found that the DBW 14, Raj 3765, HD 2643 and Halna genotypes performed better in terms of higher membrane stability index, chlorophyll content, photosynthesis rate (Pn) and HI under stress conditions. Simarjit Kaur et al. (2013) found genotypes RAJ 4011, PBW 509 and PBW 503 to be most heat tolerant.

There have been a number of initiatives to analyse and improve aspects of quality. Mohan and Gupta (2015) found that early heading and longer grain filling was effective in increasing test weight, whilst delayed heading and shorter plant height enhanced protein and bran content. Sutapa Sarkar et al. (2014) found kernel hardness positively correlated with flour recovery and grain protein percentage with bread-making quality. Grain quality characterization of 100 high-yielding varieties (Mohan et al., 2013b) showed that Indian bread wheat varieties were generally good in test weight, chapati (Indian flat bread) quality but needed improvement in sedimentation value, flour recovery and gluten index. Flour recovery from 100 varieties sampled from different locations and regions of India during 2005–11 was found to vary across regions (Mohan et al., 2013a). They rated wheat from the peninsular zone to be best for milling yield (71%), followed by the North Western Plains (68%).

HMWGS make up about 6–8% of the total gluten content which in turn constitute about 8–15% of the dry weight of wheat flour. These HMWGS determine viscoelastic properties of wheat dough, which in turn determine end use suitability. Qualitative and quantitative variation in high- and low-molecular-weight (H and LMW) glutenins dictate end use quality in wheat. Chowdhury et al. (2006) reported declining HMW scores in recent varieties. They studied HMWGS profile of 450 varieties of bygone era and compared with those of latest cultivars. They concluded that there has been no variety with a HMWGS score of ten during the last 15 years as compared to over 65 out of 450 from the early period. Similar results of declining HMW quality were reported by Ram et al. (2015). They evaluated 240 Indian wheat cultivars released during the last several decades. They profiled the genotypes for Glu-1 and Glu-3 loci and reported that allele Glu-Bli and Glu-D1d showed highly positive effect on sedimentation volume; however, they found that the frequency of Glu-Bli had declined in Indian varieties during the last two decades and that not a single 1B/1R cultivar possessed this allele.

Investments into zinc and iron biofortification have been found to be cost effective for India, reducing associated disease burden by 20 to 51% for zinc (Stein et al., 2007) and by 19 to 58% for iron (Stein et al., 2008). Mohan et al. (2009) have studied prospects for biofortification in Indian bread wheat cultivars. They characterized 60 cultivars for protein, yellow pigment, iron, zinc, copper and manganese contents vis-à-vis grain nutrition requirements under different agro-climatic zones and production conditions. The study revealed that enhancing protein and iron content was necessary in irrigated timely sown cultivars, whereas yellow pigment improvement was required for central and peninsular zones. Phytic acid plays an important role for storing micronutrients but also makes them unavailable by chelating. The phytic acid ranged from 0.59 to 2.08% among a set of 93 wheat lines (Gupta et al., 2015). Iron on the other hand ranged from 9.97 to 45.77 mg/kg. Yashvanthkumar et al. (2014) reported iron in the range of 24.9 to 50.6 mg/kg and zinc from 25.25 to 58.8 mg/kg in a set of 100 wheat lines. They listed genotypes with higher contents of iron, zinc, manganese and copper for further use in improvement work.

Finally, there have also been initiatives to improve cultivation practice. Lobell et al. (2010) studied wheat yield data to close yield gaps in Northwest India and found two key elements, viz. sowing dates and adequate irrigation. The importance of the latter has also been emphasized by Kukal et al. (2014). Rice–wheat rotation is practiced on around 14 million hectares across IGP, which stretches across India (10.5 million ha), Pakistan (2.2 million ha), Nepal (0.5 million ha) and Bangladesh (0.8 million ha). Sustaining the productivity of this system has become a challenge (Singh et al., 2010). CA has been suggested as a way to maintain productivity as it reduces cost of cultivation and saves water compared to conventional tillage system (Singh et al., 2010; Jat et al., 2014). Dahiya et al. (2008) and Mishra et al. (2015) recommended using vermicompost, crop residues or farmyard manure to supplement chemical fertilizers to improve nutrient availability and water holding capacity of soils in rice–wheat system. Others have suggested legumes to improve soil health (Singh and Ahlawat, 2014). Varinderpal Singh et al. (2010) have also suggested using tools such as chlorophyll meters and leaf colour charts to improve the timing and quantity of fertilizer application.

9 Summary

As has been seen, the Green Revolution ushered in quantum jumps in the wheat-growing regions of Asia. However, most recent research has focussed on more incremental improvements to consolidate gains already achieved, for example, through gradual accumulation of favourable alleles and employing doubled haploids to accelerate breeding. The best possible yield at a given location requires best adapted cultivar plus best location-specific agronomy. Do we need to reconsider breeding wide adaptation cultivars for zones measuring millions of hectares or develop a more targeted approach? Are there viable options to create usable hybridity and design plant architecture with the potential to deliver more? The complex evolution that generated bread wheat needs to be recreated more precisely to enrich useful variability whilst orthology across species is opening avenues to diversify and enhance trait expressions in new ways.

10 Where to look for further information

Publications

Bhardwaj, S. C., Prasad, P., Gangwar, O. P., Khan, H. and Kumar, S. (2016), 'Wheat rust research- then and now', *Indian Journal of Agricultural* Sciences, 86 (10), 1231–44.

Bonjean, A. P., Angus, W. J. and Maarten van, G. (2014), 'The world wheat book series- Volume I. II and III, Lavoisier Tec & Doc, Paris.

Borlaug Global Rust Initiative: http://www.globalrust.org/

CGIAR Research Programme on Wheat (wheat CRP): http://www.cgiar.org/about-us/our-programs/cgiar-research-program-on-wheat/

Durable Rust Resistance in wheat: http://www.wheatrust.cornell.edu/

Fischer, R. A. and Edmeades, G. O. (2010), 'Breeding and cereal yield progress, *Crop Science*, 50, 85–98.

Global Rust Reference Centre (GRRC): http://wheatrust.org/

A global wheat rust monitoring system (rust tracker): http://rusttracker.cimmyt.org/
Grewal, S. and Goel, S. (2015), 'Current research status and future challenges to wheat production in India', *Indian Journal of Biotechnology*, 14 (4), 445–54.
Harvest Plus: http://www.cimmyt.org/project-profile/harvestplus/
International centre for agricultural research in the dry areas (ICARDA): http://www.cgiar.org/about-us/research-centers/international-center-for-agricultural-research-in-the-dry-areas-icarda/
International Maize & Wheat Improvement Centre (CIMMYT): http://www.cimmyt.org/
International wheat conference (IWC, last one in Sydney in 2015): http://www.cgiar.org/about-us/research-centers/international-center-for-agricultural-research-in-the-dry-areas-icarda/
International wheat improvement network (IWIN): http://www.cimmyt.org/international-wheat-improvement-network-iwin/
International wheat genetics symposium (IWGS, the last one was in Japan in 2013): http://12iwgs.sci.yokohama-cu.ac.jp/
Khan, M. A. and Iqbal, M. (2011), 'Breeding for drought tolerance in wheat (*Triticum aestivum* L.): constraints and future prospects', *Frontiers of Agriculture in China*, 5 (1), 31–4.
Marker assisted selection in wheat (MAS wheat): http://maswheat.ucdavis.edu/

Projects, consortia and research organizations

Reynolds, M. P. and Borlaug, N. E. (2006), 'Applying innovations and new technologies for international collaborative wheat improvement', *Journal of Agricultural Science*, 144 (2), 95–110.
Rauf, S., Zaharieva, M., Warburton, M. L.,Zhang, P. Z, Al-Sadi, A.M., Farghama K., Kozak, M. and Tariq, S.A. (2015), 'Breaking wheat yield barriers requires integrated efforts in developing countries', *Journal of Integrative Agriculture*, 14 (8), 1447–74.
Singh, Y. and Sidhu, H. S. (2014), 'Management of cereal crop residues for sustainable rice-wheat production system in the Indo-Gangetic Plains of India', *Proceedings of the Indian National Science Academy*, 80 (1), 95–114.
Wellings, C. (2010), Global status of stripe rust' BGRI 2010 technical workshop oral presentations. Full papers and abstracts. May 30–31 2010. St. Petersburg, Russia 2010, pp. 34–8.
Wheat Atlas for diverse information on wheat: http://wheatatlas.org/?AspxAutoDetectCookieSupport=1
William, M., Langridge, P., Trethowan, R., Dreisigacker, S. and Crouch, J. (2008), 'Genomics of wheat, the basis of our daily bread', in Moore, P. H. and Ming, R. (eds), *Genomics of Trpical Crop Plants*, Springer Science, New York. pp. 515–48.
Wu, W., Li, C., Ma, B., Farooq Shah, Liu, Y. and Liao, Y. (2014), 'Genetic progress in wheat yield and associated traits in China since 1945 and future prospects', *Euphytica*, 196 (2), 155–68.

11 References

Adary, A. H., Al-Shamma, A. M. (2014), 'The history of wheat breeding in Iraq', In: Eds, Bonjean, A. P. W. J., Angus, and Maarten van Ginkel, *The World Wheat Book: A History of Wheat Breeding*, Vol. 3, Lavoisier Tec & Doc, Paris, 407–38.

Adu, M. O., Sparkes, D. L., Parmar, A. and Yawson, D. O. (2011), '"Stay green" in wheat: comparative study of modern bread wheat and ancient wheat cultivars', *Journal of Agricultural and Biological Science*, 6 (9), 16–24.

Ahmad, A., Arshad, M., Naqvi, S. M. S., Ahmad, M., Sher, H., Fatima, S., Kazi, A. Z., Rasheed, A. and Mujeeb-Kazi, A. (2014), 'Exploitation of synthetic-derived wheats through osmotic stress responses for drought tolerance improvement', *Acta Physiologiae Plantarum*, 36 (9), 2453–65.

Akar, T., Yazar, S., Donmez, E., Avci, M. and Dusunceli, F. (2007), 'Current status of national winter wheat breeding in Turkey,' *Journal of Agricultural Research* (Lahore), 45 (1), 1–9.

Aleksandrova, M., Lamers, J. P. A., Martius, C. and Tischbein, B. (2014), 'Rural vulnerability to environmental changes in the irrigated lowlands of Central Asia and options for policy-makers: a review', *Environmental Science and Policy*, 41, 77–8.

Ali, S., Arpita, Sharma Singh, S. K. and Bansal, K. C. (2015), 'Comparative in vitro regeneration and transient transformation efficiency of different genotypes of wheat (*Triticum aestivum*)', *Indian Journal of Agricultural Science*, 85 (8), 1065–8.

Amandeep, K., Joshi, A. M. and Indu, S. (2011), 'Genetics of Karnal Bunt resistance in a new set of bread wheat lines', *Crop Improvement*, 38 (1), 38–44.

Amiri, R., Bahraminejad, S., Sasani, S., Jalali-Honarmand, S. and Fakhri, R. (2015), 'Bread wheat genetic variation for grain's protein, iron and zinc concentrations as uptake by their genetic ability', *European Journal of Agronomy*, 67, 20–6.

Amit, K., Sengar, R. S., Sharma, M. K. and Singh, V. K. (2015), 'Effect of plant growth regulators on in vitro callus induction and plant regeneration from mature wheat (*Triticum aestivum* L.) embryos', *Vegetos*, 28 (3), 54–61.

Arain, S., Soomro, Z. A., Mari, S. N., Sial, M. A., Munaiza Baloch and Albokari, M. M. (2015), 'Improvement of grain yield and yield associated traits in wheat genotypes through mutation breeding', *International Journal of Biology and Biotechnology*, 12 (2), 277–82.

Babar, H., Khan, A. S. and Zulfiqar, A. (2015), 'Genetic variation in wheat germplasm for salinity tolerance at seedling stage: improved statistical inference', *Turkish Journal of Agricultural and Forestry*, 39 (2), 182–92.

Badakhshan, H., Moradi, N., Mohammadzadeh, H. and Zakeri, M. R. (2013), 'Genetic variability analysis of grains Fe, Zn and Beta-carotene concentration of prevalent wheat varieties in Iran', *International Journal of Agricultural and Crop Science*, 6 (2), 57–62.

Bagherikia, S., Karimzadeh, G. and Naghavi, M. R. (2014), 'Distribution of 1aL.1RS and 1BL.1RS wheat-rye translocations in *Triticum aestivum* using specific PCR', *Biochemical Systematics and Ecology*, 55, 20–6.

Bhardwaj, S. C., Prashar, M., Kumar, Subodh Gangwar, O. P. et al., (2016), 'Patterns of physiologic diversity of *Puccinia triticina* on wheat in Indian subcontinent during 2008-2013', *Indian Journal of Agricultural Sciences*, 86 (1), 55–64.

Borlaug, N. E. (2007), 'Sixty-two years of fighting hunger: personal recollections,' *Euphytica*, 157 (3), 287–97.

Braidotti, G. (2012), 'The second green revolution', *Partners in Research for Development*. No. Summer, 23–4.

Callaway, E. (2016), 'Devastating wheat fungus appears in Asia for first time', *Nature*, 532, 421–2.

Cao, M., Chen, G., Wang, C., Zhen, S., Li, X., Zhang, W., Zeller, F. J., Hsam, S. L. K., Hu, Y. and Yan, Y. (2015), '1S^l (1B0 chromosome substitution in Chinese Spring wheat promotes starch granule development and starch biosynthesis', *Crop and Pasture Science*, 66 (9), 894–903.

Chandna, R., Gupta, S., Ahmad, A., Iqbak, M. and Prasad, M. (2010), 'Variability in Indian bread wheat (*Triticum aestivum* L.) varieties differing in nitrogen efficiency as assessed by microsatellite markers', *Protoplasma*, 242 (1/4), 55–67.

Chatrath, R., Mishra, B., Ferrara, G. O., Singh, S. K. and Joshi, A. K. (2007), 'Challenges to wheat production in South Asia', *Euphytica*, 157 (3), 447–56.

Chen, J., Tian, Y., Zhang, X., Zheng, C., Song, Z., Deng, A. and Zhang, W. (2014), 'Nightime warming will increase winter wheat yield through improving plant development and grain growth in North China', *Journal of Plant Growth Regulation*, 33 (2), 397–407.

Chen, X. and Hao, M. D. (2015), 'Low contribution of photosynthesis and water use efficiency to improvement of grain yield in Chinese wheat', *Photosynthetica*, 53 (4), 519–26.

Cheng, X., Chai, L., Chen, Z., Xu, L., Zhai, A., Peng, H., Yao, Y., You, M., Sun, Q. and Ni, Z. (2015), 'Identification and characterization of a high kernel weight mutant induced by gamma radiation in wheat (*Triticum aestivum*)', *BMC Genetics*, 16, 127.

Chowdhury, S., Sharma, R. K. and Mishra, B. K. (2006), 'The distribution of high molecular weight glutenin subunits among some old wheat varieties of India', *Crop Improvement*, 33 (2), 114–18.

CIMMYT (2011), 'Wheat: global alliance for improving food security and the livelihoods of the resource-poor in the developing world', CIMMYT.

Cui, F., Fan, X., Chen, M., Zhang, N., Zhao, C., Zhang, W., Han, J., Ji, J., Zhao, X, Yang, L, Zhao, Z, Tong, Y, Wang, T. and Li, J. (2016), 'QTL detection for wheat kernel size and quality and the response of these traits to low nitrogen stress', *Theoretical and Applied Genetics*, 129 (3), 469–84.

Dababat, A. A., Imren, M., Erginbas-Orakci, G., Ashrafi, S., et al., (2015), 'The importance and management strategies of cereal cyst nematodes, Heterodera spp., in Turkey', *Euphytica*, 202 (2), 173–88.

Dahiya, D. S., Dahiya, S. S., Lathwal, O. P., Ramesh, S. and Sheoran, R. S. (2008), 'Integrated nutrient management in wheat under rice-wheat cropping system', *Haryana Journal of Agronomy*, 24 (1/2), 51–4.

Davinder, S., Mamrutha, H. M., Gupta, V. K., Ratan, T. and Rajender, S. (2015), 'Association of SSCP variants of HSP genes with physiological and yield traits under heat stress in wheat', *Research on Crops*, 16 (1), 139–46.

Deepmala, S., Prashant, V., Sansaloni, C. P., Ortiz, C., Pierre, C. S., Payne, T., Ellis, M., Amri, A., Petroli, C. D., Wenzl, P. and Singh, S. (2015), 'Exploring and mobilizing the gene bank biodiversity for wheat improvement', *PLoS ONE*, 10 (7), e0132112.

Deng, Z., Tian, J., Chen, F., Li, W., Zheng, F., Chen, J., Shi, C. et al., (2015), 'Genetic dissection on wheat flour quality traits in two related populations', *Euphytica*, 203 (1), 221–35.

Devkota, K. P., Hoogenboom, G., Boote, K. G., Singh, U., Lamers, J. P. A., Devkota, M. and Vlek, P. L. G. (2015), 'Simulating the impact of water saving irrigation and conservation agriculture practices for rice wheat systems in the irrigated semi arid drylands of Central Asia', *Agricultural and Forest Meteorology* (214/215), 266–80.

Divya, P., Madhoolika, A. and Bohra, J. S. (2013), 'Impact of four tillage permutations in rice-wheat system on GHG performance of wheat cultivation through carbon footprinting', *Ecological Engineering*, 60, 261–70.

Dixon, J., Braun, H. J., Kosina, P. and Crouch, J. (2009), *Wheat Facts and Figures 2009*. CIMMYT, Mexico, D. F.

Dixon, J., Nalley, L., Kosina, P., Rover, R. L., Hellin, J. and Aquino, P. (2006), 'Adoption and economic impact of improved wheat varieties in the developing world', *Journal of Agricultural Science*, 144 (6), 489–502.

Dreisigacker, S., Kishii, M., Lage, J. and Warburton, M. (2008), 'Use of synthetic hexaploid to increase diversity for CIMMYT bread wheat improvement', *Australian Journal Agricultural Research*, 59 (5), 413–20.

Dzhunusova, M., Ten, D., Aubekerova, N. and Morgounov, A. (2014), 'The history of wheat breeding in Kyrgyzstan', In: Eds, Bonjean, A. P., Angus, W. J., and Maarten van Ginkel, *The World Wheat Book: A history of Wheat Breeding*, Vol. 3, Lavoisier Tec & Doc, Paris, 235–47.

Erenstein, O. (2010), 'Triangulating technology diffusion indicators: zero tillage wheat in South Asia's irrigated plains', *Experimental Agriculture*, 46 (3), 293–308.

Evans, L. T. and Fischer, R. A. (1999), 'Yield potential: its definition, measurement and significance. *Crop Science*, 34, 1544–51.

FAO (2016), FAOSTAT3.FAO.ORG, Accessed 27 July 2016.

Farshadfar, E., Rad, A. A. and Kianifar, S. (2015), 'Genetic analysis of physiological criteria of drought tolerance in bread wheat under rainfed conditions', *Biological Forum*, 7 (1), 912–19.

Gaganpreet, K., Neeraj, C., Harjot, K., Harjinder, K., Kuldeep, S. and Parveen, C. (2012), 'Characterization of wild emmer wheat germplasm for high grain protein gene GpcB1', *Crop Improvement*, 39 (1), 91–4.

Garg, M., Yanaka, M., Tanaka, H. and Tsujimoto, H. (2014), 'Introgression of useful genes from Thinopyrum intermedium to wheat for improvement of bread making quality, *Plant Breeding*, 133 (3), 327–34.

Genc, Y., Oldach, K. Taylor, J. and Lyons, G. H. (2016), 'Uncoupling of sodium and chloride to assist breeding for salinity tolerance in crops', *New Phytologist*, 210 (1), 145–56.

Ghasemi, M. and Farshadfar, E. (2015), 'Screening drought tolerant genotypes in wheat using multivariate and stress tolerance score methods', *International Journal of Biosciences*, 6(1), 326–33.

Gu, A., Hao, P., Lv, D., Zhen, S., Bian, Y., Ma, C., Xu, Y., Zhang, W. and Yan, Y. (2015), 'Integrated proteome analysis of the wheat embryo and endosperm reveals central metabolic changes involved in the water deficit response during grain development', *Journal of Agricultural and Food Chemistry*, 63 (38), 8478–87.

Gupta, R. K., Gangoliya, S. S. and Singh, N. K. (2015), 'Screening and characterization of wheat germplasm for phytic acid and iron content', *Journal of Agricultural Science and Technology*, 17 (3), 747–56.

Gupta, S. R., KC, H. B., Shah, U. S., Mandal, D. and Bhattarai, M. (2013), 'Multi-crop germplasm collection from Far Western Development Region of Nepal', *Journal of Agriculture and Environment*, 14, 55–64.

Gyanendra, S., Kulshrestha, N., Singh, B. N., Setter, T. L., (2014), 'Germplasm characterization, association and clustering for salinity and waterlogging tolerance in bread wheat (Triticum aestivum)', *Indian Journal of Agricultural Sciences*, 84 (9), 1102–10.

Hao, M, Chen, J. X., Zhang, L. Q., Luo, J. T., Yuan, Z. W., Yan, Z. H., Zhang, B., Chen, W. J., Wei, Y. M., Zhang, H. G., Zheng, Y. L. and Liu, D. C. (2013), 'The genetic study utility of a hexaploid wheat DH population with non-recombinant A- and B- genomes', *SpringerPlus*, 2, 131.

Hao, Y., Velu, G., Pena, R. J., Sukhwinder, S. and Singh, R. P. (2014), 'Genetic loci associated with high grain zinc concentration and pleiotropic effect on kernel weight in wheat', *Molecular Breeding*, 34 (4), 1893–1902.

He, L., Asseng, S., Zhao, G., Wu, D., Yang, X., Zhuang, W., Jin, N. and Yu, O. (2015), 'Impact of recent climate warming, cultivar change and crop management on winter wheat phenology across the Loess Plateau of China', *Agricultural and Forest Meteorology*, 200, 135–43.

Hovmoller, M. S., Walter, S., Bayles, R. A., Hubbard, A., Flath, K., Sommerfeldt, N., Leconte, M. et al., (2016), 'Replacement of the European wheat yellow rust populations by new races from the center of diversity in the near-Himalyan region', *Plant Pathology*, 65 (3), 402–11.

Huo, Q., Cai, X., Kang, L., Zhang, H., Song, Y. and Zhu, T. (2015), 'Estimating ammonia emissions from a winter wheat crop land in North China plain with field experiments and inverse dispersion modeling', *Atmospheric Environment*, 104, 1–10.

Ibrahim, I. H. A. L. M., Zakariya, B. F., Eman, N. I. and Wisam, M. D. (2015), 'Detection for salt tolerance character in two selected genotypes of wheat', *International Journal of Biology*, 7 91), 54–6.

Inokuma, T., Vrinten, P., Shimbata, T., Sunohara, A., Ito, H., Saito, M. et al. (2016), 'Using the hexaploid nature of wheat to create variability in starch characteristics', *Journal of Agricultural and Food Chemistry*, 64 (4), 941–7.

Jagdish, K. S. V., Kishor, P. B. K., Bahuguna, R. N., Wiren, N. von and Sreenivasulu, N. (2015), 'Staying alive or going to die during terminal senescence-an enigma surrounding yield stability', *Frontiers in Plant Science*, 6 (November), 1070.

Jat, M. L., Bijay-Singh and Gerard, B. (2014a), 'Nutrient management and use efficiency in wheat systems of South Asia', *Advances in Agronomy*, 125, 171–259.

Jat, R. K., Sapkota, T. B., Singh, R. G., Jat, M. L. et al., (2014b), 'Seven years of conservation agriculture in a rice-wheat rotation of Eastern Gangetic Plains of South Asia: yield trends and economic profitability', *Field Crops Research*, 164, 199–210.

Jin, H., Wang, Z., Li, D., Wu, P., Dong, Z., Rong, C., Liu, X. et al. (2015), 'Genetic analysis of chromosomal loci affecting the content of insoluble glutenin in common wheat', *Journal of Genetics and Genomics*, 42 (9), 495–505.

Joshi, A. K., Azab, M. Mosaad, M., Moselthy, M. et al. (2011), 'Delivering rust resistant wheat to farmers: a step towards increased food security', *Euphytica*, 179 (1), 187–96.

Joshi, A. K., Mishra, B., Chatrath, R., Ferrara, G. O. and Singh, R. P. (2007), 'Wheat improvement in India: present status, emerging challenges and future prospects', *Euphytica*, 157 (3), 431–46.

Kazi, A. G., Awais, R., Tariq, M. and Abdul, M. K. (2012), 'Molecular and morphological diversity with biotic stress resistance of high 1000 grain weight synthetic hexaploid wheats', *Pakistan Journal of Botany*, 44 (3), 1021–8.

Karabayev, M., Morgounov, A., Braun, H. J., Wall, P., Sayre, K., Zelenskiy, Y., Zhapayev, R., Akhmetova, A., Dvurechenskii, V., Iskandarova, K., Friedrich, T., Fileccia, T. and Guadagni, M. (2014), *Journal of Agricultural Science and Technology*, 4 (10), 761–5.

Khalikulov, Z., Sharma, R. C., Amanov, A. and Morgounov, A. (2014), 'The history of wheat breeding in Uzbekistan', In: Eds., Bonjean, A. P., W. J., Angus and Maarten van Ginkel, *The World Wheat Book: A History of Wheat Breeding*, Vol. 3, Lavoisier Tec & Doc, Paris, 250–81.

Khush, G. S., Lee, S. C., Cho, J. and Jeon, J. (2012), 'Biofortification of crops for reducing malnutrition', *Plant Botechnology Reports*, 6 (3), 195–202.

Kim, K. H., Jung, J. U., Kang, C. S., Shin, S. H., Park, J. C. et al. (2014), 'Mapping of QTL for pre-harvest sprouting resistance and seed dormancy in a doubled haploid population', *Plant Breeding and Biotechnology*, 2 (3), 247–56.

Knight, S., Kightley, S., Bingham, I., Hoad, S., Philpott, H. and Lang, B. (2012), 'The impact of changes to agronomic practices on farm yield trends in wheat', *Aspects of Applied Biology*, 117, 105–12.

Kobayashi, F., Wu, J., Kanamori, H., Tanaka, T., Katagiri, S., Karasawa, W., Kaneko, S., Watanbe, S., Sakaguchi, T. et al., (2015), 'A high-resolution-physical map integrating an anchored chromosome with the BAC physical map of wheat chromosome 6B', *BMC Genomics*, 16 (595).

Kukal, S. S., Yadvinder-Singh, Jat, M. L. and Sidhu, S. S. (2014), 'Improving water productivity of wheat-based cropping systems in South Asia for sustained productivity', *Advances in Agronomy*, 127, 157–258.

Kulwal, P. L., Mir, R. R., Kumar, S. and Gupta, P. K. (2010), 'QTL analysis and molecular breeding for seed dormancy and pre-harvest sprouting tolerance in bread wheat', *Journal of Plant Biology*, 37 (1), 59–74.

Kumar, G., Prasad, A., Pullabhotla, S. K. and Hemant (2012), 'Supply and demand for cereals in Bangladesh, 2010–2030', IFPRI-Discussion Papers, #1186, vi+28pp.

Kumar, J., Sivasamy, M., Jayaprakash, P., Vikas, V. K. et al., (2015), 'Pathotype surveillance and pre-emptive breeding against black and brown rusts of wheat in Nilgiri hills', *Indian Phytopathology*, 68 (2), 139–44.

Latafat, P. and Khalil, I. H. (2011), 'Improvement of agronomic traits in spring wheat cultivars released in NWFP during 1958 to 2000', *Sarhad Journal of Agricultural*, 27 (1), 51–7.

Li, C., Bai, G., Carver, B. F., Chao, S. and Wang, Z. (2015a), 'Single nucleotide polymorphism markers linked to QTL for wheat yield traits', *Euphytica*, 206 (1), 89–101.

Li, G., Lang, T., Dai, G., Li, D., Li, C., Song, X. and Yang, Z. (2015b), 'Precise identification of two wheat-*Thinopyrum intermedium* substitution reveals the compensation and rearrangement between wheat and *Thinopyrum* chromosomes', *Molecular Breeding*, 35 (1), 1.

Li, J, Ye. X, An, B. Y., Du, L. and Xu, H. (2012), 'Genetic transformation of wheat: current status and future prospects', *Pant Biotechnic Reports*, 6 (3), 183–93.

Li, J., Wan, H. S. and Yang, W. Y. (2014a), 'Synthetic hexaploid wheat enhances variation and adaptive evolution of bread wheat in breeding processes', *Journal of Systematics and Evolution*, 52 (6), 735–42.

Li, K, Yang, X, Liu, Z, Zhang, T, Lu, S. and Liu, Y. (2014b), 'Low yield gap of winter wheat in the North China Plain', *European Journal of Agronomy*, 59, 1–12.

Li, T., Zhang, D., Zhou, X., Bai, G., Li, L. and Gu, S. (2016), 'Fusarium head blight resistance loci in a stratified population of wheat landraces and varieties', *Euphytica*, 207 (3), 551–61.

Li, Z., Ren, T., Yan, B., Tan, F., Yang, M. and Ren, Z. (2016), 'A mutant with expression deletion of gene Sec-1 in a 1RS.1BL line and its effect on production quality of wheat', *PLoS ONE*, 11 (1), e0146943.

Liang, X. N., Zhen, S. M., Han, C. X., Wang, C. et al., (2015), 'Molecular characterization and marker development for hexaploid wheat-specific HMW glutenin subunit 1By18 gene', *Molecular Breeding*, 35 (12), 221.

Lobell, D. B., Ortiz-Monasterio, J. I. and Lee, A. S. (2010), 'Satellite evidence for yield growth opportunities in Northwest India', *Field Crops Research*, 118 (1), 13–20.

Lobell, D. B., Schlenker, W. and Costa-Roberts, J. (2011), 'Climate trends and global crop production since 1980', *Science*, 333 (6042), 616–20.

Lopes, M. S., Reynolds, M. P., Manes, Y., Singh, R. P., Crossa, J. and Braun, H. J. (2012), 'Genetic yield gains and changes in associated traits of CIMMYT spring bread wheat in a "historic" set representing 30 years of breeding', *Crop Science*, 52 (3), 1123–31.

Lu, M., Wang, L., Zhang, J., Sun, S., Li, Y., Du, W. et al., (2014), 'Molecular cytogenetic identification of a novel 1AL.1RS translocation line with powdery mildew resistance', *Genetics and Molecular Research*, 13 (4), 10678–89.

Ma, M., Zhao, H. X., Li, Z. J., Hu, S. W., Song, W. N. and Liu, X. L. (2016), '*TaCYP78A5* regulates seed size in wheat (*Triticum aestivum*)', *Journal of Experimental Botany*, 67 (5), 1397–410.

Maciel, J. L. N. (2011), '*Magnaporthe oryzae*, the blast pathogen: current status and options for its control', *CAB Reviews: Perspectives in Agriculture, Veterinary Science, Nutrition and Natural Resources*, 6 (50), 1–8.

Mahpara, S., Hussain, S. T. and Farooq, J. (2014), 'Drought tolerance studies in wheat (*Triticum aestivum* L.)', *Cercetari Agronomice in Moldova*, 47 (4), 133–40.

Manes, Y., Gomes, H. F., Puhl, L., Reynolds, M., Braun, H. J. and Trethowan, R. (2012), 'Genetic yield gains of the CIMMYT international semi-arid wheat yield trials from 1994 to 2010', *Crop Science*, 52 (4), 1543–52.

Mehrabi, R., Kamali, S., Majidi, E. and Khodarahmi, M. (2014), 'Evaluation of CIMMYT synthetic hexaploid wheats for resistance to septoria blotch', *Crop Breeding Journal*, 4 (1), 23–33.

Mir, R. R., Kumar, N., Jaiswal, V., Girdharwal, N. et al., (2012), 'Genetic dissection of grain weight in bread wheat through quantitative trait locus interval and association mapping', *Molecular Breeding*, 29 (4), 963–72.

Miri, H. R. (2009), 'Grain yield and morpho-physiological changes from 60 years of genetic improvement of wheat in Iran', *Experimental Agriculture*, 45 (2), 149–63.

Mishra, A. N., Kaushal, K., Dubey, V. G. and Prasad, S. V. S. (2015b), 'Genetic diversity for adult plant resistance to leaf rust in bread wheat (*Triticum aestivum* L.)', *Indian Journal of Genetics and Plant Breeding*, 75 (2), 167–73.

Mishra, V. K., Srivastava, S., Bhardwaj, A. K., Sharma, D. K., Singh, Y. P. and Nayak, A. K. (2015a), 'Resource conservation strategy for rice-wheat cropping systems on partially reclaimed sodic soild of the Indo Gangetic region, and their effects on soil carbon', *Natural Resources Forum*, 39 (2), 110–22.

Misra, B. K., Sharma, Rajiv K. and Nagarajan, S. (2004), 'Plant Breeding: a component of public health strategy', *Current Science*, 86 (9), 1210–15.

Mochida, K. and Shinozaki, K. (2013), 'Unlocking Triticeae genomics to sustainably feed the future', *Plant and Cell Physiology*, 54 (12), 1931–50.

Moghaddam, M. E., Kamali, M. R. J., Anet, Z., Roshani, M. and Ghodsi, M. (2014), 'Temporal variation in phenological characteristics, grain yield, and yield components of spring bread wheat (*Triticum aestivum* L.) cultivars released in Iran between 1952 and 2009', *Crop Breeding Journal*, 4 (1): 57–64.

Mohan, D. and Gupta, R. K. (2015), 'Relevance of physiological efficiency in wheat grain quality and the prospects of improvement', *Physiology and Molecular Biology of Plants*, 21 (4), 591–6.

Mohan, D., Gupta, R. K. and Indu, S. (2013a), 'Harnessing flour recovery in the Indian wheat', *Indian Journal of Agricultural Sciences*, 83 (7), 781–7.

Mohan, D., Gupta, R. K. and Ajay, V. (2013b), 'Characterization of popular bread wheat cultivars of India for grain quality and the stable genetic resource', *Indian Journal of Genetics and Plant Breeding*, 73 (1), 14–22.

Mohan, D., Gupta, R. K. and Mishra, B. (2009), 'Improving nutritional quality of Indian bread wheat (*Triticum aestivum*)', *Indian Journal of Agricultural Sciences*, 79 (1), 29–34.

Muhammad, F., Talat, M., Ghulam, S., Naeem, A., Kazi, A. G. and Abdul Mujeeb Kazi. (2015), 'Assessment of D-genome based genetic diversity in drought tolerant wheat germplasm', *International Journal Agricultural and Biology*, 17 (4), 791–6.

Muhammad, I., Tariq, M., Ahmad, A., Muhammad, B., Awais, R. and Mujeeb-Kazi, A. (2015), 'Characterization of D-genome diversity for tolerance to boron toxicity in synthetic hexaploid wheat and in silico analysis of candidate genes', *Acta Physiologiae Plantarum*, 37 (2), 17.

Muhammad, I., Muhammad, R., Arif, M. S., Shahzad, S. M. et al., (2015), 'Spatial distribution of pollutant emissions from crop residue burning in the Punjab and Sindh provinces of Pakistan: uncertainities and challenges', *Environmental Science and Pollution Research*, 22 (21), 16475–91.

Mujeeb-Kazi, A., Kazi, A. G., Dundas, I., Rashid, A., Ogbonnaya, F., Kishii, M., Bonnett, D., Wang, R. R. C., Xu, S., Chen, P., Mohammad, T., Hadi, B. and Sumaira, F. (2013), 'Genetic diversity for wheat improvement as a conduit to food security', *Advances Agronomy*, 122, 179–257.

Muminjanov, H., Otambekova, M. and Morgounov, A. (2014), 'The history of wheat breeding in Tajikistan', In: Eds., Bonjean, A. P., W. J., Angus and Maarten, van Ginkel, *The World Wheat Book: A history of Wheat Breeding*, Vol. 3, Lavoisier Tec & Doc, Paris, 283–309.

Muthamilarasan, M., Parida, S. K. and Prasad, M. (2014), 'Advances in wheat genomics and its potential in ensuring food security in the scenario of climate change', *Proceedings of the Indian National Science Academy*, 80 (2), 325–31.

Nagarajan, S., Kogel, K. H. and Zadocks, J. C. (2014), 'Epidemiological analysis of the damage potential of Pgt-Ug99 in Central East, North East Africa; Iran and Punjab (India)', *Indian Phytopathology*, 67 (1), 26–32.

Naik, B. K., Vinod, Sharma, J. B., Sivasamy, M., Prabhu, K. V. et al., (2015), 'Molecular mapping and validation of the microsatellite markers linked to the Secale cereal derived leaf rust resistance gene Lr45 in wheat', *Molecular Breeding*, 35 (2), 61.

Najafi-Mirak, T. (2012), 'Evaluation of resistance to sunn pest (*Eurygaster integriceps* Put.) in wheat and triticale genotypes', *Crop Breeding Journal*, 2 (1), 43–8.

Nazim, H., Azra, Y., Muhammad, A. and Hamid, N. (2014), 'Exploring the role of row spacing in yield improvement of wheat cultivars', *Pakistan Journal Agricultural Science*, 51 (1), 25–31.

Niaz, A., Iqbal, M., Yasmin, S., Ahmad, I. and Asif, M. (2013), 'Genetic variation at loci controlling quality traits in spring wheat', *Pakistan Journal Agricultural Sciences*, 50 (4), 637–47.

Niroula, R. K. and Bimb, H. P. (2009), 'Overview of wheat X maize system of crosses for dihaploid induction in wheat', *World Applied Science Journal*, 7 (8), 1037–45.

Niu, N., Arief, V. N., DeLacy, I. H., Lush, D., Sheppard, J., Zhang, GaiSheng and Dieters, M. J. (2010), 'Genetic gain in yield and protein over two cycles of a wheat recurrent selection program', *Breeding Science*, 60 (3), 181–6.

Ortiz, R., Sayre, K. D., Govartes, B., Raj Gupta, Subbarao, G. V., Ban, T., Hodson, D., Dixon, J. M., Ortiz-Monasterio, J. I. and Reynolds, M. (2008), 'Climate change: can wheat beat the heat? *Agriculture, Ecosystems and Environment*, 126 (1/2), 46–58.

Osman, M., He, X., Singh, R. P., Duveiller, E., Lillemo, M., Pereyra, S. A. et al., (2015), 'Phenotypic and genotypic characterization of CIMMYT's 15th international Fusarium head blight screening nursery of wheat', *Euphytica*, 205 (2), 521–37.

Pal, D., Bhardwaj, S. C., Sharma, D., Kumari, S., Patial, M. et al., (2015), 'Assessment of genetic diversity and validating rust resistance gene sources using molecular markers in wheat (*Triticum aestivum* L.)', *SABRAO Journal of Breeding and Genetics*, 47 (2), 89–98.

Pandey, A., Khan, M. K., Thomas, G., Hakki, E. E., Kayis, S. A., Hamurcu, M., Gezin, S., Topal, A. and Akkaya, M. S. (2015), 'Estimation of Indian and Turkish hexaploid wheat population structure employing molecular markers,' *Notulae Botanicae, Horti Agrobotanici Cluj-Napoca*, 43 (1), 70–80.

Pask, A. J. D. and Reynolds, M. P. (2013), 'Breeding for yield potential has increased deep soil water extraction capacity in irrigated wheat', *Crop Science*, 53 (5), 2090–104.

Patpour, M., Nazari, K., Ogbonnaya, F., Alavi, S. M. and Mousavi, A. (2014), 'Phenotypic and molecular characterization of resistance to stem ryst in wheat cultivars and advanced breeding lines from Iran and Syria', *Crop Breeding Journal*, 4 (1), 1–14.

Peng, J. H., Sun, D. F., Peng, Y. L. and Nevo, E. (2013), 'Gene discovery in Triticum dicoccoides, the direct progenitor of cultivated wheats', *Cereal Research Communication*, 41 (1), 1–22.

Peng, Y., Yu, K., Zhang, Y., Islam, S., Sun, D. and Ma, W. (2015), 'Two novel y-type high molecular weight glutenin genes in Chinese wheat landraces of the Yangtze-river basin', *PLoS ONE*, 10 (11), e0142348.

Pradeep, K., Uppal, R. S., Ram, H. and Kaur, H. (2014), 'Grain quality of bread wheat as affected by varieties and nutrient application of nitrogen, manganese and copper', *Indian Journal of Fertilizers*, 10 (10), 48–53.

Pradeep, S., Sindhu, S., Manoj, S., Ajay, V., Tyagi, B. S. and Indu, S. (2014), 'Assessing genetic variation for heat tolerance in synthetic wheat lines using phenotypic data and molecular markers', *Australian Journal of Crop Science*, 8 (4), 515–22.

Prashar, M., Bhardwaj, S. C., Jain, S. K. and Gangwar, O. P. (2015), 'Virulence diversity in *Puccinia striiformis* f.sp. *tritici* causing yellow rust on wheat (*Triticum aestivum*) in India', *Indian Phytopathology*, 68 (2), 129–33.

Qin, X, Zhang, F, Liu, C., Yu, H., Cao, B, Tian, S, Liao, Y. and Siddique, K. H. M. (2015), 'Wheat yield improvements in China: past trends and future directions', *Field Crops Research*, 177, 117–24.

Rajbir, Y., Singh, S. S., Neelu, J., Singh, G. P. and Prabhu, K. V. (2010), 'Wheat production in India: technologies to face future challenges', *Journal of Agricultural Sciences*, 2 (2), 164–73.

Rajender, S., Ratan, T., Davinder, S., Vinod, T. and Indu, S. (2015), 'Variability in yield trials of TILLING population of bread wheat (*Triticum aestivum* L.)', *Journal of Applied and Natural Science*, 7 (1), 443–6.

Ram, S., Sharma, S. and Sharma, I. (2015), 'Allelic diversity of HMW and LMW glutenins in Indian wheats wheats and their relationship with sedimentation volume and mixograph parameters', *Cereal Research Communications*, 43 (3), 492–503.

Rao, B. B., Chowdary, P. S., Sandeep, V. M., Pramod, V. M. and Rao, V. U. M. (2015), 'Spatial analysis of the sensitivity of wheat yields to temperature in India', *Agricultural and Forest Meterology*, 200, 192–202.

Rasheed, A., Xia, X. C., Mahmood, T., Quraishi, U. M., Aziz, A., Bux, H., Mahmood, Z., Mirza, J. I., Mujeeb-Kazi, A. and He, Z. H. (2016), 'Comparison of economically important loci in landraces and improved wheat cultivars from Pakistan', *Crop Science*, 56 (1), 287–301.

Reynolds, M. P. and Borlaug, N. E. (2006a), 'Applying innovations and new technologies for international collaborative wheat improvement,' *Journal of Agricultural Science* 144 (2), 95–110.

Reynolds, M. P. and Borlaug, N. E. (2006b), 'Impacts of breeding on international collaborative wheat improvement', *Journal of Agricultural Science*, 144 (1), 3–17.

Reynolds, M. P., Foulkes, J., Furbank, R., Griffiths, S., King, J., Murchie, E., Parry, M. and Slafer, G. (2012), 'Achieving yield gains in wheat', *Plant Cell and Environment*, 35 (10), 1799–1823.

Rizvi, S. J. H., Manan, A. R., Habibi, A., Sharma, R. K., Obaidi, Q., Azmatyar, M. H., Qayum, A., Aria, A. Z., Wahed, A. and Osmanzai, M. (2014), 'The history of wheat breeding in Afghanistan', In: Eds., Bonjean, A., P. W. J., Angus and Maarten van Ginkel, *The World Wheat Book: A History of Wheat Breeding*, Vol. 3, Lavoisier Tec & Doc, Paris, 439–54.

Ruby, C. and Altaf, A. (2015), 'Nitrogen stress induced alterations in the leaf proteome of two wheat varieties grown at different nitrogen levels', *Physiology and Molecular Biology of Plants*, 21 (1), 19–33.

Safrian, A. and Abdolshahi, R. (2014), 'Genetic evaluation of different characteristics of bread wheat (*Triticum aestivum*) in cyclic drought stress conditions', *Seed and Plant Impact Journal*, 30 (4), 683–701.

Sayyah, S. S., Ghobadi, M., Mansoorifar, S. and Zebarjadi, A. R. (2015), 'The yield of wheat genotypes associated with yield components under irrigated and drought stress after anthesis', *Archives of Agronomy and Soil Science*, 61 (12), 1743–55.

Sehgal, S. K., Kaur, S., Gupta, S., Sharma, A., Kaur, R. and Bains, N. S. (2011), 'A direct hybridization approach to gene transfer from *Aegilops tauschii* coss. to *Triticum aestivum* L.', *Plant Breeding*, 130 (1), 98–100.

Shahbaz, M. and Ashraf, M. (2013), 'Improving salinity tolerance in cereals', *Critical Reviews in Plant Sciences*, 32 (4), 237–49.

Sharma, R. C., Crossa, J., Velu, G., Huerta-Espino, J., Vargas, M., Payne, T. S. and Singh, R. P. (2012), 'Genetic gains for grain yield in CIMMYT spring bread wheat across international environments', *Crop Science*, 52 (4), 1522–33.

Sharma, R. (2013), 'Does low yield heterosis limit commercial hybrids in wheat?' *African Journal of Agricultural Research*, 8 (5), 6663–9.

Sharma, R. K., Osmanzai, M., Singh, R. P., Braun, H. J. and Sharma, R. C. (2014), 'Identification of high yielding genotypes through evaluation of international nurseries', *Cereal Research Communication*, 42 (2), 303–13.

Sharma, R. K., Osmanzai, M. and Ward, R. (2011), 'Wheat Research: a crucial pre-requisite to food security in Afghanistan', *Afghanistan Research & Evaluation Unit (AREU)*, No. 28, 2–5.

Sharma, R. K., Singh, P., Vinod, Joshi, A. K., Bhardwaj, S. C., Bains, N. and Singh, S. (2013), 'Protecting South Asian Wheat Production from stem rust (Ug99) epidemic', *Journal of Phytopathology*, 161, 299–307.

Sharma, R., Sonder, K. and Sika, G. (2015), 'Potential impact of climate change trends on wheat production and mitigation strategies in Afghanistan', *Journal of Agricultural Science*, 7 (4), 41–7.

Sharma, R. K. and Jagshoran (2000), 'Contribution of exotic germplasm in wheat breeding in India', *Indian Journal of Plant Genetic Resources*, 13 (1), 75–9.

Sharma, R. K. and Mishra, S. K. (1999), 'Wheat breeding in India: past achievements, current status and future prospects', *Annals of Agricultural Research*, 20 (2), 137–43.

Shewry, P. R. (2009), 'Darwin review: Wheat', *Journal of Experimental Botany*, 60, 1537–53.

Sheibanirad, A., Mirlohi, A., Mohammadi, R., Ehsanzadeh, P. and Sayed-Tabatabaei, B. E. (2014), 'Cytogenetic and crossability studies in hulled wheat collected from Central Zagros in Iran', *Plant Systematics and Evolution*, 300 (8), 1895–901.

Shivani, N., Singh, V. P., Ajay, A., Rajkumar, D. and Shiv, R. (2015), 'Assessment of terminal heat tolerance ability of wheat genotypes based on physiological traits using multivariate analysis', *Acta Physiologiae Plantarum*, 37 (12), 257.

Simarjit, K., Sohu, V. S. and Mavi, G. S. (2013), 'Climate change resilience in a set of diverse Indian and exotic genotypes of bread wheat', *Crop Improvement*, 40 (1), 50–7.

Singh, P. K., Zhang Y., He, X., Singh, R. P., Ramesh Chand, Mishra, V. K., Malaker, P. et al., (2015), 'Development and characterization of the 4th CSISA-spot blotch nursery of bread wheat', *European Journal of Plant Pathology*, 143 (3), 595–605.

Singh, R. J. and Ahlawat, I. P. S. (2014), 'Effects of transgenic cotton based cropping systems and their fertility levels on succeeding wheat crop', *Communications in Soil Science and Plant Analysis*, 45 (18), 2385–96.

Singh, R. P., Huerta-Espino, J., Sharma, R., Joshi, A. K. and Trethowan, R. (2007), 'High yielding spring bread wheat germplasm for global irrigated and rainfed production systems', *Euphytica*, 157 (3), 351–63.

Singh, S., Aggarwal, P. K. and Kumar, S. (2006), 'Physiological analysis of growth and productivity in wheat cultivars', *Indian Journal of Plant Physiology*, 11 (1), 57–62.

Singh, S. K., Singh, A. M., Jain, N., Singh, G. P., et al., (2013), 'Molecular characterization of vernalization and photoperiod genes in wheat varieties from different agro-climatic zones of India', *Cereal Research Communications*, 41 (3), 376–87.

Singh, S. N., Sah, A. K., Om, P., Singh, R. K. and Singh, V. K. (2010), 'Assessing the impact of zero tilled wheat growing in rice (*Oryza sativa* L.)- wheat (*Triticum aestivum* L.) cropping systems: the case of central Uttar Pradesh in the Indo-Gangetic Plain', *Outlook on Agriculture*, 39 (3), 197–202.

Sohail, Q., Manickavelu, A. and Ban, T. (2015), 'Genetic diversity analysis of Afghan landraces (*Triticum aestivum*) using DArT markers', *Genetic Resources and Crop Evolution*, 62 (8), 1147–57.

Sommer, R., Glazirina, M., Yuldashev, T., Otarov, A., Ibraeva, M., Martynova, L. et al., (2013), 'Impact of climate change on wheat productivity in Central Asia', *Agriculture, Ecosystems and Environment*, 178, 78–99.

Stein, A. J., Meenakshi, J. V., Qaim, M., Nestel, P. et al., (2008), 'Potential impacts of iron biofortification in India', *Social Science & Medicine*, 66 (8), 1797–808.

Stein, A. J., Nestel, P., Meenakshi, J. V., Qaim, M. et al., (2007), 'Plant Breeding to control zinc deficiency in India: how cost effective is biofortification?', *Public Health Nutrition*, 10 (5), 492–501.

Subhani, G. M., Javed, A., Abid, S., Makhdoom, H., Abid, M. (2015), 'Nutritional diversity in spring wheat with chronological perspective and its association with grain yield', *Australian Journal of Crop Science*, 9 (5), 363–71.

Suman, L., Saharan, M. S., Indu, S. and Megha, C. (2015), 'Decision support system for resistant wheat varieties', *International Journal of Agriculture and Crop Science*, 8 (2), 114–17.

Sun, D. and He, X. (2014), 'The future way for wheat', *Molecular Plant Breeding*, 5, 5.

Sutapa, S., Singh, A. M., Ahalawat, A. K., Mridul, C. et al., (2014), 'Genetic variability for grain and bread making quality traits in wheat (*Triticum aestivum* L.) and their correlations validated through multi-environmental trials', *Indian Journal of Genetics and Plant Breeding*, 74 (3), 309–16.

Tadesse, W., Inagaki, M., Tawkaz, S., Baum, M. and Ginkel, van, M. (2012), 'Recent advances and application of doubled haploids in wheat breeding', *African Journal of Biotechnology*, 11 (89), 15484–92.

Tadesse, W., Morgounov, A. I., Braun, H. J., Akin, B., Keser, M., Kaya, Y., Sharma, R. C., Rajaram, S., Singh, M., Baum, M. and Ginkel, van, M. (2013), 'Breeding progress for yield in winter wheat genotypes targeted to irrigated environments of the CWANA region', *Euphytica*, 194 (2), 177–85.

Tadesse, W., Nachit, M. M., Abdalla, O. and Rajaram, S. (2014), 'Wheat breeding at ICARDA: Achievements and Prospects in the CWANA Region', In: Eds., Bonjean, A. P., W. J., Angusand Maarten van Ginkel, *The World Wheat Book: A History of Wheat Breeding*, Vol. 3, Lavoisier Tec & Doc, Paris, 371–405.

Takahashi, F., Tilbrook, J., Trittermann, C., Berger, B., Roy, S. J., Seki, M., Shinojaki, K. and Tester, M. (2015), 'Comparison of leaf sheath transcriptome profiles with physiological traits of bread wheat cultivars under salinity stress', *PLoS ONE*, 10 (8), e0133322.

Tang, Y., Rosenwarne, G. M., Li, C., Wu, X., Yang, W. and Wu, C. (2015), 'Physiological factors underpinning grain improvements of synthetic-derived wheat in southwestern China', *Crop Science*, 55 (1), 98–112.

Tian, Y., Chen, J., Chen, C., Deng, A., Song, Z., Zheng, C., Hoogmoed, W. and Zhang, W. (2012), 'Warming impacts on winter wheat phenophase and grain yield under field conditions in Yangtze Delta Plain, China', *Field Crops Research*, 134, 193–9.

Tian Z. W., Fan Y. H., Yin M., Wang F. R., Cai J., Jiang D. and Dai T. B. (2015), 'Genetic improvement of root growth and its relationship with grain yield of wheat cultivars in the Middle-Lower Yangtze River', *Acta Agronomica Sinica*, 41 (4), 613–22.

Tomar, R. S. S., Deshmukh, R. K., Naik, K. B., Tomar, S. M. S. and Vinod (2014), 'Development of chloroplast specific microsatellite markers for molecular characterization of alloplasmic lines and phylogenetic analysis in wheat', *Plant Breeding*, 133 (1), 12–18.

Trethowan, R. and Turner, M. (2009), 'New targets in wheat improvement', CAB Reviews: perspectives in Agriculture, Veterinary Science', *Nutrition and Natural Resources*, 4 (9), 1–14.

Umesh, G., Ratan, T., Gupta, R. K., Sarvjeet, K. and Ashok, C. (2015), 'Allelic variation of functional markers for high molecular weight glutenin genes in Indian wheat (*Triticum aestivum* L.) cultivars and their correlation with bread loaf volume', *Indian Journal of Plant Physiology*, 20 (1), 97–102.

Uzair-Hashmi, Shafqat S., Khan, F., Majid, M., Hussain, H., Kazi, A. G., John, R. and Parvaiz Ahmad. (2015), 'Plant exomics: concepts, applications and methodologies in crop improvement', *Plant Signalling and Behaviour*, 10 (1), e976152.

Van Ginkel, M. and Ogbonnaya, F. (2007), 'Novel genetic diversity from synthetic wheats in breeding cultivars for changing production conditions', *Field Crops Research*, 104, 86–94.

Vandana, J., Vijay, G., Saloni, M., Priyanka, A. et al., (2015), 'Identification of novel SNP in promoter sequence of TaGW2–6A associated with grain weight and other agronomic traits in wheat (*Triticum aestivum* L.)', *PLoS ONE*, 10 (6), e0129400.

Velu, G., Ortiz-Monasterio, I., Singh, R. P. and Payne, T. (2011a), 'Variation for grain micronutrients concentration in wheat core-collection accessions of diverse origin', *Asian Journal of Crop Science*, 3 (1), 43–8.

Velu, G., Singh, R. P., Huerta- Espino, J., Pena, R. J., Arun, B., Mahendru-Singh, A., Mujahid, M. Y., et al., (2012), 'Performance of biofortified spring wheat genotypes in target environments for grain zinc and iron concentrations', *Field Crops Research*, 137, 261–7.

Velu, G., Singh, R., Huerta-Espino, J., Pena, J. and Ortiz-Monasterio, I. (2011b), 'Breeding for enhanced zinc and iron concentration in CIMMYT spring wheat germplasm', *Czech Journal of Genetics and Plant Breeding*, 47, S174–7.

Vishwakarma, M. K., Mishra, V. K., Yadav, P. S., Saxena, R. R. and Balasubramanian, A. (2015), 'Molecular validation of quality trait linked markers for exploring the wheat germplasm of Eastern Gangetic Plains of India', *Current Trends in Biotechnology and Pharmacy*, 9 (1), 68–74.

Vrinderpal, S., Bijay, S., Yadvinder, S., Thind, H. S. and Gupta, R. K. (2010), 'Need based nitrogen management using the chlorophyll meter and leaf colour chart in rice and wheat in South Asia: a review', *Nutrient Cycling in Agroecosystems*, 88 (3), 361–80.

Wang, F., He, Z, Sayre, K., Li, S., Si, J., Feng, B. and Kong, L. (2009), 'Wheat Cropping systems and technologies in China', *Field Crops Research*, 111 (3), 181–8.

Wang, M., Wang, S. and Xia, G. (2015a), 'From genome to gene: a new epoch for wheat research?', *Trends in Plant Science*, 20 (6), 380–7.

Wang, Y., Du, Y., Yang, Z., Chen, L., Condon, A. G. and Hu, Y. (2015b), 'Comparing the effects of GA responsive dwarfing genes Rht13 and Rht8 on plant height and some agronomic traits in common wheat', *Field Crops Research*, 179, 35–43.

Wang, Y., Wang, C., Quan, W., Jia, X., Fu, Y., Zhang, H. et al., (2016), 'Identification and mapping of PmSE5785, a new recessive powdery mildew resistance locus, in synthetic hexaploid wheat', *Euphytica*, 207 (3), 619–26.

Wei, L., Muranty, H. and Zhang, H. (2011), 'Advances and prospects in wheat eyespot research: contributions from genetics and molecular tools', *Journal of Phytopathology*, 159 (7/8), 457–70.

Whitford, R., Fleury, D., Reif, J. C., Garcia, M., Okada, T., Korzun, V. and Langridge, P. (2013), 'Hybrid breeding in wheat: technologies to improve hybrid wheat seed production', *Journal Experimental Botany*, 64 (18), 5411–28.

Wulff, B. B. H. and Moscou, M. J. (2014), 'Strategies for transferring resistance into wheat: from wide crosses to GM cassettes', *Frontiers in Plant Science*, 5, 692.

Wu, W., Li, C, Ma, B., Farooq, S., Liu, Y. and Liao, Y. (2014), 'Genetic progress in wheat yield and associated traits in China since 1945 and future prospects', *Euphytica* 196 (2), 155–68.

Xu, Z., Gongbuzhaxi, Wang, C., Xue, F., Zhanf, H. and Ji, W. (2015), 'Wheat NAC transcription factor TaNAC29 is involved in response to salt stress', *Plant Physilogy and Biochemistry*, 96, 356–63.

Yadamsuren, M., Noov, B. and Serchin, B. (2014), 'The history of wheat breeding in Mongolia', In: Eds., Bonjean, A. P., W. J., Angus, and Maarten van Ginkel, The World Wheat Book: A History of Wheat Breeding, Vol. 3, Lavoisier Tec & Doc, Paris, 215–33.

Yang, X., Wu, L., Zhu, Z., Ren, G. and Liu, S. (2014), 'Variation and trends in dough rheological properties and flour quality in 330 Chinese wheat varieties', *Crop Journal*, 2 (4), 195–200.

Yashavanthkumar, L. M., Naik, V. R. and Biradar, S. S. (2014), 'Identification of micronutrient dense genotypes on the basis of biochemical characteristics in wheat (*Triticum* spp.)', *Karnataka Journal of Agricultural Sciences*, 27 (3), 285–8.

Yasmeen, F., Khurshid, H. and Ghafoor, A. (2015), 'Genetic divergence for high-molecular weight glutenin subunits (HMWGS) in indigenous landraces and commercial cultivars of bread wheat of Pakistan', *Genetics and Molecular Research*, 14 (2), 4829–39.

Ye, X. L., Lu, Y. Q., Liu, W. H., Chen, G. Y., Han, H. M., Zhang, J. P., Yang, X. M., Gao, A. and Li, L. (2015), 'The effects of chromosome 6P on fertile number of wheat as revealed in wheat-Agropyron cristatum chromosome 5A/6P translocation lines', *Theoretical and Applied Genetics*, 128 (5), 797–811.

You, G. X., Sun, G. Z., Zhang, X. Y. and Xiao, S. H. (2015), 'Cold hardiness and its relationship with the *VRN1* genotypes in wheat varieties in the Yellow-Huai-Hai River valley region of China', *Acta Agronomica Sinica*, 41 (4), 557–64.

Yu, L., Wang, X. and Liu, J. (2011), 'Analysis on relationship between overcast and rainy weather, drought, flood and wheat yield', *Meterological and Environmental Research*, 2 (9), 47–8.

Yu, S., Lonh, H., Deng, G., Pan, Z., Liang, J and Yu, M. (2015), 'Location of the powdery mildew resistance gene Pm07j126 in wheat (*Triticum aestivum* L.)', *Euphytica*, 205 (3), 691–8.

Zhang, F., Gao, S., Zhao, Y., Zhao, X., Liu, X. and Xiao, K. (2015a), 'Growth traits and nitrogen assimilation-associated physiological parameters of wheat (*Triticum aestivum* L.) under low and high N conditions', *Journal of Integrative Agriculture*, 14 (7), 1295–308.

Zhang, J. and Li, L. (2014), 'The analysis on Chine's biofortified iron rich wheat nutrition intervening Chinese people's disease burden of iron deficiency anaemia', *Journal of Agricultural Science and Technology* (Beijing), 16 (6), 132–42.

Zhang, J., Zhang, J., Liu, W., Han, H., Lu, Y., Yang, X., Li, X. and Li, L. (2015b), 'Introgression of *Agropyron cristatum* 6P chromosome segment into common wheat for enhanced thousand grain weight and spike length', *Theoretical and Applied Genetics*, 128 (9), 1827–37.

Zhang, L., Yu, Y., Zhang, M. and Hu, X. (2015c), 'Cloning of wheat GPX gene and construction of its plant expression vector', *Guizhou Agricultural Sciences*, 43 (4), 31–4.

Zhang, R., Hou, F., Feng, Y., Zhang, W., Zhang, M. and Chen, P. (2015d), 'Characterization of a *Triticum aestivum-Dasypyrum villosum* T2VS.2DL translocation line expressing a longer spike and more kernels traits', *Theoretical and Applied Genetics*, 128 (12), 2415–25.

Zhao, H., Zhang, W., Wang, J., Li, F., Cui, F., Ji, J., Wang, D. and Li, J. (2013), 'Comparative study on drought tolerance of wheat and wheat-*Aegilops biuncialis*6U^b addition lines', *Journal of Food, Agriculture and Environment*, 11 (3/4), 1046–52.

Zhao, W. C., Gao, X., Dong, J., Zhao, Z. J. et al., (2015), 'Stripe rust resistance and dough quality of new wheat-*Dasypyrum villosum* translocation lines T1DL.1V#3S and T1DS.1V#3L and the location of HMW-GS genes', *Genetics and Molecular Research*, 14 (3), 8077–83.

Zhao, Y., Li, X., Zhang, S., Wang, J., Yang, X., Tiang, J., Hai, Y. and Yang, X. (2014), 'Mapping QTLs for potassium deficiency tolerance at the seedling stage in wheat (*Triticum aestivum* L.)', *Euphytica*, 198 (2), 185–98.

Zhen, S., Han, C., Ma, C., Gu, A., Zhang, M., Shen, X., Li, X. and Yan, Y. (2014), 'Deletion of the low molecular weight glutenin subunit allele Glu-A3a of wheat (*Triticum aestivum* L.) significantly reduces dough strength and bread making quality', *BMC Plant Biology*, 14, 367.

Zheng, J, Yang, Z, Madgwick, P., Carmo-Silva, E., Parry, M. A. J. and Hu, Y. (2015), '*TaER* expression is associated with transpiration efficiency traits and yield in bread wheat', *PLoS ONE*, 10 (6), e0128415.

Zhou, B. W., Sanz-Saez, A., Elazab, A., Shen, T. M. et al., (2014), 'Physiological traits contributed to the recent increase in yield potential of winter wheat from Henan Province, China', *Journal of Integrative Plant Biology*, 56 (5), 492–504.

Zhu, Q, Weng, Q, Zhang, G, Zhang, J., Jiu, L., Yu, Y, Niu, N, Wang, J. and Ma, S. (2014), 'Mitochondrial DNA AFLP and SCAR market linked to the male sterile lines with *Aegilops Kotschyi* cytoplasm in wheat (*Triticum aestivum* L.)', *Journal of Agricultural Biotechnology*, 22 (8), 958–65.

Chapter 12

Improving wheat production in the Central and West Asia and North Africa (CWANA) region

W. Tadesse, A. Amri, M. Sanchez-Garcia, M. El-Bouhssini, M. Karrou, S. Patil, F. Bassi and M. Baum, International Center for Agricultural Research in the Dry Areas, Morocco; and T. Oweis, International Center for Agricultural Research in the Dry Areas, Jordan

1 Introduction
2 Major wheat production environments
3 Challenges to wheat production
4 Improved technologies to enhance wheat production
5 Intensification of wheat technologies
6 Future trends
7 Where to look for further information
8 References

1 Introduction

Wheat is the principal staple food in most countries of the Central and West Asia and North Africa (CWANA) region, accounting for 45% of the region's per capita calorie intake with an average wheat consumption of about 200 kg/capita/year, which is the highest in the world. The CWANA region is a vast geographic area extending west to east from the Atlas Mountains in Morocco to the fertile irrigated Indus valley in Pakistan, and from the highland, high-rainfall areas of Ethiopia in the south to the temperate and dry northern Kazakhstan. As expected, this vast geographic area is characterized by large variations in agro-ecology, farming systems, moisture, temperature, soil types and cultural practices. Accordingly, the region harbours all kinds of wild and cultivated wheat types of different growth habits. During the period 1961–2013, the area of wheat production increased from 26.9 to 54 million ha, while total production and yield (t/ha) increased from 22 to 122 million tons and from 1.1 to 2.6 t/ha, respectively (Fig. 1).

The increase in production is mainly due to the adoption of improved wheat varieties originating from the International Maize and Wheat Improvement Center (CIMMYT) and the International Center for Agricultural Research in the Dry Areas (ICARDA), utilization of inputs, better agronomic practices, increased area of production and favourable policies.

http://dx.doi.org/10.19103/AS.2016.0004.39

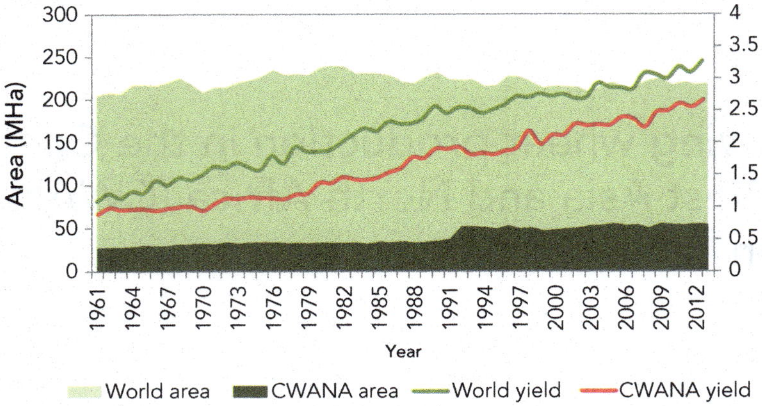

Figure 1 Trends in wheat area, production and yield at the global and CWANA level, 1961–2013; FAO, 2015.

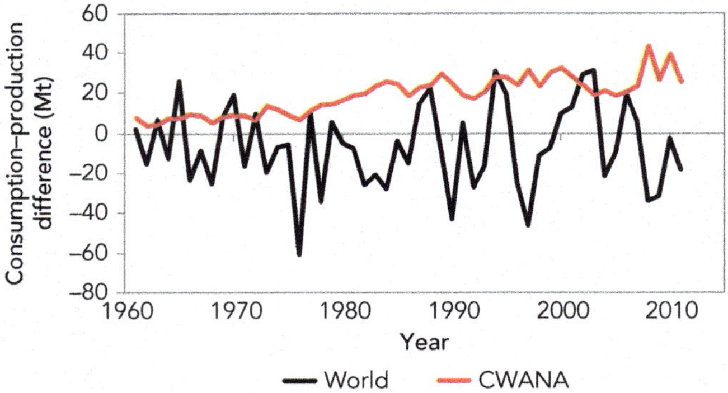

Figure 2 Wheat consumption and production difference at global and CWANA levels, 1961–2011; FAO, 2015.

The most important wheat-growing countries of the CWANA region (see Table 1 for a detailed list) with the area of cultivation in decreasing order are: Kazakhstan (13.7 Mha), Pakistan (8.9 Mha), Turkey (7.9 Mha), Iran (6.8 Mha), Morocco (3.0 Mha), Afghanistan (2.4 Mha), Algeria (1.8 Mha), Ethiopia (1.6 Mha), Syria (1.5 Mha), Uzbekistan (1.4 Mha) and Egypt (1.3 Mha). As indicated in Fig. 2, there is a huge difference between production and consumption in the CWANA region. There is not sufficient production at regional level to cope with the increasing demand for wheat consumption.

Most of the countries in the CWANA region, except Kazakhstan, Syria, Pakistan and Turkey, are not self-sufficient in wheat production, and consequently, wheat is their single most important imported commodity. Among North African countries, Egypt is the largest

Table 1 Average area, production and yield of wheat from 2009 to 2013 in wheat-growing countries of the CWANA region

Country	Region	Area (million ha)	Production (million tons)	Yield (t ha^{-1})
Afghanistan	Central Asia	2.45	4.64	1.89
Algeria	North Africa	1.79	2.97	1.66
Egypt	North Africa	1.33	8.47	6.35
Ethiopia	North Africa/Sub-Saharan Africa	1.60	3.26	2.03
Iran (Islamic Republic)	West Asia	6.82	13.42	1.97
Iraq	West Asia	1.36	2.90	2.11
Jordan	West Asia	0.02	0.02	1.16
Kazakhstan	Central Asia	13.70	14.64	1.07
Kyrgyzstan	Central Asia	0.36	0.81	2.20
Lebanon	West Asia	0.04	0.12	3.38
Libya	North Africa	0.15	0.16	1.03
Morocco	North Africa	3.05	5.62	1.84
Pakistan	Central Asia	8.88	24.05	2.71
Saudi Arabia	West Asia	0.18	1.01	5.60
Sudan	North Africa/sub-Saharan Africa	0.23	0.39	1.74
Syrian Arab Republic	West Asia	1.51	3.49	2.32
Tajikistan	Central Asia	0.33	0.89	2.70
Tunisia	North Africa	0.65	1.32	2.00
Turkey	West Asia	7.92	20.84	2.63
Turkmenistan	Central Asia	0.72	1.59	2.20
Uzbekistan	Central Asia	1.42	6.67	4.70
Yemen	West Asia	0.13	0.24	1.81
	North Africa	1.40	3.71	2.57
	West Asia	2.25	5.26	2.62
	Central Asia	3.98	7.61	2.50
	CWANA	54.52	117.20	2.55
	World	219.52	684.02	3.12

FAO, 2015.

importer with 9 million tons of wheat imported every year. According to Shiferaw et al. (2013), the demand for wheat is growing at 5.6 and 2.2%/year in Central Asia and North Africa, respectively. In the year 2050, the population in CWANA is expected to increase from the current 0.9 billion to 1.4 billion, and the demand for wheat will rise from the present 164 million tons to 268 million tons, calling for more research and development efforts to increase wheat productivity at regional and global levels to meet the needs of the increasing population.

CWANA also hosts some of the most damaging and virulent races of diseases and pests. Farmers are typically required to apply fungicide and pesticides multiple times during their cropping season to avoid near complete losses. These chemical products are often at a premium in the small rural markets of CWANA and their availability is far from certain. Furthermore, farmers are rarely trained in the deployment of integrated pest management (IPM) practices; the results of IPM malpractices can be the rapid insurgence of novel races with acquired resistance to the pesticides.

The increasing cost of importing grains, combined with the heavy investments in chemical inputs, and the constant risk of the rise of new virulent races clearly show that the current status of wheat production in CWANA is not sustainable in the long term. To cope with such a daunting challenge, ICARDA and its partners have identified, as the primary goals for sustainability, (i) the deployment of novel cultivars, which include genetic resistances, yield stability and increased yield potential; (ii) training and deployment of novel agronomical practices to close the productivity gap with reduced inputs per unit of grains and (iii) integrated approaches to pest and disease management to prevent the surge of new races and control the ones present.

2 Major wheat production environments

According to Rajaram et al. (1995), a mega-environment (ME) is a region with similar climate, moisture regimes, soil types, growth habit, plant performance, prevalent diseases and insect pests. Thus, based on moisture availability, cropping systems and temperature regimes, the wheat production area in the CWANA region can be classified under five distinct large MEs (E), namely: (E1) favourable irrigated and high rainfall spring wheat environment of CWANA that covers over 13.0 million hectares; (E2) semi-arid (Mediterranean rain-fed) spring wheat environment with about 11.0 million hectares; (E3) favourable irrigated winter/facultative wheat environment that covers an area of 5.4 million hectares; (E4) semi-arid rain-fed winter/facultative wheat environment that has an area of approximately 12.7 million hectares and (E5) the high latitude (spring planted) spring wheat environment with approximately 8.0 million hectares, all in northern Kazakhstan. The area under wheat cultivation in North Africa, West Asia, Central Asia and the Caucasus and the Nile Valley subregions is characterized by large variations in agro-ecology, farming systems, moisture, temperature, soil types, races of pests and cultural practices. In these subregions, most of the rain-fed wheat farmers, whose livelihoods solely depend on income from wheat grain, fodder and straw production, live in areas with less than 350 mm of rainfall per annum. Hence, the adverse effects of the harsh climate, poor soil conditions, crop diseases and insects resulted generally in low and highly variable wheat production. Because of these complexities, it is prudent that international breeding programmes take the ME approach and use the shuttle breeding and multilocation yield trial strategy to develop high-yielding genotypes with wide and specific adaptations.

3 Challenges to wheat production

3.1 Abiotic and biotic stresses

A significant proportion of wheat production in the CWANA region is predominantly carried out under rain-fed conditions, where the distribution of rainfall is erratic and sometimes causes problems with germination, often resulting in terminal moisture stress. Water is becoming scarce even in areas where wheat production is dependent on the use of supplementary irrigation. This is further accentuated by the increasing incidence of adverse climatic events through climate change, characterized by rising temperature (heat), less and erratic rainfall (drought) or sometimes excessive rainfall (flooding), and when combined with virulent pests and diseases, it makes agricultural productivity less predictable. Wheat is affected by many diseases such as rusts (*Puccinia* spp.), septoria (*Septoria tritici*), tan spot (*Pyrenophora tritici-repentis*), fusarium (*Fusarium* spp.), powdery mildew, the bunts and smuts, take-all and root rots. Yield loss estimates for each disease vary from country to country and season to season; however, the three rusts (stem, yellow leaf), each affecting 50 Mha worldwide (Consortium Research Program (CRP) wheat), are the most damaging ones. Of all the insects attacking wheat crop, Hessian fly, Sunn pest, wheat stem sawfly and Russian wheat aphids cause significant economic losses. Hessian fly (*Mayetiola destructor*) damage can result in a total loss of the crop if high infestations occur in the early stages of crop development. In Morocco, bread wheat and durum wheat yield losses due to Hessian fly have been estimated annually at 36 and 32%, respectively, amounting to about US$200 million/annum (Lhaloui et al., 1992). Sunn pest (*Eurygaster integriceps* Puton) affects some 15 million ha of wheat in West and Central Asia. With the current effects of climate change, it is anticipated that new pests and diseases will emerge as already exemplified in the recent epidemics of stripe/yellow rust across the CWANA region and the *Ug99* stem rust epidemic in East African countries (Solh et al., 2012). Climate change is also affecting the quality of wheat as increasing CO_2 may negatively affect N content and protein quality and content and increasing temperatures can negatively impact grain size. Modelling results indicate that wheat production will be affected by climate change more in the developing countries than in the developed ones (Braun et al., 2010).

3.2 Limited availability and high price of inputs

As has been clearly witnessed by the effects of the Green Revolution, increasing yield per unit area has been achieved mainly through the application and utilization of inputs such as improved varieties, irrigation, fertilizers and pesticides. However, the costs of these inputs are also rising and are sometimes becoming unaffordable to poor farmers in developing countries. Rising energy costs, for example, have contributed to higher fertilizer prices, because the costs of natural gas used to produce ammonia are increasing, and higher transportation costs have also indirectly contributed to higher input prices.

Such rapid increases in input costs offset wheat production and pose a disincentive to wheat producers.

3.3 Yield gap, stagnating yield and increasing population

After the quantum leap in wheat production attributed to the Green Revolution, wheat yields have recently been rising by only 1.1% per year, a level that falls far short of the demand of a global population that is growing at 1.5% or more, annually. According to some estimates, global wheat production must increase at least by 1.6% annually to meet a projected yearly wheat demand of 760 million tons by 2020. In the year 2050, the world population is estimated to be 9 billion, while the demand for wheat is estimated to reach more than 900 million tons (Dixon et al., 2009). The current annual production level of about 126 million tons of wheat on a total area of 54 million hectares is far below the regional demand of about 164 million tons. Such imbalance between demand and production has led to the importation of 44 million tons of wheat at a cost of 15 billion dollars during the 2011 season alone. Occasionally, a price hike in wheat can also cause political instability as has been seen with the price hikes in 2008. As mentioned earlier, in the year 2050, the CWANA population is expected to increase from the current 0.9 billion to 1.4 billion, and the demand for wheat will reach 268 million tons.

Fulfilling this demand is challenging amid reports of yield stagnation in major wheat-growing regions of the world and in the face of the expected adverse effects of climate change. Yield stagnation is a complex issue, which might be the result of a combination of factors, such as approaching a genetic ceiling in wheat improvement, declining soil fertility, unfavourable policies and marketing, biotic and abiotic stresses associated with climate change and other factors. Some authors have attributed yield stagnation to the genetic ceiling in India and Europe (Nagarajan, 2005), while others have reported the presence of genetic gain in both spring wheat (Manes et al., 2012; Sharma et al., 2012) and facultative winter wheat (Tadesse et al., 2013). However, it is evident that the potential of new cultivars has not been fully utilized in most of the developing countries due to poor agronomic management, application of incomplete packages of inputs, reduced incentives and unstable market prices. Recent yield gap analysis at a global level in wheat indicated that the difference between potential yield and farm yield in most countries, although narrowing, ranges between 26 and 69% with an average of 48% (Fischer et al., 2014). However, at the regional level, especially in most developing countries of the CWANA region, there is still a huge yield gap. According to Pala et al. (2011), average yield gaps in Morocco ranged from 98 to 207% in rain-fed areas and from 51 to 89% in irrigated areas. Similarly, yield gaps of 82–125% and 61–201% have been reported in Syria and Turkey, respectively.

4 Improved technologies to enhance wheat production

4.1 Modern varieties (MVs)

The CRP wheat breeding programme of the Consultative Group of International Agricultural Research (CGIAR) centres (CIMMYT and ICARDA) apply both conventional and molecular breeding approaches to develop high-yielding and widely adapted germplasm with resistance/tolerance to the major biotic and abiotic constraints prevailing at global levels in general, and in developing countries in particular. Some of these strategies and techniques include the classification of ME, assembling targeted crossing blocks, shuttle breeding, utilization of doubled haploids, marker-assisted selection, genomic selection, key location yield trials, distribution of germplasm to the National Agricultural Research System (NARS) through international nurseries and partnership and capacity building of NARS. As water is becoming ever scarcer, even in irrigated areas, wheat germplasm development at the CGIAR centres is based on the identification of genotypes with disease resistance, high-yield potential and water-use efficiency, so that wheat genotypes targeted for irrigated areas can cope with temporary drought periods. Similarly, this approach minimizes yield losses and maximizes yield gains during drought as well as good seasons for rain-fed production systems.

The continuous supply of improved germplasm from the International Wheat Improvement Network (IWIN), an alliance of NARS, CIMMYT, ICARDA and advanced research institutes, for nearly half a century has led to major gains during and after the Green Revolution and has also enabled developing countries to have a sustained increase in wheat production and productivity, thereby improving food security and farmers' livelihoods (Payne, 2004; Reynolds and Borlaug, 2006; Trethowan et al., 2007). The success in terms of increased wheat production within the CGIAR has been remarkable, with more than 50% of the wheat varieties grown in the CWANA region attributable to CIMMYT/ICARDA programmes.

The impact of wheat research has been witnessed not only by farmers, governments, policy makers and professionals but also by donors. According to the World Bank (2008), wheat is the major crop with the highest percentage of farmers in the developing world that cultivate modern and improved cultivars.

4.1.1 Breeding for drought tolerance

Identification and development of wheat varieties combining high-yield potential with water-use efficiency can help stabilize yield gains in the face of climate change. The development of wheat varieties with early vigour and cold tolerance has been a major target of wheat breeders in dry land areas, as early and complete canopy establishment shades the soil and reduces evaporative loss from the soil surface, thereby significantly improving the water productivity of wheat. High-yielding and drought-tolerant spring wheat genotypes have been identified from the ICARDA's breeding programme, and their agronomic performance has been assessed at the Sids Research Station in Egypt (31.5N; 34.5E) for yield potential, at the Wad Medani Station in Sudan (14.4N; 33.6E) for heat tolerance or in the case of durum wheat at the Sylla Station in Mauritania (16.13N; 13.58W) and at the Marchouch Station in Morocco (33.6N; 6.7W), as indicated in Table 2.

4.1.2 Development and deployment of high-yielding and disease-resistant wheat varieties

Among the major diseases known in wheat, stem/black rust and stripe/yellow rust are the most damaging and can cause very significant wheat yield losses. Developing and deploying genetically resistant varieties adapted to target-growing environments is the most economical and environmentally friendly strategy for controlling rust diseases of wheat, particularly for resource-poor farmers. However, because of the coevolution of

Table 2 Performance of elite bread wheat (IBW) or durum wheat (IDYT) genotypes across key locations in Egypt, Sudan (or Mauritania) and Morocco, 2013/14

Variety	Marchouch (t/ha)	Sids (t/ha)	Wad Medani (t/ha)	Average (t/ha)	% of the check (Attila-7)
IBW-IMAR	7.11	11.62	3.23	7.32	119%
IBW-AMAL	6.49	12.72	2.50	7.24	118%
IBW-HAMID	7.13	9.56	3.77	6.82	111%
IBW-WAHID	7.31	10.16	2.92	6.80	111%
IBW-BRIVAN	8.98	7.81	3.05	6.61	108%
IBW-FARID	6.39	10.23	3.03	6.55	106%
IBW-OMAR	6.96	9.72	2.65	6.45	105%
IBW-AKID	6.50	9.87	2.94	6.44	105%
IBW-WIDAD	6.51	8.46	3.87	6.28	102%
IBW-TARTUS	5.89	9.39	3.40	6.23	101%
IDYT-MARGHERITA[a]	6.52	13.54	3.06[b]	7.71	120%
IDYT-ICAVERVE[a]	6.58	12.17	3.33[b]	7.36	115%
Sids-1 (check)	5.36	10.88	2.35	6.20	101%
Attila-7 (check)	6.19	8.40	3.86	6.15	100%
Pastor-2 (check)	6.24	7.49	3.42	5.72	93%
Commercial check[a]	5.20	11.05	2.99[b]	6.41	

[a]Durum wheat entries compared to the best commercial check in each country.
[b]Data from Sylla Kaedi Station, Mauritania.

the host and pathogen, the deployment of individual resistance genes often leads to the emergence of new virulent pathogen mutants, and hence the 'boom and bust cycle' performance of varieties continues. A new race of stem rust first identified in 1999 in Uganda (*Ug99*) became a threat to the global wheat industry as it overcame many of the known and most commonly used stem rust resistance genes, such as *Sr36*, *Sr31* and *Sr24*. Similarly, the breakdown of stripe rust resistance gene *Yr9* in varieties derived from 'Veery' lineage in the 1980s and *Yr27* in the 2000s in major mega-cultivars derived from the 'Attila' cross, such as MH97 (Pakistan), Kubsa (Ethiopia), PBW343 (India) and others such as Achtar in Morocco, Hidab in Algeria and many other cultivars in the CWANA region, has caused significant wheat production losses of up to 70% (Solh et al., 2012).

The CGIAR centres (CIMMYT and ICARDA) have carried out intensive screening and germplasm evaluation activities against the *Ug99* race of stem rust at Njoro, Kenya and at Debre Zeit, Ethiopia through the Borlaug Global Rust Initiative programme. High-yielding varieties of CGIAR origin with resistance to *Ug99* and other diseases such as stripe rust and septoria have been released recently and deployed in many countries (Table 3).

Table 3 Recently released *Ug99* stem rust-resistant wheat varieties of CGIAR origin

Variety name	Type	Cross/pedigree	Selection history	Year of release	Country
KINGBIRD	SBW	TAM-200/TUI/6/PAVON-76//CAR-422/ ANAHUAC-75/5/BOBWHITE/CROW// BUCKBUCK/PAVON-76/3/YECORA-70/4/TRAP-1	CMSS99M00216-S-040M-030Y-0030M-16Y-2M-0Y	2007	Kenya
DANPHE	SBW	KIRITATI//2*PBW-65/2*SERI.1-B	CGSS02B00119T-099B-099Y-099M-099Y-099M-24WGY-0B	2008	Ethiopia
HOGGANA	SBW	PYN/BAU//MILAN (= ETBW 5780)	CMSW94WM00188S-0300M-0100Y-0100M-15Y-8M-0Y-0IAP-0QTAP-0YT	2011	Ethiopia
SHORIMA	SBW	UTQE96/3/PYN/BAU//MILAN	ICW02-00330-4AP/0TS-0AP-030AP-10KUL-030KUL-0AP/0KUL-0DZ/0AP	2011	Ethiopia
KARIM	SBW	T.AEST/SPRW//CA8055/3/BACANORA86	ICW92-0477-1AP-1AP-4AP-1AP-0AP	2011	Iran
GOUMRIA-3	SBW	VEE#7/KAUZ	ICW94-0029-0L-1AP-1AP-7AP-0APS-0AP-0SD	2013	Sudan
NORMAN	FWW	OR F1.158/FDL//BLO/3/SHI4414/CROW	ICWH860291-3AP-1AP-0AP-1AP-0AP	2007	Tajikistan
GIZIL BUGDA	FWW	SAULESKU41/SADOVO1	TCI950295-3AP-0AP-0E-1YE-0YE-1YM-0YM	2009	Azerbaijan
NARC-2011	SBW	OASIS/ SKAUZ//4*BACANORA-88/3/2*PASTOR[3811];	CMSS00Y01881T-050M-030Y-030M-030WGY-33M-0Y-0S	2011	Pakistan
MUQAWIM 09	SBW	OASIS/ SUPER-KAUZ//4*BACANORA-88/3/2*PASTOR	CMSS00Y01881T-050M-030Y-030M-030WGY-33M-0Y-0S	2009	Afghanistan
SUPER 152	SBW	PFAU/SERI.1B//AMAD/3/WAXWING	CGSS02Y00153S-099M-099Y-099M-46Y-0B	2011	India
MISR 1	SBW	OASIS/SKAUZ//4*BCN/3/2*PASTOR	CMSS00Y01881T-050M-030Y-030M-030WGY-33M-0Y-0S	2010	Egypt
MANGUDO	SDW	MRF1/STJ2/3/1718/BT24//KARIM	ICD01-0251-T-9AP-TR-4AP-0AP	2012	Ethiopia
MUKIYE	SDW	STJ3//BCR/LKS4/3/TER3	ICD99-0091-T-3AP-AP-6AP-AP	2012	Ethiopia
Miki3	SDW	STJ3//BCR/LKS4	ICD94-0994-C-10AP-0AP-2AP-0AP-9AP-0TR	2013	Lebanon

4.1.3 Breeding for quality

In most developing countries, apart from grain yield, disease resistance and drought/heat tolerance, grain quality has not been a strong criterion for variety selection. However, some NARS are critically looking for high-quality varieties suited for the preparation of a range of end products. Varieties such as Anza, Bezostaya, HD1220 and Pavon-76 are still dominantly grown in some countries not only because of their wide adaptation, high-yield potential and stability, but also because of their high protein content and excellent bread-making quality. With this understanding, the wheat breeding programmes at ICARDA and CIMMYT routinely undertake evaluation of germplasm for quality traits following international standard grain quality procedures. Most of the currently available elite genotypes for both irrigated and rain-fed environments are excellent in quality, with protein levels of 12–16%. Most of these genotypes have the 5+10 (Glu-D1), 7+8 (Glu-B1) and 2* (Glu-A1) glutenin subunit alleles, which are known to be highly correlated with desired protein quality. Recently, through the Harvest Plus (H+) Challenge programme, progress has been made in the development of bio-fortified wheat with improved zinc (Zn) and iron (Fe) content in wheat grains in addition to protein content and quality. However, because of the negative correlation of yield and these quality traits, it may be difficult to combine high-yield potential with high micronutrient grain content.

4.1.4 Seed production and distribution

MVs and seed technology play a significant role in increasing agricultural production and productivity, ensuring food and nutritional security and improving livelihoods of farming communities. A robust seed system provides farmers with sufficient quantity of seed of appropriate quality at the right place and time, and at reasonable cost. There are relatively well-functioning wheat seed sectors across most developing countries, although the seed sector is dominated by the public sector, both in terms of agricultural research (varietal releases) and seed delivery (quantity supplied). The diversification of the national seed sector led to the emergence of the private sector in wheat seed delivery in countries like Egypt, India, Morocco, Pakistan and Turkey (Bishaw, 2004). However, as wheat is a self-pollinated crop, farm-saved seed continues to dominate the landscape. As a result, the varietal replacement rate measured by the average age of varieties in farmers' fields and the average annual seed renewal rate measured by farmers' regular purchase of certified seed remains low, with over ten years and less than 20%, respectively. These figures are still worse in less favourable rain-fed areas and less accessible remote regions. In recent years, the emergence of rust threats and food price spikes triggered renewed interest in food security in general and wheat seed supply in particular. For the rapid deployment of stem rust-resistant wheat varieties in Egypt, Ethiopia and Pakistan, ICARDA, supported by a United States Agency for International Development (USAID) project from 2009 to 2012, developed a strategy to fast-track testing and release of rust-resistant wheat varieties, massive extension and demonstration, accelerated seed multiplication of early generation seeds (breeder, prebasic and basic seed), large-scale certified seed production by linking to existing public and/or private seed enterprises and on-farm seed production with farmer groups. In total, two varieties in Egypt, eight varieties in Ethiopia and five varieties in Pakistan were released and substantial quantities of seed were produced and distributed under an accelerated scheme within the shortest possible period of time. In Egypt, 10 760 t of certified seeds were produced, which was sufficient to plant

7% of the total wheat area in Egypt; in Ethiopia, 27 000 t were produced, sufficient to plant 10% of Ethiopia's total wheat area and in Pakistan, 42 750 t of certified seed were produced, sufficient to plant 5% of the total wheat area in that country in 2011/13. The popularization and promotion schemes of rust-resistant varieties have been reaching over 5000 farmers every year in each of the three target countries.

4.2 Sustainable nutrient management

Researchers and policy makers must consider several soil and plant nutrient management options to sustain soil fertility, which includes Integrated Nutrient Management (INM) and involves the conjunctive use of fertilizers and organic matter. The Green Revolution, which was based on the intensive use of high-yielding varieties of wheat, coupled with other inputs like chemical fertilizers and irrigation water, was successful in boosting food supply and reducing food prices. Critics of the Green Revolution, however, argue that the excessive application of chemical fertilizers has caused an imbalance in soil health by way of reducing soil fertility and increasing soil alkalinity. The challenge remains for scientists and farmers to work together to maintain soil health while increasing wheat production to meet the growing demand. Organic manures are traditionally used to supply nutrients to plants. However, the amount of organic manure required to fertilize a unit area of land and other operational problems discourage their use. Organic manure provides nutrients in an efficient way as well as improving the soil conditions. The application of farmyard manure (FYM) helps to increase the yield and nutrient uptake by wheat. The incorporation of mustard/taramira + FYM and FYM at 10 t/ha has shown a significant increase in the grain yield of wheat across years (Regar et al., 2005). Eight years of research on INM in a rice–wheat system at Jabalpur (India) revealed that the conjunctive use of 5 t/ha FYM and 6 t/ha green manure with 90 kg N/ha not only sustained the productivity of wheat, but also saved nearly 90–100 kg/ha of fertilizer nitrogen (N) (Singh et al., 2001). Blanket fertilizer N recommendations for wheat production lead to low N-use efficiency due to field-to-field variability in soil N supply and seasonal variability in wheat yield. To achieve high N-use efficiency, a site-specific N management strategy using GreenSeeker™ optical sensor is the best option. High N-use efficiency in irrigated wheat can be achieved by replacing blanket fertilizer recommendations by GreenSeeker optical sensor-based N management strategies consisting of applying moderate amounts of fertilizer N at planting and crown root initiation stages and GreenSeeker optical sensor-guided fertilizer N dosages at Feekes 5–6 or 7–8 stages of wheat. The utilization of information communication technologies, including computers, internet, geographical information systems and mobile phones, as well as traditional media such as radio or TV, is playing a major role in large-scale deployment of site-specific recommendations to farmers and extension workers. Decision support systems (DSS) are now being progressively used to facilitate the application of improved nutrient management practices in farmers' fields. A recently developed DSS 'Nutrient Expert for Wheat' synthesized the wheat on-farm research data into a simple delivery system that enables wheat farmers to rapidly implement site-specific nutrient management for their individual fields.

4.3 Sustainable crop rotation systems

In the irrigated parts of the CWANA region, such as Pakistan, Egypt, Turkey and Uzbekistan, rice–wheat or cotton–wheat rotation systems are commonly practised. In such rotation systems, wheat is planted during the cooler and drier winter season after rice or cotton, which is normally grown in the warm, subhumid monsoon summer months. The late harvesting time of cotton pushes the wheat planting late into December, which in turn exposes wheat to heat stress during grain filling in late April and May. The biggest challenge for this rotation process comes from cotton residues, which immobilize key nutrients such as nitrogen, sulphur, phosphorus and potassium.

The legumes–wheat rotation system is the most common and sustainable system in the rain-fed environments of the CWANA region. Legumes play an increasingly important role in rain-fed wheat production environments, especially in soils with low N content, by enriching soil through biological N fixation, enhancing water-use efficiency and breaking the cycle of weeds, pests and diseases, which affect wheat production. Grain legume species and varieties growing in the same location differ significantly in dry matter production, N accumulation, N_2-fixation, N-balance and residue quality (Evans et al., 2001). These differences may be the main factors in determining the residual N contribution to subsequent crops. Among the legumes, faba bean is the best nitrogen fixer with a mean of 100 kg N/ha, followed by groundnut and soybean (Smil, 1999). Many trial experiments have shown that the introduction of legumes, such as faba beans, chickpea and field bean, in wheat-based cropping is a viable strategy for reducing the application of inorganic fertilizer and thus reducing the costs for resource-poor small- and medium-scale farmers.

In the central plateau of Turkey, highlands of eastern Algeria and west Iran, the wheat-fallow system has been practised for many years. The practice is a definite improvement over continuous cropping as it stabilizes yields and provides farmers with a reliable income from year to year. However, the wheat-fallow cropping system is an inefficient user of annual precipitation in regions where water is the major limiting factor. Production of wheat can be enhanced in this cropping system and more intensive rotations can be sustained by maximizing water storage and building soil residue to trap snow, absorb more rain, reduce runoff and minimize evaporation, as well as perform complete weed control at critical times in the production cycle.

4.4 Conservation agriculture

Conservation agriculture (CA), which has its roots in the universal principles of providing permanent soil cover, minimum soil disturbance and crop rotation, is now considered the principal road to sustainable agriculture, which enables increased farm productivity, while conserving the natural resource base. The decreased soil disturbance with ground cover keeps biological activity and organic matter decomposition near the surface and helps maintain a soil structure that allows greater infiltration of water associated with reduced soil water evaporation rates. Residue retention can also improve the water-holding capacity and long-term nutrient cycling. Moussadek (2012) found relatively low hydraulic conductivity and absorptivity values under no-tillage, compared to conventional tillage systems in the Zaers region of Morocco. CA practices enhance soil organic matter levels and nutrient availability by growing green manure/cover crops and returning their residues back to the soil. Thus, arable land under CA is more productive and for much longer periods of time. Increasing soil organic carbon (SOC) content showed positive effects on soil physical properties, including increased stable aggregates (Mrabet et al., 2012). Moussadek et al. (2014) indicated that after five years of continuous no-tillage, the topsoil

(0–30 cm) SOC values increased by 10% in the Vertisol, 8% in the Cambisol and 2% in Luvisol, compared to conventional tillage.

CA improves soil structure and protects the soil against erosion and nutrient losses by maintaining a permanent soil cover and minimizing soil disturbance. In central Morocco, a reduction of 50% in soil losses was observed with no-till, having less than 2 Mg ha^{-1} of residue mulch, compared to conventional till on a Vertisol (Moussadek et al., 2011). On a Mollisol, no-till reduced the runoff volume by 30–50% and sediment loss by 50–70% compared with disk ploughing. In comparison with chisel ploughing, no-till reduced runoff volume and sediment loss by, respectively, 24–53% and 43–65%. This will increase water-holding capacity and crop productivity particularly in drier seasons.

Farmers using CA technologies typically report higher wheat yields (up to 45–48% higher) with less water, fertilizer, pesticides, capital and labour inputs, thereby resulting in higher overall farm profits. Studies of no-tillage systems in Brazil and Paraguay indicated a 10–70% saving of the conventional labour input, depending on the farming system and conventional tillage practice. Other benefits include a more even distribution of labour across the year, more timely operations, a reduction in drudgery and opportunities for livelihood diversification (Pieri et al., 2002). Reducing the time that draft animals spend in one farmer's field means that there is more time available for tilling additional land (if available) or hiring out their services to others. CA has proved particularly beneficial for many small holder farmers who do not have access to animal or mechanical tillage since it enables them to carry out all their operations on time and precisely, increasing productivity and yield potential. On the other hand, the increasing demand of straw and crop residue for animal feed becomes a major challenge for the proper expansion of CA.

4.5 Improved wheat water management

Water is the most important input for crop production and hence needs to be utilized efficiently and properly using better strategies and methods in both rain-fed and irrigated environments. The seasonal water requirement of wheat ranges from 450 to 650 mm. Wheat can tolerate lower water supplies, but usually yields are lower when exposed to water stress. Generally, spring wheat is more sensitive to stress than winter wheat and the flowering stage of growth for both types is the most sensitive to water shortage. The upper limit of wheat yields is 10–12 t/ha and wheat water productivity is about 2 kg/m^3 of net water use, which is only attainable under favourable conditions and good management (Passioura and Angus, 2010). Recently, wheat yield levels of up to 15 t/ha have been reported in the United Kingdom. In rain-fed environments, the use of drought-tolerant wheat varieties with better agronomic managements such as dry planting, conservation agriculture, control of weeds and diseases will help to increase water-use efficiency and productivity. Application of supplementary irrigation compensates for lower rainfall and less moisture storage and alleviates soil water stress during dry spells.

In northern Syria, applying supplementary irrigation (SI) one to three times (100–300 mm) in the spring increased wheat yields from 2 to 6 t/ha and water productivity from 0.5 to 2.0 kg/m^3 (Oweis et al., 1999). Research in the central Anatolian highlands of Turkey and western Iran has shown that early wheat sowing by applying 50–70 mm of SI may increase wheat yields by over 2 t/ha. Finally, more inputs such as fertilizers can be applied and high-yielding varieties can be used, when favourable soil moisture is secured with SI (Ilbeyi et al., 2006).

In raised bed planting, wheat is grown in elevated strips of 1–1.25 m width between large furrows. In Egypt, since water is provided free of charge, ICARDA and the national research institutes failed to convince the farmers not to over-irrigate the wheat. A programme was initiated to adapt the raised bed practice within an integrated package at farmers' fields in the Nile Delta. The setup of the farrows at wide spaces encouraged the farmers to reduce water application because farrows cannot take more water, as is the case with close spacing in the traditional practice. The result of several years of research was that farmers automatically cut on the amount of water application by over 30% without any loss in yield. Additional improvement in the package included better fertility, weed control and landscape of the crop. The overall package increased yield by 25% and water-use efficiency by over 50%. Furthermore, a revolution in water saving and crop productivity was achieved with the adoption of a newly developed implement that simultaneously opens the furrows and sows the seeds into the beds. The simplicity of the operation and reduced fuel cost were highly appreciated by the smallholder farmers in Egypt's Nile Delta. In just three years, farmers have shifted to the use of these new machines along with the developed package on a total of 42 000 ha. This package is now being out-scaled to irrigated wheat areas in Eretria, Ethiopia, Iraq, Morocco, Nigeria and Sudan (ICARDA, 2013).

4.6 Integrated pest management

IPM uses a combination of cultural, physical, chemical and biological methods to control wheat pests by minimizing economic, health and environmental risks. The use of resistant varieties and crop rotations and the one-time application of fungicides have been found to control most wheat diseases. Development and deployment of resistant varieties against the *Ug99* race of stem rust and the *Yr27* virulent race of yellow rust have been carried out in many countries, including Egypt, Iran, India, Pakistan, Nepal and Bangladesh. Through joint efforts between the National Institute of Agricultural Research Morocco, ICARDA and Kansas State University, ten resistance genes to Hessian fly have been identified in bread wheat. In addition, over a hundred other sources of resistance have been identified in wild relatives of wheat (El-Bouhssini et al., 2012). Three bread wheat varieties ('Massira' in 1996; 'Arrihane' and 'Aguilal' in 1998) have been released in Morocco. Despite similar efforts on durum wheat, only a single source of resistance has been identified to Hessian fly in Morocco. However, resistant lines were developed through the introgression of genes from bread wheat and wild relatives (*T. araraticum*) into durum wheat. Six resistant cultivars ('Irden', 'Chaoui', 'Marwane', 'Amria', 'Nassira' and 'Faraj') of durum wheat, with good grain quality and adaptation to Mediterranean dry lands, were released in Morocco in 2003 and 2009.

IPM is also recommended to control the most important insect pests of wheat, such as Hessian fly, Sunn pest, Sawfly and Russian wheat aphid. Early planting and use of resistant wheat varieties are effective in controlling Hessian fly. The IPM package for the control of Sunn pest has multiple components: changing spraying policy from indiscriminate aerial spraying to localized ground sprays based on economic insect population thresholds, devolution of control to farmers, enhancement/conservation of egg parasitoids through the use of flowering medicinal plants alongside wheat fields and the use of entomopathogenic fungi in overwintering sites. Participatory approaches such as farmers' field schools have been used to disseminate IPM options for both Hessian fly and Sunn pest. Some elements of the Sunn pest package are being used by farmers on over 3 million hectares in several countries in West Asia. Use of insect-resistant wheat varieties (e.g. with solid-stem), swathing, tillage, delayed planting, crop rotation and the use of parasitoids have been recommended as effective IPM options in controlling wheat stem sawfly.

5 Intensification of wheat technologies

Agricultural extension (also known as agricultural advisory services), along with innovation, plays a crucial role in promoting agricultural productivity, increasing food security, improving rural livelihoods and promoting agriculture as an engine of pro-poor economic growth. Many countries in the CWANA region such as Pakistan, Turkey, Iran, Syria and Egypt have launched credit systems market, information networks and extension services in order to demonstrate and diffuse wheat technology packages, including improved varieties, fertilizers, herbicides, fungicides, irrigation and other technologies.

In Syria, for example, the utilization of agricultural technologies and inputs (improved varieties, fertilizer, irrigation, etc.) through the government-led extension system has helped to increase wheat production especially after 1992 (Fig. 3). Jamal (2014) has clearly indicated the major role played by the National Research System and several international and regional organizations, especially ICARDA, in making cooperative efforts in the matter of building capacity and joint research to release the famous Sham series of bread and durum wheat varieties, which are adapted to both rain-fed and irrigated environments in Syria.

Wheat system intensification by non-governmental organizations has also played a significant role in demonstrating and diffusing wheat technologies, increasing wheat productivity and bolstering food security. Recently, ICARDA, in collaboration with national programmes, has implemented a project on the intensification of wheat technologies in Arab countries.

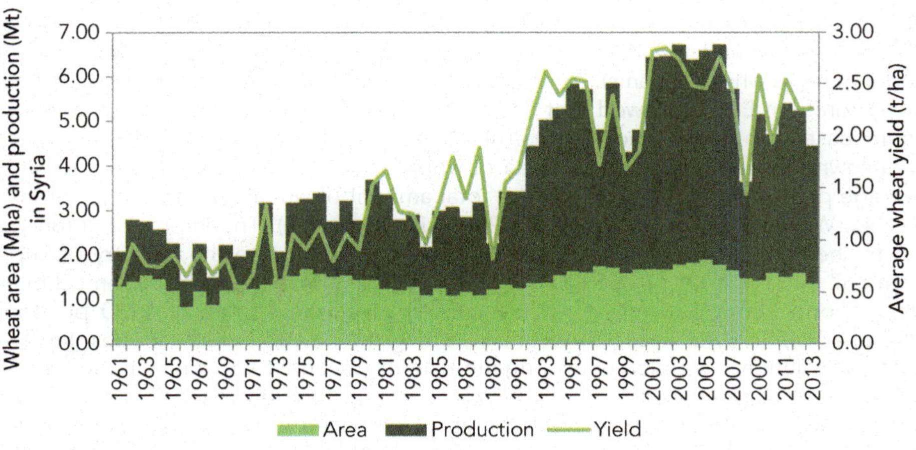

Figure 3 Wheat production, yield and area from 1961 to 2013 in Syria (FAO 2014).

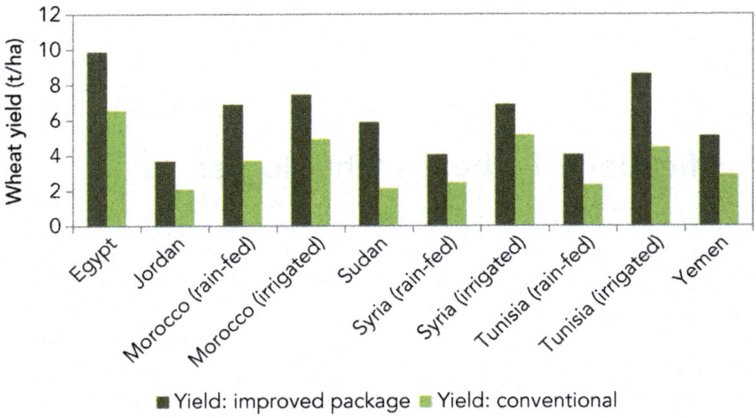

Figure 4 Grain yield of wheat under improved and conventional practices in selected countries, 2011–13.

The project seeks to improve wheat production and yields in wheat-based agricultural systems in the nine countries of Algeria, Egypt, Iraq, Jordan, Morocco, Sudan, Syria, Tunisia and Yemen (Fig. 4).

Different approaches were followed in each country to disseminate improved wheat packages (e.g. improved varieties, fertilizers, irrigation frequency and raised-bed management). Yield results of participating farmers in all countries showed that increases under all production systems (i.e. irrigated, supplementary irrigated or rain-fed) were achievable through the use of improved technologies as compared to the use of farmers' own practices. An average increase of 27% was achieved in the fields of participating farmers across all countries involved. Taken separately, the maximum average increase per hectare was achieved in Sudan (68% under irrigated wheat systems) and the minimum average increase was recorded in Morocco (8% under rain-fed wheat systems). An impact assessment study conducted in the El Sharkia governorate in Egypt showed that after the implementation of the project, there was an increase in the total amount of certified seeds sold to farmers (+21%), in the total area sown in wheat (+8%), in the area of wheat grown under raised beds (RB), in the average productivity (+16%) and in the total amount of wheat sold to the government (+36%). Wheat yield increased from 6.2 to 7.2 t/ha (+16%). The improvement in yield led to the increase in Al-Sharkia's total wheat amount sold to the Ministry of Supply from 557 030 t in 2009 to 755 496 t in 2013 (+36%). The increase has been estimated to be worth about $36 million at the conservative wheat price of $180 per tonne (Halila, 2014). In Syria, research results including the new released varieties, fertilizers recommendations, supplemental irrigation methods and total agricultural practices packages were demonstrated with pioneer farmers through extension service, with the help of ICARDA and other organizations. Demonstrations through field days were conducted and the new technologies were presented to farmers. Such an approach has enabled the country to achieve self-sufficiency in wheat and even to export the surplus production (Jamal, 2014). Similarly, systems of wheat intensification (SWI) have been implemented in Nepal by Mercy Corps, a non-governmental organization, under the European Union-Food and Agriculture Organization of the United Nations Food Facility Program. SWI practices using modern wheat varieties raised wheat yields by 91%, reaching a yield of 6.5 tonne/ha, compared to the control (3.4 t/ha) (Khadka and Raut, 2011). The sustainability and continuity of intensification programmes depend on many factors, including profitability, access to market, inputs, information, capacity building, subsidies and conducive policies.

6 Future trends

6.1 Exploring adaptive genes from genetic resources

Both biotic (rusts, sptoria, insects) and abiotic (drought, heat, salinity) factors are important limiting factors for wheat production in the CWANA region. Genetic resources remain a major source of genes to overcome these major biotic and abiotic stresses and for improving further the quality and nutritional attributes. CIMMYT and ICARDA gene banks hold the largest collection of wheat genetic resources available to all partners around the world. ICARDA gene bank holds 'in-trust' around 41 300 accessions in its active collection. This is a unique collection as it has more than 80% of landraces and the richest *Aegilops* collection. Following the Focused identification of Germplasm Strategy approach (Street et al., 2008), resistance sources for Sunn pest, Russian wheat aphid, powdery mildew, drought and salinity have been identified (El-Bouhssini et al., 2009; Endresen et al., 2011; Bari et al., 2012). Although wheat was among the few crops that benefited from wide crosses, more coordinated pre-breeding and breeding efforts are needed in the future both at the international and national levels to develop and deploy high-yielding wheat varieties with resistance/tolerance to the major biotic and abiotic stresses.

6.2 Adopting climate-smart wheat technologies

Climate change affects wheat production by increasing both abiotic stresses (heat, drought, cold and flood) and biotic stresses (aggressive diseases and insect pests). With the current climate change effects, it is anticipated that new pests and diseases will emerge as already exemplified in the recent epidemics of yellow rust across CWANA region and *Ug99* stem rust epidemic in East African countries (Solh et al., 2012). Climate change is also affecting the quality of wheat as higher temperatures result in shrivelled wheat grains. Modelling results indicate that wheat production will be affected more in the developing countries than in the developed countries. It is, therefore, important to develop climate-smart wheat varieties combining high-yield potential with resistance/tolerance to drought and heat stresses, in order to cope with the increasing effect of climate change and thereby to produce more grains per drop of water. Furthermore, climate-smart technologies such as improved water-harvesting techniques, deficit irrigation, conservation agriculture, fertigation and site-specific nutrient management need to be implemented and promoted in order to enhance sustainable wheat production across the CWANA region.

6.3 Improving integrated pest-management practices

IPM is a sustainable approach to managing pests such as weeds, diseases and insects by combining cultural, physical, chemical and biological tools in a way that minimizes economic, health and environmental risks. This approach has the potential to substantially reduce the use of chemicals, while providing a high level of effective pest management. However, IPM programmes require a higher degree of commitment and awareness creation from policy makers and extension agents in order to implement and promote IPM packages at farm level. Recent reports indicate that one-time application of fungicides even for the resistant varieties has increased wheat yield and quality. As yellow rust is the major biotic constraint in the CWANA region, fungicide application should be part of the wheat production package to improve wheat production.

6.4 Establishing strong seed systems

Efficient and effective seed delivery systems are critical for new wheat varieties to reach farmers and bring meaningful impact. Many national seed systems in the CWANA region are weak and operate under heterogeneous agro-ecological farming system, crop and market environments. The rate of variety replacement is very low in most of the NARS in the CWANA region. They also face a broad range of constraints such as poor policy and regulatory frameworks; inadequate institutional and organizational arrangements; deficiencies in production, processing and quality assurance infrastructure; poorly trained personnel and limiting technical and managerial capacities, compounded by farmers' difficult socio-economic circumstances. It is, therefore, crucial to assist and strengthen NARS in managing a strong seed system through capacity development, establishment of fast-track variety release systems, participatory demonstrations and accelerated seed multiplication of newly released wheat varieties to ensure speedy replacement of existing vulnerable commercial varieties.

6.5 Improving grain quality segregation for added value

With increasing incomes of many global consumers, the demand for specific wheat quality attributes and products is increasing. This creates a differentiation of wheat products in markets, based on visible or non-visible characteristics, and opens the possibility of adding value to the wheat industry, creating extra employment along value chains and increasing farm gate prices. The willingness to pay premium price for high-quality wheat encourages farmers to grow wheat varieties with known quality profiles and reasonably good yield potential. In this regard, it is noteworthy that it is possible to improve the nutritional value of wheat by introgressing genes for high-quality proteins and micronutrients such as iron and zinc. This effort, however, requires investment in research, institution capacity building for varietal classification, grading into different wheat classes and raising public awareness.

6.6 Promoting enabling policies and working environments

For sustainable wheat production, favourable and conducive government policies play an important role. Government subsidies to agricultural inputs such as improved seed, irrigation water and chemicals encourage farmers to adopt improved wheat technologies and increase wheat production. Furthermore, the creation of adequate infrastructures and marketing systems is of paramount importance for having a successful and competent wheat industry at national and regional levels.

6.7 Strengthening regional and international networks

The IWIN and currently the CRP wheat, an alliance of international centres (CIMMYT and ICARDA), NARS, universities and regional institutions, have been successful and efficient networks for distributing new wheat genotypes (Payne, 2004; Dixon et al., 2009; Byerlee and Dubin, 2010) globally. Such a network could be strengthened by widening partnerships and collaboration, in order to develop, disseminate and market more productive, stress-tolerant and nutritive wheat varieties and to perfect and promote production practices based on the principles of conservation agriculture, which boost yields while conserving or enhancing critical resources like soil and water.

7 Where to look for further information

The best single source of information is the International Research Center for Agricultural Research in the Dry Areas (ICARDA) (www.icarda.org). Founded in 1977, and with 600 staff in over 30 countries, ICARDA focuses on West and Central Asia, North Africa and the Caucasus.

ICARDA is committed to the improvement of livelihoods of the resource-poor in dry areas by enhancing food security and alleviating poverty through agricultural research, and the equitable use and conservation of natural resources. ICARDA is active in areas such as breeding new varieties, improved practices for farming and natural resource management (such as water harvesting and conservation agriculture) as well as diversification for improved incomes and security.

A single important research initiative is the CGIAR Research Program on Dryland Systems (http://drylandsystems.cgiar.org). The Research Program involves eight CGIAR centres including ICARDA, the International Crop Research Institute for the Semi-Arid Tropics (ICRISAT) (www.icrisat.org) (which focuses on sub-Saharan Africa and South Asia), the International Maize and Wheat Improvement Center (CIMMYT) (www.cimmyt.org) (which focuses on breeding new varieties) and the International Water Management Institute (IWMI) (www.iwmi.cgiar.org). The Research Program supports innovation in areas such as wheat-legume cropping systems and integrated watershed management which improve food and income security, resilience and better management of natural resources.

8 References

Bari, A., Street, K., Mackay, M., Endresen, D. T. F., De Pauw, E. and Amri, A. (2012), Focused Identification of germplasm strategy (FIGS) detects wheat stem rust resistance linked to environmental variables. *Genetic Resources and Crop Evolution* [online first]. doi:10.1007/s10722-011-9775-5 [Published online: 3 December 2011].

Bishaw, Z. (2004), *Wheat and Barley Seed Systems in Ethiopia and Syria*. Wageningen, the Netherlands: Wageningen University.

Braun, H.-J., Atlin, G. and Payne, T. (2010), Multi-location testing as a tool to identify plant response to global climate change. In M. P. Reynolds (ed.), *Climate Change and Crop Production*. London: CABI International, pp. 115–38.

Byerlee, D. and Dubin, H. J. (2010), Crop Improvement in the CGIAR as a global success story of open access and international sharing. *International Journal of the Commons* 4, 452–80.

CRP Wheat Phase I: the Consultative Group of International Agricultural Research (CGIAR) Research Program on Wheat (WHEAT).

Dixon, J., Braun, H. J., Kosina, P. and Crouch, J. H. (eds) (2009), *Wheat Facts and Futures 2009*. Mexico, DF (Mexico): CIMMYT, p. 95.

El-Bouhssini, M., Ogbonnaya, F. C., Chen, M., Lhaloui, S., Rihawi, F., Dabbous, A. (2012), Sources of resistance in primary synthetic hexaploid wheat (Triticum aestivum L.) to insect pests – Hessian fly, Russian wheat aphid and Sunn pest in the fertile crescent. *Genetic Resources and Crop Evolution*, doi.10.1007/s10722-012-9861-3.

El-Bouhssini, M., Street, K., Joubi, A., Ibrahim, Z. and Rihawi, F. (2009), Sources of wheat resistance to Sunn pest, *Eurygaster integriceps* Puton, in Syria. *Genetic Resources and Crop Evolution* 56(8), 1065–9.

Endresen, D. T. F., Street, K., Mackay, M., Bari, A. and De Pauw, E. (2011), Predictive association between biotic stress traits and ecogeographic data for wheat and barley landraces. *Crop Science* 51, 2036–55.

Evans, J., McNeill, A. M., Unkovich, M. J., Fettell, N. A. and Heenan, D. P. (2001), Net nitrogen balances for cool-season grain legume crops and contributions to wheat nitrogen uptake: A review. *Animal Production Science* 41(3), 347–59.

FAO (2014), FAOSTAT, http://faostat.fao.org. Accessed 11 March 2015. FAO, Rome, Italy.

FAO (2015), FAOSTAT, http://faostat.fao.org. Accessed 14 September 2015. FAO, Rome, Italy.

Fischer, R. A., Byerlee, D. and Edmeades, G. O. (2014), *Crop Yields and Global Food Security: Will Yield Increase Continue to Feed the World?* ACIAR Monograph No. 158. Canberra: Australian Centre for International Agricultural Research, pp. xxii, 634.

Halila, H. (2014), Enhancing food security in Arab countries project. In Abdul Karim Sadik, Mahmoud El Solh and Najib Saab (eds), *Food Security in Arab Countries*. Annual Report of Arab Forum for Environmental Development.

ICARDA (2013), ICARDA Annual Report. Beirut, Lebanon.

Ilbeyi, A., Ustun, H., Oweis, T., Pala, M., and Benli, B. (2006), Wheat water productivity in a cool highland environment: Effect of early sowing with supplemental irrigation. *Agricultural Water Management* 82, 399–410.

Jamal, M. (2014), Enhancing wheat self-sufficiency in Syria. In Abdul Karim Sadik, Mahmoud El Solh and Najib Saab (eds), *Food Security in Arab Countries*, Annual Report of Arab Forum for Environmental Development.

Khadka, R. B. and Raut, P. (2011), System of Wheat Intensification (SWI): A new concept on low input technology for increasing wheat yield in marginal land. System of Rice Intensification website. Report on a SWI project in Far Western Nepal by European Union Food Facility Project, Forum for Awareness and Youth Activity Nepal and Safety Nets Supporting Agricultural Productivity (SNAP) Mercy Corps Nepal.

Lhaloui, S., Buschman, L., El-Bouhssini, M., Amri, A., Hatchett, J. H., Keith, D., Starks, K. and El Houssaini, K. (1992), Control of Mayetiola species (Diptera: Cecidomyiidae) with carbofuran in bread wheat, durum wheat and barley, with yield loss assessment and its economic analysis. *Awamia* 77, 55–73.

Manes, Y., Gomez, H. F., Puhl, L., Reynolds, M., Braun, H. J. and Trethowan, R. (2012), Genetic yield gains of the CIMMYT international semi-arid wheat yield trials from 1994 to 2010. *Crop Science* 52(4), 1543–52.

Moussadek, R. (2012), Impacts de l'agriculture de conservation sur les propriétés et laproductivité des vertisols du Maroc Central. PhD thesis, University of Ghent, Belgium.

Moussadek, R., Mrabet, R., Dahan, R., Douaik, A., Verdoodt, A., Van Ranst, E. and Corbeels, M. (2011), Effect of tillage practices on the soil carbon dioxide flux during fall and spring seasons in a Mediterranean Vertisol. *Journal of Soil Science and Environmental Management* 2(11), 362–9.

Moussadek, R., Mrabet, R., Dahan, R., Zouahri, A., El Mourid, M. and Van Ranst, E. (2014), Tillage system affects soil organic carbon storage and quality in central Morocco. *Applied and Environmental Soil Science* 2014, 1–8.

Mrabet, R., Moussadek, R., Fadlaoui, A. and van Ranst, E. (2012), Conservation agriculture in dry areas of Morocco. *Field Crops Research* 132, 84–94.

Nagarajan, S. (2005), Can India produce enough wheat even by 2020. *Current Science* 89(9), 1467–71.

Oweis, T., Pala, M. and Ryan, J. (1999), Management alternatives for improved durum wheat production under supplemental irrigation in Syria. *European Agronomy Journal* 11, 255–66.

Pala, M., Oweis, T., Benli, B., De Pauw, E., El Mourid, M., Karrou, M., Jamal, M. and Zencirci, N. (2011), Assessment of wheat yield gap in the Mediterranean: Case studies from Morocco, Syria and Turkey. Aleppo, Syria: International Center for Agricultural Research in the Dry Areas (ICARDA), pp. iv +36.

Passioura, J. B. and Angus, J. F. (2010), Improving productivity of crops in water-limited environments. *Advances in Agronomy* 106, 37–75.

Payne, T. (2004), The international wheat improvement network (IWIN) at CIMMYT. www.cimmyt.org.

Pieri, C., Evers, G., Landers, J., O'Connell, P. and Terry, E. (2002), No-Till Farming for Sustainable Rural development, Agriculture and Rural Development Working Paper. Washington, DC: World Bank/International Bank for Reconstruction and Development.

Rajaram, S., van Ginkel, M. and Fischer, R. A. (1995), CIMMYT's wheat breeding mega-environments (ME). In *Proceedings of the 8th International Wheat Genetic Symposium*, 19–24 July 1993, Beijing, China.

Regar, P. L., Rao, S. S. and Vyas, S. P. (2005), Crop-residue management for increased wheat (*Triticum aestivum)* production under saline soils of arid fringes. *Indian Journal of Agriculture Sciences* 75(2), 83–6.

Reynolds, M. and Borlaug, N. E.. (2006), Impacts of breeding on international collaborative wheat improvement. *Journal of Agricultural Science* 144, 3–17.

Sharma, R. C., Crossa, J., Velu, G., Huerta-Espino, J., Vargas, M., Payne, T. S., Singh, R. P. (2012), Genetic gains for grain yield in CIMMYT spring bread wheat across international environments. *Crop Science* 52, 1522–33.

Shiferaw, B., Smale, M., Braun, H. J., Duveiller, E., Reynolds, M. and Muricho, G. (2013), Crops that feed the world 10. Past successes and future challenges to the role played by wheat in global food security. *Food Security* 5(3), 291–317.

Singh, M., Singh, V. P. and Reddy, K. S. (2001), Effect of integrated use of fertilizer N and FYM or green manure on transformation of NK and S and productivity of rice-wheat system on a Vertisols. *Journal of the Indian Society Soil Science* 49, 430–5.

Smil, V. (1999), Nitrogen in crop production: An account of global flows. *Global Biogeochemical Cycles* 13(2), 647–62.

Solh, M., Nazari, K., Tadesse, W. and Wellings, C. R. (2012), The growing threat of stripe rust worldwide. Borlaug Global Rust Initiative (BGRI) conference, Beijing.

Street, K., Mackay, M., Zuev, E., Kaur, N., El-Bouhssini, M., Konopka, J. and Mitrofanova, O. (2008), Diving into the gene pool: A rational system to access specific traits from large germplasm collections. In R. Appels, R. Eastwood, R. Lagudah, E. Langridge and M. Mackay (eds),

Proceedings of the 11th International Wheat Genetics Symposium, Brisbane, Australia, pp. 28–31.
Tadesse, W., Morgounov, A. I., Braun, H. J., Akin, B., Keser, M., Yuksel, K., Sharma, R. C., Rajaram, S., Singh, M., Baum, M., van Ginkel, M. (2013), Breeding progress for yield and adaptation of winter wheat targeted to irrigated environments at the International Winter Wheat Improvement Program (IWWIP). *Euphytica* 194, 177–85.
Trethowan, R. M., Reynolds, M. P., Ortiz-Monasterio, J. I. and Ortiz, R. (2007), The genetic basis of the green revolution in wheat production. *Plant Breeding Reviews* 28, 39–58.
World Bank (2008), *World Development Report*. Washington, DC: The World Bank.

Index

Abiotic stresses
on durum wheat 202–204
and wheat cultivation, in Asia 267–269
and wheat production, in CWANA 291
and wheat variety selection 15–16
AGFACE. *see* Australian Grains FACE (AGFACE)
Agri-transfer
farmer-managed certified seed production 252
farmer-managed trials and demonstrations 251–252
overview 250–251
wheat production handbook 252
Agronomists, and ICM
challenges 151–153
conservation agriculture 145–148
genotype, environment and management (GxExM) 148–151
APSIM crop modelling system 126
Asia, wheat cultivation in
and abiotic stresses 267–269
and biotic stresses 266–267
and genetic improvement 262–264
improving breeding techniques 264–266
improving cultivation practices 270–271
improving quality aspects 269–270
in India 271–274
overview 261–262
Australian Grains FACE (AGFACE) 115–116

Biophysical limitations, and wheat cultivation
addressing 246–248
climatic issues 242
pests and diseases 243
soil fertility 242–243
varieties and seeds 243–244
Biotic stresses
and wheat cultivation, in Asia 266–267
and wheat production, in CWANA 291
and wheat variety selection 15–16
BMS. *see* Breeding Management System (BMS)
Breeding
for drought tolerance 293
of durum wheat
doubled haploids 215–216
marker-assisted selection (MAS) 216–218
for quality 296
Breeding Management System (BMS) 127

CA. *see* Conservation agriculture (CA)
Central and West Asia and North Africa (CWANA), wheat production in
adaptive genes from genetic resources 303
challenges to
abiotic and biotic stresses 291
limited availability and high price of inputs 291–292
yield gap 292
climate-smart wheat technologies 303
conservation agriculture (CA) 298–299
and grain quality segregation 304
integrated pest management (IPM) 300–301, 304
intensification of 301–303
major 290–291
modern varieties (MVs)
breeding for drought tolerance 293
breeding for quality 296
high-yielding and disease-resistant wheat varieties 293–295
overview 292–293
seed production and distribution 296–297
overview 287–290
policies and working environments 304–305
regional and international networks 305
seed delivery systems 304
sustainable crop rotation systems 297–298
sustainable nutrient management 297
wheat water management 299–300
Climate change, and wheat crop modelling
global climate models (GCMs) 117–118
other impact factors 120–121
rainfall 118–119
scaling 118
temperature 119–120
Climate-smart wheat technologies 303
Conservation agriculture (CA) 145–148, 298–299
Conservation tillage (CT) wheat
agronomic and economic impacts of 45–49
constraints to
developments in 53–55
heterogeneous technology impacts 50–51
institutional and policy hurdles 51–53
limited information availability and access 49
diffusion, factors affecting 43–45
overview 41–43
Consumer demand, and wheat variety selection 16–17
Conventional wheat *vs.* organic wheat 187–188
Crop establishment, and root development
elongation and growth 27
embryo size and seed mass 33–35
overview 25–26
research contributions to 35–36

resource acquisition and yield 28
root system architecture 28
seed rate 31
seminal and nodal root axes 26–27
and sowing
depth of 31–33
time 28–30
wetness and dryness 33
Crop management, and wheat variety selection 14–15
Crop rotations
organic wheat and spelt
concentration 169–170
conditions of cultivation 178–179
importance of 169
reducing pests, diseases and weeds 170–172
for supplying nutrients 172–174
types of 174–178
Cultivars and traits 122

Domestication, and wheat variety selection 5
Doubled haploids 215–216
Drought tolerance, and WUE 78–79
Durum wheat
abiotic stresses on 202–204
breeding technologies
doubled haploids 215–216
marker-assisted selection (MAS) 216–218
diseases of
Fusarium head blight (FHB) 208–210
leaf spotting complex 207–208
wheat rusts 205–207
end-use quality of 212–215
insect pests of 210–212
overview 199–200
yield and agronomic traits 200–202

Embryo size and seed mass 33–35
End-use quality, of durum wheat 212–215
ET. *see* Evapotranspiration (ET)
Evapotranspiration (ET) 70
Extrapolation, and wheat crop modelling
field 111–113
global scale 113–114
region 113
time 111
treatments 111

FACE. *see* Free-air carbon dioxide enrichment (FACE)
Farm biodiversity, and organic wheat
vs. conventional wheat 187–188
vs. other organic crops 188
wheat management 188–189
Farmer(s)
-managed certified seed production 252
-managed trials and demonstrations 251–252
and wheat crop modelling 126–127
wheat variety selection by
participatory varietal selection (PVS) 17–18
seed cost 18
Farm machinery and labour 244
FHB. *see Fusarium* head blight (FHB)
Free-air carbon dioxide enrichment (FACE) 115
Fusarium head blight (FHB) 208–210

GCMs. *see* Global climate models (GCMs)
Gender and food security 100–101
Genetic improvement, and wheat cultivation 262–264
Genomic selection 10–11
Genotype, environment and management (GxExM) 148–151
Global climate models (GCMs) 117–118
Grain quality segregation, and wheat production 304
GxExM. *see* Genotype, environment and management (GxExM)

Hot serial cereal (HSC) 116
HSC. *see* Hot serial cereal (HSC)

India, wheat cultivation 271–274
Insect pests, of durum wheat 210–212
Integrated Breeding Platform 127
Integrated crop management (ICM)
and agronomists
challenges 151–153
conservation agriculture 145–148
genotype, environment and management (GxExM) 148–151
overview 143–145
for wheat stem sawfly (WSS) 153–158
and decision support system 157–158
harvest management strategies 156–157
overview 153–155
planting and in-crop strategies 156
pre-crop strategies 155
Integrated pest management (IPM) 300–301, 304
Intensification, of wheat production 301–303
Intercropping, of organic wheat
with combinable crops 179–181
with forage crops 181–182
IPM. *see* Integrated pest management (IPM)
Irrigation
deficit 76–78
and drainage 124
scheduling 75–76

Leaf spotting complex 207–208

Marker-assisted selection (MAS) 216–218
and wheat variety selection 8–10

Modern varieties (MVs)
breeding for drought tolerance 293
breeding for quality 296
high-yielding and disease-resistant wheat varieties 293–295
overview 292–293
seed production and distribution 296–297
MVs. *see* Modern varieties (MVs)

Natural resource management (NRM)
cultivars and traits 122
irrigation and drainage 124
managing seasonal variability 124–126
nitrogen fertilization 122–123
sowing 121
spatial variability 124
Natural selection, of wheat variety 4
Nitrogen application rate 114–115
Nitrogen fertilization 122–123
NRM. *see* Natural resource management (NRM)

Organic wheat and spelt
breeding and variety choice 182–183
crop rotations
concentration 169–170
conditions of cultivation 178–179
importance of 169
reducing pests, diseases and weeds 170–172
for supplying nutrients 172–174
types of 174–178
examples 184–187
and farm biodiversity
vs. conventional wheat 187–188
vs. other organic crops 188
wheat management 188–189
intercropping
with combinable crops 179–181
with forage crops 181–182
mixtures and populations 183–184
overview 167–169

Participatory varietal selection (PVS) 17–18
Phenotyping 11–12
Plant breeders, and wheat variety selection
biotic and abiotic stresses 15–16
consumer demand 16–17
crop management practices 14–15
environment and location 12–14
Post-harvest wheat losses, in Africa
crop production information 90–91
demographics 88–90
description 91–98
gender and food security 100–101
information flow and training requirements 99–100
methods 88
overview 85–88
prevention of 99
PVS. *see* Participatory varietal selection (PVS)

Rainfall, and wheat crop modelling 118–119
Resource acquisition and yield 28
Root system architecture 28

Scaling, and wheat crop modelling 118
Seasonal variability, managing 124–126
Seed cost, and wheat variety selection 18
Seed delivery systems 304
Seed mass and embryo size 33–35
Seed production and distribution 296–297
Seed rate 31
Seminal and nodal root axes 26–27
Smallholder wheat cultivation
agri-transfer
farmer-managed certified seed production 252
farmer-managed trials and demonstrations 251–252
overview 250–251
wheat production handbook 252
biophysical limitations
addressing 246–248
climatic issues 242
pests and diseases 243
soil fertility 242–243
varieties and seeds 243–244
overview 237–241
socio-economic limitations
addressing 248–249
farm machinery and labour 244
finance 245
information and advice 244–245
land 244
post-harvest handling 244
women in 245–246
Socio-economic limitations, and wheat cultivation
addressing 248–249
farm machinery and labour 244
finance 245
information and advice 244–245
land 244
post-harvest handling 244
Soil and water conservation 74–75
Soil fertility 242–243
Sowing 121
and crop establishment
depth of 31–33
time 28–30
Spatial variability 124
SRA. *see* Strategic Research Agenda (SRA)
Strategic Research Agenda (SRA) xix–xx
themes of
agronomy and crop management xxvi–xxvii

controlling wheat diseases and pests xxii–xxiii
enabling technologies and methods xxix–xxxi
ensuring high quality and safe wheat supply xxviii–xxix
genetic resources xxxiii–xxxiv
improving tolerance of wheat to abiotic stress xxiii–xxv
knowledge exchange and education xxxiv–xxxv
nutrient use efficiency xxv–xxvi
shared platforms and technologies xxxii–xxxiii
wheat field potential growth xx–xxi
Sustainable crop rotation systems 297–298
Sustainable nutrient management 297

Temperature, and wheat crop modelling 119–120

Water management
and evapotranspiration (ET) 70
irrigation
deficit 76–78
and irrigation scheduling 75–76
overview 63–68
soil and water conservation 74–75
water-use efficiency (WUE) 70–71
and drought tolerance 78–79
wheat yield-ET relationship 71–73
winter wheat yield 68–69
yield determination 70
Water stress experiment 115
Water-use efficiency (WUE) 70–71
and drought tolerance 78–79
Wheat breeding
for drought tolerance 293
durum
doubled haploids 215–216
marker-assisted selection (MAS) 216–218
for quality 296
Wheat crop modelling
and climate change
global climate models (GCMs) 117–118
other impact factors 120–121
rainfall 118–119
scaling 118
temperature 119–120
decision support for farmers 126–127
experiments for testing
Australian Grains FACE (AGFACE) 115–116
free-air carbon dioxide enrichment (FACE) 115
hot serial cereal (HSC) 116
nitrogen application rate 114–115
water stress experiment 115
extrapolation
field 111–113
global scale 113–114
region 113
time 111
treatments 111
natural resource management (NRM)
cultivars and traits 122
irrigation and drainage 124
managing seasonal variability 124–126
nitrogen fertilization 122–123
sowing 121
spatial variability 124
overview 107–110
Wheat cultivation, in Asia
and abiotic stresses 267–269
and biotic stresses 266–267
and genetic improvement 262–264
improving breeding techniques 264–266
improving cultivation practices 270–271
improving quality aspects 269–270
in India 271–274
overview 261–262
Wheat diseases, durum
Fusarium head blight (FHB) 208–210
leaf spotting complex 207–208
wheat rusts 205–207
The Wheat Initiative, xix–xx
Wheat management, and organic wheat 188–189
Wheat production, in CWANA
adaptive genes from genetic resources 303
challenges to
abiotic and biotic stresses 291
limited availability and high price of inputs 291–292
yield gap 292
climate-smart wheat technologies 303
conservation agriculture (CA) 298–299
and grain quality segregation 304
integrated pest management (IPM) 300–301, 304
intensification of 301–303
major 290–291
modern varieties (MVs)
breeding for drought tolerance 293
breeding for quality 296
high-yielding and disease-resistant wheat varieties 293–295
overview 292–293
seed production and distribution 296–297
overview 287–290
policies and working environments 304–305
regional and international networks 305
seed delivery systems 304
sustainable crop rotation systems 297–298
sustainable nutrient management 297

wheat water management 299–300
Wheat rusts 205–207
Wheat stem sawfly (WSS)
and ICM 153–158
and decision support system 157–158
harvest management strategies 156–157
overview 153–155
planting and in-crop strategies 156
pre-crop strategies 155
Wheat variety selection
and domestication 5
by farmers
participatory varietal selection (PVS) 17–18
seed cost 18
genomic selection 10–11
and marker-assisted selection (MAS) 8–10
natural selection 4
overview 3
and phenotyping 11–12
by plant breeders
biotic and abiotic stresses 15–16
consumer demand 16–17
crop management practices 14–15
environment and location 12–14
in post-Mendelian era 6–7
in pre-Mendelian era 5–6
selection by humans 4–5
Wheat water management 299–300
Winter wheat yield 68–69
WSS. *see* Wheat stem sawfly (WSS)

Yield/agronomic traits, of durum wheat 200–202
Yield Prophet 126

Lightning Source UK Ltd.
Milton Keynes UK
UKOW06n1202200717
305705UK00001B/2/P

9 781786 760203